AF518375

TREE & TREELIKE VEGETABLES

THE AUTHORS

Dr. Sanyat Misra is a Forestry graduate from UAS, Bangalore; M.Sc. (Ag.) Hort. from GBPUAT, Pantnagar (Udham Singh Nagar), Uttarakhand; and Ph.D. from Dr. YSP UHF, Nauni, Solan (H.P.). He served ICAR-NBPGR (New Delhi and Shimla) for 3 years and in 2004 joined BAU, Kanke (Ranchi) as Assistant Professor *cum* Jr. Scientist. At present, he is Head, Deparment of Horticulture looking after administrative, extension, teaching and research activities of the Department. He had been Chairman in case of 7 M.Sc. students. To his credit, he has some 45 research publications, 11 books and 3 bulletins.

Sanjay Mishra is a Lucknow-based botany graduate *cum* software professional who to his credit has some 8 books out of which 2 are exclusively on computer science. It is his untiring efforts that this book has reached to this stage.

Dr. R.L. Misra got retired from Division of Floriculture and Landscaping, ICAR-IARI, New Delhi-110012 in May 2008 as Principal Scientist *cum* Project Coordinator (Floriculture), All India Coordinated Research Project on Floriculture. He also worked as Officer-in-Charge, Indo-Israel Preject on Research and Development on Farms, IARI, New Delhi. As Adjunct Faculty he continued further teaching works to M.Sc. and Ph.D. students at IARI from 2010 to 2014. As Chairman, he guided 6 M.Sc. and 16 Ph.D. students in IARI. To his credit, he developed 30 gladiolus varieties, has more than 500 publications: 11 books, 9 technical bulletins, 15 Review articles, 169 research publications, 120 technical articles, 57 symposium papers and more than 140 book chapters. He is Fellow of (i) **Indian Society of Genetics and Plant Breeding** (1980), (ii) **Indian Society of Ornamental Horticulture (2009),** (iii) **Horticultural Society of India** (2018), and (iv) **Promotion of Horticultural Science, Bengaluru** (2022). Also he is HSI Gold Medal Awardee (2013). **Indian Society of Ornamental Horticulture,** New Delhi in 2018 on his outstanding floricultural works, and **The Orissa Horticultural Society,** Bhubaneswar in 2023 on his outstanding horticultural works conferred Lifetime Achievement Awards. In 2007, he was also conferred with Urivi Vikram Charitable Trust Award in New Delhi.

TREE & TREELIKE VEGETABLES

Authors

Sanyat Misra

Sanjay Mishra

R.L. Misra

2025

Daya Publishing House®

A Division of

Astral International Pvt. Ltd.

New Delhi – 110 002

ISBN 978-93-5919-986-3

Published by : **Daya Publishing House®**
A Division of
Astral International Pvt. Ltd.
– ISO 9001:2015 Certified Company –
4736/23, Ansari Road, Darya Ganj
New Delhi-110 002
Ph. 011-43549197, 23278134
e-mail: info@astralint.com
Website: www.astralint.com

Printed at : **Replika Press Pvt. Ltd.**

Preface

Moringa (drumstick) is a very common vegetable used in every state of South India where its pods after peeling are used as vegetables, as well as cooked with pulses. It is also used in preparing the *sambhar* ~ a popular South Indian dish used specially with *dosa*. Apart from its fruits, new leaves and flowers are also used as vegetables as these are very nutritive. In Mahararashtra, Goa, Gujarat and in a part of Madhya Pradesh too, *Moringa* is used similar to South Indian states. This is so important in South India that every family maintains a few trees for their day-to-day use. In other parts of India too, people have developed its taste for preparing the vegetables with young or nonfibrous pods or in various other ways similar to South Indian people.

Jackfruit has its origin in South India but there it is relished usually as ripened fruits where only its flesh around the seed commonly known as bulb/ lobe is eaten though seeds are used as vegetables or eaten as roasted chestnuts, but in northern India apart from its ripened fruits, mostly its raw fruits are used as a tasty vegetable on certain auspicious occasions. In North India, its raw fruits for vegetables are available all the 12 months due to continuous supply from South India during off season as in South India it continues fruiting throughout the year though in North India or elsewhere it usually fruits from April-May to August-September only (raw only from April to July looking into prevailing temperatures). **Breadfruit** in India is found growing only at or in the vicinity of sea-shores, and is usually consumed as cooked vegetable.

Banana is a tree-like herbaceous perennial though in no case it is a tree, *vis-a-vis* can be grown only in the tropical and subtropical climatic conditions of

India oftenly for its ripened fruits, however, its mature raw fruits are occasionally used as cooked vegetables though there are specific varieties for the purpose. There are certain specific banana varieties for preparing chips of high quality. For cultivation, its water requirement is very high so it is grown only where water is in plenty. However, on certain occasions the peeled fruits are used for preparing vegetables by frying with *Carum copticum* seeds. It is highly relished when cooked as *pakorah* (an oil-fried preparation mixed with gram flour paste and spices). Just emerged **bamboo** shoots are also only sometimes cooked as vegetables and for pickling but only here and there in the country. Though we have never tasted but it is said that its cooked vegetables are one of the best delicacies. Bamboo may be tree-like quite tall but the fact is this that it is a perennial grass and not a tree. We thought to include even **papaya** but it is purely a fruit and not a vegetable though sometimes it is used as vegetable and for pickling only under the forced circumstances but with no taste as a vegetable, however, only in case of upset stomach it is used as a medicinal vegetable that is why it did not get space in this book.

Mahua is a dry region huge indigenous tree found in north and central India, being very common in Uttar Pradesh, Madhya Pradesh, Maharashtra, Bihar, Odisha, Andhra Pradesh and Gujarat. This tree is very well known to the rural folk since the time immemorial. It is of high economic value whose leaves are used as fodder, stems and thick branches as timber, thin branches as fuel wood, flowers fresh or dried for various sweet preparations without addition of sugar, boiled with dry gram or peas to be used as staple food, as well as with curd, popped ones for various other dry preparations, *vis-a-vis* for country liquor ('*daaroo*') distillation. Fully developed raw fruits, after peeling and extracting the seeds, are cooked as delicious vegetables, but after ripening the pulp is generally of no use though seeds are used for extraction of oil ~ used for cooking as well as for soap industry. Its oil contains linoleic (a major unsaturated fatty acid) which is used in reducing the cholesterol content in blood serum.

All these **fruits** belonging to different nature and taste though as vegetables scarsely used but are of high value so all these under one umbrella have been christened as '**Tree & Treelike Vegetables**'.

Curry leaf is though not a vegetable but its green or dry leaves add most amicable taste & flavour to the prepared food, especially vegetables, salad, chutney, curries, pulses, pulao, biryani, puddings, pakauras, snack dishes and many other preparations. Its addition to the food items increases the value of the food. It was being previously used only in the dishes of South Indian and adjoining states but now because of its taste and flavour is being used everywhere in the country.

Kachnar thrives in the arid landscapes of North and Central India, notably in states like Uttar Pradesh, Madhya Pradesh, Maharashtra, Bihar, Odisha, Andhra Pradesh and Gujarat, where it plays an integral role in the local culture and economy. Its leaves serve as nutritious fodder for livestock, while its sturdy stems and branches find purpose as valuable timber and fuel wood. The delicate flowers, whether fresh or dried, are a culinary delight used for pickling, making raayta, in sweet preparations without added sugar as well as in staple dishes alongside pulses. Mature fruits yield seeds rich in linoleic acid, utilized for cooking and soap making, contributing to better cholesterol management. This multifaceted utility, from nourishing livestock to enriching cuisine and providing sustainable resources, underscores Kachnar's significance in sustaining rural livelihoods and preserving traditional practices.

Positive comments are welcome for further improvement of the book so for this we may be contacted on phone no. 8076235469.

Sanyat Misra

Sanjay Mishra

R.L. Misra

Contents

1

Banana

Family: Musaceae.

Synonyms of genus *Musa* L.: *Karkandela* Raf. (POWO, 2024a).

Ancestral species: *Musa acuminata* Colla (seedless) and *Musa balbisiana* Colla (seeded) (POWO, 2024b; 2024c). **Hybrid species**: *Musa × paradisiaca* L. ~ for the hybrid *M. acuminata × M. balbisiana* (POWO, 2024d).

Synonyms of *Musa acuminata* Colla: *Musa acuminata* subsp. *acuminata*; *Musa acuminata* var. *chinensis* Häkkinen & H.Wang; *Musa acuminata* subsp. *errans* (Blanco) R.V.Valmayor; *Musa acuminata* subsp. *halabanensis* (Meijer) M.Hotta; *Musa acuminata* subsp. *malaccensis* (Ridl.) N.W.Simmonds; *Musa acuminata* subsp. *microcarpa* (Becc.) N.W.Simmonds; *Musa acuminata* subsp. *siamea* N.W.Simmonds; *Musa acuminata* var. *sumatrana* (Becc.) Nasution; *Musa acuminata* var. *tomentosa* (Warb. ex K.Schum.) Nasution; and *Musa acuminata* subsp. *truncata* (Ridl.) Kiew (POWO, 2024b).

Synonyms of *Musa balbisiana* Colla: *Musa balbisiana* var. *bakeri* (Hook.f.) Häkkinen; *Musa balbisiana* var. *balbisiana*; *Musa balbisiana* var. *brachycarpa* (Backer) Häkkinen; *Musa balbisiana* var. *dechangensis* (J.L.Liu & M.G.Liu) Häkkinen; *Musa balbisiana* var. *elavazhai* A.Joe, Sreejith & M.Sabu; *Musa balbisiana* var. *liukiuensis* (Matsum.) Häkkinen; and *Musa balbisiana* var. *sepa-athiya* Borborah, Borthakur & Tanti (POWO, 2024c).

Synonyms of *Musa × paradisiaca* L.: *Karkandela × malabarica* Raf.; *Musa × acutibracteata* M.Hotta; *Musa × alphurica* Miq.; *Musa × aphurica* Rumph. ex Sagot; *Musa × arakanensis* F.W.Ripley ex Blechynden; *Musa × bacoba* Rottb.; *Musa*

× *berteroi* Colla; *Musa* × *bidigitalis* De Wild.; *Musa* × *carolinae* Sterler; *Musa* × *champa* Baker; *Musa* × *chapara* Perr.; *Musa* × *chiliocarpa* Backer ex K.Heyne; *Musa* × *consociata* Nakai; *Musa* × *corbieri* A.Chev.; *Musa* × *corniculata* Lour.; *Musa* × *dacca* Horan.; *Musa* × *decrescens* De Briey ex De Wild.; *Musa* × *decrescens* var. *pembuki* De Briey ex De Wild.; *Musa* × *decrescens* var. *rubromaculata* De Briey ex De Wild.; *Musa* × *decrescens* var. *viridis* De Briey ex De Wild.; *Musa* × *discolor* Planch.; *Musa* × *dulcissima* Nakai; *Musa* × *emasculata* De Briey ex De Wild.; *Musa* × *emasculata* var. *kiala* De Briey ex De Wild.; *Musa* × *emasculata* var. *kimbende* De Briey ex De Wild.; *Musa* × *emasculata* var. *lomba* De Briey ex De Wild.; *Musa* × *emasculata* var. *zengani* De Briey ex De Wild.; *Musa* × *humilis* Perr.; *Musa* × *ingrata* Nakai; *Musa* × *jaheri* Nakai; *Musa* × *maculata* Jacq.; *Musa* × *megalocarpa* Nakai; *Musa* × *mensaria* Moench; *Musa* × *mirabilis* Nakai; *Musa* × *nigra* Perr.; *Musa* × *odorata* Lour.; *Musa* × *oleracea* Vieill.; *Musa* × *pallida* Nakai; *Musa* × *paradisiaca* f. *dongila* De Briey ex De Wild.; *Musa* × *paradisiaca* f. *funu-nua* De Briey ex De Wild.; *Musa* × *paradisiaca* f. *kilola* De Briey ex De Wild.; *Musa* × *paradisiaca* f. *seluka* De Briey ex De Wild.; *Musa* × *paradisiaca* f. *tuba* De Briey ex De Wild.; *Musa* × *paradisiaca* subsp. *normalis* Kuntze; *Musa* × *paradisiaca* subsp. *sapientum* (L.) Kuntze; *Musa* × *paradisiaca* var. *acicularis* G.Forst.; *Musa* × *paradisiaca* var. *bende* De Briey ex De Wild.; *Musa* × *paradisiaca* var. *bilul* De Briey ex De Wild.; *Musa* × *paradisiaca* var. *champa* (Baker) K.Schum.; *Musa* × *paradisiaca* var. *cinerea* Blanco; *Musa* × *paradisiaca* var. *coarctata* G.Forst.; *Musa* × *paradisiaca* var. *compressa* Blanco; *Musa* × *paradisiaca* var. *coriacea* G.Forst.; *Musa* × *paradisiaca* var. *corniculata* G.Forst.; *Musa* × *paradisiaca* var. *dacca* (Horan.) K.Schum.; *Musa* × *paradisiaca* var. *exsicca* G.Forst.; *Musa* × *paradisiaca* var. *fatua* G.Forst.; *Musa* × *paradisiaca* var. *glaberrima* Blanco; *Musa* × *paradisiaca* var. *glauca* Blanco; *Musa* × *paradisiaca* var. *kitebbe* De Briey ex De Wild.; *Musa* × *paradisiaca* var. *lacatan* Blanco; *Musa* × *paradisiaca* var. *longa* Blanco; *Musa* × *paradisiaca* var. *lunaris* G.Forst.; *Musa* × *paradisiaca* var. *magna* Blanco; *Musa* × *paradisiaca* var. *martabarica* (Baker) K.Schum.; *Musa* × *paradisiaca* var. *maxima* Blanco; *Musa* × *paradisiaca* var. *mensaria* (Baker) K.Schum.; *Musa* × *paradisiaca* var. *mensaria* G.Forst.; *Musa* × *paradisiaca* var. *odorata* (Lour.) K.Schum.; *Musa* × *paradisiaca* var. *oleracea* (Vieill.) K.Schum.; *Musa* × *paradisiaca* var. *papillosa* G.Forst.; *Musa* × *paradisiaca* var. *punctata* G.Forst.; *Musa* × *paradisiaca* var. *purpurascens* G.Forst.; *Musa* × *paradisiaca* var. *regia* G.Forst.; *Musa* × *paradisiaca* var. *rubra* (Firminger ex Baker) K.Schum.; *Musa* × *paradisiaca* var. *sanguinea* (Welw. ex Baker) K.Schum.; *Musa* × *paradisiaca* var. *suaveolens* Blanco; *Musa* × *paradisiaca* var. *subrubea* Blanco; *Musa* × *paradisiaca* var. *ternatensis* Blanco; *Musa* × *paradisiaca* var. *tetragona* G.Forst.; *Musa* × *paradisiaca* var. *tombak* Blanco; *Musa* × *paradisiaca* var. *ulnaris* Blanco; *Musa* × *paradisiaca* var. *violacea* (Baker) K.Schum.; *Musa* × *paradisiaca* var. *violacea* Blanco; *Musa* × *paradisiaca* var. *viridis* De Briey ex De

Wild.; *Musa × paradisiaca* var. *vittata* (G.W.Ackermann ex Rodigas) K.Schum.; *Musa × polycarpa* Nakai; *Musa × prematura* Nakai; *Musa × protractorachis* De Wild.; *Musa × purpureotomentosa* De Wild.; *Musa × sapidisiaca* K.C.Jacob; *Musa × sapientum* f. *dubia* King; *Musa × sapientum* L.; *Musa × sapientum* subsp. *paradisiaca* (L.) Baker; *Musa × sapientum* var. *americana* N.G.Teodouro; *Musa × sapientum* var. *angao* Quisumb.; *Musa × sapientum* var. *baca* Quisumb.; *Musa × sapientum* var. *binutig* N.G.Teodouro; *Musa × sapientum* var. *canara* N.G.Teodouro; *Musa × sapientum* var. *canaya* Quisumb.; *Musa × sapientum* var. *champa* (Baker) Baker; *Musa × sapientum* var. *cinerea* (Blanco) N.G.Teodouro; *Musa × sapientum* var. *compressa* (Blanco) N.G.Teodouro; *Musa × sapientum* var. *cubensis* N.G.Teodouro; *Musa × sapientum* var. *dacca* (Horan.) Baker; *Musa × sapientum* var. *daryao* N.G.Teodouro; *Musa × sapientum* var. *dinalaga* Quisumb.; *Musa × sapientum* var. *dool* Quisumb.; *Musa × sapientum* var. *dubia* (King) A.M.Cowan & Cowan; *Musa × sapientum* var. *eda* Quisumb.; *Musa × sapientum* var. *fieleto* De Briey ex De Wild.; *Musa × sapientum* var. *flabellata* Quisumb.; *Musa × sapientum* var. *galatayan* Quisumb.; *Musa × sapientum* var. *garangao* N.G.Teodouro; *Musa × sapientum* var. *glaberrima* (Blanco) N.G.Teodouro; *Musa × sapientum* var. *glauca* (Blanco) N.G.Teodouro; *Musa × sapientum* var. *grandis* N.G.Teodouro; *Musa × sapientum* var. *humilis* Merr.; *Musa × sapientum* var. *inarnibal* N.G.Teodouro; *Musa × sapientum* var. *kinamay* Quisumb.; *Musa × sapientum* var. *lacatan* (Blanco) N.G.Teodouro; *Musa × sapientum* var. *longa* (Blanco) N.G.Teodouro; *Musa × sapientum* var. *martabarica* Baker; *Musa × sapientum* var. *mensaria* Baker; *Musa × sapientum* var. *odorata* (Lour.) Baker; *Musa × sapientum* var. *oleracea* Baker; *Musa × sapientum* var. *padilat* Quisumb.; *Musa × sapientum* var. *pamotion* Quisumb.; *Musa × sapientum* var. *paradisiaca* (L.) Baker; *Musa × sapientum* var. *pelipia* Quisumb.; *Musa × sapientum* var. *principe* Quisumb.; *Musa × sapientum* var. *putian* Quisumb.; *Musa × sapientum* var. *raines* Quisumb.; *Musa × sapientum* var. *regia* (G.Forst.) Baker; *Musa × sapientum* var. *rubra* Firminger ex Baker; *Musa × sapientum* var. *sanguinea* Welw. ex Baker; *Musa × sapientum* var. *sarocsoc* Quisumb.; *Musa × sapientum* var. *satama* De Briey ex De Wild.; *Musa × sapientum* var. *sision* Quisumb.; *Musa × sapientum* var. *suaveolens* (Blanco) N.G.Teodouro; *Musa × sapientum* var. *ternatensis* (Blanco) N.G.Teodouro; *Musa × sapientum* var. *tombak* (Blanco) N.G.Teodouro; *Musa × sapientum* var. *tudlong* N.G.Teodouro; *Musa × sapientum* var. *tuldoc* N.G.Teodouro; *Musa × sapientum* var. *violacea* (Blanco) N.G.Teodouro; *Musa × sapientum* var. *violacea* Baker; *Musa × sapientum* var. *vittata* (G.W.Ackermann ex Rodigas) Hook.f.; *Musa × trichocarpa* Nakai; *Musa × vittata* G.W.Ackermann ex Rodigas; and *Musa balbisiana* var. *vittata* (G.W.Ackermann ex Rodigas) M.R.Almeida (POWO, 2024d).

Other common names: **English**: Banana, Dessert banana, Plantain, Cooking banana; **French**: *Banane, bananier*; **German**: *Banane, Bananen, Bananenpisang,*

Bananenstaude; **Spanish & Latin American**: *Banano* (plant), *Plátano, Platanero* (plantain), *Guineo* (dessert banana); **Japanese**: *Chotda banana* (plant), *Banana no tsubomi* (flower); **Indonesian**: *Getang, Djantoong* (plant), *Jantung pisang* (flower), *Pisang*; **Malaysian**: *Djantoong* (plant), *Jantung pisang* (flower); **Philippines**: *Saging*; and **Mexican**: *Barbaro, Zapolete* (Nelson *et al.*, 2006).

Names in Pacific islands: *Aga* (ripe banana) (Chamorro); *Chotda* (Chamorro, Guam, Northern Marianas); *Fa'i* (Samoa); *Leka, Jaina* (Indian derivation) (Fiji); *Mai'a* (Hawai'i); *Maika, Panama* (New Zealand: Maori); *Meika* (Cook Islands); *Meika, mei'a* (French Polynesia); *Siaine* (introduced cultivars), *Hopa* (native) (Tonga); *Sou* (Solomon Islands); *Te banana* (Kiribati); *Uchu* (Chuuk); *Uht* (Pohnpei); *Usr* (Kosrae); and Fe'i banana cultivars have a host of common names in different islands (Nelson *et al.*, 2006). **Indian Hindi name:** Kela.

1.1 Introduction

Banana is an elongated edible fruit, botanically a berry, produced by several kinds of large tree-like herbaceous flowering plants in the genus *Musa* (Merriam-Webster, 2024; Morton, 2013). The term 'banana' encompasses a rich diversity of species and hybrids originating from the tropical regions of Indomalaya and Australia. Likely first domesticated in New Guinea, bananas have played a pivotal role in human societies throughout history. Notably, the genus *Musa* (both bananas and plantains) of extraordinary significance yields the 4th most essential food globally ~ following rice, wheat and maize.

In some countries, bananas utilized for culinary purposes are distinctively referred to as 'plantains', setting them apart from the sweeter dessert bananas. The *Musa* species demonstrate remarkable versatility, adapting to various environments and serving diverse human needs. It provides staple foods in tropical regions to contributing cold-hardy fibre and ornamental plants. *Musa* has been integral to human diets since recorded history.

Within Pacific societies, *Musa* species occupy a central role, serving as a source for food, beverages, fermentable sugars, medicines, flavourings, cooked foods, silage, fragrance, rope, cordage, garlands, shelter, clothing, smoking material, and fulfilling numerous ceremonial and religious functions. Particularly well-suited for traditional Pacific island agroforestry, interplanting in diversified systems, and plantation-style cultivation in full sun—except for atoll islands—bananas and plantains offer versatile agricultural solutions. While predominantly consumed locally in the Pacific region, these fruits have also gained prominence in the worldwide export market (Nelson *et al.*, 2006). *Musa* genus stands as a testament to its enduring importance ~ not only as a global food source but as a multifaceted resource deeply ingrained in cultural, agricultural and economic landscapes.

Banana is the second most significant fruit crop in India after mango. The enduring popularity of bananas stems from their year-round availability, affordability, diverse range of varieties, delightful taste, and their noteworthy nutritional *cum* medicinal values ~ making them a cherished favourite among people of all social classes. Furthermore, bananas boast considerable export potential. Engaging in hi-tech cultivation practices for banana proves to be an economically viable enterprise. This approach not only results in increased productivity but also enhances the quality of the produce and accelerates crop maturity. The superior quality of the cultivated bananas positions them to command premium prices in the market. Overall, the cultivation of bananas stands as a promising venture, combining economic viability with advancements in agricultural technology (nhb.gov.in, 2024).

In Indian context, the delineation between bananas and plantains is absent. Instead, all cultivars are assessed based on their suitability either for dessert consumption or for cooking. The classification revolves around the intended use of the fruit, with no inherent separation between the two categories.

1.2 Etymology

Etymology of the genus name *Musa* can be traced back to its Arabic origin, believed to be derived from the term 'mouz' possibly bestowed in honour of Antonius Musa (63–14 BC) who served as the physician to Octavius Augustus, the inaugural emperor of Rome. The name 'banana' has its roots in the Arabic word 'banan' meaning finger, and is believed to have been applied in Guinea (West Africa) during the introduction of the fruit by the Portuguese. Subsequently, this appellation found its way to the New World, signifying the global dissemination of both the plant and its distinctive nomenclature.

1.3 Origin and Distribution

Musa species, indigenous to tropical Indomalaya and Australia, are believed to have undergone their initial domestication in New Guinea, as supported by various studies (Simmonds, 1956; Argent, 1976; Lentfer and Boyd 2003; Nelson *et al.*, 2006; Morton, 2013). The plants found in 135 countries (ProMusa, 2013) are predominantly cultivated for their fruits though they also serve additional purposes such as fibre production, preparing banana wine & beer and occasionally as ornamental plants. The edible banana, with its roots traced back to the hot and humid tropical regions of SE Asia is thought to have originated at diverse centres with India standing out as one of them. As of today, the cultivation of bananas has proliferated across warm tropical zones worldwide thriving between latitudes 30°N and 30°S of the equator. This widespread cultivation has transformed bananas into a globally consumed fruit with diverse varieties adapted to different climates and regions across the tropical belt (nhb.gov.in, 2024; kvk.icar.gov.in, 2024). Noteworthy regions where bananas are grown include Bangladesh, the Caribbean Islands, the Canary Islands, Florida, Hawaii, Egypt, Israel, Ghana, Congo, South Africa, Fiji, Taiwan, Indonesia, the Philippines, South China, Queensland and Sri Lanka. The adaptability and widespread cultivation of bananas highlight their significance in a multitude of climates and geographical locations (kvk.icar.gov.in, 2024).

1.4 Production

Total annual world production is estimated at 86 million tonnes of fruits. Notably, India and China emerged as the world's foremost banana producers in 2017 ~ jointly contributing to approximately 38% of the total global production. By 2023, India had reached an impressive annual banana yield of nearly 30.5 million tonnes, surpassing China's output which was approximately 20 million tonnes (ProMusa, 2013; worldpopulationreview.com, 2024).

Banana production by 2024 in various countries

Country	Banana production
India	30.5 million tonnes
China	12mt
Indonesia	7.3mt
Brazil	6.8mt
Ecuador	6.6mt
Philippines	6mt
Guatemala	4.3mt
Angola	4mt
Tanzania	3.4mt
Colombia	2.9mt

Source: (worldpopulationreview.com, 2024)

1.5 Banana and Plantain

In the United States, the classification of yellow fruits is relatively straightforward, with 'bananas' commonly designated for raw consumption and 'plantains' for cooking purposes. However, this distinction is not always so clear and can be subject to inversion. On a global scale, there is a lack of a precise differentiation between bananas and plantains. In the Americas and Europe, the term banana typically refers especially to Cavendish group where fruits are soft, sweet and for dessert purposes, as well as meant for major exports. As of 2019, these bananas ranked as the most consumed fresh fruit in the U.S. In contrast, the firmer and starchier fruits from *Musa* cultivars are identified as plantains.

Beyond the global overview, regional nuances in nomenclature add complexity. Spanish-speaking regions employ a diverse set of terms including *banana china, banano enano, cambur, cachaco, colicero, cuatrofilos, carapi, curro, guineo, murrapo, mampurro, patriota, platano, platano de seda, platano enano, suspiro* and *zambo*. In Brazil, Portuguese names such as *banana maca, banana de São Tomé* and *banana da Prata* are used. French-speaking areas may utilize terms like *bananier nain, bananier de Chine, figue, figue banane* and *figue naine*. German-speaking regions have their own designations including *echte banane, feige* and *feigenbaum*. Meanwhile, Sudan uses the term '*baranda*'. These regional variations underscore the complexity of fruit nomenclature and highlight the diverse ways in which cultures and languages across the world distinguish and categorize these yellow fruits (Morton, 2013).

The nomenclature of plantain varieties reveals a rich tapestry of diverse designations across various regions. In Surinam, these fruits are acknowledged as '*banaan*' while in Panama the name shifts to '*banano macho*'. The Caribbean islands of Haiti, Guadeloupe and Martinique name these as '*banane*' or '*bananier*'. Specifically, in Haiti the nuanced terms like '*banane misquette*', '*banane musquee*' or '*pie banane*' are used. Geographical expanses further contribute to this linguistic mosaic, with Brazil employing '*bananeira de terra*', Costa Rica opting for '*banano indio*', Mexico designating them as '*barbaro*', Honduras as '*butuco*' and Venezuela as '*parichao*'.

Guyana, Jamaica and Trinidad prefer the term '*plantain*', while in Cuba, Puerto Rico and the Dominican Republic variations like '*platano*', '*platano burro*' or '*platano hembra*' are employed. Intricate distinctions emerge including '*platano macho*' in Cuba & Panama, '*platano de la isla*' in Peru and '*topocho*' or '*yapuru*' in Venezuela. Mexico contributes its own unique identifier, recognizing them as '*zapolote*'. This expansive range of nomenclatures reflects the cultural diversity and linguistic intricacies associated with the cultivation and consumption of plantains across different regions (Morton, 2013).

An alternative categorization of bananas involves distinguishing between dessert bananas and cooking bananas, with plantains identified as a specific subgroup within the category of cooking bananas (OGTR, 2008). Dessert bananas, exemplified by triploid cultivars derived solely from *M. acuminata*, are contrasted with plantains, which are triploid cultivars resulting from the hybridization between *M. acuminata* and *M. balbisiana*, specifically falling under the plantain subgroup of the AAB Group (Stover and Simmonds, 1987; Qi *et al.*, 2000). The cultivation practices among small farmers in Colombia display a broader spectrum of banana cultivars compared to large commercial plantations. A comprehensive study of these cultivars reveals that they could be categorized into at least three distinct groups based on their characteristics: dessert bananas, non-plantain cooking bananas and plantains. However, it's important to note that there were instances of overlapping between the categories of dessert and cooking bananas (Gibert *et al.*, 2009).

1.6 Species

Certain species within the genus *Musa* exhibit diverse uses, reflecting their unique characteristics and regional significance. For instance, *M. basjoo* Sieb. & Zucc. hailing from Japan and *M. ornata* Roxb. native to the region spanning Pakistan to Myanmar primarily serve as ornamental plants or sources of fibre. In contrast, *M. textilis* Nee, originating from the Philippines, is specifically cultivated for its robust fibres ~ highly prized for manufacturing strong ropes and delicate tea bags.

The Abyssinian banana, *Ensete ventricosum* Cheesman (formerly known as *E. edule* Horan and *Musa ensete* Gmel.), finds cultivation in Ethiopia for a dual purpose ~ fibre production and the derivation of staple foods from its young shoots, stem base and corm. This multifaceted utilization underscores its economic and nutritional importance within the local context.

Contemporary edible banana varieties have undergone significant development, stemming from two primary species ~ *Musa acuminata* Colla and *Musa balbisiana* Colla, along with their natural hybrids. These progenitors were originally discovered in the lush rainforests of SE Asia. The cultivation of bananas, tracing its roots back to the 7th century AD, gradually extended beyond SE Asia, reaching regions as distant as Egypt and Africa (nhb.gov.in, 2024). Edible-fruited bananas, typically seedless are prominently associated with *M. acuminata* Colla species. The hybrid *M.* × *paradisiaca* L. syn. to *M.* × *sapientum* L., results from the crossing of *M. acuminata* and *M. balbisiana*. *M. balbisiana* found in southern Asia and the East Indies bears a seeded fruit but is highly valued for its disease resistance. This attribute positions it as a crucial parent in breeding programmes aimed at enhancing disease resistance in edible bananas, highlighting the species' pivotal role in the cultivation of this globally significant fruit (Simmonds, 1956; Argent, 1976; Lentfer and Boyd 2003; Nelson *et al.*, 2006; Morton, 2013).

The taxonomic classification of bananas commonly referred to as fehi (=fe'i) is currently marked by uncertainty. The specific names *M. fehi* Bertero ex Vieill. & *M. troglodytarum* L. are in a state of pending resolution, awaiting final decisions from taxonomists to establish their definitive categorization. The lingering taxonomical ambiguity underscores the need for further investigation and consensus within the scientific community regarding the precise classification of these banana varieties (Morton, 2013).

1.6.1 Indicative listing of the possible species in *Musa**

Below is an enumeration of species that could be categorized within the *Musa* genus. It is important to acknowledge that the *Musa* genus comprises over 50 species, yet the provided list includes species that are subject to uncertainty or doubt.

Chromosome no.	Section	'Minor' section	Species	Main distribution
2n=2x=14	*Ingentimusa*		*M. ingens*	Papua New Guinea
2n=2x=20	*Callimusa*	*Callimusa*	*M. angcorensis*[a]	Cambodia
			M. bauensis	Borneo
			M. borneënsis	Borneo
			M. campestris	Indonesia, Borneo
			M. coccinea	China, Indochina, Indonesia,Vietnam
			M. exotica	Vietnam
			M. flavida	Borneo
			M. gracilis	Malaysia
			M. lawitiensis	Borneo
			M. paracoccinea	China, Vietnam
			M. pigmaea	Borneo
			M. salaccensis	Indonesia
			M. splendida[a]	Vietnam
			M. suratii	Borneo
			M. violascens	Malaysia
		Australimusa	*M. alinsanaya*	Philippines
			M. angustigemma[b]	Papua New Guinea
			M. beccarii[c]	Borneo
			M. boman	Papua New Guinea
			M. bukensis	Papua New Guinea
			***M. fitzalanii*[a]**	**Australia**
			M. hirta	Indonesia, Borneo
			M. insularimontana[a]	Taiwan
			M. jackeyi	**Australia**
			M. johnsii	Papua New Guinea
			M. lolodensis	Papua New Guinea
			M. maclayi	Papua New Guinea
			M. monticola	Borneo
			M. muluensis	Borneo
			M. peekelii	Papua New Guinea, Philippines
			M. textilis	Philippines,Brunei, Moluccas
			M. tuberculata	Borneo
			Fe'I cultivars[d]	New Guinea

Chromosome no.	Section	'Minor' section	Species	Main distribution
2n=2x=22	*Musa*	*Musa*	***M. acuminata***[e]	India, Indonesia, Malaysia, Philippines, Sri Lanka, Thailand, Vietnam, **Australia**
			M. balbisiana	Philippines, Bhutan, China, India, Vietnam, Papua New Guinea, Sri Lanka
			M. basjoo	Japan, China
			M. cheesmanii	India
			M. flaviflora	Bangladesh, Bhutan, India
			M. halabanensis	Indonesia
			M. itinerans	China, India, Thailand,Vietnam
			M. nagensium	China, India, Thailand
			M. ochracea	
			M. schizocarpa	Indonesia, Papua New Guinea
			M. sikkimensis	Bangladesh, Bhutan, India, Thailand
			M. griersonii	Bhutan
		Rhodochlamys	*M. aurantiaca*	India
			M. laterita	India, Myanmar, Thailand
			M. mannii	India
			M. ornata	Bangladesh, Myanmar, India, Borneo
			M. rosea	Bangladesh, Myanmar
			M. rubra	Myanmar, India
			M. sanguine	Northern India, Myanmar, China
			M. thomsonii[a]	India
			M. velutina	Northern India

[There is still uncertainty about the taxonomy of *Musa* and this table may overestimate the number of species and the number of sections. The species are grouped into 3 sections although the existence of *Ingentimusa* is arguable. The possible sub-division into 5 sections is indicated by the inclusion of the column headed 'minor' section, though it should be noted that this is not a recognized classification terminology. There is also uncertainty about which section some species belong in]

[a]These species are very poorly known and may not exist

[b]These species have been classified as subspecies of *M. peekelii*

[c]*M. beccarii* is placed by some in Australimusa and others in Callimusa

[d]The Fe'i bananas are widely distributed throughout the Pacific islands, from the Moluccas to Hawaii and Tahiti and their domestication is thought to have occurred independently of other bananas and plantains. Their origin is controversial.

[e]*M. acuminata* is regarded by some to comprise a number of subspecies, namely *banksii*, *burmannica*, *burmannicoides*, *errans*, *malaccensis*, *macrocarpa*, *truncata*, *siamea*, *zebrina*

*Information in the table was compiled from a number of sources (Simmonds and Weatherup, 1990; Constantine and Rossel, 2001; Ross, 1987; Horry *et al.*, 1997; Pillay and Tripathi, 2007; Wong *et al.*, 2002; Constantine and Rossel, 2001)

Source: OGTR, 2008

1.7 Botany

1.7.1 Size and growing habit

Young banana plant

By Nepali Keto 62 Talk to me - Own work, CC BY-SA 4.0, https://commons.wikimedia.org/w/index.php?curid=49875707

Banana rhizome

By Obsidian Soul - Own work, CC0, https://commons.wikimedia.org/w/index.php?curid=14997197

The banana plant, usually misidentified as a 'tree', is a substantial perennial herb, reaching heights between 2 to 9 metres. This monocotyledonous plant emerges from robust, subterranean rhizomes, often referred to as corms. The plant features a succulent and highly juicy stem, more accurately termed a pseudostem, comprising a cylindrical arrangement of leaf-petiole sheaths. Around the primary plant, offspring shoots, known as suckers, emerge, forming a cluster or stool. The eldest sucker takes over as the main plant when it bears fruit and eventually dies. This succession process perpetuates indefinitely, ensuring a continuous life cycle for the banana plant (Nelson *et al.*, 2006; Morton, 2013).

1.7.2 Leaves

Constantine and Rossel, 2001; Horry *et al.*, 1997; Nelson *et al.*, 2006; OGTR, 2008; Pillay and Tripathi, 2007; Ross, 1987; Simmonds and Weatherup, 1990; Wong *et al.*, 2002 mention that entire aboveground structure of the plant deviates from the conventional concept of a true woody trunk found in other tree species. Instead, it is characterized by a unique feature known as a 'false trunk' ('false stem'), scientifically termed a pseudostem. This pseudostem is formed by the fusion of leaves and their petiole bases. Functioning as a supportive framework, the pseudostem bears a canopy composed of 4 to 5 to as many as 20 leaves (Nelson *et al.*, 2006).

The leaves of the plant are notable for their tender, smooth texture and oblong or elliptic shape, each attached to the stem by fleshy stalks. Arranged in a spiral fashion, these leaves gradually unfurl at the pace of one per week during warm weather. As the plant matures, the leaves extend both upward and outward, reaching dimensions of up to 2.75 metres in length and 60 centimeters in width. Their colouration varies, presenting as entirely green, green with maroon splotches or featuring a dual colour scheme with green on the upper side and red-purple hues beneath (Morton, 2013).

1.7.3 Flowers

Plant showing fruit and inflorescence

By Daniela Kloth - Own work, GFDL 1.2, https://commons.wikimedia.org/w/index.php?curid=73500454

Female flowers with petals at the tip of ovary

By By Ruestz (Own work) [GFDL (http://www.gnu.org/copyleft/fdl.html), CC-BY-SA-3.0 (https://creativecommons.org/licenses/by-sa/3.0/) or CC-BY-SA-2.5-2.0-1.0 (https://creativecommons.org/licenses/by-sa/2.5-2.0-1.0)], via Wikimedia Commons - File:M._acuminata_x_balbisiana.JPG, CC BY 3.0, https://commons.wikimedia.org/w/index.php?curid=21458471

The inflorescence, an altered growing point, emerges as a terminal spike extending from the apex of the stem's heart. Initially, it manifests as a sizeable, elongated-oval and tapering bud enveloping in a vibrant purple hue. Upon unfolding, it reveals a composition of slender, nectar-rich, tubular and serrated white flowers arranged in whorled double rows along the floral stalk. Each cluster is ensconced by a substantial, waxy and hood-like bract, characterized by an exterior purple shade and an interior deep-red hue. Typically, the bract undergoes elevation from the first set of flowers within a span of 3-10 days, though in cases of weakened plants, this process might extend to 10 or 15 days.

Partially opened inflorescence

By Ruestz - Own work, CC BY-SA 3.0, https://commons.wikimedia.org/w/index.php?curid=824363

The lower sections of the inflorescence are predominantly occupied by female flowers, spanning 5 to 15 rows, while hermaphrodite or neutral flowers may appear above them, with male flowers positioned in the uppermost rows. Though certain varieties maintain an upright inflorescence, more commonly it starts bending downward shortly after unfolding. Approximately one day post the blooming of flower clusters, the male flowers along with their bracts are shed, rendering the upper portion of the stalk mostly bare except for the very tip where an unopened bud usually harbours the final-formed male flowers.

However, deviations exist such as in case of mutants like 'Dwarf Cavendish' which exhibits the persistence of male flowers and bracts. These withered structures endure, filling the space between the fruits and the terminal bud, creating a distinctive characteristic in contrast to the typical life cycle of the inflorescence (Morton, 2013).

1.7.4 Fruits (Nelson *et* al., 2006; Morton, 2013)

As the nascent fruits emerge from the female flowers, their appearance resembles slender green fingers. The protective bracts are eventually shed allowing the fully developed fruits within each cluster to form a distinctive 'hand' of bananas. As the weight increases, the stalk bends causing the entire bunch to assume an upside-down position. The quantity of 'hands' varies depending on the species and variety.

Fruits of four cultivars

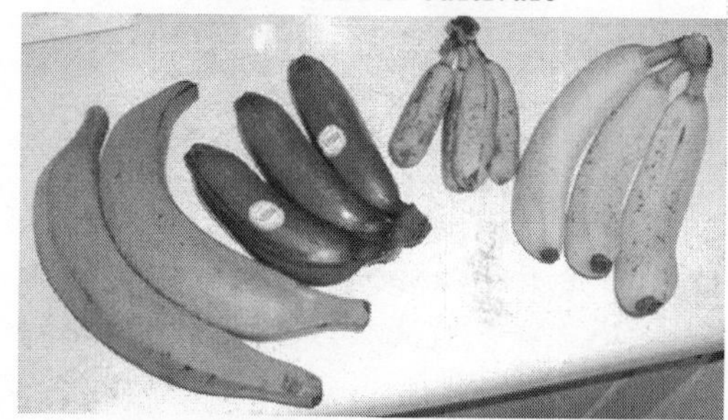

By User TimothyPilgrim on en.wikipedia - Taken by TimothyPilgrim., CC BY-SA 3.0, https://commons.wikimedia.org/w/index.php?curid=1101389

The transformation of the fruit technically classified as a 'berry' unfolds with a shift in colour from a deep-green hue to vibrant yellow, red or in certain variations a striking green-and-white-striped pattern. The dimensions of the bananas can range from 3 to 40cm in length and 2 to 8cm in diameter. The shapes may vary from oblong, cylindrical and blunt to distinctly 3-angled, somewhat curved and hornlike.

The peel of the fruit exhibits diversity ranging from thin and tender to thick and leathery, with colours spanning silver, yellow, green or red. Notably, the thin-peeled varieties are more susceptible to damage during transport compared to those with a comparatively thicker peel as observed in the Fe'i variety 'Karat' and others.

Banana flesh which is technically ivory-white but transforms to yellow or salmon-yellow, undergoes a remarkable textural evolution. In its unripe state, the flesh may be firm, astringent or may even possess a somewhat gummy consistency due to latex. As the banana ripens, the texture transitions to tender and slippery, soft and mellow, rather dry and mealy or starchy. The flavour profile is equally diverse, ranging from mild and sweet to subacid, often exhibiting a distinct apple undertone.

1.7.5 Seeds

Fruits of wild-type bananas with numerous large, hard seeds

Wild types of bananas may exhibit a nearly solid interior filled with black, hard and rounded or angled seeds ranging from 3-16mm in width, accompanied by scant flesh. When present the seeds exhibit variations in shape and morphology across different species. For instance, *Musa balbisiana* which serves as a parent to many commercially grown edible banana varieties, features ovoid dark brown seeds approximately 4mm in length, characterized by a conspicuous white *cum* powdery endosperm (Nelson *et al.*, 2006).

In contrast, common cultivated banana types are typically seedless, with only minute vestiges of ovules visible as brown specks in the slightly hollow or faintly pithy centre, particularly evident in overripe fruits. While there are occasional instances of cross-pollination by wild types resulting in the development of seeds in typically seedless varieties like 'Gros Michel', this phenomenon is not observed in the Cavendish type (Morton, 2013).

1.7.6 Rooting habit

Musa plants exhibit a prolific network of fibrous roots, typically numbering between 200 and 500. In conditions characterized by well-drained, deep and fertile soils, these roots have the capacity to extend vertically to a depth of 1.5 metres, while laterally these can span up to 4.9 metres. However, in regions with dry, shallow or rocky soils, *Musa* roots may face challenges in competitive growth. Nevertheless, under optimal conditions, *Musa* demonstrates an average to good competitive ability.

1.8 Genetics

Musa species are categorized based on their ploidy which refers to the number of chromosome sets in their genome and presence of relative proportion of *Musa acuminata* (A) and *Musa balbisiana* (B). The most common cultivated banana varieties are triploid hybrids (AAA, AAB, ABB) while diploids (AA, AB, BB) and tetraploids (AAAA, AAAB, AABB, ABBB) are much rarer and mainly experimental hybrids (Nelson *et al.*, 2006).

Sections 'Callimusa' and 'Rhodochlamys' primarily comprise non-parthenocarpic species that are valued for ornamental purposes rather than their nutritional fruits (Pillay and Tripathi, 2007). In contrast, the majority of cultivated sweet bananas and plantains fall under Section *Musa*. These are triploid varieties that originated from the evolution of two wild diploid species, *M. acuminata* with the genome designation 'AA', and *M. balbisiana* with the genome designation 'BB' (Simmonds and Shepherd, 1955).

The development of homogenomic triploid hybrids (2n=3x) with the AAA genotype took place within *M. acuminata*, leading to creation of cultivars predominantly consisting of sweet bananas (OGTR, 2008). Notably, the transmission of chloroplast DNA in *Musa* spp. occurs maternally while mitochondrial DNA is transmitted paternally. Phylogenetic studies utilising this feature suggest that the origin of edible banana cultivars is notably linked to two sub-species of *M. acuminata*, *viz. M. acuminata* subsp. *banksii* and *M. acuminata* subsp. *errans* (Horry *et al.*, 1997).

Crossbreeding between diploid and triploid types of *M. acuminata* with *M. balbisiana* resulted in the formation of heterogenomic triploid hybrids, primarily comprising plantains (AAB genotype) and other cooking bananas (ABB genotype). Additionally, tetraploid (2n=4x) and various diploid combinations also exist (Pillay *et al.*, 2004). Hybrids of *M. acuminata* and *M. balbisiana* are denoted as *Musa* × *paradisiaca* (OGTR, 2008), with *M. paradisiaca* initially assigned as the name by Linnaeus, who did not recognize it as a hybrid. The utilization of isozymes and molecular markers has verified the multi-specific origin of edible bananas (Visser, 2000). Furthermore, studies involving restriction polymorphisms of chloroplast and mitochondrial DNA indicate that species within the section Rhodochlamys may constitute a secondary genepool for the improvement of cultivated bananas (Nwakanma *et al.*, 2003).

Cultivated *Musa* species typically yield sterile or minimally fertile fruits, exhibiting parthenocarpy wherein fruit pulp develops without pollination which results into seedless fruits. While sterility and parthenocarpy enhance the appeal of banana fruits but pose challenges to breeding programmes. Over millennia, natural somatic mutations, hybridization and selective cultivation have led to substantial genetic diversity within cultivated bananas thereby resulting in over 1000 varieties globally. Particularly in the Pacific, notably in Papua New Guinea and the Solomon Islands, there is a rich diversity of banana varieties but concerns have been raised about the increasing rarity of certain varieties which threatens overall banana diversity (Nelson *et al.*, 2006).

Breeding programmes have faced obstacles due to fertility issues among desirable parent plants such as the virtually sterile Cavendish subgroup. Recent progress in developing useful cultivars has been hindered by susceptibility to significant diseases and pests in both naturally occurring and artificially bred hybrids (Nelson *et al.*, 2006).

Edible bananas worldwide belong to the Eumusa section of the Musa genus except for the Fe'i group which falls under Australimusa section. Fe'i bananas are characterized by erect bunches, pink-red to purple sap and deep yellow or orange-coloured fruit pulp (Nelson *et al.*, 2006).

1.9 Classification and Cultivar Groups

Simmonds and Shepherd (1955) advocated for a more systematic approach to taxonomic naming, proposing that genome nomenclature would be more fitting for designating taxa. Their suggestion involved appending a letter combination after the generic name, indicating both the ploidy and genome sets, followed by the specific cultivar or cultivar group name as illustrated in the table below. Cultivars within each subgroup exhibit minimal genetic diversity and are interconnected through somatic mutations as elucidated by Horry *et al.* (1997).

1.9.1 Examples of genome nomenclature for some common banana cultivars

Genome Group	*Subgroup*	*Example of common cultivar name*[a]
AA Diploid *Musa acuminata*, both wild banana plants and cultivars	Sucrier	'Sucrier'; 'Lady's Finger'[b]
AB Diploid cultivars of *Musa × paradisiaca*	Ney Poovan	Lady's Finger[b]
AAA Triploid *Musa acuminata*, both wild banana plants and cultivars	Cavendish	'Giant Cavendish' (e.g. 'Williams', 'Mons Mari')
		'Grande Naine'
		'Dwarf Cavendish[c]
		'Extra Dwarf Cavendish' (e.g. 'Dwarf Parfitt')
	Gros Michel	'Gros Michel'
	Red	'Red Dacca'
AAB Triploid cultivars of *Musa × paradisiaca*. Contains the Plantain subgroup composed of "true" plantains or African Plantains. The Group's centre of diversity is C&W Africa where a large number of cultivars were domesticated following the introduction of ancestral plantains from Asia, possibly 2000–3000 years ago.	Maoli-Popoulu	'Pacific Plantain'
	Mysore (cooking and dessert bananas)	'Mysore'
	Plantain	'French', 'Pisang Ceylan'
	Pome	'Lady's Finger'[b]
	Silk	'Sugar'

Genome Group	*Subgroup*	*Example of common cultivar name*[a]
ABB Triploid cultivars of *Musa* × *paradisiaca*	Bluggoe	'Bluggoe' (orinoco and "burro"), 'Mondolpin'
	Ney Mannan	'Blue Lubin', 'Blue Java' (Ice Cream, Ney Mannan, Ash plantain, Pata Hina, Dukuru, Vata)
	Pisang Awak	'Ducasse', 'Kluai Namwa Khom'
AABB Tetraploid cultivars of *Musa* × *paradisiaca*		Kalamagol banana, Pisang Awak (Ducasse)
AAAB Tetraploid cultivars of *Musa* × *paradisiaca*		FHIA-01[d] ('Goldfinger'), FHIA-18 ('Bananza')
ABBB Tetraploid cultivars of *Musa* × *paradisiacal*		Tiparot banana
AAAA Tetraploid *Musa acuminata* - both wild bananas and cultivars		'Bodles Altafort', 'Golden Beauty'
BB Diploid *Musa balbisiana*, wild bananas		

[a] Common names vary depending on the country

[b] 'Lady ('s) Finger' has been used to name at least four distinct AA, AB and AAB clones; genome group AAB subgroup Pome represents the cultivar that makes up about 4% of Australian production

[c] Dwarf Cavendish is the most widely distributed clone of edible banana worldwide

[d] FHIA=varieties bred at the Fundación Hondureñade Investigación Agrícolain Honduras

Source: OGTR, 2008; Wikipedia, 2024; Molina and Roa, 2000.

Edible bananas are intricately classified into various main groups and subgroups. The first group highlighted by Simmonds is the diploid *M. acuminata* group **'Sucrier'**, which is distributed across regions such as Malaya, Indonesia, the Philippines, southern India, East Africa, Myanmar, Thailand, the West Indies, Colombia and Brazil. Characterized by dark-brown sheaths, yellowish leaves with minimal wax, small bunches and petite, thin-skinned *cum* sweet fruits. Cultivars within this group hold particular significance in New Guinea.

Among the notable cultivars in this group is the **'Lady Finger'**, also known as 'Dte' or 'Fig' and referred to in Spanish as 'Dedo de Dama', 'Datil', 'Nino', 'Bocadillo', 'Manices', 'Guineo Blanco' or 'Cambur Titiaro'. This banana plant reaching a height of up to 7.5m, features a slender pseudotrunk and a robust root system providing resistance against strong winds. With outer sheaths displaying streaks or patches of reddish-brown, the bunch comprises of 10-14 hands each carrying 12-20 fingers. The fruit measuring 10-12.5cm in length boasts thin, light-yellow skin and sweet flesh. Resilient to drought, Panama disease and the black weevil, it is however susceptible to Sigatoka leaf spot disease. The 'Lady Finger' is prevalent in Latin America and holds commercial value in Queensland and New South Wales.

Within the second category, the group is exemplified by the extensively cultivated **'Gros Michel'** which basically originates from Myanmar, Thailand, Malaya, Indonesia and Sri Lanka. This widely grown variety serves as a representative specimen of this category showcasing its geographical roots and distribution across various regions. Introduced to Martinique in the early 19th century by a French naval officer, it subsequently made its way to Jamaica, Fiji, Nicaragua, Hawaii and Australia. Characterized by its large, tall stature and long bunches of sizable yellow fruits, 'Gros Michel' was once the primary commercial cultivar in Central Africa, Latin America and the Caribbean. However, its extensive susceptibility to Panama disease has led to its gradual phasing out. Over time, 'Gros Michel' has given rise to various named sports or mutants.

The **Cavendish** subgroup comprises several significant banana cultivars each with distinctive characteristics (Morton, 2013):

'Dwarf Cavendish': Initially identified in China, this cultivar is widely cultivated particularly in the Canary Islands as well as in East & South Africa. The plant ranging from 1.2-2.1m in height features broad leaves on short petioles, demonstrating hardiness and resistance to wind. The medium-sized fruit of good quality has thin skin and requires careful handling during shipment. Notably, this cultivar is easily recognizable due to the retention of male bracts and flowers.

'Giant Cavendish' (syn. 'Mons Mari', 'Williams', 'Williams Hybrid' or 'Grand Naine'): With an uncertain origin, this cultivar closely resembles the 'Gros Michel' and has replaced the 'Dwarfs' in various regions. Found in Colombia, Australia, Martinique, many Hawaiian plantations and to some extent in Ecuador, it has become the commercial banana of Taiwan. The plant reaches 2.7-4.9m in height. The pseudostem displays dark brown splashes, the bunch is elongated and cylindrical and the fruits are larger and less delicate than those of the 'Dwarf'. Male bracts and flowers are shed, leaving a space between the fruits and the terminal bud.

'Pisang Masak Hijau' (syn. 'Bungulan'): Originating from the Philippines, Indonesia and Malaya, this triploid Cavendish clone is erroneously named 'Lacatan' in Jamaica. It replaced 'Gros Michel' due to its resistance to Panama disease though it is susceptible to Sigatoka leaf spot. Its tall and slender plant is prone to wind damage. Fruits ripen unevenly in winter, bruise easily and are prone to spoilage in storage. Though no longer grown commercially in Jamaica and the Windward Islands, the fruits are commonly used for cooking in Jamaican households. Some experts suggest that 'Pisang Masak Hijau' may have been the primary source for all members of the Cavendish group.

'Robusta': Similar to the so-called 'Lacatan', this cultivar has largely replaced 'Gros Michel' in Jamaica and the Windward Islands and is grown in Central America, Brazil, eastern Australia, Samoa and Fiji. It is shorter, thick-stemmed and less susceptible to wind. It is resistant to Panama disease but is prone to Sigatoka. 'Robusta' has become a commercial choice in various regions.

'Valery': Another triploid Cavendish clone which closely resembles 'Robusta'. It is cultivated as a successor to 'Robusta' and is already more widely grown than 'Lacatan' for export. However, in cooking trials, it has received low ratings as the flesh hardens and develops a waxy texture.

The Banana Breeding Research Scheme in Jamaica has successfully developed numerous tetraploid banana clones with enhanced disease resistance, some of which exhibit dessert qualities comparable to the renowned 'Lacatan' and 'Valery'. One notable example is **'Bluggoe'** known by various local names which serves as a cooking banana and boasts exceptional resistance to both Panama disease and Sigatoka. 'Bluggoe' produces distinct hands of large, nearly straight, starchy fruits, holding significant importance in regions such as Myanmar, Thailand, southern India, East Africa, the Philippines, Samoa and Grenada.

'**Ice Cream**' banana, originating from Hawaii ('Cenizo' in Central America and the West Indies; 'Krie' in the Philippines), is a relative of 'Bluggoe'. The plant, growing up to 3-4.5m features a light pink leaf midrib and a flower stalk that can reach several feet in length. The bunch comprises only 7 to 9 hands with the fruit measuring 17.5-22.8cm in length and up to 6.25cm in thickness. Initially bluish with a silvery bloom when young, the fruit ripens to a pale yellow. The flesh sweetish in taste can be consumed raw or cooked.

'**Mysore**', also known as 'Fillbasket' and 'Poovan' stands out as the predominant banana type in India, constituting 70% of the total crop. While it is sparingly grown in regions like Malaya, Thailand, Ceylon and Myanmar, its significance in the New World is primarily observed in Trinidad where it is cultivated as shade for cacao. 'Mysore' is a robust and large plant immune to Panama disease and largely resistant to Sigatoka. It bears large compact bunches of medium-size bearing plumpy, thin-skinned and bright yellow fruits with a subacid flavour.

Additional commercial cultivars such as **'Salembale'** and **'Rasabale'** are less suitable for canning due to their starchy taste and weak flavour. On the other hand, **'Pachabale'** and **'Chandrabale'** are locally preferred varieties for canning. Naik had identified 34 cultivars as particularly important among the diverse range grown in South India.

The **'Silk,'** 'Silk Fig' or 'Apple' banana renowned as the most popular dessert banana in the tropics is widely distributed but not grown on a large scale. The plant standing at 3-3.6m is moderately vigorous, highly resistant to Sigatoka but susceptible to Panama disease. Each bunch comprises only 6-12 hands, each with 16-18 fruits. The plump bananas are slightly curved, astringent when unripe and pleasantly subacidic when fully ripe emitting an apple scent. If left on the bunch until fully developed, the thin skin splits lengthwise but the fruit remains firm and keeps well.

The **'Red'**, 'Red Spanish', 'Red Cuban', 'Colorado' or 'Lal Kela' bananas believed to have originated in India exhibiting robust growth taking 18 months from planting to harvest. Highly resistant to disease, the plant features purplish-red pseudostem, petiole, midrib and fruit peel which turns orange-yellow when fully ripe. The compact bunch may contain over 100 fruits with thick peel and strong-flavoured flesh. In the 'Green Red' mutant, the plant is variegated green *cum* red reaching a height of 8.5m with 45cm pseudostem thickness. Bunch bears 4-7 hands with thick and 12.5-17.5cm long fruits.

The **'Fehi'** or 'Fe'i' group from Polynesia stands out for its erect bunches and the purplish-red or reddish-yellow sap used for ink and dyeing. These plants reaching up to 10.9m tall feature wide leaves of 50-75cm length. The bunches contain about 6 hands of orange or coppery, thick-skinned and starchy fruits with good flavour when boiled or roasted. It is often cultivated as ornamental in Hawaii.

In a separate category, 'I.C. 2' or **'Golden Beauty'** bred at the Imperial College of Tropical Agriculture in Trinidad in 1928 by crossing the 'Gros Michel' with a wild *Musa acuminata*, is resistant to Panama disease and highly resistant to Sigatoka. Despite small bunches and short fruits, it ships and ripens well, making it suitable for export in Honduras and cultivation in Hawaii, Samoa, and Fiji.

The **'Orinoco,'** 'Horse,' 'Hog,' or 'Burro' banana is a medium-tall and sturdy plant known for its hardiness. The bunch comprises only a few hands of very thick, 3-angled fruits, about 15cm long. The flesh, with a salmon tint, is edible raw when fully ripe but is enhanced when cooked whether fried, baked or otherwise similar to plantains.

Between 1976 and 1979, trials were conducted at the Campo Agricola Experimental Station in Tecoman (Mexico), involving five clones of the 'Giant

Cavendish' and nine other cultivars ('Robusta A', 'Robusta B', 'Cocos A', 'Cocos B', 'Golden Beauty', 'Enano Nautia', 'Enano Gigante', 'Enano' and 'Valery'). Though 'Enano Gigante' was the most widely grown cultivar in the region, tests revealed that 'Enano Nautia' and 'Golden Beauty' produced heavier bunches of superior quality fruit despite 'Enano Gigante' having a greater number of bunches and the highest yield per ground area. Clones 1, 2, 3 and 4 of 'Giant Cavendish' and 'Cocos B' exhibited tall growth, low yields and poor-quality fruit.

Among plantains various forms exist, some with pink, red or dark-brown leaf sheaths and others featuring coloured midribs or splotches on leaves or fruits. Typically large and vigorous, these plants are resistant to Panama disease and Sigatoka but susceptible to borers. Major subgroups include 'French plantain' and 'Horn plantain' with the former having persistent male flowers. Usually large and angled fruits are borne in few hands and are vital food sources in southern India, East Africa, tropical America and the West Indies. Leading commercial plantains include the tall **'Maricongo'** and the **'Common Dwarf'**. A dwarf mutant, the 'Plantano Enano of Puerto Rico' ('banane cochon' of Haiti) also exists. Ordinary plantains in Mexico are known as 'cuadrado', 'chato' and 'topocho'. Key commercial cultivars resistant to Black Sigatoka are 'Pelipita' and 'Saba' though lacking the culinary quality of 'Harton', 'Dominico-Harton', 'Currare' and 'Horn'. The fertile 'Laknau' resembles 'Horn' but is of inferior quality, opening up possibilities for hybridization with 'Pelipita' and 'Saba.'

In Florida, the most commonly grown banana and plantain cultivars include the 'Dwarf Cavendish', 'Apple' and 'Orinoco' bananas along with the 'Macho' plantain. 'Red' and 'Lady Finger' bananas are occasionally cultivated in sheltered locations. Worldwide, there are five major collections of banana and plantain clones, with United Brands maintaining a collection of 470 cultivars and 100 species at La Lima (Honduras).

1.9.2 Cultivars grown in India

In the Indian context, bananas are systematically categorized into dessert and culinary types for commercial purposes. The culinary varieties are characterized by their starchy fruits, often utilized in their mature unripe state as vegetables. Noteworthy banana cultivars in India encompass a diverse range including Dwarf Cavendish, Robusta, Monthan, Poovan, Nendran, Red banana, Nyali, Safed Velchi, Basrai, Ardhapuri, Rasthali, Karpurvalli, Karthali and Grand Naine among others.

Grand Naine, an imported variety originating in Israel is gaining significant traction and is poised to potentially emerge as the most favoured banana variety. This popularity can be attributed to its remarkable tolerance to abiotic stresses and the production of high-quality bunches. The fruit itself develops an appealing uniform yellow colouration with an enhanced shelf life and overall superior quality compared to other cultivars.

Important banana varieties cultivated in different states of India:

State	Varieties grown
Andhra Pradesh	Amritpant, Chakrakeli, Dwarf Cavendish, Karpoora Poovan, Monthan, Rasthali, Robusta, Thellachakrakeli and Yenagu Bontha
Assam	Bharat Moni, Bhimkol, Borjahaji (Robusta), Chinia (Manohar), Chini Champa, Digjowa, Honda, Jahaji (Dwarf Cavendish), Jatikol, Kanchkol, Kulpait, Malbhog, Manjahaji
Bihar	Alpon, Chinia, Chini Champa, Dwarf Cavendish, Gauria, Kothia, Malbhig, Muthia
Gujarat	Basrai, Dwarf Cavendish, Gandevi Selection, Harichal, Harichal (Lokhandi), Lacatan, Robusta, G-9, Shrimati
Jharkhand	Basrai, Singapuri
Karnataka	Elakkibale, Dwarf Cavendish, Monthan, Poovan, Robusta, Rasthali
Kerala	Monthan, Nendran (Plantain), Palayankodan (Poovan), Rasthali, Red Banana, Robusta
Madhya Pradesh	Basrai
Maharashtra	Basrai, Dwarf Cavendish, Grand Naine, Lal Velchi, Rajeli Nendran, Red Banana, Robusta, Safed Velchi, Shreemanti
Orissa	Champa, Dwarf Cavendish, Patkapura (Rasthali), Robusta
Tamil Nadu	Karpuravalli, Nendran, Matti, Monthan, Peyan, Poovan, Rasthali, Red Banana, Robusta, Sakkai, Virupakshi
West Bengal	Champa, Dwarf Cavendish, Giant Governor, Kanthali, Mortman, Singapuri

(nhb.gov.in, 2024; kvk.icar.gov.in, 2024)

Dwarf Cavendish (AAA): Its 'Basrai', the leading commercial variety is widely cultivated for table and processing purposes in Maharashtra, Gujarat, Bihar, West Bengal, Tamil Nadu, Karnataka and Andhra Pradesh. Its plant has a dwarf stature so reducing susceptibility to wind damage. Despite a rather poor keeping quality, the bunch size, *vis-à-vis* the fruit length and size are commendable with an average bunch weight of 15-25kg. Its thick rind retains the greenish colouration even when ripe. The Gandevi selection known as 'Hanuman' or 'Padarre' is gaining importance, yielding larger bunches of 55-60kg but with a long crop duration. It performs well under light soils with high inputs, successful especially with high-density planting and drip irrigation. It is highly susceptible to Sigatoka leaf spot disease in humid tropics which limits its commercial cultivation.

- **Karpuravalli (ABB):** It is popular for table purposes in medium-rich soils. It is commercially cultivated in central and southern districts of Tamil Nadu and Kerala. Occasionally it seeds depending on seasonal variability. It is a tall and robust plant with sweet ash-coated golden yellow fruits. It is highly susceptible to wilt disease though tolerant to leaf spot and is well-suited for drought, salt-affected areas and for low-input conditions.
- **Monthan (ABB):** This variuety is widely cultivated for processing. It is a fairly tall and robust plant bearing bunches of 18-20kg after 12 months. Its fruits are bold, stocky, knobbed and pale green. Its new clones like 'Kanchi Vazhai' and 'Chakkia' are gaining popularity in Tamil Nadu. It is immune to 'banana bunchy top virus' (BBTV), is salt-tolerant and maintains a normal bunch mass even under marginal conditions though is highly susceptible to *Fusarium* wilt.
- **Nendran (AAB):** It is popular in Kerala for both consumption and processing and is gaining rapid cultivation in Tamil Nadu. It displays considerable diversity in plant features. Bunches have 5-6 hands weighing about 12-15kg. Fruits have a distinct neck with thick green skin turning buff yellow on ripening. Remains starchy even after ripening. Highly susceptible to 'banana bract mosaic virus (BBMV), nematodes and borers.

Cv. Nendran

By ARUNKUMAR P.R - Own work, CC BY-SA 3.0, https://commons.wikimedia.org/w/index.php?curid=20981988

- **Ney Poovan (AB):** A choicest diploid cultivar under commercial mono-cultivation, especially in Karnataka and Tamil Nadu. It is grown in backyards in Kerala, shifting to large-scale cultivation. Slender plant bearing bunches of 15-30kg after 12-14 months having highly fragrant, tasty, powdery and firm fruit. Its dark green fruits turn golden yellow with excellent keeping quality. It is tolerant to leaf spot but susceptible to *Fusarium* wilt and BBMV.
- **Pachanadan (AAB):** This variety is popular in Tamil Nadu for its cooling effects in hot tracts during summer. It thrives in marginal soils without yield reduction. It is well suited as an intercrop in coconut/arecanut gardens. Its bunch weight ranges from 12 to 15kg. It is tolerant to leaf spot and BBTV diseases but is susceptible to wilt.

- **Poovan (Mysore AAB):** Widely grown throughout the country with location-specific ecotypes. It is cultivated as a perennial crop especially in Tamil Nadu and is commercially used for the leaf industry. Its slightly acidic fruits with a sour-sweet aroma turn attractive golden yellow on ripening. It produces medium-sized bunch with closely packed fruits that have good keeping quality but are highly susceptible to BBMV and 'banana streak virus' (BSV) which may lead to reduced yield.
- **Rasthali (Silk AAB):** A medium-tall variety commercially grown in Tamil Nadu, Andhra Pradesh, Kerala, Karnataka and Bihar which is a highly prized cultivar for table purposes. It is known for its unique fruit quality. Fruits are yellowish green during development, turning pale to golden yellow after ripening. Fruit is tasty with a good aroma though has longer crop duration so requires covering of the bunches. It is severely susceptible to *Fusarium* wilt.
- **Red Banana (AAA):** Highly prized in Kerala and Tamil Nadu with commercial cultivation in specific regions of Karnataka, Andhra Pradesh and W&C India. It is known as 'Lal Velchi' in Bihar and 'Chandra Bale' in Karnataka. Its pseudostem, petiole, midrib and fruit rind are purplish-red. Its plants are quite robust which bear bunches weighing 20-30kg under good management practices. Its orange-yellow fruits are sweet and aromatic. It is highly susceptible to bunchy top, *Fusarium* wilt and nematodes.
- **Robusta (AAA):** It is a semi-tall variety primarily grown in Tamil Nadu, parts of Karnataka, Andhra Pradesh and Maharashtra for table purposes. It is a high yielding variety with large bunches and well-developed fruits. The dark green fruits turn bright yellow upon ripening. Despite its sweet taste and good aroma, its poor keeping quality limits long-distance transportation. It is susceptible to Sigatoka leaf spot disease in humid tropics.
- **Safed Velchi Musa (AB Group):** It fruits in 13 months and is considered a high-quality fruit for table purpose. This variety is cultivated in Thane and Nasik districts of Maharashtra and under the shade of arecanut gardens in South Kanara districts (Karnataka). Its medium-sized plants with slender yellowish-green pseudostem are recognized by reddish petiole margin, large fruits, thin papery rind and white firm sweet flesh. Its average bunch weight is about 12kg with around 150 fruits per bunch.

- **Virupakashi (AAB):** An elite variety in South India, it is grown for table purposes in Palani and Shevroy hills of Tamil Nadu. Its hardy plants are vigorous with a characteristic flavour at higher elevations. It is well suited for growing as a shade plant for young coffee in mixed cultivation. Ecotypes like 'Kali', 'Sirumalai' and 'Vannan' are well-suited for plains. Its perennial cultivation is aggravated due to attack of BBTV.

1.10 Nutritional Value

Banana is important to acknowledge that slight variations in nutritional content may arise from differences in cultivation practices, ripeness levels and measurement methodologies. The information presented here serves as a broad overview of the nutritional profile of bananas, providing insights into their potential health benefits. For a more comprehensive understanding of the nutrient composition, refer to the following nutritional data compiled from diverse sources.

Morton (2013) through undergoing a sequence of choppings of the leaves, pseudostems, fruit stalks and peels; their fermentation; *vis-a-vis* drying processes reported that resulting product is a meal that exhibits nutritional qualities surpassing those found in alfalfa presscake. This byproduct which has demonstrated enhanced nutritional value is under consideration for utilization as organic fertilizer in Somalia. Furthermore, in Malaya the incorporation of pseudostems into pig diets has shown a correlation with reduced susceptibility to liver and kidney parasites when compared to pigs on alternative diets. The comprehensive analysis of banana peels reveals the presence of several beneficial compounds including beta-sitosterol, stigmasterol, campesterol, cycloeucalenol, cycloartanol and 24-methylene cycloartanol. Notably, the primary constituents include 24-methylene cycloartanol palmitate and an unidentified triterpene ketone. These insights highlight the multifaceted potential of banana waste materials not only as a source of organic manure but also as a reservoir of bioactive compounds with potential applications in various fields. As per USDA (2020) reports, raw bananas, excluding the peel boast a composition that comprises 75% water, 23% carbohydrates, 1% protein and minimal fat content. A 100g serving provides 89 calories, representing 31% of the recommended Daily Value (DV) for vitamin B6 in the United States. It also contains moderate levels of vitamin C, manganese and dietary fibre while other micronutrients are present in negligible amounts as referred in the table.

Food value per 100g of edible portion*

	Banana				Plantain			
	Ripe	Green	Dried	Flour (green)	Ripe	Ripe (cooked)	Green	Dried (green)
Calories	65.5-111	108	298	340	110.7-156.3	77	90.5-145.9	359
Moisture	68.6-78.1g	72.4g	19.5-27.7g	11.2-13.5g	52.9-77.6g	79.8g	58.7-74.1g	9.0g
Protein	1.1-1.87g	1.1g	2.8-3.5g	3.8-4.1g	0.8-1.6g	1.3g	1.16-1.47g	3.3g
Fat	.016-0.4g	0.3g	0.8-1.1g	0.9-1.0g	0.1-0.78g	0.10g	0.10-0.12g	1.4g
Carbohydrates	19.33-25.8g	25.3g	69.9g	79.6g	25.50-36.81g	18.1g	23.4-37.61g	83.9g
Fibre	0.33-1.07g	1.0g	2.1-3.0g	3.2-4.5g	0.30-0.42g	0.2g	0.40-0.48g	1.0g
Ash	0.60-1.48g	0.9g	2.1-2.8g	3.1g	0.63-1.40g	0.7g	0.63-0.83g	2.4g
Calcium	3.2-13.8mg	11mg		30-39mg	5.0-14.2mg		10.01-12.2mg	50mg
Phosphorus	16.3-50.4mg	28mg		93-94mg	21.0-51.4mg		32.5-43.2mg	65mg
Iron	0.4-1.50mg	0.9mg		2.6-2.7mg	0.40-0.11mg		0.56-0.87mg	1.1mg
B-Carotene	0.006-0.151mg				0.11-1.32mg		0.06-1.38mg	45mg
Thiamine	0.04-0.54mg				0.04-0.11mg		0.06-0.09mg	0.10mg
Riboflavin	0.05-0.067mg				0.04-0.05mg		0.04-0.05mg	0.16mg
Niacin	0.60-1.05mg				0.48-0.70mg		0.32-0.55mg	1.9mg
Ascorbic acid	5.60-36.4mg				18-31.2mg		22.2-33.8mg	1mg
Tryptophan	17-19mg				8-15mg		7-10mg	14mg
Methionine	7-10mg				4-8mg		3-8mg	
Lysine	58-76mg				34-60mg		37-56mg	

*Derived from various analyses made in Cuba, Central America and Africa.

Table source: Morton (2013)

Contrary to popular belief, bananas are not exceptionally rich in potassium per typical serving, offering only 8% of the recommended DV for potassium which is considered a relatively low level (see nutrition table). When compared to other foods, the potassium content of raw dessert bananas (326mg/100g) is moderate. Several foods surpass bananas in potassium content including ground flaxseed (793mg/100g), dry-roasted almonds (684mg/100g), peanut butter (654mg/100g), raw baby spinach (582mg/100g), white button mushrooms (393mg/100g) and whole grain oats (350mg/100g). Raw yellow plantains contain 487mg potassium per 100g while dehydrated dessert bananas (banana powder) boast a higher potassium content at 1490 mg/100g.

Bananas, raw (Daily Value)	
Nutritional value per 100g	
Energy	371 kJ (89 kcal)
Carbohydrates	22.84 g
Sugars	12.23 g
Dietary fibre	2.6 g
Fat	0.33 g
Protein	1.09 g
Vitamins	**Quantity (%DV†)**
Thiamine (B1)	0.031 mg (3%)
Riboflavin (B2)	(0.073 mg) 6%
Niacin (B3)	(0.665 mg) 4%
Pantothenic acid (B5)	(0.334 mg) 7%
Vitamin B6	(0.4 mg) 31%
Folate (B9)	(20 μg) 5%
Choline	(9.8 mg) 2%
Vitamin C	(8.7 mg) 10%
Minerals	**Quantity (%DV†)**
Iron	(0.26 mg) 2%
Magnesium	(27 mg) 8%
Manganese	(0.27 mg) 13%
Phosphorus	(22 mg) 3%
Potassium	(358 mg) 8%
Sodium	(1 mg) 0%
Zinc	(0.15 mg) 2%
Other constituents	**Quantity**
Water	74.91 g
Units: μg = micrograms, mg = milligrams, IU = International units	
†Percentages are roughly approximated using US recommendations for adults.	

Source: USDA, 2020

1.11 Allergen

Those individuals who have a latex allergy may potentially undergo an adverse reaction upon the consumption of bananas (Taylor and Erkek, 2004). This cross-reactivity is attributed to the presence of proteins in bananas that share similarities with those found in latex. Therefore, individuals with a latex allergy should exercise caution when consuming bananas as they may be more susceptible to experiencing allergic symptoms such as itching, swelling, hives or in severe cases anaphylaxis. It is advisable for individuals with known latex allergies to consult with a healthcare professional for personalized advice on managing and navigating potential risks associated with banana consumption.

1.12 Phytochemicals

The banana peel contains a variety of bioactive compounds including beta-sitosterol, stigmasterol, campesterol, cycloeucalenol, cycloartanol and 24-methylene cycloartanol. The predominant constituents are 24-methylene cycloartanol palmitate and an unidentified triterpene ketone (Morton, 2013). Both the peel and flesh of the banana are abundant sources of valuable phytochemicals, encompassing polyphenols, flavonoids, fatty acids, carotenoids and phytosterols (Someya *et al.*, 2002; Vilela *et al.*, 2014; Mathew and Negi, 2017; Qamar and Shaikh, 2018; Ajijolakewu *et al.*, 2021). Major polyphenolics include catechin, epicatechin, gallic acid, cinnamic acid, chlorogenic acid and protocatechuic acid while banana phytosterols comprise stigmasterol, cycloeucalenone, beta-sitosterol and cycloartenol (Pereira and Maraschin, 2015).

1.12.1 Lipophilic compounds

Major lipophilic compounds in hexane extract (HE) as tentatively identified through GC–MS are:

NO	Compound	RT (min)	Area (%) [a]
1	Benzoic acid, 4-ethoxy-, ethyl ester	22.303	0.15 ± 0.12
2	Benzene, 1,2,3-trimethoxy-5-(2-propenyl)-	22.833	0.17 ± 0.03
3	Phenol, 2,6-dimethoxy-4-(2-propenyl)-	24.044	0.58 ± 0.05
4	Palmitic acid, methyl ester	31.02	0.35 ± 0.05
5	Palmitic acid	31.767	7.76 ± 0.73
6	Ethyl 9,12-hexadecadienoate	32.014	0.25 ± 0.01
7	Ethyl palmitate	32.355	4.46 ± 0.18
8	Linoleic acid, methyl ester	34.261	0.20 ± 0.01
9	9,12,15-Octadecatrienoic acid	34.373	0.21 ± 0.01
10	Linolenic acid	35.096	2.84 ± 0.38
11	Oleic Acid	35.155	1.94
12	Z-6,17-Octadecadien-1-ol acetate	35.467	2.97 ± 0.01

NO	Compound	RT (min)	Area (%) [a]
13	Ethyl linolenate	35.585	2.42 ± 0.03
14	Ethyl Oleate	35.685	0.77 ± 0.03
15	Butyl palmitate	35.902	0.15 ± 0.01
16	Ethyl stearate	36.032	0.16 ± 0.01
17	Ethyl linoleolate	39.879	0.09 ± 0.01
18	α-Tocopherol	52.342	0.32 ± 0.03
19	Campesterol	55.389	2.53 ± 0.14
20	Stigmasterol	56.089	5.50 ± 0.06
21	γ-Sitosterol	57.466	50.28 ± 0.93
22	(E)-24-Propylidene cholesterol	57.772	6.82 ± 0.17

[a]Data expressed as mean ± standard deviation (SD) ($n = 3$).

Source: Kim *et al.*, 2022

1.12.2 Antioxidant compounds in banana fruits

The banana fruit emerges as a rich source of antioxidant compounds, offering potential health benefits attributed to its diverse phytochemical profile. Reactive oxygen species (ROS) and reactive nitrogen species (RNS), including hydroxyl radicals, superoxide ions, nitric oxide radicals, singlet oxygen and hydrogen peroxide have been implicated in the onset of various disorders such as arthritis, diabetes, arteriosclerosis, age-related macular degeneration, cancer, inflammation, genotoxicity and Alzheimer's disease. The exact mechanisms remain elusive but the interaction of these ROS and RNS species with biomolecules like lipids, proteins and DNA is thought to contribute to the development of these conditions (Shukla *et al.*, 2009; Septembre-Malaterre *et al.*, 2016).

Kandasamy and Aradhya (2014) shed light on the antioxidant potential of the banana rhizome, showcasing its richness in polyphenolic compounds. Notably, anthocyanins derived from banana bracts, identified by Pazmino-Duran *et al.* (2001) including cyanidin-3-rutinoside were proposed for use as natural colourants. The inhibition of polyphenol oxidase (PPO) during banana fruit ripening at room temperature was observed with the addition of heat-treated onion extract, and Maillard reaction products (MRP) significantly impacted banana PPO activity (Lee, 2007). The phytochemistry and pharmacology of wild banana (*Musa acuminata* Colla) have been reviewed by Mathew and Negi (2017) and they suggested the use of banana pulp and peel for the development of drugs and use in functional foods.

Beyond banana pulp, both the pseudostem and banana fruit peel have emerged as significant sources of antioxidants. Aziz *et al.* (2011) conducted a comprehensive analysis, revealing higher polyphenol content, flavonoids, dietary fibre, lignin, hemicellulose, cellulose, antioxidant capacity and free-radical scavenging capacity in native banana pseudostem flour compared to tender core pseudostem flour. Pereira and Maraschin (2015) and Singh *et al.* (2016) emphasized the abundance of bioactive compounds in bananas such as carotenoids, flavonoids, phenolics, amines, vitamin C and vitamin E ~ all possessing antioxidant activities that contribute to various health benefits.

Recent reviews by Vu *et al.* (2018) highlighted the phenolic compounds in banana peel and their potential health benefits, proposing the utilization of this valuable by-product in the food and pharmaceutical industries. Furthermore, Anyasi *et al.* (2018) when conducted a meticulous analysis of unripe banana flour, it revealed presence of essential minerals and phenolic compounds including flavonoids such as epicatechin and 3-O-rhamnosyl-glucoside. The cultivars examined exhibited varying concentrations of these compounds with potassium emerging as the highest essential mineral content, underlining the nutritional richness of bananas.

1.12.3 Carotenoids

Carotenoids constitute a diverse class of compounds encompassing approximately 600 members. Within this family, certain carotenoids serve as precursors for vitamin A while others exhibit potent antioxidant capabilities, effectively scavenging reactive oxygen species (ROS). In the context of banana fruit, noteworthy carotenoids include α-carotene, β-carotene and β-cryptoxanthin which possess provitamin A activity. Concurrently, lycopene and lutein also found in bananas exhibit robust antioxidant capacities. Among these carotenoids, α-carotene, β-carotene and β-cryptoxanthin contribute to provitamin A activity while lycopene and lutein are known for their antioxidant properties (Erdman *et al.*, 1993). Lycopene has been associated with protection against prostate cancer in men whereas lutein offers human health benefits by acting as an inhibitor of age-related macular degeneration (Davey *et al.*, 2006). A comprehensive analysis conducted by Davey *et al.* (2009) involved examining 171 different genotypes of *Musa* spp. for provitamin A carotenoids and 47 genotypes for two minerals (Fe & Zn). The study revealed substantial variability in provitamin A content among various cultivars but minimal variability in iron and zinc levels irrespective of soil type and growing conditions. The workers proposed the adoption of cultivars with high provitamin A and trace mineral content as a strategic approach to enhance nutritional health and address micronutrient deficiencies prevalent in *Musa*-consuming populations.

Yellow- and orange-fleshed banana cultivars, as reported by Englberger *et al.* (2006), are recognized for their heightened trans-β-carotene content. A detailed overview of carotenoid content in various banana cultivars is presented in the accompanying table. Scientific literature suggests that the consumption of fruits rich in carotenoids can fortify the immune system and mitigate the risk of diseases such as cancer, type II diabetes and cardiovascular disorders (Krinsky and Johnson, 2005).

Certain banana cultivars particularly those abundant in provitamin A carotenoids hold the potential to address vitamin A deficiency especially in economically challenged populations worldwide (Fungo and Pillay, 2013). Encouraging the cultivation and consumption of these nutrient-rich banana varieties emerges as a promising strategy for alleviating nutritional deficiencies and enhancing the overall well-being of vulnerable populations.

Carotenoid content of different banana cultivars (μg/100g)

Cultivar	Trans-α carotene	Trans-β carotene	Total carotenoid content	Reference
Hung Tu ripe pulp	1849 FW	5653 FW	7760 FW	Beatrice *et al.*, 2015
To'o ripe pulp	2055 FW	5267 FW	7765 FW	
Sepi ripe pulp	4728 FW	5611 FW	10,067 FW	
Apantu ripe pulp	3287 FW	6387 FW	10,056 FW	
Bungaoisan ripe pulp	779 FW	857 FW	1675 FW	
Aibwo	2358 FW	5945 FW	9400 FW	Englberger *et al.*, 2006
Fagufagu	1524 FW	3428 FW	5054 FW	
Ropa	3682 FW	1324 FW	5218 FW	
Gatagata	79 FW	695 FW	774 FW	
Toraka Parao	250 FW	526 FW	776 FW	
Red banana peel and pulp	ND	123 and 29.6 DW	200 and 250 DW	Arora *et al.*, 2008

Source: Adopted and modified from Singh *et al.*, 2016.

1.12.4 Phenolic compounds

Phenolics constituting the predominant bioactive compounds in banana fruit play a crucial role in conferring antioxidant properties that contribute to various health benefits. The extensive repertoire of identified phenolics in bananas includes gallic acid, catechin, epicatechin, tannins and anthocyanins.

Notably, the banana rhizome, particularly in South India stands out as a rich source of phenolics, making it both a dietary staple and a traditional medicinal resource (Kandasamy and Aradhya, 2014). Russell *et al.* (2009) expanded the list of banana phenolics, detecting compounds such as ferulic, sinapic, salicylic, gallic, p-hydroxybenzoic, vanillic, syringic, gentisic and p-coumaric acids with ferulic acid standing out as the predominant component, constituting 69% of cinnamic acids.

The banana peel emerges as a noteworthy reservoir of phenolic compounds as revealed by Tsamo *et al.* (2015) who analyzed pulp and peel from various banana cultivars. Hydroxycinnamic derivatives, particularly ferulic acid-hexoside were identified as major constituents in plantain pulp, exhibiting considerable variation among cultivars. In plantain peel, rutin emerged as the most abundant flavonol glycoside underscoring the diverse phenolic profile present in different parts of banana. Consequently, both banana pulp and peel emerge as valuable sources of health-promoting phenolics. Among the detected flavonoids in bananas ~ quercetin, myricetin, kaempferol and cyanidin contribute to health benefits by acting as effective scavengers of free radicals, reactive oxygen species (ROS) and reactive nitrogen species (RNS) (Kevers *et al.*, 2007). The health-promoting attributes of these phenolics extend beyond their antioxidant properties encompassing antibacterial, antiviral, anti-inflammatory, antiallergenic, antithrombotic and vasodilatory activities (Cook and Sammon, 1996). Sulaiman *et al.* (2011) conducted a comprehensive analysis of eight banana cultivars in Malaysia, determining total phenolic and mineral contents in both pulp and peel. Peel extracts exhibited higher total phenolics and total antioxidant activities than the pulp with potassium emerging as the primary mineral element followed by phosphorus, magnesium and sodium. This study underscores the significance of banana peel as a potent source of phenolics and antioxidants.

Bananas showcase a diverse array of phenolic compounds across different parts of fruit with the rhizome, pulp and peel all contributing significantly to the overall phenolic content. These bioactive compounds not only confer antioxidant benefits but also exhibit a range of health-promoting activities reinforcing the nutritional and medicinal value of bananas.

1.13 Health Benefits of Bioactive Components

Banana flesh is recognized for its rich content of bioactive antioxidant compounds essential for cellular defense against free radicals and the mitigation of oxidative damage (Qamar and Shaikh, 2018). The antioxidant potential is attributed to phenolic compounds, vitamins, catecholamines and carotenoids present in bananas (Vu *et al.*, 2019).

1.13.1 List of bioactive compounds and their uses

Bioactive compound	Health benefits	Reference source
Tannic acid	Applied as medicinal agents for the treatment of burns	Siang (1983)
Catechin	Resistance of LDL to oxidation, brachial artery dilation increased plasma antioxidant activity and fat oxidation	Williamson and Manach (2005)
Gallic acid	Antioxidant and potential hepatoprotective effects	Rasool *et al.* (2010)
Cinnamic acid	Is a precursor to the sweetener aspartame by means of enzyme catalysed amination to phenylalanine	toppr, 2024
p-Coumaric acid	Antioxidant properties and potentially reducing the risk of stomach cancer	Ferguson *et al.* (2005)
Gallocatechin gallate	Cholesterol reduction	Ikeda *et al.* (2003)
Quercetin	Promotes overall cardiovascular health by encouraging blood flow	Perez-Vizcaino and Duarte (2010)
Ferulic acid	Antioxidant, antimicrobial, anti-inflammatory, antiallergic, anticarcinogenic, modulation of enzyme activity, antiviral and vasodilatory actions	Kumar and Pruthi (2014)
Trans-α carotene	Precursor to vitamin A	Li *et al.* (2011)
Trans-β carotene	Reduces the risk of CVD and cancer	Li *et al.* (2011)
Violaxanthin	Used as a food colourant	
Cryptoxanthin	Food colourant might reduce the risk of lung cancer	DeLorenze *et al.* (2010)
Serotonin	Might contribute to feelings of well-being and happiness	Young (2007)
Dopamine	Reduces the plasma oxidative stress and enhances the resistance to oxidative modification of LDL	Yin *et al.*, (2008)
Catecholamines	Increases blood pressure, glucose levels and heart beat rate	Kuklin and Conger (1995)
β-Sitosterol	Potential to reduce blood cholesterol levels and benign prostatic hyperplasia (BPH)	Wilt *et al.* (1999)
Campesterol and stigmasterol	Reduces the absorption of cholesterol in the human intestines	Choudhary and Tran (2011)

Source: Sidhu and Zafar (2018).

1.13.2 Health benefits of phenolics

The health benefits associated with phenolics, particularly the flavonoid leucocyanidin found in the aqueous extract of unripe banana pulp, have been a subject of considerable scientific interest. Lewis *et al.* (1999) identified leucocyanidin as a predominant component, highlighting its noteworthy anti-ulcerogenic activity. This discovery suggests a potential therapeutic avenue for the treatment of gastric diseases with emphasis on the therapeutic potential of various flavonoids, especially leucocyanidin analogues.

Understanding the structure–activity relationship of flavonoids further underscores their health-promoting properties. Heim *et al.* (2002) demonstrated that the antioxidant capacity, free radical scavenging and chelating action of flavonoids are intricately linked to specific functional groups in their nuclear structure. The antioxidant and chelating properties of flavonoids contribute significantly to their health benefits as highlighted by Rice-Evans *et al.* (1996) who associated these properties with antimutagenic and antitumoral activities.

The multifaceted impact of flavonoids extends to enzyme inhibition, particularly oxygenases like prostaglandin synthase which is crucial in eicosanoid synthesis. Havsteen (2002) observed that flavonoids inhibit hyaluronidase activity thereby aiding in the maintenance of connective tissue proteoglycans. This inhibition potentially prevents the spread of bacterial infections or tumour metastases.

An additional layer to the health benefits of flavonoids lies in their ability to prevent the oxidation of the body's natural water-soluble antioxidants such as ascorbic acid (Korkina and Afanas'ev, 1997). The preferential oxidation of flavonoids serves as a protective mechanism preserving the efficacy of other antioxidants within the body.

Interestingly, the by-products of banana processing particularly banana peel have gained attention for their potential health-promoting qualities. Zhang *et al.* (2005) highlighted the environmental challenge posed by disposal of banana peel and other by-products. Banana peel is reported to be a rich source of high-value antioxidant phytochemicals including anthocyanins, delphinidin and cyanidins (Seymour, 1993). Rebello *et al.* (2014) conducted a comprehensive study revealing that banana peel extract is a rich source of total phenolics (29mg/g as GAE). The extract also contains various antioxidant compounds such as highly polymerized prodelphinidins (~3952mg/kg), flavonol glycosides, B-type procyanidin dimers and monomeric flavan-3-ols (~126mg/kg) underscoring the potential health benefits of harnessing these compounds from banana by-products.

In summary, the health benefits associated with phenolics particularly flavonoids present in banana pulp and peel offer a promising avenue for therapeutic applications ranging from gastric disease treatment to antioxidant-rich dietary supplements. These findings contribute to the broader understanding of the nutritional and medicinal potential of banana-derived compounds.

1.13.3 Health benefits of biogenic amines

The health benefits of biogenic amines derived from banana peel and pulp are multifaceted and underscore the physiological impact of these compounds on human well-being. Biogenic amines, specifically catecholamines are produced through the decarboxylation of amino acids or the amination of aldehydes and ketones. Within the banana, notable catecholamines include dopamine, serotonin, epinephrine and norepinephrine present in substantial amounts across various banana cultivars (Ponchet *et al.*, 1982). In animals, these biogenic amines are reported to work as neurotransmitters for the hormonal regulation of glycogen metabolism (Kimura, 1968).

In the realm of human consumption, serotonin found in the banana pulp at concentrations ranging from 8 to 50µg/g contributes to feelings of well-being and happiness upon consumption. Dopamine and norepinephrine abundant in bananas play crucial roles in influencing mood and emotional stability (Buckley, 1961; Waalkes *et al.*, 1958). The dopamine content in different banana varieties, such as yellow banana (*M. acuminata*), red banana (*Musa sapientum*) and plantain varies with reported concentrations of 42, 54 and 5.5µg/g, respectively (Feldman *et al.*, 1987).

The potential health benefits extend beyond mood regulation as dopamine a neurotransmitter abundant in the peel and pulp of commercially ripened *Musa* Cavendish has been reported in concentrations ranging from 80 to 560mg/100g and 2.5 to 10mg/100g, respectively (Kanazawa and Sakakhibara, 2000). This suggests the possibility of utilizing banana by-products, especially those rich in dopamine for developing pharmaceutical formulations aimed at preventing neurodegenerative diseases like Parkinson's.

The presence of tryptophan in banana peel serving as one of the precursors for dopamine synthesis, sparks considerable interest in exploring the potential role of this amino acid in mitigating neurodegenerative diseases like Parkinson's. The utilization of banana peel, a by-product of the food-processing industry, in the development of pharmaceutical formulations is an avenue that holds promise. This concept is rooted in the understanding that enhancing dopamine levels, facilitated by tryptophan could have neuroprotective effects. Despite the therapeutic potential, it is crucial to acknowledge findings that report an increase in dopamine content from unripe to ripened stages, both in the peel and pulp of bananas. Romphophak *et al.* (2005) and Gonzalez-Montelongo *et al.* (2010) have documented this phenomenon independently. Their observations highlight a dynamic change in dopamine levels during the ripening process. However, an intriguing aspect emerges during the over-ripening stage where a decline in dopamine concentration is noted.

While the antioxidant properties of banana peel are evident, with water extracts showing a remarkable 65-70 per cent suppression of linoleic acid autooxidation after 5 days of incubation, dopamine emerging as a potent natural antioxidant. Comparative studies with other antioxidants including ascorbic acid revealed reduced glutathione and phenolic acids such as gallocatechin gallate which reveals dopamine's higher antioxidant activity *in vitro* as demonstrated by the DPPH assay (Kanazawa and Sakakhibara, 2000). Banana peel extracts rich in dopamine, L-dopa and catecholamines exhibit significant antioxidant capacity showcasing their potential as functional foods (Gonzalez-Montelongo *et al.*, 2010).

Moreover, biogenic amines including dopamine are implicated in enhancing resistance of plants to pathogens by interacting with phytohormones particularly through auxin oxidation. This interaction plays a pivotal role in influencing plant growth and development, highlighting the broader ecological significance of these compounds in the plant kingdom (Newman *et al.*, 2001; von Roepenack-Lahaye *et al.*, 2003). Overall, the diverse roles of biogenic amines in bananas, from promoting human well-being to influencing plant physiology, underscore the potential applications and significance of these compounds in various fields.

1.13.4 Health benefits of phytosterols

Phytosterols ~ naturally occurring plant sterols have garnered considerable attention from food manufacturers due to their potential health benefits. These compounds structurally similar to cholesterol play a significant role in promoting heart health. By virtue of their resemblance to cholesterol, phytosterols compete for absorption in the gut, thereby contributing to a reduction in blood cholesterol levels (Marangoni and Poli, 2010). Studies have indicated that a daily intake of 3 grams of phytosterols can lead to a marked reduction in LDL cholesterol levels.

Banana peels contain various phytosterols, including stigmasterol, β-sitosterol, campesterol, 24-methylene cycloartenol, cycloeucalenol and cycloartenol (Knapp and Nicholas, 1969). Health professionals now recommend incorporating plant sterols into the diet particularly for individuals who cannot tolerate cholesterol-lowering statin drugs (Ostlund *et al.*, 2003). Notably, bananas both in the peel and pulp are rich sources of phytosterols (Akihisa *et al.*, 1986). Villaverde *et al.* (2013) reported phytosterol content in unripe bananas ranging from 2.8 to 12.4g/kg DW with *Musa balbisiana* cultivars like 'Dwarf Red' and 'Silver' exhibiting higher phytosterol amounts compared to *M. acuminata* cultivars.

Furthermore, the lipophilic extract of ripe banana pulp from various cultivars of both *M. acuminata* and *M. balbisiana* species has been found to be rich in beneficial compounds. This extract includes omega-3 and omega-6 fatty acids, phytosterols, long-chain aliphatic alcohols and α-tocopherol, offering established nutritional and health benefits (Vilela *et al.*, 2014). Therefore, bananas not only serve as a flavourful and versatile fruit but also contribute to a heart-healthy diet through their phytosterol content.

1.14 Uses

1.14.1 Fruits as food

Bananas whether consumed in their raw, green, partially-ripe or fully ripe form stand as one of the most substantial sources of dietary calories on a global scale. Beyond their caloric contribution, bananas are notably rich in potassium. Varieties with yellow and orange flesh also boast high levels of provitamin A and other carotenoids. The presence of provitamin A carotenoids, especially beta-carotene is crucial for guarding against vitamin A deficiency and anemia, given their involvement in iron metabolism. Furthermore, carotenoid-rich foods including bananas may offer protective effects against diabetes, heart disease and certain cancers ~ a matter of increasing concern in Pacific region.

Recent attention has been directed towards carotenoid-rich bananas in Pohnpei with particular emphasis on the 'Karat' ~ a Fe'i banana variety featuring an upright bunch. Studies have revealed remarkable levels of beta-carotene reaching up to 8508µg/100g in certain Fe'i bananas. Other carotenoid-rich varieties in Pohnpei include 'Kudud' (known internationally as 'Sucrier'), boasting 315µg of beta-carotene per 100g ~ more than 10 times the content found in the common Cavendish banana (21µg/100g) (Nelson *et al.*, 2006).

Banana or plantain flour derived either domestically through sun-drying and pulverizing unripe fruit slices or commercially through methods such as spray-drying or drum-drying serves various culinary purposes. This flour when mixed with wheat flour can produce cupcakes. Puerto Rican dishes like 'pasteles' and 'alcapurias' feature pastry made from plantain flour or a combination of plantain with cassava or cocoyam. Plantain cultivars such as 'Saba', 'Tundoc' and 'Latundan' are deemed highly suitable for flour production (Morton, 2013).

The production and popularity of fried green plantain and banana chips have witnessed growth globally in the past 25 years. These snack items often found in retail groceries alongside potato chips are commonly made from 'Carinosa' and 'Bungulan' bananas while in Puerto Rico 'Guayamero Alto' and 'Congo Enano' are preferred choices for chip production (Morton, 2013).

Dried bananas ('banana figs') involve peeling firm-ripe bananas by splitting them lengthwise, sulphur treated and oven-dried to achieve a moisture content of 18-20%. Individually wrapped and packed these exportable dried bananas can be stored for up to one year. These serve as snacks or are minced and used in fruit cakes and bakery products. Varieties like the 'Dwarf Cavendish' in India and the true 'Lacatan' or 'Higo' in the Philippines are favoured for drying (Morton, 2013).

In SE Asia, the terminal male bud of wild *M. balbisiana* banana is marketed and often boiled for consumption as a vegetable. While considered somewhat astringent, the male buds of cultivated bananas are occasionally consumed similarly. In various cuisines, the flowers may be separated from the bud and prepared separately, featuring in dishes such as curries in Malaya and consumed with palm oil in W. Tropical Africa (Morton, 2013).

In India, an ash solution derived from burned banana leaves and pseudostems is utilized as a seasoning salt for vegetable curries which provides essential minerals such as potassium, magnesium, phosphorus, calcium and sodium (Morton, 2013). Dried and finely ground green plantains have historically served as a substitute for coffee (Morton, 2013).

1.14.2 Leaf vegetable

The culinary potential of banana plants extends beyond their iconic fruits. In particular, the new shoots of young plants emerge as a delectable leafy green suitable for cooking. The core of the banana pseudostem comprising approximately 10-15% of the entire structure emerges as a culinary resource. This core contains essential components boasting 1% starch, 0.68% crude fibre and 1% total ash. In India, it is a common practice to cook and consume banana pseudostem core as a vegetable but usually only in eastern & NE Indian parts but usually not in North India. Furthermore, this versatile core finds its way into canned dishes often paired with potatoes and tomatoes in a flavourful curry sauce. To enhance preservation and flavour, circular slices approximately 1.25cm thick undergo a meticulous treatment involving citric acid and potassium metabisulfite before being candied (Morton, 2013).

Beyond the pseudostem, the banana plant offers yet another culinary delight ~ the edible leaf buds. These tender buds are not only palatable but also nutritious, contributing to the overall versatility of the banana plant in gastronomy. One specific variety, *M. acuminata* ssp. *zebrina* known for its striking red variegation stands out not only for its ornamental beauty but also for its gustatory appeal. The leaves of this particular cultivar are noted for the exceptional taste ~ adding an extra layer of culinary interest to an already aesthetically pleasing plant. This dual-purpose characteristic makes *M. acuminata* ssp. *zebrina* a noteworthy choice for both ornamental and culinary purposes (Nelson *et al.*, 2006).

In essence, the banana plant transcends its identity as a mere fruit-bearing tree offering a spectrum of edible components that contribute to diverse and flavourful culinary experiences. This utilization aligns with cultural practices in regions like India, where various parts of the banana plant have found a place in traditional cuisines.

1.14.3 Floral culinary uses and traditional practices

The diminutive male flowers concealed within the banana bud hold culinary significance, especially in certain varieties like 'Saba' and 'Dippig', as observed in the Philippines and among the Filipino community in Hawai'i. In these cultures, the male flowers located inside the protective bud are harvested and prepared as a delectable dish. Entire bud can be transformed into a vegetable delicacy provided the appropriate banana variety is chosen. However, it is imperative to subject the bud to multiple rounds of boiling in different water batches to mitigate its inherent bitterness.

Historically, the traditional uses of banana flowers extend beyond mere gastronomic applications. Specifically, within the Iholena, Maoli and Pōpō'ulu subgroups, the male flowers yield a thick nectar. This natural sweetener, resembling a semi-jelled 'honey' was employed in pre-contact Hawai'i as a nourishing source for infants. The cultural practices surrounding the utilization of this floral 'honey' underscore the banana plant's multifaceted contribution to the traditional diets of these communities (Nelson *et al.*, 2006).

In sharp contrast to the male flowers which find their way into culinary creations the semi-jelled 'honey' emerges as a delightful gift from the banana plant catering to those with a penchant for sweetness. This unique floral offering with its cultural and culinary significance exemplifies the diverse roles played by the banana plant in both traditional practices and contemporary culinary exploration. The exploration of banana flowers extends beyond their botanical existence, intertwining with cultural heritage, culinary innovation and traditional remedies.

1.14.4 Beverages

The spectrum of beverage production extends beyond traditional boundaries encompassing a diverse range that includes alcohol, beer, vinegar and wine ~ all derived from fruits. Although such practices were more prevalent in historical contexts, even in contemporary times the production of vinegar endures as a craft in certain regions. An exemplar of this enduring tradition can be observed in modern French Polynesia where vinegar is crafted in limited quantities by small-holding farmers. These farmers utilize the 'Yangambi km5' banana variety belonging to the AAA group within the Ibota subgroup. Originally introduced for livestock purposes this particular banana cultivar has demonstrated exceptional versatility emerging as a preferred choice for its edibility and its excellence in fermenting into vinegar (Nelson *et al.*, 2006). This insightful exploration not only underscores the resilience of traditional practices but also sheds light on the adaptive and innovative use of agricultural resources in contemporary agricultural landscapes.

1.14.5 Medicinal

Medicinal applications of flowers, fruits and roots persist in select Pacific cultures albeit many of these traditional practices have waned over time. Among these cultures, historical records from Hawai'i indicate the utilization of banana stalks which were mashed and employed as poultices for treating sprains or broken bones. The root sap from specific banana varieties served as a remedy for ailments such as thrush ~ a childhood mouth irritation, and was also applied to address skin warts. In the context of traditional medicinal practices in French Polynesia, banana root sap found application in treating various health concerns. Pounded banana peels ~ a common therapeutic method across many Pacific cultures have recently garnered attention for their inherent antibiotic properties. This traditional technique involves utilizing banana peels as a type of wound 'bandage' underscoring the historical wisdom encapsulated in these indigenous healing practices. While the prevalence of these medicinal applications has diminished over time, the recognition of the therapeutic potential within bananas and their components offers a unique intersection between traditional knowledge and modern scientific inquiry. The mention of these authorities helps anchor the historical authenticity of these practices within the specified cultural contexts (Nelson *et al.*, 2006).

1.14.6 Ceremonial and religious significance

The ceremonial and religious significance of bananas is deeply rooted in various Pacific cultures, with traditions and rituals that highlight the symbolic and sacred nature of this fruit. The multifaceted importance of bananas in religious practices is exemplified through the following aspects (Nelson *et al.*, 2006):

i) **Offerings to the Gods:** In the Pacific including Hawai'i, bananas hold a central role as offerings. Tradition recounts that all mai'a (native bananas) were planted by the gods Kane and Kanaloa. As a result, bananas were regularly presented to them as 'gifts of the land' at heiau (places of worship) and other sacred locations.

ii) **Taboos and Royalty:** The consumption of certain banana varieties particularly those from the Maoli Group (excluding 'Maoli Ka'ualau') was restricted for females. These varieties were considered food fit for male ali'i (royalty) and priests in Hawai'i. Bananas with red or gold pseudostems, petioles or fruit at various stages of development were deemed royal or sacred.

iii) **Use of leaves and stalks in ceremonies:** Banana leaves were utilized to cover altars not only in Hawai'i but also across Oceania. Banana stalks were symbolically linked to manhood. In specific circumstances Maoli bananas were incorporated into ceremonies where banana offerings served as substitutes for human sacrifices in Hawai'i.

iv) **Embodiment of Deities:** Bananas were revered as the embodiment of Kanaloa, Tangaroa or Ta'aroa, primary deities in Hawai'i, Society Islands, Samoa and possibly Fiji.

v) **Primordial sustenance and ancestral spirits:** Wild bananas were considered primordial sustenance for the gods in Polynesia. Furthermore, bananas were employed in ceremonial feeding practices dedicated to ancestral spirits in Hawai'i. The 'Iholena Lele' variety known for its delicious cooking bananas with salmon-fleshed fruit was particularly planted around heiau for this purpose.

vi) **Love magic and cultural symbolism:** Bananas especially the Maoli varieties were associated with 'love magic' in Hawai'i. Throughout the Pacific, Maoli and Pōpō'ulu-type bananas were intricately linked with sexuality. This cultural connection is still evident in the society and Marquesas Islands where these bananas are bestowed with 'specially appropriate' names. The prohibition of Maoli bananas for women in old Hawai'i was likely influenced by these cultural associations.

vii) **Rituals in childbirth:** Banana plants played a role in rituals associated with the birth of children, especially among royal families in the Marquesas Islands and Tahiti.

These diverse ceremonial and religious practices involving bananas showcase the intricate cultural tapestry woven around this fruit across various Pacific societies. Authorities mentionings in the original article are retained to ensure the accuracy and credibility of the information.

1.14.7 Other uses

The multifaceted utility of the banana plant extends far beyond its well-known role as a source of fruit. The diverse applications of *Musa* species and hybrids as highlighted by Nelson *et al.* (2006) include:

i) **Flavouring:** Ashes derived from burning banana leaves and pseudostems serve as flavour enhancers in curries and can be utilized as a salt substitute, particularly in India.

ii) **Cut flowers:** Certain *Musa* species and hybrids distinguished by their vibrant floral bracts and flowers find application in ornamental landscapes and tropical flower arrangements.

iii) **Fuel wood/cooking:** In the Pacific, banana trunks play a role in lining underground ovens, providing steam for cooking. Banana leaves are also employed to cover food, ensuring cleanliness during the cooking process.

iv) **Fodder and silage:** Both fruits and stems of the banana plant are converted into silage, serving as valuable feed for cattle. The underground parts are also utilized in feeding pigs and cattle as observed in regions like Marquesas.

v) **Canoe/boat/raft making:** Historically, banana stalks were employed as canoe rollers in Hawaii and this practice continues in certain remote Micronesian and Polynesian communities such as Yap and Tuvalu.

vi) **Fibre/weaving/clothing:** Banana leaves and plant fibres are utilized to craft thread and cloth. Specific varieties like *Musa basjoo* and *M. textilis* are still used to create high-quality clothing and fabrics. The outer pseudostem sheath, especially in traditional Polynesian varieties, was historically employed in weaving pandanus mats, contributing to intricate patterns.

vii) **Rope/cordage/string:** Leaf fibres particularly from *Musa textilis* contribute to the production of string, thread, cordage and rope with applications such as 'Manila hemp' and 'Taiwan hemp'.

viii) **Wrapping/parcelization:** Banana leaves serve as versatile packing material and are used to create leis, garlands and parcels. These also find application in traditional preservation methods such as lining pits for breadfruit in Micronesia.

ix) **Thatch/roofing/mats:** Banana leaves are utilized for constructing house roofs, wall linings, thatching and weaving mats.

x) **Resin/gum/glue/latex:** Banana sap has been historically employed as dye in various cultures with some varieties producing bright magenta sap. The sap is also known for its staining properties used as an indelible black dye.

xi) **Body ornamentation/garlands:** Leaf sheath fibres are employed for stringing leis, garlands and for tying and plaiting clothing. In some cases, twisted strips of banana trunk are preferred by lei makers exemplified by the ancient Hawaiian variety 'Maoli Manai-'ula'.

xii) **Fishing:** Specific banana varieties were historically used to feed sharks believed to ensure good fishing in Hawaii.

xiii) **Tannin/dye:** Banana sap serves various purposes including dyeing and tanning in India. In some Pacific islands, ashes derived from banana plant combustion are also utilized for tattooing.

xiv) **Eating utensils:** Banana leaves serve as eco-friendly dining plates and tablecloths in various cultures including Polynesia and India.

xv) **Agricultural:** Banana leaves find application in agriculture where these are used to wrap root balls of seedlings or plants before transplanting.

xvi) **Protection:** Banana leaves offer practical uses as umbrellas and temporary raincoats providing protection against rains.

xvii) **Smoking:** Banana leaves serve as tobacco paper particularly in the SW Pacific and Micronesia, contributing to local smoking practices.

1.15 Cultivation

(nhb.gov.in, 2024; kvk.icar.gov.in, 2024; Nelson *et al.*, 2006; Morton, 2013)

1.15.1 Agro-climatic requirements

Banana, being inherently a tropical crop thrives within a temperature range of 15-35°C coupled with a relative humidity of 75-85%. It exhibits a particular affinity for tropical humid lowlands and demonstrates adaptability across a broad altitude spectrum, flourishing from sea level up to elevations reaching 2000 metres above mean sea level. The cultivation of this crop in India spans diverse climatic conditions ranging from humid tropical regions to arid mild subtropics, achieved through the careful selection of suitable varieties.

It is crucial to note that bananas are sensitive to chilling temperatures, and injuries can occur when exposed to conditions below 12°C. Furthermore, the crop faces vulnerability to high wind velocities exceeding 80km/h which can inflict damages.

The significance of a well-defined monsoon season is paramount for robust vegetative growth, typically spanning four months from June to September, accompanied by an average rainfall of 650-750mm. This climatic pattern plays a pivotal role in sustaining the vigorous growth of banana plants. In areas of higher altitudes, banana cultivation is confined to a select few varieties such as the 'Hill banana,' emphasizing the importance of tailoring cultivation practices to specific regional climatic nuances. This nuanced understanding of the preferred climatic conditions and associated challenges provides a foundation for effective banana cultivation practices, ensuring optimal yields and quality.

1.15.2 Growing and potential belts

The state-wise growing belts are given in the following table:

State	Growing belts
Andhra Pradesh	East Godavari, West Godavari, Kurnool, Cuddapah
Assam	Goalpara, Nagaon, Sonitpur, foothills of Garo hills
Gujarat	Surat, Vadodara, Anand, Kheda, Junagadh, Narmada, Bharuch
Jharkhand	Ranchi, Sahebganj
Karnataka	Bengaluru, Chitradurga, Shioroga, Hassan, Chikka Mangloor
Kerala	Thiruvananthapuram, Kollam, Pathanamthitta, Alappuzha, Kottayam, Idukki, Ernakulam, Thrissur, Palakkad, Malappuram, Kozhikode, Wynadu, Kannur, Kasargod
Madhya Pradesh	Khandwa, Badwani, Khargaon, Dhar
Maharashtra	Jalgaon, Ahmednagar, Buldhana, Pune, Wardha, Dhule, Nanded, Parbani, Nandurbar, Satara, Sangli, Osmanabad, Buldhana, Akola, Yeothmal, Amravati, Thane, Kulara, Alibag
Odisha	Ganjam, Puri, Khurda, Gajpati, Cuttack, Dhenkanal, Angul, Sundargarh, Sambalpur, Bargarh, Deogarh, Koraput, Keonjhar, Raygada, Mayurbhanj
Tamil Nadu	Thoothukudi, Tiruchirapalli, Coimbatore, Tirunelveli, Karur, Erode, Kanniyakumari
West Bengal	Hooghly, Nadia, North 24 Parganas

(nhb.gov.in, 2024)

1.15.3 Soil

Banana thrives best in deep nutrient-rich loamy soils that exhibit a pH range of 6.5-7.5. The ideal soil conditions for bananas include well-draining properties, ample fertility and consistent moisture availability. Unsuitable for banana cultivation are saline soils and calcareous substrates. Optimal banana-friendly soil strikes a balance between acidity and alkalinity presenting a neutral pH. Furthermore, it should boast the high organic matter content, particularly rich in nitrogen with sufficient phosphorus levels and an abundance of potash. This combination of soil attributes provides an environment conducive to the healthy growth and development of banana plants, fostering robust yields.

1.15.4 Propagation

Banana and plantain propagation primarily relies on vegetative division with seeds being a considerably less common method and typically reserved for purposes such as banana breeding, cultivation of ornamental varieties, and preservation of wild species. Notably, the utilization of seeds in banana and plantain reproduction is infrequent. The prevailing trend in recent years within commercial plantations is the adoption of tissue culture as a standard practice. This method offers distinct advantages, particularly in ensuring the initiation of plant growth with disease-free planting material. It is noteworthy that edible banana varieties are predominantly seedless. However, exceptions exist as exemplified by certain cultivars like 'Pisang Awak' which may produce seeds under specific conditions particularly when growing in close proximity to a fertile pollen source. This nuanced aspect underscores the importance of understanding the exceptional cases where seed development may occur in typically seedless edible banana varieties.

1.15.4.1 Vegetative propagation

Vegetative propagation is the exclusive method employed for cultivation of commercial bananas due to their seedless nature. The reproductive cycle involves a unique underground stem known as the rhizome which harbours multiple buds. These buds upon sprouting give rise to their individual pseudostems and distinctive bulbous rhizomes commonly referred to as suckers. Two main types of suckers utilized in banana propagation are 'sword suckers' and 'water suckers'.

Sword suckers exhibit a well-developed base with narrow and sword-shaped leaf blades during early stages of growth. Suckers aged 2-4 months are typically chosen for propagation. Apart from suckers, other viable planting materials include whole or bits of rhizomes. For instance, the Basrai variety in Jalgaon (Maharashtra) follows a propagation method primarily relying on dormant rhizomes. Post-harvest the rhizomes are dug out of the soil and stored in a cool and dry place for approximately two months. During this resting period, the lower part of the pseudostem naturally sheds, revealing a prominent heart bud. Selection criteria for rhizomes involve favouring conical shapes while discarding flat ones. The ideal weight of rhizomes for planting ranges between 500 to 750g with an optimal age of 3-4 months.

Crucially, the size of rhizome has implications for subsequent growth and development of banana plant. Smaller rhizomes tend to produce larger fruits with delayed flowering whereas larger rhizomes initiate early flowering but result in smaller-sized fruits or bunches. Given the inherent genetic instability of bananas, it is imperative to select suckers and rhizomes from healthy plants that exhibit desirable bunch qualities and high-yielding capabilities, preferably with a minimum of 10 hands in a bunch. This meticulous selection process ensures the propagation of banana plants with the desired genetic traits for optimal fruit yield as well as for commercial cultivation.

1.15.4.2 Tissue culture

In contemporary agricultural practices, the propagation of banana plants has evolved with the integration of tissue culture techniques. This advanced method is increasingly employed for cultivation of specific varieties such as 'Shrimanti', 'Gross Michael' and 'Grand Naine'. Through tissue culture, disease-free plantlets featuring 3 to 4 leaves are meticulously produced and supplied in pots, facilitating the establishment of secondary nurseries.

The cultivation process commences with careful placement of these plantlets in shaded environments, typically at 50% shade, to ensure optimal acclimatization. Over the initial six weeks, a gradual reduction in shade exposure takes place as the plants progressively adapt and harden. After this acclimatization period of about six weeks, the plants demonstrate robust resilience, rendering them independent of shade requirements.

Following the successful completion of the secondary nursery phase lasting around two months, the banana plants are deemed ready for transplantation into designated field pits. This meticulous approach, combining tissue culture and well-managed nursery practices not only ensures the production of disease-resistant plantlets but also contributes to the overall health and vigour of the banana plants throughout their life cycle.

1.15.5 Land preparation

Before initiating the banana planting process it is advisable to cultivate a green manuring crop such as *Sesbania* (dhaincha), sunhemp or cowpea. This practice serves to enhance soil fertility and structure. The land preparation phase involves ploughing the soil meticulously typically between 2 to 4 times followed by levelling to create an optimal planting surface. To further refine the soil structure, a rotavator or harrow is employed to break down clods and for achieving fine tilth.

In the course of soil preparation, it is crucial to incorporate a basal dose of farmyard manure (FYM). Approximately 50 tonnes per hectare of FYM are recommended and this application is performed before the final harrowing. The FYM is meticulously mixed into the soil ensuring even distribution and thorough integration. This comprehensive approach to soil preparation sets the stage for an enriched and conducive environment for successful banana cultivation.

1.15.6 Planting season

The cultivation of tissue culture bananas offers flexibility in planting schedules, enabling year-round planting to align with market demand, with the exception of periods characterized by excessively low or high temperatures. It is advisable to fine-tune the planting schedule to circumvent adverse conditions particularly high temperatures and drought during critical emergence phase of bunches, typically occurring approximately 7-8 months post-planting. It is crucial to note that the optimal planting time may vary between long-duration cultivars and short-duration varieties, necessitating a discerning approach to planting schedules for different banana cultivars. This strategic adjustment not only accounts for the distinct growth and maturation periods of various cultivars but also aims to mitigate potential challenges associated with environmental factors, ensuring a more resilient and productive banana cultivation system.

Thus, the important seasons for planting followed in different states of India are:

State	Planting time
Maharashtra	*Kharif*: June–July; *Rabi*: October–November
Tamil Nadu	February–April and November-December
Kerala	Rainfed: April-May; Irrigated crop: August- September

(nhb.gov.in, 2024)

1.15.7 Spacing

Historically, banana growers have adhered to a traditional planting pattern, spacing crops at a density of 1.5×1.5 metres. Despite the high density, this conventional approach has proven suboptimal as it results in compromised plant growth and diminished yields due to intense competition for sunlight. Recognizing the need for tailored cultivation practices particularly in regions characterized by distinct climatic conditions such as North India, coastal belts and areas experiencing high humidity with temperatures dropping to 5-7°C, a more prudent approach is recommended. In these specific regions, it is advised to adopt a wider planting distance of not less than 2.1×1.5m. This strategic adjustment aims to optimize sunlight exposure, mitigate competition among plants and enhancing overall banana crop performance acknowledging the nuanced environmental factors inherent to these geographical locations.

Season	Spacing
Kharif	1.5×1.5m, 2× 2m or 2.5×2.5m
Rabi	1.5×1.2m, 1.5×1.37m

(nhb.gov.in, 2024)

Banana planting is executed employing the patta double line method, a strategic approach designed for optimal plant spacing and efficient agricultural operations. Under this method, distance between two planting lines is maintained within the range of 0.9-1.2 metres. Simultaneously, the plant-to-plant distance is set between 1.2 to 2 metres. This specific spacing configuration offers several advantages including enhanced accessibility for intercultural operations and a notable reduction in the cost associated with drip irrigation.

Recent experimental endeavours have shed light on the significance of planting distances in influencing banana quality and bunch weight. Achieving robust banana quality and substantial bunch weight have been observed by adopting a planting distance of 1.8×1.8 metres. However, to attain maximum yield, an optimal plantation configuration involves maintaining a closer proximity specifically at 1.2×1.5 metres.

This careful consideration of planting distances reflects a nuanced approach to banana cultivation, balancing factors such as plant accessibility, irrigation efficiency and yield optimization. By strategically aligning these elements farmers can cultivate bananas with superior quality and substantial yields contributing to the overall success and sustainability of banana plantations.

High density planting: Cultivation technique of high-density planting has gained prominence with plantations accommodating a range of 4444 to 5555 plants per hectare. Remarkably, the resulting yields have been documented at an impressive scale often reaching 55-60 tonnes per hectare or even surpassing these figures. Cultivators commonly adopt the square or rectangular system of planting which involves placing 3 suckers per pit at a specified spacing.

For Cavendish varieties, the practice involves planting at a spacing of 1.8×3.6 metres amounting to 4,600 plants per hectare. On the other hand, Nendran varieties are typically planted using 2×3 metres spacing, accommodating 5000 plants per hectare. This meticulous approach to planting configuration underscores the precision and adaptability required in banana cultivation with farmers employing strategies tailored to the specific requirements of different banana varieties. The implementation of such practices not only optimises land usage but also plays a pivotal role in achieving robust yields, contributing to the overall success of banana farming endeavours.

1.15.8 Planting method

1.15.8.1 Pit method

The pit planting method commonly employed in garden cultivation systems involves the meticulous preparation of planting pits to ensure optimal growth conditions for banana plants. The pits with dimensions of 0.5×0.5×0.5 metres are excavated particularly in areas with ridges and furrows. To enhance the fertility of the pits a mixture of topsoil, 10kg of well-decomposed FYM, 250 grammes of neemcake and 20 grammes of carbofuran is used.

After the pit preparation, an essential step is to leave the pits open for a period of 15-20 days. This practice serves multiple purposes as it allows exposure to solar radiation which effectively eliminates insects and soil-borne diseases. This duration permits aeration of the soil so enhancing its overall quality. In regions with saline-alkali soil characterized by a pH above 8, the pit mixture is customized by incorporating organic matter and gypsum to address specific soil conditions.

Planting process involves carefully situating the banana suckers in the centre of each pit with subsequent compaction of the surrounding soil. The plants are positioned in such a way that the pseudostem remains approximately 2cm below the ground level. Gentle pressing of the soil around the plant is performed emphasizing the importance of avoiding deep planting. Immediate irrigation of the field follows the planting process to ensure that newly established banana plants receive the necessary moisture for early establishment and growth. This comprehensive approach to pit planting sets the stage for optimal banana cultivation practices, for promoting soil fertility, preventing the diseases and for improving over all plant vigour.

Digging pits of dimensions 0.5m×0.5m×0.5m represents the traditional approach for planting banana rhizomes. Despite its adherence to the desired planting depth, this method is characterized by its labour-intensive nature and associated high costs. The primary advantage lies in the elimination of the need for earthing up given that the planting occurs at the specified depth. However, due to the substantial labour and expense involved, this practice has witnessed a decline in popularity and is not widely adopted in current banana cultivation practices.

1.15.8.2 Furrow method

This cultivation technique is widely employed involving the use of a tractor or ridger to create furrows measuring 20-25cm in depth spaced at a distance of 1.5 metres. Rhizomes are then strategically planted within these furrows. However, to ensure optimal growth conditions in this method, regular earthing up becomes necessary. This entails the periodic addition of soil to cover the exposed rhizomes, promoting their protection and facilitating a conducive environment for successful development of the crop.

In the agricultural landscape of India, distinct planting practices are employed in different regions. In the states of Gujarat and Maharashtra, the annual planting system is characterized by the implementation of furrow planting. This method involves creating furrows or shallow trenches in the soil, providing an organized and systematic arrangement for planting crops. On the other hand, in the wetland cultivation areas of the Cauvery delta region in Tamil Nadu a different approach is adopted. Trench planting takes precedence in this region, offering a well-suited technique for the unique conditions of wetland farming. Trench planting involves the creation of deeper furrows or trenches in the saturated soil, optimizing water management and facilitating the growth of crops in the Cauvery delta's distinct agricultural environment. These regional variations in planting practices underscore the adaptability of agricultural techniques to local climatic and soil conditions, showcasing the diversity and ingenuity present in India's farming practices.

1.15.9 Manure and fertilizers

Banana cultivation demands a substantial quantity of nutrients where soil contributes it partially. On a nationwide scale in India, the recommended nutrient composition per banana plant includes 10kg of FYM, 200-250g of nitrogen, 60-70g of phosphorus and 300g of potassium. Specifically, for a metric ton of yield, the banana crop requires 7-8kg of nitrogen, 0.7-1.5kg of phosphorus and 17-20kg of potassium. Traditionally, farmers tend to prioritize urea over phosphorus and potash, applying urea in 3-4 split doses.

A strategic nutrient management approach involves top-dressing with approximately 100 grammes of nitrogen per plant divided into three equal doses at 60, 90 and 120 days after planting while 100 grammes of potash and 40 grammes of phosphorus are crucial and applied at the time of planting. It is recommended to apply full dose of phosphorus and potassium at planting, distributing nitrogen in three equal doses in shallow rings about 8-10 centimetres deep.

Enhanced yields are achieved by applying 150g of nitrogen during vegetative phase and 50g during reproductive phase. Combining 25% of nitrogen as FYM and 1kg of neemcake proves beneficial. An effective practice is to apply 25% of nitrogen in organic form, 75% in inorganic form, coupled with the cultivation of green manure crops.

Phosphorus requirements are comparatively low with superphosphate being the primary source followed by the application of rock phosphate (50-95g/plant at planting). In acidic soils, the recommendation is of triple superphosphate or diammonium phosphate. Phosphorus is applied in a single dose at planting, and the quantity of P_2O_5 varies based on soil type ~ ranging from 20 to 40g/plant.

Potassium plays a crucial role in banana cultivation due to its involvement in vital functions so its continuous supply is essential during the finger-filling stage. The recommended application includes 100g of potassium in two splits during the vegetative phase and another 100g in two splits during the reproductive phase. Cultivar-dependent recommendations suggest 200-300g of K_2O, with plantains requiring higher potassium levels than other cultivar groups. Muriate of potash is commonly used but in soils with pH above 7.5, however, potassium sulphate is advantageous.

Calcium, influencing yield through interaction with N, P and K is addressed by using dolomite (Mg_2CO_3) and limestone ($CaCO_3$) as soil amendments in acidic soils. In cases of acute magnesium deficiencies, foliar application of $MgSO_4$ is found effective. While sulphur deficiency has been reported in some instances, it is not a widespread concern for bananas. Sulphur uptake is active during suckering to shooting stage though after that sulphur supply comes from leaves and pseudostem.

1.15.9.1 Fertigation

To prevent nutrient losses inherent in traditional fertilization methods such as nitrogen leaching, volatilization, evaporation and phosphorus and potassium fixation in the soil, a strategic approach involves the utilization of water-soluble or liquid fertilizers administered through drip irrigation ~ a technique commonly known as fertigation. This innovative method not only addresses the challenges associated with conventional fertilizers but also yields substantial benefits.

Fertigation has demonstrated its efficacy by delivering a remarkable 25-30% increase in crop yield compared to traditional fertilization practices. This heightened productivity is attributed to the enhanced efficiency in nutrient utilization and absorption facilitated by precise application of water-soluble fertilizers directly to root zone through drip irrigation system.

Furthermore, the adoption of fertigation brings about operational advantages. Labour and time savings are notable benefits as the process of applying liquid fertilizers through drip irrigation is more automated and streamlined compared to traditional methods. This efficiency is particularly advantageous in large-scale agricultural operations, contributing to overall resource optimization.

Another key advantage lies in the uniform distribution of nutrients throughout the field. Fertigation ensures that plants receive a consistent and balanced supply of essential elements, promoting homogeneous growth and development. This uniform nutrient distribution not only maximizes crop yield but also minimizes the risk of nutrient deficiencies or excesses in different parts of the field.

In summary, the adoption of fertigation represents a sophisticated and effective strategy to overcome the limitations of conventional fertilization practices. The observed increase in yield, coupled with operational efficiencies and precise nutrient management, underscores fertigation as a valuable and sustainable approach for modern agriculture.

1.15.9.2 Micronutrients

Application of a combined foliar spray consisting of $ZnSO_4$ at 0.5%, $FeSO_4$ at 0.2%, $CuSO_4$ at 0.2% and H_3BO_3 at 0.1%, administered at intervals of 3, 5 and 7 months post-planting has demonstrated notable efficacy in enhancing both the yield and quality of banana crops. This strategic application of micronutrients addresses crucial nutritional requirements during key developmental stages, fostering robust growth, improving fruit development and positively influencing the overall productivity of banana plants.

The incorporation of zinc sulphate contributes to enhanced enzyme activities and metabolic processes so promoting better nutrient uptake and utilization. Ferrous sulphate ~ a source of iron plays a pivotal role in chlorophyll synthesis, ultimately influencing photosynthetic efficiency and plant vigour. Copper sulphate ~ an essential micronutrient supports various enzymatic functions vital for plant development. Boric acid supplementation aids in the regulation of physiological processes ~ contributing to improved fruit setting and quality.

Carefully timed foliar application at 3, 5 and 7 months after planting aligns with critical growth stages ensuring that the banana plants receive necessary micronutrient support at key developmental milestones. This comprehensive approach not only addresses potential nutrient deficiencies but also optimizes the physiological processes governing banana growth, ultimately resulting in a quantifiable enhancement of both yield and fruit quality. Employing this tailored foliar application regimen underscores a proactive and targeted strategy for maximizing the agricultural output of banana cultivation.

1.15.10 Irrigation

Bananas being the succulent herbaceous evergreens bearing shallow roots exhibit a heightened demand for water to optimize productivity. The recommended water requirement for banana cultivation ranges between 1,800 and 2,000mm per annum. Precision in irrigation scheduling is crucial for effective cultivation, with winter necessitating irrigation intervals of 7-8 days and summer requiring more frequent at intervals of 4-5 days. It is noteworthy that during rainy season irrigation is applied judiciously to avoid excess water as this could lead to root zone congestion, disrupting air exchange within soil pores and adversely impacting plant establishment and growth. In totality, approximately 70-75 irrigations are administered throughout the banana crop cycle.

To enhance water use efficiency and foster sustainable banana production, the implementation of an efficient irrigation system such as drip irrigation is of paramount importance. Traditional methods like furrows, basins and trench systems are commonly employed. However, the adoption of advanced techniques like drip irrigation coupled with mulching has demonstrated notable improvements. Reports indicate a substantial 58% reduction in water consumption accompanied by a remarkable increase in yield ranging from 23-32% under drip irrigation systems. Beyond water conservation, this method also facilitates precise fertilizer application through fertigation, contributing to overall resource optimization in banana cultivation.

1.15.10.1 Drip irrigation

Drip irrigation presents a valuable approach in cultivating bananas, offering several advantages for emhancing optimal plant growth and yield. Through this irrigation system, a delicate balance between soil-air and soil-water content is maintained thus fostering an early and robust bunch development. This method facilitates the accelerated maturation of raw bunches which results in a significant reduction of the maturation period by 30-45 days. Concurrently, the overall yield experiences a notable increase ranging between 15 to 30 per cent.

Moreover, the adoption of drip irrigation contributes to resource efficiency with a substantial conservation of water resources reaching up to 58-60% in comparison to conventional irrigation methods. This water-saving aspect is particularly crucial in sustainable agriculture practices. Reduced water requirement corresponds to a decline in weed growth so further diminishing the need for intercultural operations and the associated costs. The precision of drip irrigation also allows for the targeted application of water-soluble fertilizers thus optimizing nutrient management for banana plants.

To operationalize drip irrigation effectively, a specific schedule is recommended. Initially, from planting until the 4th month a rate of 15 liters of water per plant per day is advised. From 5th month until the shooting stage this rate is increased to 20 liters per plant per day. Subsequently, from the shooting stage until 15 days prior to harvest a further adjustment sees the application rate elevated to 25 liters per plant per day.

Two distinct methods are commonly employed in drip irrigation for bananas:

i) **Single line system:** Plant spacing of 1.5×1.5 metres is maintained with each plant connected to one lateral line and one dripper.

ii) **Double line system:** In this approach, lines are spaced 1 metre apart while the distance between plants within a line is set at 1.5 metres. Furthermore, the gap between two double lines is established at 1.8 metres. A single lateral line and one dripper serve two plants. Alternatively, the distance between two lines may be adjusted at 2.1×2.4 metres.

By integrating drip irrigation with these specified techniques, banana cultivation not only experiences enhanced growth and yield but also exemplifies resource efficiency and sustainable agricultural practices.

1.15.11 Intercultural operations

To maximize the productivity of the crop, a series of intercultural operations are followed ensuring meticulous care and attention throughout the cultivation process. These operations include:

i) **Weeding management:** As a pre-planting spraying in the field of Glyphosate @ 2 litres/ha is done, and afterwards when weeds are seen, 4-5 weedings as needed during growth period of the crop are undertaken. This proactive approach helps safeguard the crop from competition for resources which fosters its unhindered growth.

ii) **Harrowing and earthing up:** The field should undergo harrowing 3-4 times to ensure the soil remaining loose and conducive to healthy root development. Earthing up of the soil 25-30cm around the base should be executed at 3-4 months after planting. This method not only maintains soil friability but also ensures strategic positioning of the drip line about 5-7.5cm away from the plant. This approach not only aids in protecting the plants from wind-induced damage but also minimizes production losses thus contributing to the overall resilience of the crop.

1.15.12 Desuckering

Desuckering is a vital practice in banana cultivation aimed at mitigating internal competition with the main plant. This operation involves the systematic removal of unwanted suckers extending up to 7-8 months after planting, especially the smaller ones on a regular basis. By executing this practice diligently, growers optimize the allocation of nutrients and resources to the main banana plant which fosters its healthy growth and development. Timely and consistent desuckering not only enhances the overall productivity of the banana plantation but also contributes to the quality of the harvested fruits.

1.15.13 Propping for bunch support

The inherent weight of a mature banana bunch can lead to imbalancing of the plant which may result in the potential lodging of the bearing plant. Such lodging not only jeopardizes the structural integrity of the plant but may also have detrimental effects on both production and fruit quality. So to counteract this, a proactive measure involves providing support through a carefully implemented propping technique. In this, two bamboo supports are strategically positioned against the stems on the side where the plant is leaning. The formation of a triangular support structure created by these bamboos effectively redistributes the weight of the bunch so restoring proper balance of the plant. This not only mitigates the risk of lodging but also plays a crucial role in sustaining both production levels and fruit quality.

Beyond the structural benefits, propping contributes to the uniform development of the banana bunch. By providing additional support to natural inclination of the plant to lean under the weight of the fruit, the propping system ensures that each fruit within the bunch receives equal exposure to sunlight and nutrients. In turn this promotes consistent growth and ripening so ultimately enhancing the overall quality of the harvest. Not only the physical stability of the plant is safeguarded through judicious placement of bamboo supports but subsequent positive impact on production and fruit uniformity underscores the importance of this cultivation practice in ensuring a thriving banana crop.

1.15.14 Bunch protection and spraying practices

i) **Bunch covering for enhanced fruit quality:** The economical practice of covering banana bunches with dried leaves from the plant serves multiple purposes. This technique not only shields the bunch from direct exposure to sunlight but also contributes to overall enhancement of fruit quality. However, it is crucial to note that during rainy season this practice should be avoided to prevent potential issues.

ii) **Sleeving for comprehensive fruit protection:** To safeguard banana fruits against various external elements including dust, spray residues, insects and birds bunches are sleeved. This protective measure involves use of transparent and perforated polythene sheets. The level of ventilation ranging from 2% during the cool season to 4% during summer can be adjusted accordingly. This approach may be complemented by the application of neem cake (1kg/ha) which not only increases the temperature around the developing bunch but also promotes early maturity.

iii) **Monocrotophos spray for thrips control:** Thrips infestation poses a threat to banana crops leading to discolouration of the fruit skin so deteriorating its visual appeal. For effective control of thrips a targeted approach involves spraying Monocrotophos (0.2%) after the emergence of all the banana hands. This strategic application helps mitigating the impact of thrips on the crop so ensuring the development of visually appealing and marketable fruits.

Implementing these bunch covering and spraying practices not only addresses potential environmental challenges but also contributes to the overall health, quality and marketability. It underscores the importance of adopting a holistic and adaptable approach based on seasonal considerations and specific pest threats to maximize yield and fruit quality in banana cultivation.

1.15.15 Dehandling of false hands of bunch

Dehandling involves removal of false hands from a banana bunch. It is a crucial practice for enhancing the overall quality of the produce. This entails promptly eliminating incomplete hands that do not meet the desired standards for export-quality bananas shortly after blooming. Dehandling contributes to the improvement of the weight of the remaining hands in the bunch. This selective removal allows the plant to allocate more nutrients to the viable hands. This practice also optimizes the banana finger size as this way remaining fingers attain more uniform and desirable size to meet proper export standards. This practice also facilitates an improved skin-to-pulp ratio in the banana fingers. Elimination of substandard hands enhances overall fruit composition which results in a more favourable balance between outer skin and inner pulp. This is particularly significant for meeting the stringent criteria set by export standards.

In essence, the strategic dehandling of false hands is a proactive measure that contributes to the overall excellence of banana production. It not only ensures adherence to quality benchmarks but also promotes a more efficient allocation of resources which ultimately leads to a superior and market-ready banana harvest.

1.15.16 Mulching

Use of wheat and banana straws at 12.5kg per plant as mulching materials has proven to be a valuable practice in banana orchards. This has several advantages such as effective conservation of soil moisture and increase of the bunch weight. Application of mulch occurs strategically at the commencement of summer, *i.e.* in February. This layer serves as a multifaceted tool, addressing critical aspects of banana cultivation such as:

i) **Bunch weight enhancement:** Use of mulch in the specified quantity demonstrates a positive impact on weight of banana bunches. This enhancement can be attributed to insulation provided by mulch layer which aids in nutrient retention, root protection and temperature regulation.

ii) **Soil moisture conservation:** Mulching plays a pivotal role in mitigating water loss from the soil through evaporation. By forming a barrier between the soil and the atmosphere, the mulch layer reduces water evaporation thereby fostering optimal soil moisture levels. This is particularly crucial in banana orchards where maintaining consistent soil moisture is essential for sustained plant health and productivity.

iii) **Seasonal timing:** The application of mulch at the initiation of summer, *i.e.* in February aligns with growing season's demands. This timing ensures that mulch is in place to offer its protective and moisture-conserving benefits during period when environmental stressors such as elevated temperatures and increased evaporation are more pronounced.

In conclusion, the integration of wheat straw and banana straw as mulching materials in banana orchards emerges as a proactive and effective cultivation practice. Beyond its direct impact on bunch weight and soil moisture conservation this contributes to the overall resilience of banana plants fostering a conducive environment for healthy growth and improved yields.

1.15.17 Intercropping

Intercropping practices with banana require careful consideration due to the plant's superficial root system which is susceptible to damage from cultivation activities. While intercropping is generally discouraged, there are specific recommendations for short-duration crops with growth cycles of 45-60 days such as *mung*, cowpea and *Sesbania* which can be strategically utilized as green manuring to enhance soil fertility. During initial 3-4 months of banana cultivation intercropping can be done with certain crops such as legumes, sunnhemp, *Sesbania*, beetroot, elephant foot yam, ginger or turmeric. These crops can be strategically integrated without causing significant harm to the delicate banana root system. However, it is imperative to exercise caution when considering intercropping, especially with cucurbitaceous vegetables as these are known carriers of viruses that can adversely affect banana plants. Therefore, cultivation of cucurbitaceous vegetables in conjunction with bananas is to be avoided to prevent transmission of viral infections.

In specific coastal regions such as Karnataka, Kerala and Andhra Pradesh an alternative intercropping strategy involves cultivating bananas within coconut and arecanut plantations, particularly with tall banana cultivars. This approach leverages the unique ecological conditions of coastal areas to create symbiotic relationships between banana plants and existing plantation ecosystems which contribute to overall agricultural sustainability in these regions.

1.15.18 Growth regulators

Utilizing growth regulators strategically plays a crucial role in optimizing the quality and yield of banana bunches. To enhance the grade of bunches a recommended approach involves the application of 25ppm 2,4-D following the emergence of the last hand. This application not only contributes to the overall improvement of bunch quality but also aids in the removal of seeds in specific varieties such as Poovan and CO-1.

Further enhancement of yield can be achieved through a well-timed application of growth regulators. Spraying CCC (chlormequat chloride) at 1000ppm is advised at the 4th and 6th months after planting. Plantozyme application at a rate of 2ml/l is recommended at the 6th and 8th months after planting which promotes robust growth and thereby increased yield potential.

Upon the full development of the banana bunch a strategic combination of potassium dihydrogen phosphate at 0.5% and urea at 1% can be applied through spraying. Alternatively, a 2,4-D solution at a concentration of 10ppm may be employed. This post-development treatment is designed to optimize banana size and elevate overall fruit quality.

1.15.19 Miscllaneous farm operations

Various essential farm operations are involved in the cultivation of bananas to ensure optimal growth and productivity which include:

i) **Removal of dry leaves:** Regular maintenance involves the removal of dry leaves and retaining only healthy green leaves. This practice supports the overall health of the banana plants by minimizing the risk of disease and providing better exposure to sunlight.

ii) **Winter care:** During winter months when temperatures drop below the optimal range, *i.e.* 10^0C, growth of banana plants may be adversely affected. To counter this, strategic measures such as night-time irrigation or use of smoke generated by inducing controlled fires can be employed to mitigate the impact on plant growth.

iii) **Neem cake application:** A recommended practice during winter months involves the application of neem cake at a rate of 1kg per plant which will aid in facilitating the formation of robust banana bunches and most productive harvest.

iv) **Wind protection:** To shield banana plantations from potentially damaging effects of strong winds it is advisable to plant tall vegetation along the farm borders. This natural windbreak serves as a protective barrier preventing wind-induced stress and potential damage to banana plants.

v) **Support structures:** Providing adequate support to banana plants is crucial for preventing structural damage especially when plants bear the weight of developing fruit. Commonly, farmers use bamboo or eucalyptus poles strategically placed to offer support and stability ensuring upright growth and reducing the risk of breakage.

These farm operations collectively contribute to the overall well-being of banana crops: promoting healthy growth, protecting against environmental challenges and optimizing the conditions for a successful harvest. Implementing these practices underscores the importance of thoughtful and proactive management in banana cultivation.

1.15.20 Harvesting

Banana harvesting is a nuanced process influenced by factors like fruit maturity, market preferences and the need for long-distance transportation.

The ideal harvesting stage is determined by maturity of fruits with the common practice being harvesting at 75-80% maturity for transportation over extended distances. Given the climacteric nature of the fruit, it attains its consumable stage through ripening operations after harvest.

Banana plants typically become ready for harvest within 12-15 months of planting, with the primary harvesting season spanning from September to April. The maturity of bunches occurring 90-150 days post-flowering depends on various factors including the variety, soil conditions, weather and elevation. Harvesting is meticulously timed with the recommendation to harvest bunches when the fingers of second hand from the top are 3/4 rounded. A sharp sickle is employed for cutting the bunch 30cm above the first hand. In certain cases harvest may be delayed up to 100-110 days after the opening of the first hand.

Post-harvest handling is crucial emphasizing the use of well-padded trays or baskets for collecting bunches and avoiding exposure to light as this expedites ripening and softening. For local consumption, hands are often retained on stalks and sold to retailers.

Dwarf varieties exhibit quicker maturation and are ready for harvesting in 11-14 months while tall varieties may require 14-16 months. Following harvest only the leaves are removed and plant system is retained for ratoon crop development. This not only enhances food supply but also allows for a 15% reduction in irrigation needs. To ensure premium banana quality, maintaining just 7 to 8 berries in a bunch is recommended.

The potential for multiple harvests is realized through ratoon cropping with the first ratoon crop being ready in 8-10 months after the main crop harvest and the second ratoon crop following 8-9 months after the second crop. Over a span of 28-30 months, three harvests ~ comprising one main crop and two ratoon crops can be taken. The yield of bananas is contingent on various factors such as variety, plant density and overall management practices.

1.15.21 Yield

Variety-wise average yield of banana (t/ha):

Varieties	Average yield (t/ha)
Basrai, Rasthali	40-50
Shrimanti	70
Grand Naine	65
Ardhapuri , Meanyham	55
Hirsal, Safed Velchi, Red banana, Lal Velchi	45
Poovan	40-50

Varieties	Average yield (t/ha)
Monthan	30-40
Dwarf Cavendish , Robusta Champa and Chini Desi	50-60
Nendran	30-35

1.15.22 Post harvest management

Post-harvest management of bananas involves a systematic process encompassing grading, storage, packing, transportation and marketing. Each stage is crucial in maintaining the quality and ensuring the optimal market value of the produce.

i) **Grading:** Grading plays a pivotal role in determining the marketability of bananas. The criteria for grading include size, colour and maturity. Small and large fruits are segregated to facilitate uniform ripening. In the grading process, efforts are made to eliminate immature, overripe and damaged or diseased fruits. Harvesting is typically conducted early in the season capturing a pre-mature stage to meet the demands of the early market. Ethylene application emerges as an effective method to expedite ripening without compromising fruit quality and flavour.

ii) **Storage:** Mature green bananas can be stored for varying durations based on storage conditions. In ethylene-free air, storage can extend up to three weeks while in a controlled atmosphere at 14°C this duration can stretch up to six weeks.

iii) **Packing:** The choice of packaging significantly influences the shelf life and market value of bananas. Wooden or cardboard boxes, bamboo baskets and sometimes old gunny bags wrapped with banana leaves are utilized. Dehanded fruits packed in polythene bags with perforations exhibit enhanced shelf life whereas those without perforations are susceptible to fungal infections due to build up of high humidity. For export purposes, high-quality bananas undergo meticulous processing including washing, treatment with solutions to remove latex, grading based on length and girth and packing in plastic corrugated fibreboard cartons.

iv) **Transportation:** Road transport *via* trucks and lorries is the predominant mode for transporting bananas ensuring easy access from orchards to markets. Rail wagons are employed for distant markets to facilitate efficient transportation.

v) **Marketing:** The marketing phase involves various intermediaries such as wholesalers and commission agents. Private traders dominate the sector, overseeing approximately 95% of the trade and often providing credit to farmers at relatively high-interest rates. Cooperatives contribute to the remaining 5% of the trade volume, emphasizing a balance in the market dynamics.

In summary, the comprehensive post-harvest management approach outlined above is essential for preserving banana quality, meeting market demands and ensuring a smooth and profitable flow within the banana supply chain.

1.16 Nutrient Deficiencies

Banana plants are susceptible to various nutrient deficiencies, each manifesting distinct symptoms that can impact plant health and overall yield. Here's an in-depth exploration of common nutrient deficiencies and recommended control measures:

i) **Nitrogen deficiency**

- **Symptoms:** Pale green leaves; reddish-pink midribs, petioles and leaf sheaths; rosette appearance; reduced bunch weight and compromised fruit quality.
- **Control:** Application of urea (300g/plant) followed by irrigation.

ii) **Phosphorus deficiency**

- **Symptoms:** Stunted growth, poor root development, sawtooth marginal chlorosis in old leaves, leaf curling, breaking of petioles and bluish-green colouration in younger leaves.
- **Control:** Application of DAP (50g/plant) followed by irrigation.

iii) **Potassium deficiency**

- **Symptoms:** Orange-yellow colouration in old leaves, scorching along margins, reduced total leaf area, curving of midribs, delayed flower initiation and reduced yield.
- **Control:** Spraying of 1% potassium sulphate on leaves.

iv) **Calcium deficiency**

- **Symptoms:** Deformation or absence of leaf lamina (spike leaf), marginal leaf necrosis and thickening of veins.
- **Control:** Application of lime (50g/plant) followed by irrigation.

v) **Magnesium deficiency**

- **Symptoms:** Yellow discolouration in mid-blade and midrib, green leaf margins, purple mottling of petioles, marginal necrosis and separation of leaf sheaths.
- **Control:** Application of magnesium sulphate (25g/plant) followed by irrigation.

vi) **Sulphur deficiency**

- **Symptoms:** Yellow or white appearance of young leaves, necrotic patches on leaf margins, stunted growth and small or choked bunches.
- **Control:** Application of complex fertilizer (20:20:0:15 NPKS) @ 20g/ plant followed by irrigation.

vii) **Manganese deficiency**

- **Symptoms:** Narrow green edge at leaf margins and comb-tooth appearance spreading along veins.
- **Control:** Spraying manganese sulphate (0.5%) on leaves.

viii) **Zinc deficiency**

- **Symptoms:** Smaller lanceolate leaves, anthocyanin pigmentation on leaf undersides, alternating chlorotic and green bands in unfurled leaves.
- **Control:** Spraying of zinc sulphate (0.5%) on leaves.

ix) **Iron deficiency**

- **Symptoms:** Yellow or white younger leaves.
- **Control:** Spraying iron sulphate (0.5%) along with urea (1%) on leaves.

x) **Copper deficiency**

- **Symptoms:** Chlorosis in young and old leaves and umbrella-like appearance.
- **Control:** Spraying of copper sulphate (0.5%) on leaves.

xi) **Boron deficiency**

- **Symptoms:** Reduced leaf area, leaf curling, lamina deformation, white stripes on young leaf lamina, thickening of secondary veins and inhibited root and flower formation.
- **Control:** Application of borax salt (25g/plant) in the soil around the root zone.

Implementing these control measures can help address and mitigate nutrient deficiencies, and ensuring the health and productivity of banana plants. Regular monitoring and timely interventions are crucial for optimal crop development.

1.17 Physiological Disorders - Impact on Fruit Quality

Like many other crops, bananas are susceptible to various physiological disorders that can significantly impact their growth, development and overall health. Among these abiotic fruit disorders stand out as they not only reduce fruit quality but may also render the fruit unmarketable. Here, we delve into several common physiological disorders affecting banana plants and exploring effective management strategies:

i) **Chilling injury**

- **Description:** Necrotic flecking just beneath the surface of the green skin and pulp which results from exposure to temperatures below 13°C.
- **Management:** Prevent refrigeration at temperatures below 13°C to mitigate chilling injury.

ii) **Fused fingers**

- **Description:** Fusion of banana fingers due to a genetic mutation or defect, particularly observed in Cavendish varieties.
- **Management:** Remove affected plants and their suckers in commercial farms and utilize fused hands as animal feed.

iii) **Maturity bronzing**

- **Description:** Reddish-brown to brown discolouration of mature green bananas due to periodic water stress or calcium deficiency.
- **Management:** Ensuring proper irrigation to prevent moisture stress in young bunches and avoiding the over-filling bunches before harvest.

iv) **Precocious/mixed ripening**

- **Description:** Premature ripening of individual fingers is possibly triggered by harvesting overmature bananas, inappropriate postharvest storage or transit temperatures.
- **Management:** Harvest at the right maturity stage; control postharvest temperatures; apply fungicide dips; and ensure well-ventilated fruit storage.

v) **Sap damage**

- **Description:** Stains on banana fruit skin caused by contact with its own sap during de-handing and packing activities.
- **Management:** Conduct careful de-handing; soak banana hands in clean water to dilute staining sap.

vi) **Senescent spots**

- **Description:** Superficial brown flecking on peels is a natural part of the ripening process for Cavendish bananas.
- **Management:** Avoid harvesting and ripening overly mature fruit to prevent senescent spots.

vii) Splitting of fingers

- **Description:** Longitudinal splits in fingers leading to separation from the hand is caused by high humidity, over-mature or over-caliper fruit.
- **Management:** Maintain humidity at 70–75% during ripening; ensure good aeration in ripening rooms and harvest fruits on time based on caliper grade.

Understanding and effectively managing these physiological disorders is crucial for sustaining banana crop health and ensuring high-quality fruit for market consumption. Implementing appropriate preventative measures and interventions can contribute to the overall success of banana cultivation.

1.18 Pests and Diseases

The banana crop is significantly vulnerable to a multitude of challenges encompassing fungal, bacterial and viral diseases as well as infestations by various insect-pests and nematodes. This collective onslaught not only diminishes banana production levels but also adversely impacts overall productivity and fruit quality. Presented below are succinct overviews of the primary pests and diseases affecting bananas coupled with recommended control measures:

Name	Symptoms	Control measures
Pest		
Rhizome weevil (*Cosmopolites sordidus*)	a) Large network creation of galleries in rhizome which weakens the plant	a) Use healthy planting material
		b) Sanitation in orchard
		c) Trapping of adult weevils using pseudostem or rhizome pieces, and
		d) Soil application of carbofuran @2g/plant
Pseudostem weevil (*Odaiporous longicolis*)	a) Small holes on pseudostem with exudation of transparent gummy substance	a) Management approach is identical to rhizome weevil
	b) Existence of tunnelling in leaf sheath and inner core of the stem	b) Secondly, injection of lime solution (Monocrotophos 150ml in 350ml water) using stem injector 1.2m above the ground level at 30° angle is recommended
	c) Abortion of bunches	c) Use longitudinal split (30cm length) or disc on stump traps @ 100/ha. Keep the split portion of tray facing the ground. Collected weevils are then killed.
Thrips (*Chaetanaphothrips signipennis* and *Heliothrips kodaliphilus*)	a) They scrap from attacked plant organs and render them brown and discoloured, especially the fruits.	Spray or inject Monocrotophos @ 0.05% on the inflorescence before unfurling of top most bract.
Fruit scarring beetle (*Besilepta subcostatum*)	a) Adults feed on tender unfolded leaves & fruits and cause scarring of skin	Sanitation spray of 0.05% moncrotophos or 0.1% carbaryl on the heart of the plants immediately after emergence of new foliage and during fruiting season is recommended.
	b) Plant loses its vigour, and quality of bunch becomes poor)	
Aphids (*Pentalonia nigronervosa*)	a) They are vectors of banana bunchy top virus (BBTV) and can be seen as congregation under leaf base of pseudostem	Spray of 0.1% monocrotophos or 0.03% phosphonidon on the leaves is effective.

Name	Symptoms	Control measures
Nematodes	a) Stunted growth	a) Apply corbofuron @40g per plant at planting and 4 months after planting.
	b) Small leaves	
	c) Cut roots	b) Use neem cake as organic manure.
	d) Purple black lesions on roots and root splitting.	c) Use marigold as trap crop.
Fungal diseases		
Panama wilt (*Furarium oxysporum*)	a) Yellowing of old leaves progressing	a) Cultivation of resistant cultivar (Cavendish group)
	b) Affected leaves collapse near petiole and hang.	b) Trim and treat the suckers in 0.1% Bavistin before planting
	c) Pseudostem splitting is common.	c) Apply bioagents like *Trichoderma* and *Pseudomonas fluorescence* with organic manure
	d) Reddish brown discolouration in cross-section of root and rhizome	d) Keep good drainage and apply a lot of organic manure in the field.
Head rot (*Erwinia carotovora*)	a) Rotting of collar region and epinasty of leaves	a) Use healthy planting material
	b) On pulling out of affected plant, the plant topples from the collar region leaving the corm with root in soil	
	c) On opening up of collar region of affected plants yellowish to reddish ooze can be seen.	b) Drench plants with 0.1% Emisan followed by another drenching after 3 months.
	d) In early stage of infection, dark brown or yellow water- soaked areas in critical region may be seen which may decay to form cavities surrounded by dark spongy tissues.	c) Avoid planting in rocks and in poorly drained soils.

Name	Symptoms	Control measures
Sigatoka leaf spot (*Mycosphaerella* spp)	a) Characterized by small lesions on the leaves; the lesions may show pale yellow to greenish yellow streaking visible from both the surfaces of leaves b) Thereafter linear brownish to blackish streaks appear.	a) Remove infected leaves and destroy them
	c) The centre of the streak eventually dries up giving appearance of eye spot.	b) Keep proper drainage and avoid waterlogging.
	d) Sometimes premature ripening is observed	c) Spray dithane M-45 (1.25 kg/ha) or Bavistine 500g/ha.
Anthracnose (*Colletotrichum musae*)	a) Characterized by dark lesions on fruits	Anthracnose can be managed through proper sanitation, pruning and fungicidal applications.
Viral diseases		
Banana bunchy top virus **(BBTV)**	a) Appearance of irregular, dark green 'morse code' streaks along secondary veins on underside of the leaves.	a) Use virus-free planting material, *i.e.* tissue culture. b) Survey and eradicate infected plants regularly.
	b) Leaf size is reduced and leaves remain abnormally erect and brittle	c) Control insect vectors especially aphids and mealy bugs. d) Indexing should be followed in case of mass multiplication
	c) Leaves are little, close to each other and bunched at the top	e) Prohibit movement of any plant part from diseased area to healthy area.
	d) The tips of the bracts in male buds are greenish.	f) Use resistant cultivar.
	e) Virus is spread through aphids.	g) Avoid growing of alternate host as mixed crop or in nearby areas.
Banana mosaic virus **(BMV)**	Chlorosis with mild chlorotic streaks along the veins occurs. These never turn necrotic as in BSV.	Elimination of affected plants and maintenance of disease-free plantation through the use of disease-free planting material, *i.e.* tissue culture seeding.

Name	Symptoms	Control measures
Banana bract mosaic virus **(BBMV)**	Presence of spindle-shaped pinkish to reddish streaks on pseudostem, midribs, petioles and lamina.	Use of disease-free planting material, *i.e.* tissue cultured seedling.
Banana streak virus **(BSV)**	Presence of inconspicuous chlorotic flecking to small lethal systematic necrosis, and includes yellow, brown and black streaking, cigar leaf necrosis, basal pseudostem splitting, internal pseudo-stem necrosis and formation of small deformed bunches.	Use of disease-free planting material, *i.e.* tissue culture seedings.

Source: Vikaspedia, 2024

Other pests

Certain other pests also pose a threat to banana crops such as wild birds, bats and rodents which sometimes become very challenging. These creatures not only consume banana fruits but also frequently build nests within the bunches, strategically positioning themselves to nourish their offspring as the bananas mature. Among these pests, rats in particular can escalate into a significant issue for communities reliant on bananas. Consequently, it is advised to harvest bananas either just before or at the initial indication of rat damage. Following harvest, a preventive measure involves suspending the bananas upside-down in an environment free from rat intrusion, ensuring a safeguarded maturation process until ready for consumption. This proactive approach serves to mitigate the potential impact of pest infestations and safeguard both the yield and quality of banana crops for local populations.

References

1. Ajijolakewu, K.A., A.S. Ayoola, T.O. Agbabiaka, F.R. Zakariyah, N.R. Ahmed, O.J. Oyedele and A. Sani, 2021. A review of the ethnomedicinal, antimicrobial and phytochemical properties of *Musa paradisiaca* (plantain). *Bull. Natl Res. Cent.*, **45**(1): 1–17.
2. Akihisa, T., N. Shimizu, T. Tamura and T. Matsumoto, 1986. (24S)-14a, 24- Dimethyl-9b, 19-cyclo-5a-cholest-25-en-3b-ol: a new sterol and other sterols in *Musa sapientum*. *Lipids*, **21**: 494–7.
3. Anyasi, T.A., A.L.O. Jideani and G.R.A. Mchau, 2018. Phenolics and essential mineral profile of organic acid pretreated unripe banana flour. *Food Res. Intern.*, **104**: 100–109.
4. Argent, G.C.G. 1976. The wild bananas of Papua New Guinea. *Notes Roy Bot. Gard., Edinb.*, **35**: 77-114.
5. Arora, A., D. Choudhary, G. Agarwal and V.P. Singh, 2008. Compositional variation in b-carotene content, carbohydrate and antioxidant enzymes in selected banana cultivars. *Intern. J. Food Sci. Techn.*, **43**: 1913–21.
6. Aziz, N.A.A., L.H. Ho, B. Azahari, R. Bhat, L.H. Cheng and M.N.M. Ibrahim, 2011. Chemical and functional properties of the native banana (*Musa acuminata* × *balbisiana* Colla cv. Awak) pseudo-stem and pseudo-stem tender core flours. *Food Chem.*, **128**: 748–53.
7. Beatrice, E., N. Deborah and B. Guy, 2015. Provitamin A carotenoid content of unripe and ripe banana cultivars for potential adoption in eastern Africa. *Afric. J. Food Compos. Anal.*, **43**: 1–6.
8. Buckley, E.H. 1961. Further studies on the biosynthesis of 3-hydroxytyramine in the peel of the banana. *Plant Physiol.*, **36**: 315–20.
9. Choudhary, S.P. and L.S. Tran, 2011. Phytosterols: perspectives in human nutrition and clinical therapy. *Curr. medic. Chem.*, **18**: 4557–67.
10. Constantine, D. and G. Rossel, 2001. The Musaceae: Anannotated list of the species of *Ensete, Musa* and *Musella*. http://www.users.globalnet.co.uk/~drc/index.htm.
11. Cook, N.C. and S. Sammon, 1996. Flavanoids chemistry, metabolism, cardioprotective effects, and dietary sources. *Nutrit. Biochem.*, **7**: 66–76.
12. Davey, M.W., I.V. Bergh, R. Markham, R. Swennen and J. Keulemans, 2009. Genetic variability in *Musa* fruit provitamin A carotenoids, lutein and mineral micronutrient contents. *Food Chem.*, **115**: 806–813.
13. Davey, M.W., J. Keulemans and R. Swennen, 2006. Methods for the efficient quantification of fruit provitamin A contents. *J. Chromatog.* A, **1136**: 176–84.

14. DeLorenze, G.N., L. McCoy, A.L. Tsai, C.P. Quesenberry Jr., T. Rice, D. Il'yasova and M. Wrensch, 2010. Daily intake of antioxidants in relation to survival among adult patients diagnosed with malignant glioma. *BMC Cancer*, **10**: 215.
15. Englberger, L., R.B. Wills, B. Blades, L. Dufficy, J.W. Daniells and T. Coyne, 2006. Carotenoid content and flesh colour of selected banana cultivars growing in Australia. *Food Nutrit. Bull.*, **27**: 281–91.
16. Erdman Jr., J.W., T.L. Bierer and E.T. Gugger, 1993. Absorption and transport of carotenoids. *Ann. New York Acad. Sci.*, **691**: 76–85.
17. Feldman, J.M., E.M. Lee and C.A. Castleberry, 1987. Catecholamine and serotonin content of foods: effect on urinary excretion of homovanillic and 5-hydroxyindoleacetic acid. *J. Amer. Dietetic Assoc.*, **87**: 1031–5.
18. Ferguson, L.R., S.T. Zhu and P.J. Harris, 2005. Antioxidant and antigenotoxic effects of plant cell wall hydroxycinnamic acids in cultured HT-29 cells. *Molec. Nutr. Food Res.*, **49**: 585–93.
19. Fungo, R. and M. Pillay, 2013. β-Carotene content of selected banana genotypes from Uganda. *Afric. J. Biotechnol.*, **10**: 5423–30.
20. Gibert, O., D. Dufour, A. Giraldo, T. Sánchez, M. Reynes, J.P. Pain, A. González, A. Fernández and A. Díaz, 2009. Differentiation between cooking bananas and dessert bananas. 1. Morphological and compositional characterization of cultivated Colombian Musaceae (*Musa* sp.) in relation to consumer preferences. *J. agr. Food Chem.*, **57** (17): 7857–69.
21. Gonzalez-Montelongo, R., M.G. Lobo and M. González, 2010. Antioxidant activity in banana peel extracts: testing extraction conditions and related bioactive compounds. *Food Chem.*, **119**: 1030–39.
22. Havsteen, B.H. 2002. The biochemistry and medical significance of the flavonoids. *Pharmacol. & Therapeutics*, **96**: 67–202.
23. Heim, K.E., A.R. Tagliaferro and D.J. Bobilya, 2002. Flavonoid antioxidants: chemistry, metabolism and structure-activity relationships. *J. nutrit. Biochem.*, **13**: 572–84.
24. Horry, J.-P., O.E. Arnaud, J.H. Crouch, R.S.B. Ferris, D.R. Jones, N. Mateo, C. Picq and D. Vuylsteke, 1997. Banana and Plantain (Chapter 6). In: *Biodiversity in Trust. Conservation and Use of Plant Genetic Resources in CGIAR Centres* (eds Fuccillo, D., L. Sears and P. Stapleton), pp. 67-81. Cambridge University Press, Cambridge, UK.
25. Ikeda, I., M. Kobayashi, T. Hamada, K. Tsuda, H. Goto, K. Imaizumi, A. Nozawa, A. Sugimoto and T. Kakuda, 2003. Heat-epimerized tea catechins rich in gallocatechin gallate and catechin gallate are more effective to inhibit

cholesterol absorption than tea catechins rich in epigallocatechin gallate and epicatechin gallate. *J. agric. Food Chem.*, **51**(25):7303-7.

26. Kanazawa, K. and H. Sakakibara, 2000. High content of dopamine, a strong antioxidant, in Cavendish banana. *J. agric. & Food Chem.*, **48**: 844–8.

27. Kandasamy, S. and S.M. Aradhya, 2014. Polyphenolic profile and antioxidant properties of rhizome of commercial banana cultivars grown in India. *Food Biosci.*, **8**: 22–32.

28. Kevers, C., M. Falkowski, J. Tabart, J.O. Defraigne, J. Dommes and J. Pincemail, 2007. Evolution of antioxidant capacity during storage of selected fruits and vegetables. *J. agric. & Food Chem.*, **55**: 8596–603.

29. Kim, D.K., M.K. Ediriweera, M. Davaatseren, H.B. Hyun and S.K. Cho, 2022. Antioxidant activity of banana flesh and antiproliferative effect on breast and pancreatic cancer cells. *Food Sci. Nutr.*, **26**[10(3)]:740-50.

30. Kimura, M. 1968. Fluorescence histochemical study on serotonin and catecholamine in some plants. *Jap. J. Pharmacol.*, **18**: 162–8.

31. Knapp, F.F. and H.J. Nicholas, 1969. The sterols and triterpenes of banana peel. *Phytochem.*, **8**: 207–14.

32. Korkina, L.G. and I.B. Afanas'ev, 1997. Antioxidant and chelating properties of flavonoids. *Adv. in Pharmacol.* (San Diego, Calif.), **38**: 151–63.

33. Krinsky, N.I. and E.J. Johnson, 2005. Carotenoid actions and their relation to health and disease. *Molec. Aspects Medicine*, **26**: 459–516.

34. Kuklin, A.I. and B.V. Conger, 1995. Catecholamines in plants. *J. Pl. Gr. Regul.*, **14**: 91–7.

35. Kumar, N. and V. Pruthi, 2014. Potential applications of ferulic acid from natural sources. *Biotechn. Rep.* (Amsterdam), **4**: 86–93.

36. kvk.icar.gov.in, 2024. Banana. https://kvk.icar.gov.in/API/Content/PPupload/k0390_41.pdf. Accessed on 29 January 2024.

37. Lee, M.K. 2007. Inhibitory effect of banana polyphenol oxidase during ripening of banana by onion extract and Maillard reaction products. *Food Chem.*, **102**: 146–9.

38. Lentfer, C. and W. Boyd (ed.), 2003. Tracing antiquity of banana cultivation in Papua New Guinea (Report). *The Australia & Pacific Science Foundation.* PBF 02-3. http://www.apscience.org.au/projects/PBF_02_3/pbf_02_3.htm. Retrieved on 28 January 2024.

39. Lewis, D.A., W.N. Fields and G.P. Shaw, 1999. A natural flavonoid present in unripe plantain banana pulp (*Musa sapientum* L. var. *paradisiaca*) protects

the gastric mucosa from aspirin-induced erosions. *J. Ethnopharmac.*, **65**: 283–8.

40. Li, C., E.S. Ford, G. Zhao, L.S. Balluz, W.H. Giles and S. Liu, 2011. Serum α-carotene concentrations and risk of death among US adults: the third national health and nutrition examination survey follow-up study. *Arch. Internal Medic.*, **171**: 507–15.

41. Marangoni, F. and A. Poli, 2010. Phytosterols and cardiovascular health. *Pharmacol. Res.*, **61**: 193–99.

42. Mathew, N.S. and P.S. Negi, 2017. Traditional uses, phytochemistry and pharmacology of wild banana (*Musa acuminata* Colla): a review. *J. Ethnopharmacol.*, **196**: 124–40.

43. Merriam-Webster, 2024. Banana. Merriam-Webster Online Dictionary. Retrieved January 4, 2013. https://www.merriam-webster.com/dictionary/banana?show=0&t=1357340585.

44. Molina, A.B. and V.N. Roa, 2000. Advancing Banana and Plantain R and D in Asia and the Pacific. International Plant Genetic Resources Institute, p. 3. ISBN 978-971-91751-3-1. Retrieved 2021-12-08.

45. Morton, J.F. 2013. Banana. In: Fruits of Warm Climates. Echo Point Books & Media, pp. 29–46. ISBN 978-1-62654-976-0. OCLC 861735500. Archived from the original on April 15, 2009 – via www.hort.purdue.edu.

46. Nelson, S.C., R.C. Ploetz and A.K. Kepler, 2006. *Musa* species (bananas and plantains) (PDF). In: Species Profiles for Pacific Island Agroforestry (edited by ERlevitch, C.R.). Hōlualoa, Hawai'i: Permanent Agriculture Resources (PAR). Musa species (bananas and plantains) (agroforestry.net). Accessed on 28 January 2024.

47. Newman, M.A., E. von Roepenack-Lahaye, A. Parr, M.J. Daniels and J.M. Dow, 2001. Induction of hydroxycinnamoyl-tyramine conjugates in pepper by *Xanthomonas campestris*, a plant defense response activated by hrp gene-dependent and hrp gene-independent mechanisms. *Molec. Pl.-Microbe Interact.*, **14**: 785–92.

48. nhb.gov.in, 2024. BANANA. https://nhb.gov.in/report_files/banana/BANANA.htm. Accessed on 29 January 2024.

49. Nwakanma, D.C., M. Pillay, B.E. Okoli and A. Tenkouana, 2003. Sectional relationships in the genus *Musa* L. inferred from the PCR-RFLP of organelle DNA sequences. *Theor. & Appl. Genet.*, **107**: 850-6.

50. OGTR, 2008. Office of the Gene Technology Regulator. The Biology *of Musa* L. (banana) (PDF). Australian Government. Archived from the

original (PDF) on December 3, 2012. Retrieved January 30, 2024. https://web.archive.org/web/20121203022848/http://www.ogtr.gov.au/internet/ogtr/publishing.nsf/content/banana-3/$FILE/biologybanana08.pdf.

51. Ostlund, R.E. Jr., S.B. Racette and W.F. Stenson, 2003. Inhibition of cholesterol absorption by phytosterol-replete wheat germ compared with phytosterol-depleted wheat germ. *The Amer. J. Clinic. Nutrit.*, **77**: 1385–9.

52. Pazmino-Duran, E.A., M.M. Giusti, R.E. Wrolstad and M.B.A. Gloria, 2001. Anthocyanins from banana bracts (*Musa* × *paradisiaca*) as potential food colourants. *Food Chem.*, **73**: 327–32.

53. Pereira, A. and M. Maraschin, 2015. Banana (*Musa* spp.) from peel to pulp: Ethnopharmacology, source of bioactive compounds and its relevance for human health. *J. Ethnopharmacol.*, **160**, 149–63.

54. Perez-Vizcaino, F. and J. Duarte, 2010. Flavonols and cardiovascular disease. *Molec. Asp. Medic.*, **31**: 478–94.

55. Pillay, M. and L. Tripathi, 2007. Banana (Chapter 15). In: Genome Mapping and Molecular Breeding in Plants, Vol. 4: *Fruits and Nuts* (ed. Kole, C.), pp. 281-301. Springer-Verlag, Berlin.

56. Pillay, M.A., A. Tenkouano, G. Ude and R. Ortiz, 2004. Molecular characterization of genomes in *Musa* and its applications (Chapter 23). In: Banana Improvement: Cellular, Molecular Biology and Induced Mutations (eds Jain, S.M. and R. Swennen). Science Publishers Inc., Enfield (NH), USA.

57. Ponchet, M., J. Martin-Tanguy, A. Marais and C. Martin, 1982. Hydroxycinnamoyl acid amides and aromatic amines in the inflorescences of some Araceae species. *Phytochem.*, **21**: 2865–9.

58. POWO, 2024a. *Musa* L. Plants of the World Online. Royal Botanic Gardens, Kew Science. Retrieved on 29 January 2024. (https://powo.science.kew.org/taxon/urn:lsid:ipni.org:names:327926-2). First published in Sp. Pl.: 1043 (1753).

59. POWO, 2024b. *Musa acuminata* Colla. Plants of the World Online. Royal Botanic Gardens, Kew Science. Retrieved on 29 January 2024. (https://powo.science.kew.org/taxon/urn:lsid:ipni.org:names:797527-1). First published in Mem. Reale Accad. Sci. Torino 25: 394 (1820).

60. POWO, 2024c. *Musa balbisiana* Colla. Plants of the World Online. Royal Botanic Gardens, Kew Science. Retrieved on 29 January 2024. (https://powo.science.kew.org/taxon/urn:lsid:ipni.org:names:797536-1). First published in Mem. Reale Accad. Sci. Torino 25: 384 (1820)

61. POWO, 2024d. *Musa × paradisiaca* L. Plants of the World Online. Royal Botanic Gardens, Kew Science. Retrieved on 29 January 2024. (https://powo.science.kew.org/taxon/urn:lsid:ipni.org:names:797595-1). First published in Sp. Pl.: 1043 (1753).

62. ProMusa, 2013. Banana-producing countries portal|Improving the understanding of banana (promusa.org). Retrieved on 28 January 2024. https://www.promusa.org/Banana-producing+countries+portal. Accessed on 28 January 2024.

63. Qamar, S. and A. Shaikh, 2018. Therapeutic potentials and compositional changes of valuable compounds from banana-A review. *Trends in Food Sci. & Techn.*, **79**, 1–9.

64. Qi, B., K.G. Moore and J. Orchard, 2000. Effect of cooking on banana and plantain texture. *J. agric. and Food Chem.*, **48** (9): 4221–6.

65. Rasool, M.K., E.P. Sabina, S.R. Ramya, P. Preety, S. Patel, N. Mandal, P.P. Mishra and J. Samuel, 2010. Hepatoprotective and antioxidant effects of gallic acid in paracetamol-induced liver damage in mice. *J. Pharm. and Pharmacol.*, **62**: 638–43.

66. Rebello, L.P.G., A.M. Ramos, P.B. Pertuzatti, M.T. Barcia, N. Castillo-Munoz and I. Hermosin- Gutierrez, 2014. Flour of banana (*Musa* AAA) peel as a source of antioxidant phenolic compounds. *Food Res. Intern.*, **55**: 397–403.

67. Rice-Evans, C.A., N.J. Miller and G. Paganga, 1996. Structure-antioxidant activity relationships of flavonoids and phenolic acids. *Free Radical Biol. & Medic.*, **20**: 933–56.

68. Romphophak, T., J. Siriphanich, Y. Ueda, K. Abe and K. Chachin, 2005. Changes in concentrations of phenolic compounds and polyphenol oxidase activity in banana peel during storage. *Food Preserv. Sci.*, **31**: 111–5.

69. Ross, E.M. 1987. Musaceae. In: *Flora of Australia*, vol. 45, *Hydatellaceae to Liliaceae* (ed. George, A.S.), pp. 16-9. Australian Government Publishing Service, Canberra.

70. Russell, W.R., A. Labat, L. Scobbie, G.J. Duncan and G.G. Duthie, 2009. Phenolic acid content of fruits commonly consumed and locally produced in Scotland. *Food Chem.*, **115**: 100–104.

71. Septembre-Malaterre, A., G. Stanislas, E. Douraguia and M.P. Gonthier, 2016. Evaluation of nutritional and antioxidant properties of the tropical fruits banana, litchi, mango, papaya, passion fruit and pineapple cultivated in réunion French island. *Food Chem.*, **212**: 225–33.

72. Seymour, G.B. 1993. Banana. In: Biochemistry of Fruit Ripening (eds Seymour, J.E. and G.A. Tucker), pp. 83-106. Chapman and Hall, NY.

73. Shukla, S., A. Mehta, J. John, S. Singh, P. Mehta and S.P. Vyas, 2009. Antioxidant activity and total phenolic content of ethanolic extract of *Caesalpinia bonducella* seeds. *Food and Chem. Toxicol.*, **47**: 1848–51.

74. Siang, S.T. 1983. Use of combined traditional Chinese and Western medicine in the management of burns. *Panminerva Medica*, **25**: 197–202.

75. Sidhu, J.S. and T.A. Zafar, 2018. Bioactive compounds in banana fruits and their health benefits. *Food Qual. and Safety*, **2**(4): 183–8. https://doi.org/10.1093/fqsafe/fyy019.

76. Simmonds, N.W. 1956. Botanical results of the banana collecting expedition 1954-5. *Kew Bull.*, **11**(3): 463-89.

77. Simmonds, N.W. and K. Shepherd, 1955. The taxonomy and origins of the cultivated bananas. *J. Linnean Soc. London* (Botany), **55**: 302-12.

78. Simmonds, N.W. and S.T.C. Weatherup, 1990. Numerical taxonomy of the wild bananas. *New Phytologist*, **115**: 567-71.

79. Singh, B., J.P. Singh, A. Kaur and N. Singh, 2016. Bioactive compounds in banana and their associated health benefits – a review. *Food Chem.*, **206**: 1–11.

80. Someya, S., Y. Yoshiki and K. Okubo, 2002. Antioxidant compounds from bananas (*Musa cavendishii*). *Food Chem.*, **79**(3): 351–4.

81. Stover, R.H. and N.W. Simmonds, 1987. Bananas (3rd ed.), p. 183. Harlow, Longman, England. ISBN 978-0-582-46357-8.

82. Sulaiman, S.F., N.A.M. Yusoff, I.M. Eldeen, E.M. Seow, A.A.B. Sajak, Supriatno and K.L. Ooi, 2011. Correlation between total phenolic and mineral contents with antioxidant activity of eight Malaysian bananas (*Musa* sp.). *J. Food Composit. and Analysis*, **24**: 1–10.

83. Taylor, J.S. and E. Erkek, 2004. Latex allergy: diagnosis and management. *Dermatol. Therapy*, **17** (4): 289–301.

84. toppr, 2024. Cinnamic Acid Formula. Chemical Formulas, toppr. https://www.toppr.com/guides/chemistry-formulas/cinnamic-acid-formula/. Accessed on 30 January 2024.

85. Tsamo, C.V.P., M.F. Herent, K. Tomekpe, T.H. Emaga, J. Quetin-Leclercq, H. Rogez, Y. Larondelle and C. Andre, 2015. Phenolic profiling in the pulp and peel of nine plantain cultivars (*Musa* sp.). *Food Chem.*, **167:**197-204.

86. USDA, 2020. Bananas, ripe and slightly ripe, raw. FoodData Central. Released in January 2020. Accessed on 29 January 2024. https://fdc.nal.usda.gov/fdc-app.html.

87. Vikaspedia, 2024. Banana. https://vikaspedia.in/agriculture/crop-production/package-of-practices/fruits-1/banana. accessed on 29 January 20224.

88. Vilela, C., S.A.O. Santos, J.J. Villaverde, L. Oliveira, A. Nunes, N. Cordeiro, C.S.R. Freire and A.J.D. Silvestre, 2014. Lipophilic phytochemicals from banana fruits of several *Musa* species. *Food Chem.*, **162**: 247–52.

89. Villaverde, J.J., L. Oliveira, C. Vilela, R.M. Domingues, N. Freitas, N. Cordeiro, C.S. Freire and A.J. Silvestre, 2013. High valuable compounds from the unripe peel of several *Musa* species cultivated in Madeira Island (Portugal). *Industrial Crops and Products*, **42**: 507-12.

90. Visser, A.V. 2000. Characterization of banana and plantain using random amplified polymorphic DNA markers. *Acta Hort.*, **540**: 113-23.

91. von Roepenack-Lahaye, E., M.A. Newman, S. Schornack, K.E. Hammond-Kosack, T. Lahaye, J.D. Jones, M.J. Daniels and J.M. Dow, 2003. p-Coumaroylnoradrenaline, a novel plant metabolite implicated in tomato defense against pathogens. *J. biol. Chem.*, **278**(44):43373-83.

92. Vu, H.T., C.J. Scarlett and Q.V. Vuong, 2018. Phenolic compounds within banana peel and their potential uses: a review. *J. funct. Foods*, **40**: 238–48.

93. Vu, H.T., C.J. Scarlett and Q.V. Vuong, 2019. Changes of phytochemicals and antioxidant capacity of banana peel during the ripening process; with and without ethylene treatment. *Scient. Hort.*, **253**: 255–62.

94. Waalkes, T.P., A. Sjoerdsma, C.R. Creveling, H. Weissbach and S. Udenfriend, 1958. Serotonin, norepinephrine and related compounds in bananas. *Science* (N.Y.), **127**: 648–50.

95. Wikipedia, 2024. List of banana cultivars. https://en.wikipedia.org/wiki/List_of_banana_cultivars. Accessed on 30 January 2024.

96. Williamson, G. and C. Manach, 2005. Bioavailability and bioefficacy of polyphenols in humans. II. Review of 93 intervention studies. *Amer. J. Clin. Nutr.*, **81**: 243S–55S.

97. Wilt, T., A. Ishani, R. MacDonald, G. Stark, C. Mulrow and J. Lau, 1999. Betasitosterols for benign prostatic hyperplasia. *Cochrane Database System. Rev.* (cited from Sidhu and Zafar, 2018 as reference no. 76 above).

98. Wong, C., R. Kiew, G. Argent, O. Set, S.K. Lee and Y.Y. Gan, 2002. Assessment of the validity of the sections in *Musa* (Musaceae) using AFLP. *Ann. Bot.*, **90**: 231-8.

99. worldpopulationreview.com, 2024. Banana Production by Country 2024. Accessed on 29, January 2024. https://worldpopulationreview.com/country-rankings/banana-production-by-country.

100. worldpopulationreview.com, 2024. Banana Production by Country 2024. https://worldpopulationreview.com/country-rankings/banana-production-by-country. Accessed on 28 January 2024.

101. Yin, X., J. Quan and T. Kanazawa, 2008. Banana prevents plasma oxidative stress in healthy individuals. *Plant Foods for Hum. Nutrit.*, **63**: 71–6.

102. Young, S.N. 2007. How to increase serotonin in the human brain without drugs. *J. Psychiatry & Neurosci.*, **32**: 394–9.

103. Zhang, P., R.L. Whistler, J.N. BeMiller and B.R. Hamaker, 2005. Banana starch: production, physicochemical properties and digestibility—a review. *Carbohydrate Polymers*, **59**: 443–58.

2

Artocarpus J.R.Forster & G.Forster

Family: Moraceae.

The generic name stems from the fusion of the Greek words '*artos*' (bread) and '*karpos*' (fruit), collectively yielding the term 'breadfruit' for its edible fruit (Sikarwar *et al.,* 2014). Breadfruit is rich in carbohydrates & minerals and overcomes malnutrition. Besides, breadfruit has bioactive compounds that provide therapeutic properties.

Tropical plants belonging to genus *Artocarpus* are renowned for their delectable and nectar-rich fruits (Jarrett, 1977). With a diverse array of approximately 50 species found in the tropical and subtropical Asian regions (Xu *et al.,* 2019), this genus exhibits significant variations in size, height, floral *cum* fruit morphology, developmental processes and functional characteristics (Zerega *et al.,* 2010; Gardner *et al.,* 2021). In their native habitats, majority of *Artocarpus* species have garnered widespread acclaim due to their abundant provision of food, therapeutic properties, wood and other essential resources (Jagtap and Bapat, 2010; Xavier *et al.,* 2014; Ragone, 2018; Buddhisuharto *et al.,* 2021). Consequently, certain species have global distribution. *Artocarpus heterophyllus* (jackfruit) and *Artocarpus altilis* (breadfruit) ~ the two most extensively disseminated and domesticated species presently thrive in tropical and subtropical climates with an oriental distribution (Zerega *et al.,* 2010; Williams *et al.,* 2017).

Contrastingly, the indigenous range of *Artocarpus* trees encompasses the expanse from the Western Ghats and SE Asia to the Oceanic Islands. Recent investigations reveal that the genus originated in Borneo, underwent divergence and subsequently expanded and diversified throughout the Miocene epoch (Williams *et al.*, 2017). However, fossils from India dating back to the Palaeocene period (Mehrotra *et al.*, 1984; Srivastava, 1998) suggest an earlier presence in the region. The seemingly unlikely Bornean origin of *Artocarpus* raises questions about the existence of Indian *Artocarpus* species, postulating potential overland or overwater dissemination over considerable distances (Williams *et al.*, 2017). Further research is imperative to unravel the intricate biogeographical history of these species.

The divergence in traits observed among these *Artocarpus* species as compared to their ancestral counterparts may be attributed predominantly to distinctions in the bioclimatic conditions of their habitats and the fauna involved in their pollination and dispersion. Understanding these divergent traits necessitates comprehensive exploration and analysis.

2.1 Botanical Characteristics and Morphological Diversity

Artocarpus plants, encompassing laticiferous trees or shrubs, exhibit a striking morphological diversity that captivates botanical enthusiasts. These plants gracefully ascend to heights ranging from 9 to 70 meters, *vis-a-vis* their leaves, twigs and stems generously yield a milky sap ~ a characteristic feature adding to their botanical allure. Within the genus *Artocarpus*, these plants display a unique monoecious nature, producing unisexual blooms that coexist harmoniously on the same plant. The intricate dance of tiny greenish female flowers on short and fleshy spikes is a spectacle to behold. Following pollination, these blooms metamorphose into syncarpous fruits whose dimensions can achieve awe-inspiring proportions. The leaf morphology within the genus is a testament to nature's artistic variability. From the small and entirely formed leaves of *Artocarpus integer* to the massive and lobed ones of *A. altilis*, the spectrum of leaf shapes is diverse. Notably, the cordate leaves of *A. altilis* culminate in long pointed ends, contributing to the overall botanical tapestry.

Artocarpus trees, renowned for their diversity in unique unisexual inflorescences and composite syncarpous fruits (Jarrett, 1977) have been a subject of fascination for researchers. The phenotypic diversity observed among the syncarps underscores the reliance of taxonomy of the genus on inflorescence morphology and structure (Zerega *et al.*, 2010). Beyond the overarching botanical characteristics, distinct species-specific traits have evolved within the genus. *A. heterophyllus* trees, reaching 15-20 or even up to 30 metres exhibit reticulate branching close to the soil. In contrast, *A. altilis* trees, towering up to 26-30 metres display moderate branching at a medium height. *Artocarpus hirsutus* ~ the wild jack standing as a testament to the diversity with towering forest trees surpassing even 50 metres and some extending even to an astonishing 70 metres with no branching until the apices. The divergence in male inflorescences among these species adds another layer to their uniqueness. *A. heterophyllus* boasts smaller cylindrical inflorescences, *A. altilis* featuring longer and thicker apices and *A. hirsutus* distinguishing itself with thin, long and filamentous stalks where male inflorescence presents entirely different from its counterparts. The differentiation extends to female inflorescences and consequently fruit morphology. The Jackfruit (*A. heterophyllus*) produces larger, multiple, low-hanging and ellipsoidal fruits reaching up to 100cm. In contrast, the breadfruit (*A. altilis*) yields numerous medium-sized, oval and starchy fruits measuring 12-20cm which hang at apices of branches at medium heights. Wild Jack (*A. hirsutus*) deviates further with multiple oval/ellipsoidal and fleshy smaller fruits at the apices of branches sheathing in orange-yellow and measuring 6-10cm at higher elevations.

Such intricately diverse phenotypic characteristics extend beyond mere botanical fascination. These hint at the existence of specialized pollinator and disperser networks and mechanisms (Jarrett, 1977; Matthew *et al.*, 2006; Jagtap and Bapat, 2010; Ragone, 2018; Buddhisuharto *et al.*, 2021). This botanical symphony unfolds in the rich tapestry of *Artocarpus* plants, inviting a deeper exploration into the evolutionary forces shaping their remarkable diversity.

2.2 Chromosome Numbers

Within the genus *Artocarpus*, the chromosomal characteristics of *Artocarpus camansi* and *A. mariannensis* is 2n=56 which is indicative of diploidy. In contrast, *A. altilis* exhibits a more intricate chromosomal pattern with 2n = 56 representing diploidy and 2n = 84 signifying triploidy. Notably, a majority of diploid cultivars in *A. altilis* display seeding though exceptions exist with two cultivars identified as having reduced seed numbers. Micronesian accessions add an intriguing dimension to the cytogenetic landscape, showcasing potential interspecific hybrids between *A. altilis* and *A. mariannensis*. Among these accessions, most manifest as seedless diploids although instances of triploid hybrids have also been observed. An interesting correlation has been established between pollen stainability and the degree of seediness (Ragone, 2001).

Despite the wealth of information available, cytological investigations within the broader context of the *Artocarpus* genus especially concerning the diverse cultivars of Pacific Island breadfruit have been relatively limited. Prior studies, however, have reported chromosomal counts of 2n=56 for various species, underscoring the need for more comprehensive cytogenetic exploration. Species such as *A. chaplasha* Roxb., *A. elastica* Reinw., *A. gomezianus* Wall. ex Trécul, *A. heterophyllus* Lam., *A. integer* (Thunb.) Merr. and *A. lakoocha* Roxb. have been among those studied (Habib, 1972; Hans, 1972; Mehra and Gill, 1974; Oginuma and Tobe, 1995). Historical records on seeded breadfruit primarily point towards diploidy with 2n=2x=56 which is the prevalent chromosomal count. Conversely, seedless cultivars commonly exhibit triploidy marked by a chromosomal count of 2n=3x=84 (Jarrett, 1959; Barrau, 1976). This intricate chromosomal diversity underscores the need for continued exploration and elucidation of the cytogenetic landscape within the genus *Artocarpus* particularly in the context of Pacific Island breadfruit cultivars.

2.3 Phytochemicals

The genus *Artocarpus* renowned for its diverse range of plant species has been a subject of extensive scientific investigation particularly with regard to its rich repertoire of bioactive compounds. Phenolic compounds including flavonoids, stilbenoids and arylbenzofurons have been identified as prominent constituents across various *Artocarpus* plants (Yen *et al.*, 2017). The phenolic constituents present in these plants exhibit a diverse array of chemical structures encompassing isoprenylflavonoids, stilbenoid and 2-arylbenzofuran derivatives, phenolic compounds featuring an oxepine ring and natural Diels-Alder type adducts. In-depth biological investigations conducted on the isolated phytochemicals provide a rational basis for the longstanding traditional utilization of *Artocarpus* plants (Mohanty and Pradhan, 2015).

Artocarpus heterophyllus, *A. altilis*, *A. communis* and *A. integer* are renowned for their diverse chemical constituents including but not limited to morin, dihydromorin, cynomacurin, artocarpin, isoartocarpin, cyloartocarpin, artocarpesin, oxydihydroartocarpesin, artocarpetin, norartocarpetin, cycloartinone, β-sitosterol, ursolic acid, betullic acid acetate and artocarpanone. Analysis of heartwood of these species revealed cellulose at 59%, glucosides at 38%, moisture at 6.7%, lipids at 0.7% and albumin at 1.7%. Essential amino acids such as arginine, cystine, histidine, leucine, lysine, methionine, threonine, tryptophan and free sugars (sucrose) are also present along with fatty acids and ellagic acid. Bark components encompass betullic acid and two novel flavone pigments, cycloheterophyllin. Triterpenic compounds including cycloartenyl acetate and cycloartenone have also been identified. Heterophylol, a phenolic compound, has been isolated from *Artocarpus heterophyllus*. Its bark contains only 3.3% tannin. Leaves and stems exhibit sapogenins, cycloartenone, cycloartenol, β-sitosterol and tannins displaying estrogenic activity. Roots contain β-sitosterol, ursolic acid, betulinic acid and cycloartenone. Additional compounds found include arcommunol C, D & E, squalene, stigmasterol, phytol, polyprenol, dihydrochalcones, hydroxyartoflavone, artogomezianone, furanocyclocommurin and prenylated stilbenes (Mohanty and Pradhan, 2015).

The exploration of phytochemical compounds within the *Artocarpus* genus has been a focal point of recent research endeavours. It's worth noting that these compounds contribute to the diverse pharmacological properties and potential health benefits associated with *Artocarpus* species including antioxidant, anti-inflammatory and estrogenic activities. Further research may uncover additional chemical constituents and biological activities which may contribute to the overall understanding of the therapeutic potential of these plants.

In-depth analyses of specific *Artocarpus* species have revealed distinctive profiles of bioactive compounds as given below:

i) *Artocarpus heterophyllus* (jackfruit) has been found to contain a noteworthy array of phenolic compounds, flavonoids, arylbenzofurans and stilbenoids which underscores its potential therapeutic significance (Burci *et al.*, 2019).

ii) Similarly, *A. altilis* (breadfruit tree) exhibits a composition rich in steroids, anthraquinone, glycoside, tannins and flavonoids which suggests its diverse range of pharmacological attributes (Sikarwar *et al.*, 2015). *A. altilis* also harbours unique chemical components including the cyclopropane-containing sterol known as cycloartenol. Isolated from the fresh fruit, cycloartenol comprises a substantial portion constituting of 12% of the non-saponifiable extract. This intriguing compound adds an extra layer of complexity to the composition of breadfruit which prompts a further exploration into its potential biological and pharmacological implications (Morton, 1987).

iii) *A. hirsutus* (wild jack) presents a pharmacologically diverse profile featuring sterols, terpenoids, tannins, saponins, flavonoids and lactones (Nayak *et al.*, 2017). The comprehensive phytochemical makeup of *A. hirsutus* implies potential applications in various therapeutic domains.

iv) *A. lakoocha* (lakucha) has unveiled the presence of flavonoids and tannins which contribute to the overall understanding of the genus's chemical diversity (Bhattacharya *et al.*, 2019). In a study, Saleem *et al.* (2018) unveiled a spectrum of bioactive constituents within lakucha fruit extract including flavonoids, tannins, terpenoids, saponins, glycosides, alkaloids and steroids. These findings align seamlessly with earlier research, specifically to a study by Kumar *et al.* (2010) which documented the presence of these primary and secondary metabolites in the pericarp of *A. lakoocha* fruit. Furthermore, Saleem *et al.* (2018) provided comprehensive evidence of the occurrence of chromatotropic acid, quercetin, gallic acid, vanillic acid, cinnamic acid, ferulic acid and kaempferol in the methanolic extract of lakucha fruits. This corroborates with existing literature such as the work of Shajib *et al.* (2013) which identified additional phytochemicals like artocarpine, artocarpetin and agglutinin in lakucha fruits. This insight into *A. lakoocha* adds to the growing body of knowledge regarding the bioactive compounds within the *Artocarpus* genus.

v) *A. camansi* (breadnut) has been found to contain polyprenol, squalene and β-sitosterol (Tsai *et al.,* 2013). Artonin E&B, heterophyllin, artobiloxanthone, cycloheterophyllin and cycloartobiloxanthone were isolated from *A. communis*. Artonin E&B showed stronger activity against cancer cells. Artonin E was also expected to be an anti-tumour promoter (Mohanty and Pradhan, 2015). This diversity of bioactive compounds within *A. camansi* suggests its potential utility in various fields such as pharmaceuticals or nutraceuticals.

Beyond their phytochemical richness, *Artocarpus* plants have been associated with diverse pharmacological activities including antioxidant, anti-inflammatory and antimicrobial properties (Sikarwar *et al.,* 2015; Yen *et al.,* 2017; Burci *et al.,* 2019). The multifaceted nature of the bioactive compounds present in *Artocarpus* species underscores the potential for further exploration and utilization of these plants in various scientific and medicinal applications.

A comprehensive compilation of diverse compounds extracted from various anatomical components of select species is presented herein below:

Artocarpus species	Plant part	Compounds	Extraction solvent	Reference
Artocarpus heterophyllus	Flowers	Carotenoids, flavonoids	Ethanol, methanol	Gupta *et al.*, 2020
	Leaves	Chromones, flavonoids, catechin, chlorogenic acid	Ethanol, methanol	Liu *et al.*, 2020; Gonzalez *et al.*, 2020
	Roots	Flavonoids (free 6- or 8-prenylated substituted flavones, free 8-geranyl substituted flavones, free 3- prenylated substituted flavones)	Ethanol	Ye *et al.*, 2019
	Stems	Chromones, flavonoids	Methanol	Liu *et al.*, 2020
	Bark	Flavonoids, phenols, saponins, (epi)catechin, glycosides	Ethanol, methanol	Ye *et al.*, 2019; Zhang *et al.*, 2017; Ilmi *et al.*, 2020; Ajiboye *et al.*, 2018
	Peels			
	Pulps			
	Seeds	Flavonoids, tannins, steroids, glycosides, saponins, anthraquinones, phenols	Ethanol, n-hexane, methanol	Zhang *et al.*, 2017; Eve *et al.*, 2020; Dwitiyanti *et al.*, 2019

Artocarpus species	Plant part	Compounds	Extraction solvent	Reference
Artocarpus altilis	Peels	Alkaloids, flavonoids, phenols, tannins, glycosides	Methanol	Marjoni *et al.*, 2018
	Bark	Prenylated stilbenoids	Ethanol	Hosek *et al.*, 2019
	Heartwood Cortex	Artocarpin, norartocarpetin, cycloartocarpin, artonol B, cyclomorusin, artoflavone A, cyclogeracommunin, andartonin M, prenylated stilbenoids, flavonoids, isocycloartobiloxanthon, furanocyclocommunin	Ethanol, methanol	Hosek *et al.*, 2019; Lan *et al.*, 2013
	Fruits	Flavonols, tannins, catechins, monoterpenes	Methanol	Soifoini *et al.*, 2018
	Leaves	Phenols, flavonoids, prenylated aurone, cycloaltilisin 7, artocarpaurone, prenylated chalcones, prenylated flavanones, triterpens, hexadecanoicacid, *cis*-13-octadecenoicacid and cinnamic acid	Ethanol, methanol	Lan *et al.*, 2013; Huong *et al.*, 2012; Ahmad *et al.*, 2020
	Root Seeds	Prenylflavonoids, triterpenoids	Ethanol	Adebajo *et al.*, 2014
Artocarpus hirsutus	Stem	Sterols, triterpenoids, flavonoids, tannins, saponins	Ethanol	Mahadeva *et al.*, 2017
	Leaves	Glycosides, alkaloids, sterols, triterpenoids, flavonoids, tannins, saponins	Acetone, ethanol, methanol	Mahadeva *et al.*, 2017; Patel *et al.*, 2016; Shanmugapriya *et al.*, 2017
	Fruits	Alkaloids, flavonoids, glycosides, saponins, terpenoids	Acetone, methanol	Shanmugapriya *et al.*, 2017

Artocarpus species	Plant part	Compounds	Extraction solvent	Reference
Artocarpus lakoocha	Flowers	Carotenoids, flavonoids	Ethanol, methanol	Gupta *et al.*, 2020
	Bark Twigs	Flavonoid (lakoochanone), lakoochanosides, catechin, moracin C, integrin, cyclocommunin, engeletin, isogemichalcone B, morachalcone A, heterophyllene B, albanin A, moracin M, artocarpesin, norartocarpin, resveratrol, artocarpanone, oxyresveratrol	Acetone	Boonyaketgoson *et al.*, 2019, 2020
	Heartwood	Oxyresveratrol	Ethanol	Hankittichai *et al.*, 2020; Aziz *et al.*, 2019
	Leaves	Flavonoids, phenols, tannins, eicosane, diethylphthalate, 9-octyl eicosane	Pentane, n-hexane, ethyl acetate, acetone, methanol	Bhattacharya *et al.*, 2019
	Fruits	Flavonoids, tannins, terpenoids, saponins, glycosides, alkaloids, steroids, quercetin, kaempferol	Methanol	Saleem *et al.*, 2018
Artocarpus camansi	Leaves Stem	Friedelinol, squalene, β-sitosterol, stigmasterol, phytol, polyprenol, cycloartenol	Acetone	Tsai *et al.*, 2013
	Peels	Steroids, triterpenoids	n-Hexane	Nasution *et al.*, 2018

Upon analysis of the aforementioned dataset, a diverse array of phytochemical compounds was identified among the five distinct *Artocarpus* species, showcasing a rich profile that encompasses flavonoids, glycosides, tannins, phenols, saponins, alkaloids, steroids and triterpenoids. The distribution of these compounds varies across different parts of the plants emphasizing the unique botanical characteristics of each species.

Flavonoids, recognized for their antioxidant properties manifest prominently in multiple components of *A. heterophyllus*, *A. altilis*, *A. hirsutus* and *A. lakoocha* with a particular abundance observed in leaves and fruits. Glycosides, known for their medicinal potential, are notably prevalent in the fruits of *A. heterophyllus*, *A. altilis*, *A. hirsutus* and *A. lakoocha*. Tannins, recognized for their astringent and antimicrobial qualities are sourced from *A. heterophyllus* seeds, *A. altilis* and *A. lakoocha* fruits, as well as *A. hirsutus* and *A. lakoocha* leaves. Phenols, with their antioxidant and anti-inflammatory attributes are detected in the stems of *A. heterophyllus* and *A. altilis*, *vis-a-vis* in the leaves of *A. lakoocha*. Saponins, known for their diverse biological activities abound in the fruits of *A. heterophyllus*, *A. hirsutus* and *A. lakoocha*. Alkaloids, recognized for their pharmacological significance are discernible in the fruits of *A. altilis*, *A. hirsutus* and *A. lakoocha*. Steroids, contributing to plant defense mechanisms are identified in *A. heterophyllus* seeds, *A. lakoocha* fruits and *A. camansi* peels. Furthermore, triterpenoids known for their anti-inflammatory and antiviral properties are sourced from *A. altilis* roots and seeds, and *A. hirsutus* stems *cum* leaves as well as *A. camansi* peels.

This comprehensive exploration not only elucidates the abundance of these phytochemicals across various *Artocarpus* species but also underscores the potential pharmacological and medicinal significance of these compounds in diverse plant parts. Such detailed insights lay the groundwork for further research on the therapeutic applications and ecological roles of these phytochemicals, fostering a deeper understanding of the intricate biochemistry within the *Artocarpus* genus.

2.4 Pharmacological Activities of *Artocarpus* Genus

The plants of *Artocarpus* have been documented for their diverse pharmacological properties such as anti-inflammatory, antioxidant, antimalarial and antifungal activities alongside cytotoxic effects and tyrosinase inhibitory activities (Liu *et al.*, 2018). This seminal work illuminates the multifaceted therapeutic potential inherent in *Artocarpus* plants. Anti-inflammatory attributes of these plants contribute to their significance in addressing inflammatory conditions, showcasing their potential as natural anti-inflammatory agents. Furthermore, their antioxidant properties underscore their capacity to neutralize harmful free radicals which positions the *Artocarpus* plants as potential candidate in oxidative stress-related disorders.

In the realm of infectious diseases, the antimalarial and antifungal activities exhibited by *Artocarpus* plants offer promising avenues for the development of novel therapeutics. The cytotoxic effects observed further accentuate their potential in cancer research and treatment as cytotoxicity is a pivotal characteristic in targeting and eliminating malignant cells. The tyrosinase inhibitory activities of *Artocarpus* plants point towards their possible application in dermatological formulations for their potential to regulate melanin synthesis. This opens avenues for research in skincare and cosmetic sciences.

The cumulative pharmacological insights derived from previous studies collectively validate the therapeutic potential of *Artocarpus* plants. As elucidated below, these findings not only provide a comprehensive understanding of their pharmacological spectrum but also pave the way for future investigations and applications in the fields of medicine, dermatology and beyond.

In recent times, there has been a significant surge in the application of ethno-medicinal knowledge for treatment of various diseases. Traditional medicine has emerged as a valuable source of potentially beneficial compounds and among the diverse array of plants utilized, *Artocarpus* species have gained particular attention. These species have been traditionally employed to address a spectrum of disorders with different parts of plants exhibiting various therapeutic activities.

Leaves of *Artocarpus* plants are traditionally used to treat conditions such as fever, boils, wounds and skin diseases. The young fruits are recognized for their astringent and carminative properties while the ripe fruits are believed to possess laxative and aphrodisiac activities, the seeds exhibit diuretic qualities and are known for their constipating effects, wood of these plants finds application as an antidiabetic, anti-inflammatory, sedative and in management of convulsions, its latex is utilized in treatment of dysopia, ophthalmic disorders, pharyngitis and as an antibacterial agent, the dried latex demonstrates marked androgenic action and aids in the healing of abscesses, snakebites and glandular swellings, and root is considered a remedy for various conditions including skin diseases, asthma, fever and diarrhoea (Somashekhar *et al.*, 2013). The widespread distribution of *Artocarpus* species positions the plant as a potent pharmaceutical tool for ethno-medicinal purposes. Extensive literature underscores the multitude of pharmacological investigations conducted on *Artocarpus* species, revealing diverse effects such as anti-inflammatory, antioxidant, antifungal, regulation of sexual behaviour, immuno-modulatory effects, antidiabetic properties, antibacterial effects, anti-cholinergic activity, chelating abilities, nutritional assessment, cosmetic applications, toxicity to cancer cells, anthelmintic effects, protease inhibition, regulation of estrogens and inhibition of melanin biosynthesis (Somashekhar *et al.*, 2013; Mohanty and Pradhan, 2015; Jagtap and Bapat, 2010).

Pharmacological properties	Extracts	Description	Reference
Antimicrobial activity	*A. heterophyllus* seed extracts	Flavonoids have shown inhibiting bacterial movement and reducing its biofilm formation in *Pseudomonas aeruginosa* and inhibiting bacterial toxin production of *Staphylococcus aureus*	Biharee *et al.*, 2020
	A. altilis leaf extract activity	Hexadecanoic acid, *cis*-13-octadecenoic acid, and cinnamic acid had antimicrobial; had shown inhibition against *Bacillus cereus* and *Escherichia coli*'s growth, respectively	Ahmad *et al.*, 2020
Antidiabetic activity	Ethanolic extract of *A. heterophyllus* seeds	Showed antidiabetic activity in diabetic rats by reducing blood glucose levels by 61.73%. The existence of saponins can lower cholesterol and blood glucose levels.	Dwitiyanti *et al.*, 2019; Ajiboye *et al.*, 2018
Anti-inflammatory activity	Methanol extract from leaves and stems of *A. heterophyllus*	Chromones and flavonoids had anti-inflammatory activity	Liu *et al.*, 2020
	A. altilis leaf extracts	Able to reduce inflammation in chronic inflammatory rats .	Palupi *et al.*, 2020
	A. lakoocha heartwood extract	Oxyresveratrol obtained from the extract work as proinflammatory mediators.	Aziz *et al.*, 2019
Antioxidant activity	Methanol extracts: peel, pulp and seed of *A. heterophyllus*	Expressed DPPH scavenging activities due to the existing phenolic and flavonoid compounds. Hydroxyl groups linked to the aromatic ring in flavonoid structure allowed them to scavenge free radicals through redox reaction.	Ajiboye *et al.*, 2016
	Stem bark extract of *A. altilis*	Has high flavonoids contents as well. The percentage inhibition of DPPH free radical by *A. altilis* stem bark extract was 62.5% at 100μg/ml.	Akanni *et al.*, 2014
	A. hirsutus seed extracts	In another study, the radical scavenging activity was 51.6 ±1.52%. The seed oils contained phenolic compounds, which performed antioxidant activity. Phenolic compounds act as an antioxidant group by donating hydrogen to free radicals.	Nagala *et al.*, 2013; Adan *et al.*, 2020

Pharmacological properties	Extracts	Description	Reference
Anticancer activity and anti-tumour promoter	Methanolic extract from leaves and stems of *A. heterophyllus*	The presence of chromones and flavonoids showed antiproliferative activity. Those compounds presented significant inhibitory effects against various human cancer cells.	Liu *et al.*, 2020
	Methanol extract of *A. altilis* pulp	The presence of quercetin was able to induce cell apoptosis in HeLa cell.	Jamil *et al.*, 2018
	Leaf extract of *A. camansi*	Showed high cytotoxic activity against breast carcinoma cells.	Tantengco and Jancito, 2015
	Artonin E and B, heterophyllin, artobiloxanthone, cycloheterophyllin & cycloartobiloxanthone isolated from *A. communis*	Artonin E demonstrated enhanced efficacy against cancer cells, specifically mouse L-1210 and colon 38. Notably, Artonin E is anticipated to function as an anti-tumour promoter akin to morusin derived from *Morus alba.* Among the tested *Artocarpus* flavonoids, artonin E displays the highest potency to inhibit tumour-necrosis factor-α (TNF-α) release. Heterophyllin and artobiloxanthone exhibited stronger inhibition compared to cycloheterophyllin and cycloartobiloxanthone. Notably, compounds demonstrating robust activity were characterized by the presence of three hydroxyl groups in the B ring, suggesting that this particular feature may be a crucial determinant of their biological activity.	Ragone, 2006; Donsing *et al.*, 2008; Pradhan *et al.*, 2013; Mohanty and Pradhan, 2015
Antifungal activity	Leaf extract of *A. heterophyllus*	Catechin and chlorogenic acid compounds showed high inhibition percentage against *Colletotrichum gloeosporioides.* Catechin exhibited antifungal activity by inhibiting cAMP synthesis and interrupting intracellular signaling pathways in fungal ~ known as the MAP kinase cascade .	Gonzalez *et al.*, 2020; Saito *et al.*, 2013
	A. altilis leaf extracts in methanol	The minimum inhibitory concentration (MIC) against *Candida albicans* was 475 mg/ml.	Riasari *et al.*, 2017

Pharmacological properties	Extracts	Description	Reference
Antiviral activity	*A. heterophyllus*, *A. altilis* and *A. camansi* leaf extracts were tested on hepatitis C virus (HCV) infected cells	Flavonoids obtained from *Artocarpus* species are effective in blocking viral transcription and translation, inhibiting viral replication and preventing virus attachment into cells.	Lalani and Poh, 2020
Antimalarial activity	Ethanolic extract of *A. altilis* leaves	Isolated prenylflavonoids and triterpenoids work as antimalarial constituents.	Adebajo *et al.*, 2014
Hepatoprotective potential	Methanol extract of *A. lakoocha*	Quercetin, gallic acid and kaempferol have been shown to act as natural antioxidants and hepatoprotective agents.	Saleem *et al.*, 2014
		Quercetin ~ a flavonoid stands out for its capacity to forestall oxidative damage, inhibiting nitric oxide formation and preventing hepatocellular necrosis.	Pavanato *et al.*, 2003
		Kaempferol ~ a flavonoid was found to mitigate oxidative stress and fortify antioxidant defense mechanisms	Wang *et al.*, 2015
		Gallic acid ~ a phenolic compound exhibited potent antioxidant activity by impeding reactive oxygen species (ROS), curbing lipid peroxidation and enhancing liver antioxidant enzymes.	Tung *et al.*, 2009
	These revelations underscore the potential therapeutic significance of lakucha fruit constituents in countering oxidative stress and promoting hepatic well-being.		

2.5 Uses

Numerous *Artocarpus* species bear delectable fruits. Among these, botanical marvels are *A. altilis* (**breadfruit**) which is renowned for its starchy versatile flesh; *A. integer* (**chempedak**) is known for its sweet and fragrant taste; *A. heterophyllus* (**jackfruit**) is esteemed as the largest fruit in the genus with a subtly sweet flavour and diverse culinary applications; *A. parvus* (**kwai muk**) is distinguished by its small round fruits and pleasant taste; *A. lakoocha* (**lakoocha**) is treasured for its delectable pulpy flesh; *A. kemando* (**pudau**) is valued for its succulent edible fruit; *A. hirsutus* (**Aanjily/jungle jack**) is a species grown for its unique fruit and high-quality timber; *A. chama* (**chaplaish**) is recognized for its distinct fruit characteristics; and the *A. odoratissimus* (**marang**) is esteemed for its aromatic and delectable flesh. Breadfruit and jackfruit stand out as widely cultivated species in tropical SE Asia where their culinary and economic significance is profound. Beyond their edible appeal, these fruits contribute to local economies and dietary diversity.

Several *Artocarpus* species serve dual purposes as these are cultivated not only for their delectable fruits but also for their valuable timber. Anjily thriving in the Western Ghats is a prime example which is esteemed for both its delectable fruit and high-quality timber. The Western Ghats serve as a vital habitat for this species where sustainable cultivation practices may contribute to the conservation of biodiversity and promotion of economic well-being. Some *Artocarpus* species such as anjily contribute to the rich tapestry of agroforestry, playing a pivotal role in enhancing environmental sustainability while meeting various human needs. With evolution of the cultural practices there has been increasing recognition of the importance of preserving and promoting the cultivation of these diverse *Artocarpus* species for their ecological, economical and cultural values.

References

1. Adan, A.A., R.A. Ojwang, E.K. Muge, B.K. Mwanza and E.N. Nyaboga, 2020. Phytochemical composition and essential mineral profile, antioxidant and antimicrobial potential of unutilized parts of jackfruit. *Food Res.*, **4**: 1125–34.
2. Adebajo, A.C., S.A. Odediran, F.A. Aliyu, P.A. Nwafor, N.T. Nwoko and U.S. Umana, 2014. *In vivo* antiplasmodial potentials of the combinations of four Nigerian antimalarial plants. *Molecules*, **19**: 13136-46.
3. Ahmad, M.N., N.U. Karim, E. Normaya, B.M. Piah, A. Iqbal and K.H. Bulat, 2020. *Artocarpus altilis* extracts as a food-borne pathogen and oxidation inhibitors: RSM, COSMO RS, and molecular docking approaches. *Sci Rep.*, **10**: 9566.

4. Ajiboye, B.O., O.A. Ojo, O. Adeyonu, O. Imiere, I. Olayide, A. Fadaka and B.E. Oyinloye, 2016. Inhibitory effect on key enzymes relevant to acute type-2 diabetes and antioxidative activity of ethanolic extract of *Artocarpus heterophyllus* stem bark. *J. Acute Dis.*, **5**: 423–9.

5. Ajiboye, B.O., O.A. Ojo, O. Adeyonu, O. Imiere, B.E. Oyinloye and O. Ogunmodede, 2018. Ameliorative activity of ethanolic extract of *Artocarpus heterophyllus* stem bark on alloxan-induced diabetic rats. *Adv. Pharm. Bull.*, **8:** 141–7.

6. Akanni, O.O., S.E. Owumi and O.A. Adaramoye, 2014. *In vitro* studies to assess the antioxidative, radical scavenging and arginase inhibitory potentials of extracts from *Artocarpus altilis, Ficus exasperata* and *Kigelia africana. Asian Pac. J. Trop. Biomed.*, **4**: 492–9.

7. Aziz, R.S., A. Siddiqua, M. Shahzad, A. Shabbir and N. Naseem, 2019. Oxyresveratrol ameliorates ethanol-induced gastric ulcer *via* downregulation of IL-6, TNF-α, NF-κB, and COX-2 levels, and upregulation of TFF-2 levels. *Biomed. Pharmacother.*, **110**: 554–60.

8. Barrau, J. 1976. Breadfruit and relatives. In: Evolution of crop plants (ed. Simmonds, N.W.), pp. 201–202. Longman, London (UK).

9. Bhattacharya, E., R. Dutta, S. Chakraborty and S.M. Biswas, 2019. Phytochemical profiling of *Artocarpus lakoocha* Roxb. leaf methanol extract and its antioxidant, antimicrobial and antioxidative activities. *Asian Pac. J. Trop. Biomed.*, **9**: 484–92.

10. Biharee, A., A. Sharma, A. Kumar and V. Jaitak, 2020. Antimicrobial flavonoids as a potential substitute for overcoming antimicrobial resistance. *Fitoterapia*, **146**: 104720.

11. Boonyaketgoson, S., Y. Du, A.L. Murillo, M.B. Cassera, D.G. Kingston and K. Trisuwan, 2020. Flavanones from the twigs and barks of *Artocarpus lakoocha* having antiplasmodial and anti-TB activities. *J-STAGE*, **68**: 671–4.

12. Boonyaketgoson, S., V. Rukachaisirikul, V. Phongpaichit and K. Trisuwan, 2019. Deoxybenzoin and flavan derivatives from the twigs of *Artocarpus lakoocha. Phytochem. Lett.*, **31**: 96–100.

13. Buddhisuharto, A. K., H. Pramastya and I. Fidrianny, 2021. An Updated Review of Phytochemical Compounds and Pharmacology Activities of *Artocarpus* genus. *Biointerface Res. Appl. Chem.*, **11**(6): 14898–905.

14. Burci, L.M., C.B. Silva, J.N. Rondon, L.M. Silva, S.F. Andrade, O.G. Miguel, J.F. Dias and D.M. Miguel, 2019. Acute and subacute (28 days) toxicity, hemolytic and cytotoxic effect of *Artocarpus heterophyllus* seed extracts. *Toxicol. Rep.*, **6**: 1304–8.

15. Donsing, P., N. Limpeanchob and J. Viyoch, 2008. Evaluation of the effect of Thai breadfruit's heartwood extract on melanogenesis-inhibitory and antioxidation activities. *J. Cosmetic Sci.*, **59** (1): 41-58.

16. Dwitiyanti, D., R.A. Rachmania, K. Efendi, R. Septiani and P. Jihadudin, 2019. *In vivo* activities and *in silico* study of jackfruit seeds (*Artocarpus heterophyllus* Lam.) on the reduction of blood sugar levels of gestational diabetes rate induced by streptozotocin. *Maced J. Med. Sci.*, **7**: 3819–26.

17. Eve, A., A.A. Aliero, D. Nalubiri, R.O. Adeyemo, S.A. Akinola, T. Pius, S. Nabaasa, S. Nabukeera, B. Alkali and I. Ntulume, 2020. *In vitro* antibacterial activity of crude extracts of *Artocarpus heterophyllus* seeds against selected diarrhoea-causing superbug bacteria. *Sci. World J.*, 9813970.

18. Gardner, E.M., M.G. Johnson, J.T. Pereira, A.S.A. Puad, A.D. Sahromi, N.J. Wickett and N.J.C. Zerega 2021. Paralogs and off-target sequences improve phylogenetic resolution in a densely sampled study of the breadfruit genus (*Artocarpus*, Moraceae). *Syst. Biol.*, **70**(3): 558–75.

19. Gonzalez, Y.V., J.A. Sanchez and M.C. Santoyo, 2020. Characterization and antifungal activity of jackfruit (*Artocarpus heterophyllus* Lam.) leaf extract obtained using conventional and emerging technologies. *Food Chem.*, **330:**127211.

20. Gupta, A.K., M.A. Rather, A. Kumar Jha, A. Shashank, S. Singhal, M. Sharma, U. Pathak, *D.* Sharma and A. Mastinu, 2020. *Artocarpus lakoocha* Roxb. and *Artocarpus heterophyllus* Lam. flowers: new sources of bioactive compounds. *Plants*, **9**(10): 1329.

21. Habib, A.F. 1972. Cytology of jack—*Artocarpus heterophyllus* Lam. *Mysore J. agric. Sci.*, **6**: 200–202.

22. Hankittichai, P., P. Buacheen, P. Pitchakarn, M. Na Takuathung, N. Wikan, D.R. Smith, S. Potikanond and W. Nimlamool, 2020. *Artocarpus lakoocha* extract inhibits LPS-induced inflammatory response in RAW 264.7 macrophage cells. *Intern. J. Mol. Sci.*, **21**(4): 1355.

23. Hans, A. S. 1972. Cytomorphology of arborescent Moraceae. *J. Arnold Arboretum*, **53**: 215–26.

24. Hosek, J., V. Lelakova, P. Bobal, H. Pizova, M. Gazdova, M. Malanik, K. Jakubczyk, O. Vesely, P. Landa, V. Temml, D. Schuster, V. Prachyawarakorn, P. Pailee, G. Ren, F. Zpurny, M. Oravec and K. Smejkal, 2019. Prenylated stilbenoids affect inflammation by inhibiting the NF-κB/AP-1 signaling pathway and cyclooxygenases and lipoxygenase. *J. Nat. Prod.*, **82**: 1839–48.

25. Huong, T.T., N.X. Cuong, L.H. Tram, T.T. Quang, L.V. Duong, N.H. Nam, N.T. Dat, P.T. Huong, C.N. Diep, P.V. Kiem and C.V. Minh, 2012. A new

prenylated aurone from *Artocarpus altilis*. *J. Asian Nat. Prod. Res.*, **14**: 923–8.

26. Ilmi, H.M., B. Elya and R. Handayani, 2020. Association between total phenol and flavonoid contents in *Artocarpus heterophyllus* (jackfruit) bark and leaf extracts and lipoxygenase inhibition. *Int. J. App. Pharm.*, **12**: 252–56.

27. Jagtap, U.B. and V.A. Bapat, 2010. *Artocarpus*: A review of its traditional uses, phytochemistry and pharmacology. *J. ethnopharmacol.* [Online], **12**(9): 143-4. Available from: www.elsevier.com / locate / jethpharm [Accessed 21st January 2014].

28. Jamil, M.M., S. Haneson, H.B. Mammam ands R.A. Wahab, 2018. *Artocarpus altilis* extract effect on cervical cancer cells. *Mater. Today*, **5**: 15559–66.

29. Jarrett, F.M. 1959. Studies in *Artocarpus* and allied genera. III. A revision of *Artocarpus* subgenus *Artocarpus*. *J. Arnold Arboretum*, **40**: 113–55; 298–326.

30. Jarrett, F.M. 1977. The syncarp of *Artocarpus* ~ a unique biological phenomenon. *Gard.*, **29**: 35–9.

31. Kumar. M.S., M.R. Kumar, A.C. Bharath, H.V. Kumar, T.P. Kekuda, K.C. Nandini, M.N. Rakshitha and H.L. Raghavendra, 2010. Screening of selected biological activities of *Artocarpus lakoocha* Roxb (Moraceae) fruit pericarp. *J. Basic Clin. Pharm.*, **1**: 239.

32. Lalani, S. And C.L. Poh, 2020. Flavonoids as antiviral agents for Enterovirus A71 (EV-A71). *Viruses*, **12**(2):184.

33. Lan, W.C., C.W. Tzeng, C.C. Lin, F.L. Yen and H.H. Ko, 2013. Prenylated flavonoids from *Artocarpus altilis*: antioxidant activities and inhibitory effects on melanin production. *Phytochem.*, **89**: 78–88.

34. Liu, X., X.D. Kuang, X.R. He, G. Ren, Y. Wang, L.Y. Xu, L.H. Feng, B. Wang and Z.W. Zhou, 2018. Prenylflavonoids from the twigs of *Artocarpus nigrifolius*. *J-STAGE*, **66**: 434–8.

35. Liu, Y.P., X.M. Yu, W. Zhang, T. Wang, B. Jiang, H.X. Tang, Q.T. Su and Y.H. Fu, 2020. Prenylated chromones and flavonoids from *Artocarpus heterophyllus* with their potential antiproliferative and anti-inflammatory activities. *Bioorg. Chem.*, **101**: 104030.

36. Mahadeva, N., N. Ananthanarayanan and M. Muhammed, 2017. Pharmacognostic evaluation of leaf and stem wood extracts of *Artocarpus hirsutus* Lam. *Pharmacogn. J.*, *9*: 887–94.

37. Marjoni, M.R., F. Sidik, F. Ovisa and Y. Sukma, 2018. Extraction of antioxidants from fruit peel of *Artocarpus altilis*. *Int. J. Green Pharm.*, **12**: 284–9.

38. Matthew, S.P., A. Mohandas, S.M. Shareef and G.M. Nair, 2006. Biocultural diversity of the endemic 'wild jack tree' on the Malabar coast of South India. *Ethnobot. Res. & Applicat.*, **4**: 025-040.

39. Mehra, P.N. and B.S. Gill, 1974. Cytological studies in Ulmaceae, Moraceae and Urticaceae. *J. Arnold Arboretum*, **55**: 663–77.

40. Mehrotra, R.C., U. Prakash and M.B. Bande, 1984. Fossil woods of *Lophopetalum* and *Artocarpus* from the Deccan Intertrappean beds of Mandla district, Madhya pradesh, *Indian J. Palaeosci.*, **324**(1-3): 310–20.

41. Mohanty, M. and C. Pradhan, 2015. A Review on Phytochemistry, Bio-Efficacy, Medicinal and EthnoPharmaceutical Importance of *Artocarpus altilis*. *Ijppr. Human*, **3**(1): 219-31.

42. Morton, Julia F. 1987. Breadfruit. Fruits of Warm Climates, pp. 50-58. Center for New Crops & Plant Products. Department of Horticulture and Landscape Architecture, Purdue University, West Lafayette, Indiana, USA, pp. 50-58. ISBN 0-9610184-1-0.

43. Nagala, S., M. Yekula and R.R. Tamanam, 2013. Antioxidant and gas chromatographic analysis of five varieties of jackfruit (*Artocarpus*) seed oils. *Drug Invent. Today*, **5**: 315–20.

44. Nasution, R., C.N. Fitrah, H. Helwati, Murniana, B. Arifin, Cutchamzurni, Y. Rizal and *M.* Marianne, 2018. Antidiabetes activities extract hexane from the peels of *Artocarpus camansi* Blanco fruit. *Asian J. Pharm. Clin. Res.*, **11**(1): 12–7.

45. Nayak, M., A. Nagarajan and M. Majeed, 2017. Pharmacognostic evaluation of leaf and stem wood extracts of *Artocarpus hirsutus* Lam. *Pharmacogn. J.*, **9**: 887–94,

46. Oginuma, K. and H. Tobe, 1995. Karyomorphology of some Moraceae and Cecropiaceae (Urticales). *J. Pl. Res.*, **108**: 313–26.

47. Palupi, D.H., D.S. Retnoningrum, M.I. Iwo and A.A. Seomardji, 2020. Leaf extract of *Artocarpus altilis* [Park.] Fosberg has potency as anti-inflammatory, antioxidant and immunosuppressant. *Rasayan J. Chem.*, **13**: 636–46.

48. Patel, J., V. Reddy and S. Kumar, 2016. Preliminary phytochemical screening and hepatoprotective activity of methanol extract of *Artocarpus hirsutus* leaves. *Int. J. Phytomedicine*, **8**. https://biointerfaceresearch.com/

wp-content/uploads/2021/04/20695837116.1489814905.pdf. https://doi.org/10.5138/09750185.1900.

49. Pavanato, A., M.J. Tuñón, S. Sánchez-Campos, C.A. Marroni, S. Llesuy, J. González - Gallego and N. Marroni, 2003. Effects of quercetin on liver damage in rats with carbon tetrachloride-induced cirrhosis. *Dig. Dis. Sci.*, **48**: 824-9.

50. Pradhan, C., M. Mohanty and A. Rout, 2013. Phytochemical screening and comparative bioefficacy assessment of *Artocarpus altilis* leaf extracts for antimicrobial activity. *Front. Life Sc.*, **6**(3–4): 71–6.

51. Ragone, D. 2001. Chromosome numbers and pollen stainability of three species of Pacific Island breadfruit (*Artocarpus*, Moraceae). *Am. J. Bot.*, **88**(4): 693-6.

52. Ragone, D. 2006. *Artocarpus altilis* (breadfruit), ver. 2.1. In: Species Profiles for Pacific Island Agroforestry (ed. Elevitch, C.R.). Permanent Agriculture Resources (PAR), Holualoa, Hawai'i. <http://www.traditionaltree.org>.

53. Ragone, D. 2018. Breadfruit–*Artocarpus altilis* (Parkinson) Fosberg. *Exot. Fruits*, 53–60. doi: 10.1016/B978-0-12-803138-4.00009-5.

54. Riasari, H., M. Ulfah, D. Prayugo and N.A. Komariah, 2017. Antibacterial and antifungal activities of various bread fruit leaves (*Artocarpus altilis* (Parkinson) Fosberg). *Int. J. Pharm. Sci.*, **8**: 1066–73.

55. Saito, H., M. Tamura, K. Imai, T. Ishigami and K. Ochiai, 2013. Catechin inhibits *Candida albicans* dimorphism by disrupting Cek1 phosphorylation and cAMP synthesis. *Microb. Pathog.*, **56**: 16–20.

56. Saleem, M., B. Ahmed, M.I. Qadir, M. Rafiq, M. Ahmad and B. Ahmad, 2014. Hepatoprotective effect of *Chenopodium murale* in mice. *Bangladesh J. Pharmacol.*, **9**: 124-8.

57. Saleem, M., A. Asiff, M.F. Akhtar and A. Saleem, 2018. Hepatoprotective potential and chemical characterization of *Artocarpus lakoocha* fruit extract. *Bangladesh J. Pharmacol.*, **13**: 90–97.

58. Shajib, M.T., M. Kawser, M.N. Miah, P. Begum, L. Bhattacharjee, A. Hossain, I.S. Fomsgaard and S.N. Islam, 2013. Nutritional composition of minor indigenous fruits: Cheapest nutritional source for the rural people of Bangladesh. *Food Chem.*, **140**: 466-70.

59. Shanmugapriya, K., S. Akanya, C.B. Anderson. N.S. Berril, P.R. Prakash and V. Suganya, 2017. Phytochemical screening of *Artocarpus hirsutus* and its antimicrobial potential. *Asian J. Pharm. Clin. Res.*, **10**: 298– 302.

60. Sikarwar, M.S., B.J. Hui, K. Subramaniam, B.D. Valeisamy, L.K. Yean and K. Balaji, 2014. A review on *Artocarpus altilis* (Parkinson) Fosberg (breadfruit). *J. Appl. Pharm. Sci.*, **4**: 91–7.

61. Sikarwar, M.S., B.J. Hui, K. Subramaniam, B.D.Valeisamy, L.K. Yean and K. Balaji, 2015. Pharmacognostical, phytochemical and total phenolic content of *Artocarpus altilis* (Parkinson) Fosberg leaves. *J. Appl. Pharm. Sci.*, **5**: 94–100.

62. Soifoini, T., D. Donno, V. Jeannoda, E. Rakotoniaina, S. Hamidou, S.M. Achmet, N.R. Solo, K. Afraitane, C. Giacoma and G.L. Beccaro, 2018. Bioactive compounds, nutritional traits and antioxidant properties of *Artocarpus altilis* (Parkinson) fruits: exploiting a potential functional food for food security on the Comoros Islands. *J. Food Qual.* Article ID 5697928, 11 pages, 2018. https://doi.org/10.1155/2018/5697928.

63. Somashekhar, M, Naira Nayeem and Basavraj Sonnad, 2013. A review on family Moraceae (Mulberry) with a focus on *Artocarpus* Species. *World J. Pharm. Pharmaceut.* Sc., **2**(5): 2614-21.

64. Srivastava, Rashmi, 1998. Fossil wood of *Artocarpus* from warkalli formation of kerala coast, India. *Phytomorphol.*, **48**(4): 391-7.

65. Tantengco, O.A. and S.D. Jancito, 2015. Cytotoxic activity of crude extracts and fractions from *Premna odorata* (Blanco), *Artocarpus camansi* (Blanco) and *Gliricidia sepium* (Jacq.) against selected human cancer cell lines. *Asian Pac. J. Trop. Biomed.*, **5**: 1037–41.

66. Tsai, P.W., K.A. Cruz, C.C. Shen, C.T. Chiou and C.Y. Ragasa, 2013. Chemical constituents of *Artocarpus camansi. Pharmacogn. J.*, **5:** 80–82.

67. Tung, Y.T., J.H. Wu, C.C. Huang, H.C. Peng, Y.L. Chen, S.C. Yang and S.T. Chang, 2009. Protective effect of *Acacia confusa* bark extract and its active compound gallic acid against carbon tetrachloride-induced chronic liver injury in rats. *Food Chem. Toxicol.*, **47**: 1385-92.

68. Wang, M., J. Sun, Z. Jiang, W. Xie and X. Zhang, 2015. Hepatoprotective effect of kaempferol against alcoholic liver injury in mice. *Am. J. Chin. Med.*, **43**: 241-54.

69. Williams, E.W., E.M. Gardner, R. Harris, A. Chaveerach, J.T. Pereira and N.J.C. Zerega, 2017. Out of Borneo: biogeography, phylogeny and divergence date estimates of *Artocarpus* (Moraceae). *Ann. Bot.*, **119**(4): 611–27.

70. Xavier, T.F., M. Kannan, L. Lija, A. Auxillia, A.K.F. Rose and S.S. Kumar, 2014. Ethnobotanical study of Kani tribes in Thodu hills of Kerala, South India. *J. Ethnopharmacol.*, **152**(1): 78–90.

71. Xu, L., X. Liu, Q. Lin, Y. Wang, X. He and Z. Zhou, 2019. Prenylflavanones and lignans from the twigs of *Artocarpus pithecogallus*. *Magn. Reson. Chem.*, **57**: 506–511.

72. Ye, J-B., G. Ren, W.Y. Li, G.Y. Zhong, M. Zhang, J.B. Yuan and T. Lu, 2019. Characterization and identification of prenylated flavonoids from *Artocarpus heterophyllus* Lam. roots by quadrupole time-of-flight and linear trap quadrupole orbitrap mass spectrometry. *Molecules*, **24**(24):4591.

73. Yen, K.H., N. Nyokat, C.J. Kutoi, A.S. Hamzah and I.F. Lim, 2017. Chemical constituents of *Artocarpus odoratissimus* from Sarawak. *J. Appl. Pharm. Sci.*, **7**: 137–141.

74. Zerega, N.J.C., M.N.N. Supardi and T.J. Motley, 2010. Phylogeny and recircumscription of Artocarpeae (Moraceae) with a focus on *Artocarpus*. *Syst. Bot.*, **35**(4): 766–82.

75. Zhang, L., Z. Tu, X. Xie, H. Wang, Z. Wang, X. Sha and Y. Lua, 2017. Jackfruit (*Artocarpus heterophyllus* Lam.) peel: a better source of antioxidants and a-glucosidase inhibitors than pulp, flake and seed, and phytochemical profile by HPLC-QTOF-MS/MS. *Food Chem.*, **234**: 303–13.

2A

Jackfruit or Jack tree (*Artocarpus heterophyllus* Lam.)

Synonyms: *Artocarpus brasiliensis* Gomes, *A. grandis* Linden ex Cazzuola & Nencioni, *A. integrifolius* var. *glaber* Stokes, *A. integrifolius* var. *heterophyllus* (Lam.) Pers., *A. maximus* Blanco, *A. nanca* Noronha, *A. philippensis* Lam., *Polyphema jaca* Lour. and *Saccus heterophyllus* (Lam.) Kuntze (POWO, 2023).

Indian names: Katahal (Hindi), Kothaal (Asomese), Chakka (Malayali), Phanas (Marathi), Ponos (Konkani), Halasa (Kannad) (Lobo, 2019), Palā (Tamil ~ 'Palāppalam' for ripe jackfruit), Panasa (Odia ~ 'Panasa Katha' for unripe jackfruit and 'Pachila Panasa' for ripe one).

Other names: Jak, Jaca, *Nangka* in Malaysia and the Philippines, *Khanun* in Thailand, *Khnor* in Cambodia, *Mak mi* or *May mi* in Laos and *Mit* in Vietnam (Morton, 1987).

Chromosome number: $2n = 4x = 56$.

2A.1 Introduction

Artocarpus heterophyllus, commonly known as jackfruit, stands out as the most prevalent species within the *Artocarpus* genus, contributing significantly to the economic landscape of SE Asia (Fathin *et al.*, 2021). Thriving exclusively in humid tropical and near-tropical climates, the jackfruit has found its niche in various regions in India, It exhibits resilience in the Himalayan foothills and adapts seamlessly across altitudes ranging from sea level to 1,500 metres in the southern regions (Morton, 1987).

Jackfruits on main stem of the tree

By Augustus Binu, CC BY-SA 3.0, https://commons.wikimedia.org/w/index.php?curid=42008455

As a member of the Moraceae family, jackfruit shares botanical ties with breadfruit, figs and mulberries. Its cultivation is characterized by relative ease, positioning it as a promising and sustainable food source. The fruit's expansive growth is noteworthy with specimens reaching remarkable sizes ~ some weighing up to 36kg or more. External appearance is marked by a spiky, greenish-yellow and thick skin, encapsulating an inner pale-yellow flesh composed of large edible bulbs/pods each harbouring a seed.

Jackfruit tree laden with clustered fruits, mostly on main trunk

By Balaram Mahalder - Own work, CC BY-SA 3.0, https://commons.wikimedia.org/w/index.php?curid=14977521

The jackfruit experience encompasses not only its physical attributes but also sensory elements. Described as having a dense *cum* sweet taste, the fruit emanates an almost intoxicating sweet aroma (Lobo, 2019). This unique combination contributes to its culinary appeal, making it a sought-after ingredient globally. Beyond its taste and aroma, jackfruit's potential as a sustainable food source is emphasized. Its adaptability to various climates and relative ease of cultivation make it an environmentally friendly option. Moreover, the use of unripe jackfruit as a meat substitute aligns with the growing trend of plant-based diets, recognized for their positive environmental impact. The global popularity of jackfruit as a meat alternative in vegetarian and vegan cuisines attests to its versatility and nutritional benefits. Its ability to absorb flavours coupled with its meaty texture, positions jackfruit as a preferred choice for those exploring plant-based alternatives.

The cultural significance of jackfruit is evident in dedicated festivals held across regions in India including Maharashtra, Karnataka, Kerala, Goa and the

North East. In smaller villages, especially during festivals, jackfruit serves as an excellent and cherished gift. This cultural connection extends to linguistic expressions such as the Kannada saying '*hasidu halasu tinnu, undu maavu tinnu*' (meaning loosely 'eat jackfruit when hungry, eat mango when full') and the Bengali sentiment associating summers with the fragrance of mango and jackfruit (Lobo, 2019).

2A.2 Etymology

The nomenclature of jackfruit (*Artocarpus heterophyllus*), unfolds a linguistic and botanical tale rooted in ancient languages. The generic name stems from the fusion of the Greek words '*artos*' (bread) and '*karpos*' (fruit), collectively yielding the term 'breadfruit' for its edible produce. Delving into specificity, the epithet '*heterophyllus*' is a Latin descriptor denoting various-leaved attributes, eloquently capturing the diversity in the sizes and shapes of its leaves. This Latin term, in turn, finds its roots in the Greek '*heteros*', meaning different.

The etymology of jackfruit itself unveils a historical journey as the term draws inspiration from the Portuguese '*jaca*' which in turn echoes the Malayali influence of '*chakka*'. This linguistic confluence found its roots during the Portuguese arrival on the Malabar Coast of Kerala in 1499, encapsulating the rich history and cultural exchanges that have shaped the language surrounding this venerable fruit (Pradeepkumar *et al.*, 2008; Silver, 2014; Lobo, 2019).

2A.3 Origin and Distribution

The jackfruit (*Artocarpus heterophyllus*) and its closely allied counterpart the 'chempedak' (*Artocarpus integer*) trace their origins back to the fertile lands of India and Malaysia. The genus *Artocarpus* showcases heightened diversity along the tropics of the Indo-Malayan region, eclipsing its Australasian counterparts. Recent taxonomic evaluations in Sri Lanka have shed light on its intricate classification and distribution (Senavirathna *et al.*, 2020).

Native to the lush landscapes of SW India, jackfruit is believed to have originated in the biodiverse rainforests of the Western Ghats. This unique tropical fruit has not confined itself to its place of origin but has ventured beyond, establishing its presence in various regions across Asia. It made its way to Bangladesh, China, Indonesia, Myanmar, Malaysia, Nepal, the Philippines, Sri Lanka, Thailand and Vietnam. Furthermore, it found its way to Latin American and East African countries through intentional introductions (Jagadeesh *et al.*, 2007; Khan *et al.*, 2010; Sindhu, 2012; POWO, 2023).

In its indigenous habitat the Western Ghats rainforests, jackfruit thrives, though has also adapted to cultivation at lower elevations throughout India,

Myanmar, Sri Lanka, southern China and the East Indies. Its presence is notable in the Philippines where it is cultivated, *vis-a-vis* naturalized. The fruit has found its way to Queensland and Mauritius albeit to a limited extent. In Africa, particularly in Kenya, Uganda and former Zanzibar, jackfruit is a common sight showcasing its adaptability. While its introduction to Hawaii dates back prior to 1888, it remains relatively rare in the Pacific islands. Similarly, in most tropical American and West Indies regions jackfruit is not as prevalent though it gained popularity in northern Brazil during mid-19th century and is more widely embraced there and in Surinam than in other parts of the New World (Morton, 1987).

Within the rich cultural tapestry of South India, jackfruit holds a revered status. In particular, India boasts the highest number of red-fleshed jackfruit locally known as 'chandra halasu'. This variant characterized by a milder and less sweet taste compared to the more common yellow fruit is especially abundant in the region. Panruthi in Tamil Nadu stands out as a jackfruit paradise where the fruit is cultivated on an extensive scale occupying over 10,000 hectares of land. The local farmers have become heavily reliant on jackfruit cultivation as a mainstay of their livelihoods (Lobo, 2019).

2A.4 Ecology

Jackfruit ~ an arboreal marvel, in its native habitats is primarily associated with evergreen forests nestled at altitudes ranging from 400 to 1200 metres. Distinguishing itself from its botanical relatives, *A. altilis* (breadfruit) and *A. integer* (chempedak), jackfruit boldly extends its arboreal domain into drier and cooler climates. This resilient species bears fruit generously up to latitudes of 30°N and 30°S in areas untouched by frost, yielding bountiful harvests within 25°N and 25°S of the equator (Bakewell-Stone, 2023).

However, jackfruit's true vitality comes to the fore in warm and humid climates flourishing beneath the 1600m altitude mark. While it exhibits diminished resilience to cold temperatures, drought and flooding, it boasts a moderate ability to withstand winds and salinity challenges. The symphony of its growth finds its rhythm in regions blessed with annual rainfall exceeding 1500mm, where dry season plays a subdued role.

Jackfruit proves itself versatile in soil preferences as it adapts to various types. Yet, it achieves its zenith in deep, well-drained, alluvial, sandy or clay loam soils with a pH range of 6-7.5. This ecological affinity showcases the intricate dance between jackfruit and its surroundings, emphasizing the nuanced interplay of altitude, climate and soil conditions that contribute to the thriving tapestry of this remarkable tree (Bakewell-Stone, 2023).

2A.4 Production

India stands as the global leader in jackfruit production holding the title of the largest producer worldwide. In 2017, jackfruit production reached 1.4 million tonnes in India leading the global output, followed by Bangladesh, Thailand and Indonesia (Sawe, 2017). Particularly prominent in South India, jackfruit holds a distinguished position in the local cuisine, ranking closely alongside mangoes and bananas in terms of total annual production. The southern region boasts an impressive landscape with over 100,000 jackfruit trees flourishing in backyards and providing shade in plantations dedicated to betelnut, coffee, pepper and cardamom. The expansive cultivation of jackfruit in India spans approximately 26,000ha as horticulturists actively encourage the planting of jackfruit trees along highways, waterways and railroads especially in South India. This strategic approach not only contributes to the country's food supply but also showcases a commitment to sustainable agricultural practices.

In neighbouring Sri Lanka more than 4,452ha land is devoted to jackfruit cultivation primarily for timber purposes. Here, the fruit serves as a highly valued by-product adding economic significance to the cultivation practices. Meanwhile, in Thailand, jackfruit enjoys widespread cultivation for its delectable fruit.

Beyond the Far East, however, jackfruit faces challenges in gaining acceptance comparable to that of breadfruit. This divergence in popularity is largely attributed to the distinct aroma of ripe jackfruit which has influenced traditional preferences favouring breadfruit in other regions. Notably, the pockets of communities with East Indian origins show a different trend as there jackfruit finds its place among their culinary preferences. These nuanced dynamics underscore the multifaceted journey of jackfruit cultivation, its diverse applications and the varying cultural perceptions surrounding this versatile fruit (Morton, 1987).

World leaders in jackfruit production

Rank	Country	Production (1000 tonnes)
1	India	1436
2	Bangladesh	926
3	Thailand	392
4	Indonesia	340
5	Nepal	18.97

Source: (Sawe, 2017).

2A.6 Botany

2A.6.1 Habit, size and stem form

Jackfruit tree is a captivating small to medium-sized, evergreen and cauliflorus species distinguished by its compact yet robust structure achieving remarkable heights between 9 and 20 metres (sometimes attaining even 30m height) with a 80-200cm diameter. With a sturdy and erect demeanor, this plant boasts a straight cylindrical trunk that stands with resilience. Its branches, extending at a lower altitude create a dense and irregularly shaped crown, forming a striking silhouette in the natural landscape (Bakewell-Stone, 2023).

Jackfruit tree trunk showing texture and colouration

By Salim_Khandoker - Own work, CC BY-SA 3.0, https://commons.wikimedia.org/w/index.php?curid=70339517

2A.6.2 Root system and bark

Mature trees have tubular roots (WFO, 2024) with a long taproot (Bakewell-Stone, 2023). The trunk may occasionally develop buttress roots to enhance stability. The bark of this tree boasts a smooth texture with a striking reddish-brown colouration. Upon injury, the bark exudes a milky sap ~ a latex not confined solely to the bark but distributed throughout the tree, rendering all its parts sticky with a white consistency (Morton, 1987; Bakewell-Stone, 2023).

2A.6.3 Foliage

A. heterophyllus showcases vibrant and glossy leaves arranged alternately in a spiral pattern along its branches. These evergreen leaves measuring 22.5cm in length exhibit a unique versatility. On mature wood these take on an oval shape occasionally displaying an oblong or deeply lobed structure on young shoots (Morton, 1987).

The leaves of this plant are characterized by a 2-3cm long petiole supporting elliptic to obovate lamina that takes on an egg-shaped form which is broadest at the apex. The dimensions of the lamina range from 5 to 20cm in length and 3.5 to 12cm in breadth, presenting a leathery texture. On young shoots, the leaves may exhibit entire margins or 3-lobed, contributing to their dynamic appearance. The upper surface of the leaves is dark green possessing a glossy finish and these appear glabrous. The base of the leaves tapers gradually displaying a wedge-shaped cuneate structure while the tips range from obtuse to subacute. Notable features include large stipules that are spathaceous and measure 5-8cm in length (WFO, 2024).

2A.6.4 Flowers

The jackfruit tree is known for its unique monoecious nature where separate male and female inflorescences coexist on the same tree. This botanical phenomenon is characterized by cauliflory wherein inflorescences emerge not from the typical axillary or terminal positions but rather from the trunk, larger branches or even the base of aged trees covered by soil (Morton, 1987; Pushpakumara, 2006; Fathin *et al.*, 2021). Female flowerheads are primarily borne on short shoots emerging from the trunk and main limbs. In contrast, male heads are not confined to specific shoots; they also manifest on shoots in the peripheral regions of the tree canopy particularly in vigorously growing trees. Male and female heads on cauliflorous shoots tend to develop almost simultaneously with male heads maturing 3 to 5 days earlier (Bakewell-Stone, 2023).

Flower buds and leaves

By B.navez - Own work, CC BY-SA 3.0, https://commons.wikimedia.org/w/index.php?curid=3563405

Within this fascinating botanical landscape, the flowers themselves are situated on a fleshy rachis (Pushpakumara, 2006). The male flowers characterized by a greenish hue occasionally include sterile specimens. These flowers feature a hairy texture with 1-1.5mm membranes as part of the perianth. Individual stamens stand erect, showcasing yellow roundish anthers. With diminutive pollen grains, approximately 60 microns in diameter, pollen distribution precedes the stamens transitioning to ash-gray and subsequently falling off after a few days. All male inflorescences eventually follow suit in detachment.

On the female side, the greenish flowers possessing even a hairy exterior and tubular perianth boast a fleshy flower-like base. Within these female flowers lies an ovary with a broad, capitate or occasionally bilobed scar. The blooming period spans from December to February or even up to March offering a visually striking display of this unique botanical process (Fathin *et al.*, 2021).

2A.6.4.1 Developmental stages of inflorescences

Jackfruit trees are capable of producing both male and female inflorescences. Initiation of inflorescences is influenced by factors such as age, climate and overall health of the tree.

2A.6.4.1.1 Footstalk initiation

Originating from the primary stem or older branches, floral buds initially enveloped by stipules emerge and undergo a gradual elongation process spanning 4-8 weeks. These buds which sometimes are referred to as footstalks are characterized as short axillary leafy twigs or buds with the potential to develop into inflorescences (Pushpakumara, 2006; Li *et al.*, 2010). Upon reaching maximum elongation the stipules unfurl revealing either male or female inflorescences. These inflorescences are pedunculated and showcase a distinctive cylindrical, ellipsoidal or pear-shaped morphology, typically measuring 10–12cm in length and 5–7cm in width (en.wikipedia.org, 2024a, b). This intricate process underscores the botanical journey from initial bud emergence to the ultimate revelation of gendered inflorescences.

Footstalk initiation (the emergence of floral bud covered with stipule) and elongation (a-c); footstalk in maximum elongation (d); the opening of stipules and emergence of inflorescences (e).

2A.6.4.1.2 Developmental stages of male inflorescences

The staminate or masculine flower cluster ~ commonly referred to as the male inflorescence, takes precedence in the reproductive cycle. It manifests initially exhibiting an arrangement of numerous small flowers clustered along a shared axis. Within this inflorescence, the male flowers play a pivotal role by generating pollen ~ a crucial element for subsequent pollination of female flowers. At anthesis, male heads become coated with sticky yellow pollen and emit a sweet scent attracting small insects such as flies and beetles. While these insects may act as pollinating agents though in India wind-induced pollination has also been reported. Subsequent to anthesis, male heads darken and detach from the tree (Bakewell-Stone, 2023).

This male inflorescence undergoes a comprehensive developmental process encompassing five distinct stages. The journey through these stages requires a substantial timeframe, spanning from 64 to 101 days to complete the intricate male developmental phases. This extended duration underscores the intricate and nuanced progression that culminates in the production of mature male flowers capable of fulfilling their essential reproductive function in the broader ecological context.

Developmental stages of male inflorescences: The emergence of male inflorescences from the opening stipules of footstalk (a); pollen maturity (pollen are sticky and yellow); anthers dehiscence; pollination and fertilization (b-c); end of pollen viability; the dry up and abscission of anthers (d); and the dry up and abscission of male inflorescences (e).

Fathin *et al.*, 2021 and Pushpakumara, 2006 have defined five phases of male inflorescence development as under:

1. **Phase 1 ~ Footstalk initiation and elongation (30-45 days):** Phase 1 marks the commencement of male inflorescence development, characterized by the initiation and elongation of the footstalk. The male inflorescences can be discerned by their longer yet slender structure compared to their female counterparts.
2. **Phase 2 ~ Maximum elongation and flower emergence (14-21 days):** In Phase 2 the male inflorescences reach their maximum elongation accompanied by the opening of stipules and the emergence of male flowers. Male flower buds appearing in a yellowish-green hue, manifest on a thinner peduncle. These inflorescences typically ascend above their female counterparts consisting of numerous small individual male flowers each adorned with a single stamen. The male flower colour transitions from light green in youth to dark green or brown when aged.
3. **Phase 3 ~ Pollen maturity (14-21 days):** Phase 3 signifies the maturity of pollen spanning 14 to 21 days. At this stage, anthers undergo dehiscence releasing mature and sticky yellowish pollen ready for fertilization within the subsequent week.
4. **Phase 4 ~ End of viability (3-7 days):** Phase 4 lasting 3 to 7 days, denotes the end of viability which is noticeable when the once-yellow anthers transform into brown or grey shades before eventual abscission. Unlike female inflorescences, male counterparts cease growth post-pollen dispersal abscissing within 4-8 weeks after stipule opening.
5. **Phase 5 ~ Mould coverage and abscission (3-7 days):** The final phase witnesses male inflorescences being enveloped by black mould signifying the conclusion of the viable phase. Drought conditions contribute to the abscission of male inflorescences at the termination of the fertilization phase ~ a process spanning 3 to 7 days.

In summary, the intricate and sequential development of male inflorescences spans five phases, each characterized by distinct physiological changes, durations and visual cues, culminating in the crucial role of pollen production for pollination of female flowers.

2A.6.4.1.3 Developmental stages of female inflorescences

Following the initiation of male inflorescence, the development of the female inflorescence referred to as the 'pistillate' or 'feminine' flower cluster unfolds in a distinct manner. Typically surpassing the male inflorescence in size, the female counterpart manifests itself on a different section of the tree. Within the female inflorescence, individual flowers carry an ovary that upon fertilization matures into the recognizable jackfruit. The entire process encompasses five distinct developmental stages spanning a duration of 92-160 days for completion. Jackfruit grown in cooler environments such as those in the temperate regions of Sri Lanka and India necessitate an extended maturation period ranging from 180 to 240 days.

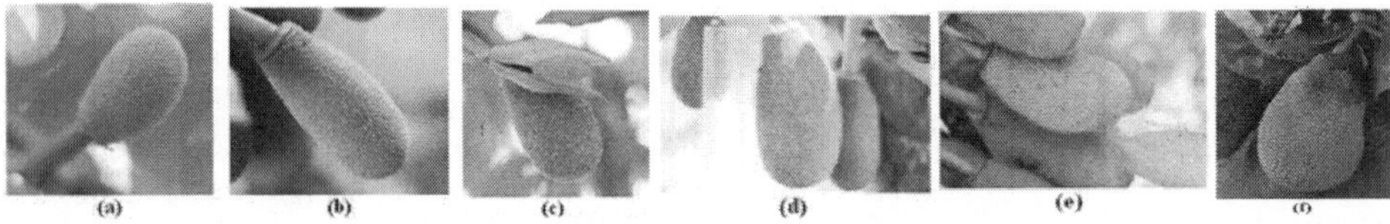

Developmental stages of female inflorescences: the emergence of female inflorescences from the opening stipules of footstalk (a); stigma receptivity (stigma are sticky, straight and white in colour); pollination and fertilization (b-c); end of stigma receptivity; the drying up and abscission of stigma; the formation of young fruits (syncarps) (d); and the development of syncarps until maturity (e-f).

As per Fathin *et al.* (2021) and Pushpakumara (2006), while developing the female flowers also undergo following five phases:

i) **Phase 1 ~ Footstalk initiation and elongation (30-45 days):** Phase 1 marks the initiation and elongation of the footstalk discernible within the initial 30 to 45 days. Female footstalks are distinguishable by their larger yet shorter dimensions and more rounded morphology compared to their male counterparts.

ii) **Phase 2 ~ Maximum elongation of footstalks, stipule opening and female inflorescence emergence (10-21 days):** Phase 2 witnesses maximum elongation of footstalks accompanied by opening of stipules and emergence of female inflorescences, unfolding over 10 to 21 days. The light green female inflorescences feature a thicker peduncle than their male counterparts. These inflorescences bear numerous small single female flowers each hosting a unilocular ovary and pistil. White and sticky pistil exhibits a yellowish and lighter stigma when receptive. The receptacles to which the flowers are attached are fleshy and edible when the fruit ripens.

iii) **Phase 3 ~ Stigma receptivity and initiation of pollination (variable duration):** Phase 3 marks the commencement of stigma receptivity characterized by liquid secretion. The pistil straight and white hosts a stigma that remains receptive for 10 to 14 days. This phase sets the stage for pollination and fertilization processes.

iv) **Phase 4 ~ End of stigma receptivity, drying and abscission (21-45 days):** Phase 4 concludes the period of stigma receptivity. The stigma dries out and undergoes abscission paving the way for the transformation into young fruits known as syncarps. This transformation spans 21 to 45 days.

v) **Phase 5 ~ Syncarp development until maturity (21-45 days):** The final phase encompasses the development of syncarps ~ a process spanning 21 to 45 days until reaching maturity. During this stage jackfruit undergoes the final maturation processes culminating in its readiness for harvest.

Fertilized female heads evolve into mature fruits over a period of some 3 months influenced by factors such as seedling type or clone. At higher altitudes or latitudes this maturation process may extend to 6 months. Unfertilized flowers transform into strap- or string-like structures, filling the interstitial spaces between developing fruitlets. A fully developed fruit may house up to 500 seeds each weighing 3-6g (Bakewell-Stone, 2023).

2A.6.5 Fruits

The fruit of the jacktree emerges from a distinctive spadix inflorescence and is formally referred to as sorosis. The jackfruit exhibits distinctive fruit morphology characterized by its ellipsoidal to roundish shape, which emerges from a robust trunk or a lengthy and thick stem. This remarkable fruit stands as a testament to nature's ingenuity as it is a multiple fruit resulting from the amalgamation of ovaries from numerous individual flowers. As the largest among tree-borne fruits, the dimensions of the jackfruit can vary significantly ranging from 20 to 90cm in length and 15 to 50cm in width with an impressive weight that can span from 4.5 to

Jackfruit flesh

By Mullookkaaran - Own work, CC BY-SA 3.0, https://commons.wikimedia.org/w/index.php?curid=19778855

20kg while in some cases reach up to 50kg. Jackfruit is known for its unique aroma which is a blend of fruity and musky scents. As the jackfruit reaches full ripeness an unopened fruit emits a potent disagreeable odour reminiscent of decayed onions. However, upon opening the pulp releases a pleasant fragrance reminiscent of a blend of pineapple and banana.

The fruit has a spiky green outer skin with a bumpy texture often referred to as the 'rind'. The skin protects the sweet, yellow and fibrous flesh inside. The rind is a resilient and adhesive covering for the compound or aggregate fruit. When ripe, this protective layer adopts vibrant shades of green or yellow and is adorned with numerous rigid cone-like points intricately attached to a substantial and rubbery and pale yellow or whitish wall.

Upon slicing open this extraordinary fruit the interior unveils a captivating arrangement. Large *cum* succulent 'bulbs' reminiscent of fully developed perianths showcase a rich yellow hue and a delightful banana flavour. These bulbs are interspersed among slender ribbons of thin and resilient undeveloped perianths (perigones) creating a visually striking contrast. At the heart of the fruit lies a central pithy core.

Nestled within the succulent banana-flavoured flesh of each bulb, there resides a velvety oval-shaped 'seed' (endocarp) exhibiting a lustrous light-brown hue. This seed is ensconced by a fragile white membrane (exocarp) forming a sophisticated and intricate composition. Beyond its aesthetic appeal, this nuanced arrangement plays a pivotal role in shaping the distinctive texture and flavour of the jackfruit underscoring the botanical intricacies that define its culinary allure.

The jackfruit's morphology is not only remarkable for its size but also for its versatility in culinary applications. It serves as a meat substitute in vegetarian and vegan cuisines owing to its fibrous texture that mimics pulled pork or shredded chicken when cooked. The intricate structure of the jackfruit along with its nutritional value makes it a significant and valuable fruit in various parts of the world (Morton, 1987).

2A.6.6 Seed

A noteworthy aspect of jackfruit is its prolific seed production with a single fruit containing anywhere from 100 to up to 500 seeds (Morton, 1987). The seeds of the jackfruit are relatively large, typically about 2 to 4cm in length and 1 to 2cm in width. They are elongated, usually flattened and somewhat bean-shaped. The colour of jackfruit seeds can vary but they are commonly light brown or beige. The surface of jackfruit seeds is often smooth with a shiny appearance. The outer seed coat may have a slight ridge or seam running along one side. The edible portion of the jackfruit seed is the starchy inner part which is usually white or pale yellow and has crisp texture. This inner part is surrounded by a thin seed coat. Like many seeds, jackfruit seeds can germinate under suitable conditions. Germination involves the sprouting of a new plant from the seed.

Jackfruit seeds

By Goldflakes - Own work, CC BY-SA 4.0, https://commons.wikimedia.org/w/index.php?curid=91815026

It is worth noting that morphology of jackfruit seeds can vary somewhat depending on the specific variety of jackfruit. Jackfruit is known for its large size and unique flavour, and both the seeds and the flesh are used in various culinary applications in many tropical regions. The seeds are often boiled or roasted before consumption.

2A.7 Maturation and Lifecycle Dynamics

Jackfruit tree is perennial and can produce fruit for 30–60 years. Depending on the latitude, jackfruits ripen from March to September, often from June to August in Central India. The tree itself can live up to 100 years (www.foreststreesagroforestry.org, 2024). The growth and development of jackfruit trees present a fascinating chronicle marked by distinct milestones and intricacies:

i) **Germination:** Seed germination is hypogeal and unlike breadfruit seedlings the cotyledons separate facilitating the unimpeded emergence of the plumule. New leaves take approximately 12-15 days to expand (Bakewell-Stone, 2023).

ii) **Growth:** The jackfruit tree grows rapidly in early years, *i.e.* up to 1.5m per year in height but, slowing to about 0.5m per year as the tree reaches maturity (Kamaluddin *et al.*, 1997). In mature trees extension growth is gradual averaging 3-5cm per month characterized by continuous and steady progression (Bakewell-Stone, 2023).

iii) **Flowering initiation and fruit development:** Under natural conditions, seed-germinated trees typically commence flowering at 7-8 years of age though grafted plants exhibit an earlier onset of yield typically starting from the 4th or 5th year or even earlier in some exceptionally early varieties like 'Singapore Jack' ('Ceylon Jack') that starts bearing from 2-3 years after planting (Priya Devi *et al.*, 2014). Full production from clonal material in Malaysia is achieved typically when the trees reach 8 to 15 years of age (Bakewell-Stone, 2023).

- **Male and female flower characteristics:** Flowerheads are primarily borne on short shoots emerging from the trunk and main limbs. However, unlike female flowerheads, male heads are not restricted to particular shoots and instead these also manifest on shoots in the peripheral regions of the tree canopy particularly in vigorously growing trees. Male and female heads on cauliflorous shoots tend to develop almost simultaneously with male heads maturing 3 to 5 days earlier (Bakewell-Stone, 2023).

- **Pollination mechanisms:** At anthesis, male heads attract small insects with sticky yellow pollen and a sweet scent serving as potential pollinators. Wind-induced pollination is also reported in India. After anthesis, male heads darken and detach from the tree (Bakewell-Stone, 2023).
- **Fruit maturation and seed development:** Fertilized female heads evolve into mature fruits over a period of 3 months or more influenced by factors such as seedling type or clone. At higher altitudes or latitudes this maturation process may extend to six months. Unfertilized flowers transform into strap- or string-like structures filling the interstitial spaces between developing fruitlets. A fully developed fruit may house up to 500 seeds each weighing 3-6g (Bakewell-Stone, 2023).

iv) **Impact of growing fruit on flowering:** The presence of a developing fruit load may exert an inhibitory effect on flowering potentially accentuating the seasonality of production (Bakewell-Stone, 2023).

v) **Flowering patterns and harvest periods:** In suitable environments, jackfruit trees exhibit the remarkable capability of flowering and fruiting throughout the year. However, a notable harvest season generally spans from April to August, *i.e.* during summer in India though September to December in Malaysia and January to May in Thailand (Bakewell-Stone, 2023).

2A.8 Flowering and Fruit Production Dynamics

Flowering frequency in jackfruit exhibits a spectrum encompassing annual (occurring once a year), biennial (once every two years), sub-annual (more than twice a year) or irregular (uncertain) patterns (Pushpakumara, 2006; Li *et al.,* 2010; Fathin *et al.,* 2021). Jackfruit is self-incompatible and displays a preference for crossmating (Gardner *et al.,* 2018; Fathin *et al.,* 2021). The variation in flowering cycles is influenced by a combination of climatic conditions (Pushpakumara, 2006; Lina and Protacio, 2015; Choudury *et al.,* 2017; Kishore, 2018) and genetic factors (Jagadeesh *et al.,* 2007; Choudury *et al.,* 2017; Nakintu *et al.,* 2019). This variability is evident in the diverse traits of flowered individuals, flower abundance, flowering initiation *cum* duration and the resulting fruit production (Pushpakumara, 2006; Fathin *et al.,* 2021).

Jackfruit's adaptability to various climatic conditions in distinct agro-ecological zones further contributes to the complexity of its phenological patterns (Kishore, 2018). Each agro-ecological zone manifests its unique flowering and fruiting characteristics, impacting the onset and duration of these crucial life stages (Lina and Protacio, 2015; Choudury *et al.*, 2017; Kishore, 2018). Interplay of genetic factors alongside climatic conditions underscores the diversity observed in jackfruit's flowering behaviour (Jagadeesh *et al.*, 2007; Li *et al.*, 2010; Choudury *et al.*, 2017; Nakintu *et al.*, 2019). Several studies have identified early and late maturing varieties (Li *et al.*, 2010), secondary flowering varieties (Nakintu *et al.*, 2019) and those that bear fruits year-round (Khan *et al.*, 2010).

Fruit production in jackfruit is a year-round phenomenon, reported across diverse geographical locations such as Australia, Indonesia, Malaysia, Thailand (Nakintu *et al.*, 2019) and China (Li *et al.*, 2010). However, the timing, duration and peak seasons exhibit variation influenced by climatic disparities among countries (Sindhu, 2012; Kishore, 2018; Nakintu *et al.*, 2019). The peak fruiting season spans from March to May in the Philippines, April to May in Thailand and September to December *cum* June to August in different climatic conditions in India (Khan *et al.*, 2010). Uganda experiences fruiting in December to January (Nakintu *et al.*, 2019) with a unique occurrence of two fruiting seasons in warmer months, *i.e.* March to April and November to December, the latter being more prolific (Balamaze *et al.*, 2019). In temperate regions like Florida, the primary fruiting season aligns with summer and fall with minimal ripening observed in winter and early spring (Fathin *et al.*, 2021). In China, jackfruit is classified into annual and biannual mature types, reflecting varieties that fruit once or twice a year (Li *et al.*, 2010).

2A.9 Slicing or Break-Opening Jackfruit

Slicing or breaking open jackfruit is a meticulous process that begins with generously coating hands with oil for ease of handling. Employing either a sizable knife or a sickle-like instrument lubricated usually with mustard oil, the jackfruit is skillfully opened. This reveals fleshy pods and seeds which are carefully extracted and separated from the inedible parts. This distinctive raw material possesses unparalleled versatility, offering myriad possibilities for culinary creations. Unlike any other raw ingredient, jackfruit stands alone in its ability to be transformed into a wide array of delectable preparations (Lobo, 2019).

2A.10 Flavour and Texture of Ripe Fruit

Jackfruit known for its distinctive characteristics boasts a unique combination of odour, flavour and texture that contribute to its popularity in culinary and cultural contexts.

i) **Odour**

- **Sweet and tropical aroma:** When ripe, jackfruit emits a sweet and tropical fragrance. The aroma is often described as a blend of pineapple, banana and mango. The enticing scent can be quite strong filling the air around the fruit.

ii) **Flavour**

- **Sweet and mild taste:** The flavour of ripe jackfruit is predominantly sweet with mild undertones. The sweetness is comparable to a combination of mango, pineapple and banana. Unripe or green jackfruit often used in savory dishes has a more neutral taste allowing it to absorb the flavours of the spices and seasonings it is cooked with.

iii) **Texture**

- **Fibrous and meaty texture:** The texture of jackfruit is one of its most remarkable features. The flesh of the fruit has a fibrous and meaty consistency resembling pulled pork or shredded chicken when cooked. This unique texture makes jackfruit a popular choice for vegetarian and vegan dishes particularly as a meat substitute.
- **Soft and juicy:** Ripe jackfruit is softer and juicier with a texture reminiscent of juicy tropical fruits. The individual bulbs that make up the fruit are soft and can be easily separated.
- **Seed texture:** When cooked jackfruit seeds have a starchy and somewhat nutty texture. They can be boiled, roasted or used in various culinary preparations.

2A.11 Advancements in Jackfruit Breeding

The realm of jackfruit breeding in India has witnessed remarkable strides marking a timeline of significant achievements and meticulous efforts to enhance the quality and yield of this versatile fruit (Morton, 1987).

i) **Horticultural Research Institute, Saharanpur, India (1961):** In 1961, HRI, Saharanpur embarked on a journey of improvement by acquiring air-layered plants of distinguished jackfruit varieties where notable selections included 'Safeda', 'Khaja', 'Bhusila', 'Bhadaiyan' and 'Handia' among others.

ii) **Fruit Experimental Station, Burliar (54 clones collection):** This research station took a comprehensive approach establishing a collection of 54 jackfruit clones sourced from various producing countries. This meticulous curation process culminated in the selection of 'T Nagar Jack' as the epitome of excellence in terms of both quality and yield.

iii) **Breeding work at Kallar (1952 onwards):** Beginning the monumental task of breeding in 1952, the Fruit Experimental Station at Kallar set forth with a vision. Their aim was to craft short, compact and many-branched precocious and highly productive trees. These trees were envisioned to bear large, yellow and high-quality fruits with a unique distribution of half in the main season and half later.

- **Hybrid success ~ 'Singapore Jack' and 'Velipala' (after 25 years of testing):** Employing a strategic approach the 'Singapore Jack' was chosen as the female parent for its ability to yield both early and late crops. The male parent 'Velipala' ~ a local forest selection contributed large fruits with superior-quality carpels regularly borne in the main summer season. After 25 years of rigorous testing, a hybrid emerged as an outstanding specimen excelling in precocity, fruit size, off-season and main season production as well as overall yield surpassing the achievements of its parental lineage. Remarkably, this hybrid had yet to be officially named as of the last reported update in 1982. In Assam, nurserymen have given names such as 'Mammoth', 'Everbearer' and 'Rose-scented' to preferred types.

2A.12 Classification

In South India, cultivated or naturally occurring jackfruits are systematically classified into two distinct types each exhibiting unique characteristics and qualities (Morton, 1987; Priya Devi *et al.,* 2014; kvk, 2024) such as:

i) **Koozha Chakka (soft carpels):** The fruits belonging to this category are characterized by small, fibrous, soft and with mushy carpels. Despite their fibrous texture, these carpels are exceptionally sweet contributing to the distinctiveness of the 'Koozha Chakka' variety. Referred to as **Rassal** in Konkani, and **Barka** or **Berka** in certain other areas, emphasizing its softness and the ease of breaking open with hands.

ii) **Koozha Pazham (firm *cum* crisp carpels):** This type holds greater commercial significance primarily due to its high quality crisp carpels known as 'varika' there. The carpels of Koozha Pazham jackfruit possess a desirable crunchiness sought after in various applications. These types are known by names like **Kapa** (in Konkani) or **Kapiya** in Goa and certain other areas which reflects the variations in texture and the method of opening ~ crisp and cut open with a knife.

   a) The specific varieties under Koozha Pazham category released from HRI, Saharanpur in 1961 are ~ **Safeda, Khaja, Bhusila, Bhadaiyan** and **Handia**. Each variety offers unique attributes contributing to the diverse jackfruit landscape. Their introduction from the institute adds to the richness of jackfruit cultivation.

b) Notable Local Selections of Koozha Pazham type in India are:

- **Gulabi (rose scented):** Recognized for its fragrant quality reminiscent of roses.
- **Champa (flavour like champak):** Known for its flavour akin to the delightful scent of champak flowers.
- **Hazar (bearing large number of fruits):** Esteemed for its prolific fruit-bearing capability, yielding a large number of fruits.

The equivalent types are known in Thailand as **Kha-nun lamoud** (soft) and **Kha-nun nang** (firm and best), and in Ceylon as **Vela** (soft) and **Varaka** or **Waraka** (firm). This shows a cross-cultural recognition of texture and quality and highlighting regional nuances in nomenclature. The Peniwaraka (honey jack) has sweet pulp and so many consider it the best of all. Kuruwaraka bears small and rounded fruits. Vela type predominates in the West Indies (Morton, 1987).

Morton (1987) draws upon the perspectives of Firminger and Dutta, providing a nuanced portrayal of distinct types:

i) **Firminger's classification**

- **Khuja**: It is characterized by the green, hard and smooth exterior. Khuja boasts juicy pulp and petite seeds.
- **Ghila**: It exhibits a rough, soft texture. Ghila features thin pulp and larger seeds but is not very juicy.

ii) **Dutta's discernment**

- **Khujja or Karcha**: This sports a pale-brown or occasionally pale-green rind. This variant possesses pulp as robust as an apple.
- **Ghila or Ghula**: It is typically light-green, occasionally brownish and this type is known for its soft pulp, offering either a sweet or acidulously-sweet taste.

iii) **Dutta's variety exploration**

- **Number of varieties**: Dutta delves into the intricate world of jackfruit detailing eight distinct varieties.
- **Hazari**: Among the named varieties, Hazari akin to Rudrakshi stands out with a relatively smooth rind. However, its flesh is noted for its inferior quality.

2A.13 Varieties

i) **Cheena** (FTG, 2024; Bruno, 2024)

- Australian Cheena variety is though classified as a jackfruit but is a natural hybrid between jackfruit and chempedak.

- Its tree has an open, low and the spreading growth habit maintainable at 2.5m with annual pruning.
- 'Cheena' consistently yields 50 to 70 kg per tree.
- Fruit is long, narrow and uniform weighing 2.4kg with 33% edible flesh.
- Green skin with blunt spines turns yellow and opens slightly upon maturity.
- Deep orange somewhat fibrous flesh has an excellent flavour and intense earthy aroma.
- Fruit preparation time is as short as 15 minutes contrasting with 1 to 2 hours for other cultivars.

ii) **Black Gold** (FTG, 2024; Bruno, 2024)

- Selected in Queensland (Australia), Black Gold is vigorous with a dense and manageable canopy.
- Maintainable at 2 to 2.5m through annual pruning yielding 55 to 90kg of fruits per tree.
- Fruits are medium-sized averaging 6.7kg with 35% edible flesh.
- Fruit has dark green exterior with sharp fleshy spines and deep orange, soft and sweet flesh.
- Skin spines do not open upon maturity which makes harvest-timing challenging.

iii) **Golden Nugget** (FTG, 2024; Bruno, 2024)

- Australian variety produces round and smaller fruits with bright yellow skin, each weighing 2.7–3.25kg.
- Flesh is dark orange with moderately firm flesh. It is ready for harvesting from May to June.
- Typically it is grown in Queensland from where it is marketed nationwide.

iv) **Cochin** (FTG, 2024; Bruno, 2024)

- It was selected in Australia. Cochin tree is low in vigour, growing 2-2.5m high though requires light annual pruning. This forms sparse, upright and narrow canopy.
- Its smooth and firm fruits with mild flesh weighing about 1.5kg each are small having spines which flatten and open with maturity of the fruits. Fruits have little latex and have 35% edible parts.
- Entire fruit, including the 'rag' can be eaten during certain periods.

v) **Honey Gold** (Bruno, 2024)

- It is a medium-sized cultivar weighing 4-5kg each. Fruits are blocky like Chompa Gob.
- It ripens in July-August with dark yellow-orange, firm, silky and sweet flesh.

vi) **Chompa Gob** (Bruno, 2024)

- It is from Singapore, producintg medium-sized blocky fruits weighing 4.5-8kg each.
- Its bright orange sweet flesh is moderately fragrant and fairly firm.

vii) **Tabouey:** (FTG, 2024; Bruno, 2024)

- This Indonesian cultivar bears medium-large fruit weighing 8-9kg each.
- Its open and rounded canopy is maintainable upto 3m high through annual pruning.
- It bears long and thin fruits tapered at the stem-end. Fruits are often misshapen and lobed with a bright yellow colour and irregular blunt spines. The fruit often split prior to harvest. The flesh is firm, pale yellow and has almost no aroma. Its flavour is mild and pleasant.
- It gives moderate cropping of 50 to 70kg per tree with mild and pleasant flavour.

viii) **Golden Pillow or Mong Tong** (FTG, 2024; Bruno, 2024)

- It is a Thai variety with uniformly round shape where fruits weigh 3-5.5kg each.
- Its trees are small and highly manageable with a height and spread of 3m.
- Its skin is intense green, spines blunt and pliant and the flesh is yellow-orange, crunchy and with sweet flesh.
- Its average fruit weight is 3.6-5.5kg with 35-40% edible flesh. The golden flesh is thick and crunchy and flavour is mild and sweet with no musky aftertaste. This cultivar is noted for precocious fruiting, bearing in the second year after planting.
- There are typically 65 to 75 seeds per fruit and with lax latex.

ix) **Dang Rasimi** (FTG, 2024)

- This originated in Thailand. Its tree is open, spreading and fast-growing but requires annual pruning.

- It is highly productive yielding 75 to 125kg fruits per tree.
- Its fruits are bright green to pale yellow bearing medium to large fruit up to 8kg each with 32% edible flesh. Flesh is firm, deep orange and sweet.
- Its skin has sharp spines that do not flatten or open with maturity.
- It has a mild sweet flavour and a pleasantly sweet aroma.

x) **Mia 1:** (FTG, 2024)

- This originated in Thailand which is vigorous, productive and delicious.
- The tree attains height and spread of 2.5m and begins to fruit within 2 to 3 years of planting.
- Fruits weighing some 9-12kg each have little latex. Skin is golden-yellow. Flesh is crisp, sweet and delicious.
- Tree can be maintained at a height and spread of 2.4m.

xi) **Lemon Gold** (Bruno, 2024)

- Australian Lemon Gold jackfruit has a block-like shape, weighing around 4.5-5kg.
- Flesh is very sweet with a firm and crisp texture.

xii) **Kun Wi Chan** (Bruno, 2024)

- This is a large and round Thai variety where a fruit may weigh 12-16kg.
- Its outer skin is bright yellow, flesh paler-yellow and is used when underripe for savory dishes.

xiii) **NS1:** (FTG, 2024; Bruno, 2024)

- This variety was selected in Malaysia. It has dense upright canopy with medium-vigour.
- It can be maintained up to 2.5-3m height through moderate annual pruning.
- It is heavy bearer producing more than 90kg fruits per tree. Its fruits are dark green and blocky with flattened spines, and the flesh is dark orange, firm and sweet.

xiv) **J-30:** (FTG, 2024; Bruno, 2024)

- This was selected in Malaysia. This is a vigorous tree forming an open and conical canopy. It can be maintained at 3m height through annual pruning.

- Yield is 50 to 60kg per year, each fruit weighing 7.6kg with edible flesh percentage of 38%. Its fruits having dark green and blunt spines hang down singly on major limbs
- Its flesh is deep orange, firm and sweet with only a slight aroma.

xv) **J-31:** (FTG, 2024; Bruno, 2024)

- This moderately vigorous tree having spreading type open canopy is a Malaysian selection which can be managed up to 2.5m high.
- It yields 42 to 60kg fruits per year.
- Its large fruits are irregular with blunt spines and different from J-30 in shape. Average weight of a fruit is about 12kg with an edible flesh percentage of 36%. Its flesh is deep yellow with thick walls, firm and sweet with strong earthy aroma.
- Off-season fruit production occurs during fall and winter.

xvi) **Vietnam Super Early** (Bruno, 2024)

- It is a biannual bearer which starts fruit bearing from second year of planting. It is a special variety for vegetables and chip making. Its small- to medium-sized fruits have lime green skins and pale yellow-orange flesh.

xvii) **Kothmale** (Bruno, 2024)

- It is a Sri Lankan variety bearing medium-large oblong fruits having lime green-yellow skins and yellow interior flesh.
- Matures between March and June with soft, sweet and juicy flesh when fully ripe.

xviii) **Brazil Jack** (Bruno, 2024)

- It is a Brazilian variety naturalized from India which in South America is known as 'jaca'.
- Its round fruits weigh about 8kg each with bright lime green skins and sugary sweet bright yellow flesh when completely ripe.

xix) **Malaysian Dwarf** (Bruno, 2024)

- This small tree bears fruits twice a year, *i.e.* once before and other after the monsoons.
- Fruits weigh about 4kg bearing olive-green skins and bright orange flesh.

xx) **Orange Crush** (Bruno, 2024)

- It thrives well in California and Florida. Average weight of a fruit is 8kg with green skin and dark orange, sweet and crunchy flesh.

xxi) Zima Pink (Bruno, 2024)

- Grown in Hawaii with dark greenish-yellow skin and orange-pink, sweet, thick and crunchy flesh.

Experts claim that best-tasting jackfruit varieties are those that have 'gold' in their name such as 'Lemon Gold', 'Black Gold' and 'Golden Nugget' and all these rate high on overall deliciousness. These three cultivars tip the scales because these combine deliciousness, sweet flavours with pleasing textures and aromas.

2A.13.1 Location-wise varieties

According to Bruno (2024) the jackfruit varieties grown in various countries and regions are:

- **India**: Kerala and Karnataka regions are rich in jackfruit cultivation where one can record even wild ones growing. Outside of these areas one should look for 'Hemachandra', 'Janagere Yellow', 'Singapuri Wada', *etc.*

a. **Tamil Nadu**: Jackfruit is the state fruit of Tamil Nadu, and the fruits grown in Chakka Gramam can weigh close to 40kg each. The most common cultivars grown here are 'Mankale Red' and 'Hemachandra'.

b. **Kerala**: Kerala's climate is ideal for growing fruit trees, and many gorgeous jackfruit varieties thrive best there. One of the most vibrant ones is the rare 'Thailand Pink' which is the primary cultivar grown in that region. 'Vietnam Super Early' is also almost at par but not quite often.

c. **Karnataka**: This SW Indian province is home to beautiful 'Mankale Red' jackfruit. It is named after the Mankale region in which it grows. One can also find 'Mijar Giant', 'Varashree Jack', 'Rudrakshi Red', 'Byrachandra Super' and just about any other red-hued jackfruit variety.

d. **Fajihabad**: NJT-1, NJT-2, NJT-3 and NJT-4.

- **USA**: Most of the jackfruits found in USA are grown in Florida or California. 'Malaysian Dwarf', 'Brazil Jack', 'Lemon Gold' and 'Black Gold' are all popular cultivars in Florida. There climate is very similar to the tropical forests in India, Malaysia and Indonesia where these trees thrive best. Hawaiians can look for 'Zima Pink', 'Black Gold' or 'Golden Nugget'. 'Orange Crush' and 'Lemon Gold' are the top jackfruit varieties grown in California.
- **Sri Lanka**: Jackfruits are known as 'kos' in Sri Lanka. There primary variety is 'Kothmale', though one can find even 'Singapuri Wada' and 'Maharagama'.

- **Malaysia**: 'J-30', 'J-31' and 'NS1' are the most common Malaysian varieties. One can find even 'Janagere Yellow' and 'Byrachandra' there as well.
- **Philippines**: 'Dang Rasimi' and 'Tinumbaga' are two of the most popular cultivars in the Philippines.

2A.13.2 Varieties grown in India

i) **Burliar-1 (T Nagar selection):** T Nagar selection stands as a noteworthy achievement of FRS, Burliar (T.N.). Burliar-1 trees are characterized by a moderate height with harmonious balance that facilitates both cultivation and harvesting. One of the distinguishing features of this cultivar is its prolific bearing which ensures a bountiful yield. The medium height of the trees coupled with their productivity makes Burliar-1 a valuable addition to the spectrum of jackfruit varieties (Priya Devi *et al.*, 2014).

ii) **Byrachandra Super Early:** it is a Bangla cultivar that took centrestage at the 2019 Jackfruit Festival in Mysore. It yields large pale green and oblong fruits which are suitable as raw (cooking), underripe and fully ripe harvests, however, fully ripe ones have sweet and musky flesh so perfect for desserts. Bright orange flesh boasts firm and chewy (Bruno, 2024).

iii) **Dang Surya:** It originated in Thailand and Malaysia which is being grown in Kerala for its honey-tasting flesh. These dwarf cultivars result in smaller fruits, rarely exceeding 4kg each. It has golden-yellow skin and peachy interior flesh (Bruno, 2024).

iv) **Gulabi:** In Tamil Nadu it is recognized as one of the three sacred trees. Gulabi jackfruit possesses light green skin which on maturity exhibits smooth spines with pale yellow, soft and sweet flesh (Bruno, 2024). It produces fruits weighing some 30kg.

v) **Gumless***:* Among the jackfruit varieties, Gumless stands out for having the lowest latex content so giving a slippery texture reminiscent of raw oyster, not like many others where due to latex the texture becomes rubbery. The flesh of full*y* ripe fruits i*s* exquisitel*y* sweet (Bruno, 2024).

vi) **Hemachandra Jack:** It is the poor man's food in Bangladesh. When underripe, it is used in stews. It medium-sized (8kg) fruits are harvested in August and September (Bruno, 2024).

vii) **Hybrid Jack:** The 'Hybrid Jack' developed at FRS, Kallar (T.N.) is a remarkable crossbreed originating from successful union of 'Singapore Jack' and 'Velippala'. This innovative hybrid showcases trees characterized by their precocious nature in bearing fruits, setting it apart in terms of early productivity. Noteworthy are the distinctive qualities found in its carpels, surpassing both the parent varieties. Its carpels are notably larger in size and boast a delightful sweetness, elevating the overall fruit quality and making it a notable addition to the spectrum of jackfruit varieties. A testament to strategic breeding efforts, this hybrid stands out as a promising cultivar with enhanced features inherited from its renowned parentage (Priya Devi *et al.*, 2014).

viii) **Janagere Yellow:** Thriving in Goa and Kerala, the 'Janagere Yellow' jackfruit produces medium-sized fruits weighing 8-9kg with yellow-beige skins and bright yellow flesh. Optimal ripeness is achieved by harvesting in June or July (Bruno, 2024).

ix) **Konkan Prolific**: It stands out as a distinguished selection introduced by Dr. BSKKV, Dapoli (Maharashtra). Its tree attains an average height of 14.7m and an expansive spread of 11.5m (north-south) with an average girth of 1.1m. This variety showcases a semi-spreading growth habit presenting a fairly large stature. The leaves are characterized by a deep green hue arranged alternately along the branches. The tree is monoecious, featuring small male flowers borne on pedicels. Female flowers surpass their male counterparts in size, and their pedicels exhibit robust thickness. The multiple type medium fruit weighs around 8-9kg. Its skin transitions from green in its immature state to a captivating greenish-yellow to brownish-yellow when fully ripe. Fruits on ripening encapsulate sweet and aromatic bulbs that offer a delightful crispiness to the palate (Priya Devi *et al.*, 2014).

x) **Mankale Red:** It is from the Mankale region (Karnataka) which exhibits an almost dark olive green skin and vibrant red inner flesh. Its fruit weighs 10-12kg having an extraordinarily crispy sweet taste even when fully ripe (Bruno, 2024).

xi) **Mijar Giant:** It is native to Dakshina Kannada region (Karnataka) which is an enormous jackfruit variety. Its rounded fruits can attain up to an impressive 30kg, boasting greenish-yellow skin adorned with flattened spines. While the pale interior flesh may lack distinctive flavour, it finds utility in the creation of papadums, chips and other savory snacks. This variety serves as an excellent vegan meat substitute (Bruno, 2024).

xii) **Pathamuttam Red (Pakamattam):** It is a red-fleshed jackfruit from Karnataka where its 2-3kg fruits are usually misshapen bearing sweet ruby-red flesh on ripening. Fruit skin is thin bearing flattening spines as the fruits ripen (Bruno, 2024).

xiii) **PLR-1 (Palur-1):** It is a high-yielding jackfruit evolved through single plant selection at the Vegetable Research Station, Palur (T.N.) as this was found in Panikkankuppam village near Panruti in the South Arcot district with special characteristic of bearing fruits twice a year, *i.e.* March to June (the usual season) and again from October to December (off season) so giving more yield. Each tree bears an impressive quantity of 60-80 fruits with an average fruit weight of 12kg, each encompassing 115-120 bulbs. These pale yellow bulbs are not only visually appealing but also boast a delightful crispness and sweetness, complemented by a TSS of 19° Brix (Priya Devi *et al.*, 2014).

xiv) **PPI-1 (Pechiparai-1):** It was developed at TNAU HRS, Pechiparai through clonal selection from Mulagumoodu Local. Its trees are medium tall with maximum bearing on tree trunk. On an average, each tree bears 107 fruits weighing 1818kg per year collectively in both the seasons. Its crisp carpels are sweet and tasty with pleasant aroma. It is suitable for commercial planting as well as for planting in home garden (Priya Devi *et al.*, 2014).

xv) **Rudrakshi Red:** Its 3-5kg fruit is of pumello-size with smooth rind, less spines and quite sweet red bulbs (kvk, 2024).

xvi) **Singapore Early:** It is ready for harvest in December, bearing red bulbs.

xvii) **Singapore Jack (Ceylon Jack):** An introduction from Sri Lanka, this is noteworthy for its exceptional early bearing capability as even seedling progenies start bearing in 3 years (18-30 months) (Morton, 1987; Priya Devi *et al.*, 2014) though in many other cases the seedling progenies start bearing the fruits only after 7-8 years of sowing. This jackfruit variant underwent extensive planting in 1949. Characterized by medium-sized fruits weighing 7-10kg each, it features small fibrous carpels renowned for their delightful sweetness. This jackfruit variety adds to its allure by providing not only in March to June-July but also a second crop from September-October to December (Morton, 1987; Priya Devi *et al.*, 2014).

xviii) **Singapuri Wada:** It is a hybrid popular in Maharashtra developed from a Singaporean cultivar and a Malaysian variety bearing medium-sized (4-14kg) oblong fruits. Its fruits bear bright yellow-green skins and very tasty flesh with gorgeously sweet flavour (Bruno, 2024).

xix) **Varashree Red:** Another entrant from Karnataka which is also cultivated successfully in Goa, the 'Varashree Red' jackfruit is of medium size (some 8kg each) with yellow-green skins and bright orange *cum* crunchy sweet flesh. Its fruits ripen in June. However, harvesting in May allows for underripe fruits suitable for savoury use (Bruno, 2024).

Some other varieties are Malaysian Dwarf, Brazil Jack, Vietnam Super Early, *etc.*

2A.14 Nutritional Value

Jackfruit is a nutritional powerhouse, boasting a diverse array of essential nutrients that contribute to its antioxidative and medicinal properties (Jagtap *et al.*, 2010). This versatile fruit is notably rich in carbohydrates, protein, potassium, calcium and various vitamins. Jackfruit stands out as an excellent source of dietary fibre encompassing both soluble and insoluble fibres. Its nutritional profile includes moderate amounts of vitamin B & C, potassium and other essential nutrients. The jackfruit pulp constitutes 25-40% of the fruit's weight. In general, fresh seeds are considered to be high in starch, low in calcium & iron and a good source of vitamins B_1 & B_2 (Morton, 1987).

Jackfruit offers a comprehensive spectrum of nutrients to consumers. Comparably, it rivals esteemed counterparts such as 'avocado' and 'olive' presenting an optimal blend of essential elements found in mother's milk. Beyond its culinary allure, this fruit unveils an additional dimension ~ the presence of the chemical 'jacalin'. Renowned for its potential in preventing conditions such as colon cancer and AIDS, jackfruit transcends its role as a delectable fruit to become a source of holistic well-being (Priya Devi *et al.*, 2014).

Its fruit has potential to offer numerous health benefits to human beings. These advantages encompass improved digestion, support for the immune system and potential anti-inflammatory properties. It is essential to acknowledge that individual responses to food can vary, highlighting the need for personalized dietary considerations. In incorporating jackfruit into one's diet, individuals may harness its nutritional richness to promote overall well-being.

As per various workers, the nutritional values of jackfruit are provided below. It is essential to note that there might be slight variation in the nutritional content due to differences in cultivation, ripeness and measurement methods. The data presented here serves as a general overview of its nutritional composition:

Composition of jackfruit (100g edible portion)		
Composition	**Young fruit**	**Ripe fruit**
Water (g)	76.2 - 85.2	72.0 - 94.0
Protein (g)	2.0 - 2.6	1.2 - 1.9
Fat (g)	0.1 - 0.6	0.1 - 0.4
Carbohydrate (g)	9.4 - 11.5	16.0 - 25.4
Fibre (g)	2.6 - 3.6	1.0 - 1.5
Total sugars (g)	-	20.6
Total minerals (g)	0.9	0.87 - 0.9
Calcium (mg)	30.0 - 73.2	20.0 - 37.0

Composition of jackfruit (100g edible portion)		
Composition	**Young fruit**	**Ripe fruit**
Magnesium (mg)	-	27.0
Phosphorus (mg)	20.0 - 57.2	38.0 - 41.0
Potassium (mg)	287-323	191-407
Sodium (mg)	3.0-35.0	2.0-41.0
Iron (mg)	0.4-1.9	0.5-1.1
Vitamin A (IU)	30	175-540
Thiamine (mg) (B1)	0.05-0.15	0.03-0.09
Riboflavin (mg) (B2)	0.05-0.2	0.05-0.4
Vitamin C (mg)	12.0-14.0	7.0-10.0
Energy (KJ)	50-210	88-410
Source: Ranasinghe *et al.* (2019)		

Nutritional value per 100g of uncooked jackfruit			
Energy	397 kJ (95 kcal)	**Minerals**	(%DV[†]) **Quantity**
Carbohydrates	23.25 g	Calcium	2% (24 mg)
Sugars	19.08 g	Iron	2% (0.23 mg)
Dietary fibre	1.5 g	Magnesium	8% (29 mg)
Fat	0.64 g	Manganese	2% (0.043 mg)
Protein	1.72 g	Phosphorus	3% (21 mg)
Vitamins	(%DV[†]) **Quantity**	Potassium	10% (448 mg)
Vitamin A equiv. beta-Carotene lutein zeaxanthin	1% (5 μg) 1% (61 μg) 157 μg	Sodium	0% (2 mg)
Thiamine (B1)	9% (0.105 mg)	Zinc	1% (0.13 mg)
Riboflavin (B2)	5% (0.055 mg)	**Other constituents**	**Quantity**
Niacin (B3)	6% (0.92 mg)	Water	73.5 g
Pantothenic acid (B5)	5% (0.235 mg)	Source: USDA, 2018	
Vitamin B6	25% (0.329 mg)		
Folate (B9)	6% (24 μg)		
Vitamin C	17% (13.8 mg)		
Vitamin E	2% (0.34 mg)		

Food Value per 100g of edible portion

Source: Morton (1987)

	Pulp (ripe-fresh)	*Seeds (fresh)*	*Seeds (dried)*
Calories	98		
Moisture	72.0-77.2 g	51.6-57.77 g	
Protein	1.3-1.9 g	6.6 g	
Fat	0.1-0.3 g	0.4 g	
Carbohydrates	18.9-25.4 g	38.4 g	
Fiber	1.0-1.1 g	1.5 g	
Ash	0.8-1.0 g	1.25-1.50 g	2.96%
Calcium	22 mg	0.05-0.55 mg	0.13%
Phosphorus	38 mg	0.13-0.23 mg	0.54%
Iron	0.5 mg	0.002-1.2 mg	0.005%
Sodium	2 mg		
Potassium	407 mg		
Vitamin A	540 I.U.		
Thiamine	0.03 mg		
Niacin	4 mg		
Ascorbic Acid	8-10 mg		

Priya Devi et al. (2014)

Constituent	Average_value
Moisture (%)	76.20
Energy (cal)	88.00
Protein (g)	1.90
Fat (g)	0.10
Fibre (g)	1.10
Carbohydrates (g)	19.80
Potassium (mg)	107.00
Calcium (mg)	20.00
Phosporous (mg)	41.00
Iron (mg)	0.56
β Carotene (mg)	175.00
Thiamine (mg)	0.03
Riboflavin (mg)	0.13
Niacin (mg)	0.40
Vitamin C (mg)	7.00

Jackfruit contains various antioxidants including phytonutrients such as flavonoids and saponins which may have health-promoting properties. The glycemic index of jackfruit is relatively low which means it has a smaller impact on blood sugar levels. This can be beneficial for individuals managing diabetes. Jackfruit is cholesterol-free which makes it a heart-healthy food option.

2A.15 Phytochemistry

Jackfruit contains many classes of phytochemicals such as carotenoids, flavonoids, volatile acids, sterols, tannins, *etc.* with varying concentrations depending on the variety, ripeness and cultivation conditions. The edible parts are replete with a diverse array of phytochemicals each contributing to the fruit's nutritional richness and potential health benefits. Following comprehensive breakdown details the various classes of phytochemicals found in jackfruit:

2A.15.1 Esters

Esters are organic compounds that contribute to the characteristic flavours and aromas of many fruits where jackfruit is not an exception. The specific esters found in jackfruit can vary but they are often responsible for the fruity and sweet notes in its scent. These esters are formed during the ripening process of the fruit as metabolic changes and enzymatic reactions take place.

It is important to note that exact composition of esters in jackfruit can depend on factors such as variety of jackfruit, ripening stage and environmental conditions. Some of the common esters found in fruits include ethyl acetate, isoamyl acetate and ethyl butyrate. These esters contribute to the overall fruity and sometimes bubblegum-like aroma associated with ripe jackfruit.

2A.15.2 Carotenoids

This class of natural pigments are responsible for imparting yellow to reddish hues. Jackfruit stands out as a notable source of carotenoids, recognized for their antioxidant properties and their potential to contribute to eye health. The carotene content within jackfruit exhibits a gradual increase during the ripening process (Ranasinghe *et al.*, 2019).

Jackfruit kernel contains a spectrum of carotenoids including β-carotene, α-carotene, β-zeacarotene, α-zeacarotene, β-carotene-5,6α-epoxide, a dicarboxylic carotenoid and crocetin (Chandrika *et al.*, 2004). Baliga *et al.* (2011) have identified key carotenoids in jackfruit including all-trans-lutein, all-trans-β-carotene, all-trans-neoxanthin, 9-cis-neoxanthin and 9-cis-vio-laxanthin.

Concentration (μg/100g fresh weight) of carotenoids of jackfruit and their vitamin A values (μg RAE/100g fresh weight) as per de Faria** et al.**, 2009:	
Carotenoids	**Concentration (μg/100g fresh weight)**
All-trans-neoxanthin	8.85
9-cis-Neoxanthin	6.87
All-trans-neochrome	0.88
All-trans-luteoxanthin	2.06
cis-Antheraxanthin	1.12
9-cis-Violaxanthin	7.05
cis-Luteoxanthin	0.34
All-trans-lutein	37.02
All-trans-zeaxanthin	0.96
All-trans-zeinoxanthin	1.72
cis-Zeinoxanthin	0.90
All-trans-α-cryptoxanthin	0.35
All-trans-β-cryptoxanthin	1.21
15-cis-β-Carotene	0.18
13-cis-β-Carotene	2.45
All-trans-α-carotene	1.24

Concentration (μg/100g fresh weight) of carotenoids of jackfruit and their vitamin A values (μg RAE/100g fresh weight) as per de Faria et al., 2009:	
Carotenoids	**Concentration (μg/100g fresh weight)**
All-trans-β-carotene	29.55
9-cis-β-Carotene	0.79
Total carotenoids	107.98
Vitamin A value	2.84

2A.15.3 Flavonoids

Flavonoids are a group of plant compounds with antioxidant properties. They are known for their potential health benefits including anti-inflammatory and anticancer properties. While the specific flavonoid content can vary among different fruits and vegetables, jackfruit is reported to contain several types of flavonoids. Some common flavonoids found in this crop include:

i) **Quercetin**: This flavonoid is known for its antioxidant and anti-inflammatory properties. It is found in various fruits and vegetables including jackfruit.

ii) **Kaempferol**: Another flavonoid with antioxidant and anti-inflammatory effects, the kaempferol is present in jackfruit and many other plant-based foods.

iii) **Catechins and epicatechin**: These are the types of flavonoid present in abundance in tea but can also be found in smaller amounts even in jackfruit. Catechins is a class of compounds with antioxidant properties. Epicatechin is a specific catechin that has been studied for its potential health benefits.

iv) **Luteolin**: Jackfruit contains luteolin ~ a flavonoid that has been studied for its potential anti-inflammatory and antioxidant effects.

v) **Isoflavones:** Jackfruit does contain a little amount of isoflavones. Isoflavones are a type of phytoestrogen which are plant-derived compounds that can mimic the effects of estrogen in the body. Soybeans and soy products are well-known sources of isoflavones but these are not as prevalent in jackfruit. The isoflavone content in jackfruit may vary based on factors such as ripeness and growing conditions. Generally fruits and vegetables contain lower levels of isoflavones compared to legumes like soybeans.

2A.15.4 Phenolic compounds

Phenolic compounds including various phenolic acids are present in jackfruit. These compounds contribute to the fruit's antioxidant activity and may have anti-inflammatory properties. These include compounds like ferulic acid and caffeic acid which also contribute to the antioxidant capacity of jackfruit.

i) **Ellagic acid:** Ellagic acid is a polyphenolic compound known for its antioxidant effects. It is found in various fruits including berries and may also be present in jackfruit.

ii) **Cinnamic acid derivatives:** Cinnamic acid and its derivatives are phenolic compounds that contribute to the flavour and potential health benefits of fruits. They may be present in jackfruit.

iii) **Gallic acid:** Gallic acid is a type of phenolic acid that is found in various plant-based foods including fruits. Gallic acid is a natural antioxidant that is believed to have various health benefits. It is commonly found in fruits, vegetables, tea and in certain nuts. In jackfruit as in many other plant-based foods, gallic acid contributes to the overall antioxidant profile of the fruit.

2A.15.5 Lignans

Jackfruit contains lignans phytochemicals with potential antioxidant and anti-cancer properties. Lignans may also contribute to heart health.

2A.15.6 Saponins

Saponins ~ glycosides with a soap-like structure are naturally occurring compounds found in various plant species including jackfruit. Saponins have been studied for their potential health benefits including anti-inflammatory and immune-modulating effects. In jackfruit saponins can be found in different parts of the plant including the seeds, leaves and bark. The concentration and types of saponins can vary depending on the specific variety of jackfruit and environmental factors.

2A.15.7 Tannins

Tannins are polyphenolic compounds present in jackfruit that may have antioxidant properties. Tannins in jackfruit are believed to be responsible for the astringent or puckering taste that some people may experience when consuming unripe or green jackfruit. As the fruit ripens the tannin levels generally decrease which leads to a sweeter taste.

i) **Hydrolyzable tannins:** These are large complex molecules that can be broken down into smaller units through hydrolysis. Ellagic acid and gallic acid are examples of hydrolyzable tannins found in some fruits including jackfruit.

ii) **Condensed tannins (proanthocyanidins):** These are polymers of flavonoid units. These are generally less soluble than hydrolyzable tannins.

The presence of proanthocyanidins contributes to the astringency of jackfruit.

2A.15.8 Terpenoids

Terpenoids are a diverse class of natural compounds found in various plants including jackfruit. Specific types and amounts of terpenoids in jackfruit may vary based on factors such as the fruit's maturity, growing conditions and genetic factors. However, some common terpenoids found in jackfruit include:

i) **Limonene:** Limonene is a cyclic terpene that contributes to the citrusy aroma found in many fruits. It is also known for its potential health benefits.

ii) **Myrcene:** Myrcene associated with a fruity and musky aroma is another terpene commonly found in jackfruit as well as in other tropical fruits.

iii) **Pinene:** Two main types of pinene ~ alpha-pinene and beta-pinene are found. These terpenes contribute to the piney and resinous aromas found in some fruits and plants.

iv) **Terpinolene:** It is a terpene with a sweet and floral aroma. It is found in various fruits and herbs and may have antioxidant properties.

v) **Linalool:** It is a terpene with a floral scent and is known for its relaxing properties. It is found in various plants including some fruits.

vi) **Caryophyllene:** It is a sesquiterpene that contributes to a spicy and peppery aroma. It is found in many plants including some fruits and herbs.

2A.15.9 Organosulphur compounds

Though jackfruit is not particularly rich in organosulphur compounds compared to certain other plants but it does contain certain sulphur-containing compounds that contribute to its nutritional profile and potential health benefits.

i) **S-allyl cysteine:** One notable organosulphur compound found in jackfruit is S-allyl cysteine which is also present in garlic. S-allyl cysteine is associated with various health benefits including antioxidant and anti-inflammatory properties. It is formed when the enzyme alliinase converts alliin into allicin and subsequent reactions lead to the formation of S-allyl cysteine. This compound is believed to contribute to the characteristic aroma and flavour of jackfruit.

ii) **Sulphur-containing amino acids:** Moreover, sulphur-containing amino acids such as cysteine and methionine may also be present in jackfruit. These amino acids play crucial role in protein synthesis and have various physiological functions in the human body.

2A.15.10 Polysaccharides

Polysaccharides are complex carbohydrates made up of long chains of sugar molecules. In jackfruit several polysaccharides can be found which contribute to its nutritional composition and potential health benefits. Some of the key polysaccharides in jackfruit include:

i) **Dietary fibre:** Jackfruit is known for its high fibre content. Fibre is a type of polysaccharide that includes non-digestible carbohydrates. It can be both soluble and insoluble. Soluble fibre like pectin can help in regulating blood sugar levels and lowering cholesterol. Insoluble fibre such as cellulose aids in digestion and promoting a healthy gut.

ii) **Starch:** Jackfruit contains a significant amount of starch which is a polysaccharide made up of glucose molecules. Starch serves as a storage form of energy in plants and can be broken down into sugars during digestion to provide a source of glucose for energy.

iii) **Resistant starch:** This is a type of starch that resists digestion in the small intestine, reaching the colon where it can be fermented by gut bacteria. Resistant starch has been associated with various health benefits including improved gut health and blood sugar regulation.

iv) **Fructans:** Jackfruit also contains fructans which are polysaccharides made up of fructose molecules. Fructans can act as prebiotics promoting growth of beneficial bacteria in the gut and contributing to digestive health.

v) **Pectin:** Pectin is a type of soluble fibre found in the cell walls of plants including fruits. It is commonly used as a gelling agent in food processing especially in production of jams and jellies. In case of jackfruit it does contain pectin but concentration may vary depending on the ripeness of fruit. As fruits ripen their pectin content tends to decrease. Therefore, unripe or green jackfruit may have a higher pectin content compared to fully ripe one. The pectin in jackfruit can have various health benefits as dietary fibre is known to support digestive health and help regulating blood sugar levels. However, it's worth noting that the primary appeal of jackfruit lies in its unique texture and flavour which make it a popular meat substitute in vegetarian and vegan dishes.

2A.15.11 Vitamins

i) **Vitamin C:** Jackfruit is a good source of vitamin C ~ an antioxidant that plays a crucial role in immune function, collagen synthesis and wound healing.

ii) **Vitamin A:** Its vitamin A is important for vision, immune function and skin health.

iii) **Vitamin B-complex:** Jackfruit provides various B-vitamins including thiamine (B1), riboflavin (B2), niacin (B3), pantothenic acid (B5), pyridoxine (B6) and folate (B9). These vitamins play roles in energy metabolism, nerve function and red blood cell formation.

iv) **Vitamin C:** Vitamin C is worth mentioning as it acts as an antioxidant and helps the body absorbing iron. Jackfruit contains a decent amount of vitamin C.

v) **Vitamin K:** Jackfruit contains vitamin K which is necessary for blood clotting and bone health.

vi) **Vitamin E:** While jackfruit is not an exceptionally high source of vitamin E, it still contributes to the overall intake of this fat-soluble antioxidant.

2A.15.12 Minerals

i) **Potassium:** It is a good source of potassium ~ an essential mineral that helps regulating blood pressure, balances fluids in the body and supports proper muscle and nerve function.

ii) **Magnesium:** It is important for muscle function, energy production and the maintenance of bone health. Jackfruit contains moderate amount of magnesium.

iii) **Manganese:** Jackfruit also contains a little iof manganese ~ a trace mineral that plays a role in bone formation, blood clotting and reducing inflammation.

iv) **Copper:** It contains a small amount of copper which involves in the formation of red blood cells and maintenance of nerve cells.

v) **Iron:** It also contains a little of iron which is important in formation of haemoglobin and transportation of oxygen in the blood.

It is important to note that concentrations of these phytochemicals and nutritional contents can vary in different parts of the jackfruit and the overall content may be influenced by factors such as varieties, ripeness and processing methods. The health benefits of these phytochemicals are an area of ongoing research and individual responses to phytochemicals can vary.

2A.16 Chemical Diversity within *Artocarpus heterophyllus*

A. heterophyllus boasts a remarkable array of chemical constituents contributing to its diverse profile. Examining its intricate chemistry unveils a spectrum of flavones, colouring matters and unique compounds such as morin, dihydromorin, cynomacurin, artocarpin, isoartocarpin, cyloartocarpin, artocarpesin, oxydihydroartocarpesin, artocarpetin, norartocarpetin, cycloartinone and artocarpanone (Rama Rao and Venkataraman, 1973). It exhibits a unique and sweet fruity fragrance. A flavour volatile analysis conducted on five different jackfruit cultivars revealed prominent compounds including ethyl isovalerate, propyl isovalerate, butyl isovalerate, isobutyl isovalerate, 3-methylbutyl acetate, 1-butanol and 2-methylbutan-1-ol (Ong *et al.*, 2008).

Analyzing the heartwood further reveals a nuanced composition with 6.7% moisture, 38.0% glucosides, 0.7% lipids, 1.7% albumin and a dominant cellulose content of 59.0% (Perkin and Cope, 1895). The overall plant structure incorporates free sugars (sucrose), fatty acids, ellagic acid and essential amino acids such as arginine, cystine, histidine, leucine, lysine, methionine, threonine and tryptophan (Pavanasasivam *et al.*, 1973).

The bark of the main trunk contains betullic acid and novel flavone pigments like cycloheterophyllin ($C_{30}H_{30}O_7$) (Chawdhary and Raman, 1997). Reports also highlight triterpenic compounds like cycloartenyl acetate and cycloartenone (Barik *et al.*, 1997) alongside the discovery of heterophylol ~ a distinctive phenolic compound with a novel skeleton (Lin and Lu, 1993).

Only 3.3% tannin is found in the bark, occasionally repurposed for cordage or cloth. Leaves and stems exhibit the presence of sapogenins, cycloartenone, cycloartenol and β-sitosterol, showcasing estrogenic activity (Om Prakash *et al.*, 2009). The root, on the other hand contains β-sitosterol, ursolic acid, betulinic acid and cycloartenone (Dayal and Seshadri, 1974).

Beyond its structural components, jackfruit seeds harbour jacalin ~ a significant protein characterized by a tetrameric two-chain lectin structure. Jacalin's specificity for the O-glycoside of the Thomsen–Friedenreich antigen, even in its sialylated form, positions it as a valuable tool for studying O-linked glycoproteins, particularly human IgA1 (Kumar *et al.*, 1982). Its unique property of being strongly mitogenic for human CD4+ T lymphocytes adds to its significance in assessing the immune status of individuals, notably those infected with HIV-1 (Pereira-da-Slva *et al.*, 2006).

Further exploration led to the isolation of two novel 2′, 4′, 6′-trioxygenated flavanones and heteroflavanones A and B from root bark. These compounds exhibit potent anti-inflammatory activity (Lu and Lin, 1993; Lin *et al.*, 1995). Three phenolic compounds ~ artocarpesin, norartocarpetin and oxyresveratrol were characterized and evaluated for their anti-inflammatory effects in murine macrophage cells (Jha *et al.*, 1997; Venkataraman, 2001).

The carotenoid composition of *A. heterophyllus* includes carotenes, β-zeacarotene, α-zeacarotene, and β-carotene-5, 6-epoxide, alongside the dicarboxylic carotenoid crocetin (Chandrika *et al.*, 2004). Further isolations from the root and bark resulted in six prenylflavonoids including two new compounds and a natural Diels-Alder-type adduct named artonin X (Chung *et al.*, 1995; Shinomiya *et al.*, 2004).

Jackfruit seed when subjected to emission and Fourier transform infrared spectra analysis, reveals the presence of previously undetected elements ~ manganese and magnesium. Specific functional groups attributed to different bands were identified in the Fourier transform infrared spectrum (Barua and Boruah, 2004).

2A.17 Phytochemical Examination Across Various Components

Phytochemical analysis conducted by various researchers has revealed a diverse array of compounds present in different components of the jackfruit tree, highlighting its rich chemical profile. The following key findings have been reported:

- Barik *et al.* (1994) isolated tetracyclic triterpenoids, including 9,19-cyclolanost-3-one-24,25-diol and 9,1-cyclolanost-3-one-24,2Sdiol along with cycloartenone and cycloartenol.
- Yuan *et al.* (2017) identified flavonoids such as artoheteroids A-D, morin, artocarmin A, albanin A, euchrenone A, norartocarpanone and steppogenin from the roots of *A. heterophyllus.*
- Arung *et al.* (2010) discovered a novel prenylated flavonoid 3-prenyl luteolin (1) in wood extract.
- Wang *et al.* (2017) obtained an 2-arylbenzofuran derivative (artocarstilbene B) and a benzaldehyde derivative [(E)-3,5-dihydroxy-4-(3-methylbut-1-enyl)benzaldehyde] from the leaves of *A. heterophyllus.*
- From stems and leaves, Liu *et al.* (2020) isolated seven prenylated chromones and five prenylated flavonoids including two new prenylated chromones, artoheterophines A and B; five known prenylated chromones as well as five known biogenetically related prenylated flavonoids, *viz.* cnidimol D, ficuformodiol B, harperamone, perforatin B, 5,7-dihydroxy-8- [(2*E*)-4-hydroxy-3-methylbut-2-enyl]-2-methyl-4H-1-benzopyran-4-one, 2-(4-hydroxy- phenyl)-8-(3-methyl-but-2-enyl)-chroman-4-one, bracteflavone B, dinklagin C, 6-(3-methyl-(*E*)-1-butenyl) chrysin and 5,7,3′,5′-tetramethoxy-6-*C*-prenylflavone.

- Zheng *et al.* (2014) isolated new phenolic compounds ~ artoheterophyllins E-J, 4-geranyl-2',3,4',5-tetrahydroxy-cis-stilbene and 5-methoxymorican M. Two additional natural compounds ~ 2,3-dihydro-5,7-dihydroxy-2-(2-hydroxy-4-methoxyphenyl)-4H-benzopyran-4-one & 6-[(1S,2S)-1,2-dihydroxy-3-methylbutyl]-2-(2,4-dihydroxyphenyl)-5-hydroxy-7-methoxy-3-(3-methyl-2-buten-1-yl)-4H-1-benzopyran-4-one were isolated from ethanol extract of *A. heterophyllus* wood.
- Previous studies by Zheng *et al.* (2008) identified furanoflavone,7-(2,4-dihydroxyphenyl)-4-hydroxy-2-(2-hydroxypropan-2-yl)-2, 3-dihydrofuro(3, 2-g)chromen-5-one (artocarpfuranol), along with 14 known compounds including dihydromorin, steppogenin, norartocarpetin, artocarpanone, artocarpesin, artocarpin, cycloartocarpin, cycloartocarpesin, artocarpetin, brosimone I, cudraflavone B, carpachromene, isoartocarpesin and cyanomaclurin from *A. heterophyllus* wood.
- Zheng *et al.* (2009) isolated phenolic compounds from the ethanol extract of the twigs of *A. heterophyllus* including an isoprenylated 2-arylbenzofuran derivative (artoheterophyllin A) and three isoprenylated flavonoids (artoheterophyllin B, artoheterophyllin C and artoheterophyllin D).
- Fu *et al.* (2020) purified artocarheterones A-E from the ripe fruits of *A. heterophyllus.*
- Septama *et al.* (2020) identified dihydromorin and norartocarpetin in the heartwoods of *A. heterophyllus.*
- Om Prakash *et al.* (2009) & Rama Rao and Venkataraman (1973) stated that jackfruit possesses compounds like morin, dihydromorin, cynomacurin, artocarpin, isoartocarpin, cyloartocarpin, artocarpesin, oxydihydroartocarpesin, artocarpetin, norartocarpetin, cycloartinone, betulinic acid, artocarpanone and heterophylol.

These findings collectively emphasize the diverse array of phytochemicals present in various parts of jackfruit tree, underscoring its potential significance in pharmaceutical and nutritional applications.

2A.18 Beneficial Effects of Jackfruit on Human Health

Jackfruit is associated with a myriad of potential health benefits as evidenced by diverse array of high-value compounds it contains. The flesh and seeds of jackfruit are considered as a cooling and nutritious tonic. Recognized for its antibacterial, antifungal, antidiabetic, anti-inflammatory and antioxidant activities, jackfruit stands out as a versatile and nutritious addition to a healthy diet (Jagtap *et al.*, 2010; Shanmugapriya *et al.*, 2011).

Jackfruit contains carotenoids. Beyond their role as colourants, carotenoids display provitamin A activity and are associated with positive impacts on numerous chronic degenerative diseases. Studies by Krinsky *et al.*, 2003; Stanner *et al.*, 2004; Coyne *et al.*, 2005; Stahl and Sies, 2005 and de Faria *et al.*, 2009 suggest the potential benefits of carotenoids in mitigating conditions like cancer, inflammation, cardiovascular disease, cataract and age-related macular degeneration.

2A.18.1 Antioxidants

Jackfruit boasts a rich repertoire of phytonutrients, particularly carotenoids which exhibit potent antioxidant properties as highlighted by Baliga *et al.* (2011) and Ranasinghe *et al.* (2019). Antioxidants characterized by their ability to impede, retard or even prevent the oxidation process (Halliwell, 1997) play a crucial role in shielding the body and biomolecules from the detrimental effects of excessive free radicals.

Excess free radicals are implicated in various harmful processes including tissue injury, protein oxidation, DNA damage and in elicitation of proinflammatory responses (Xu and Touyz, 2006). Jagtap *et al.* (2010) have noted a positive correlation between the antioxidant activities of jackfruit flesh extracts and their total phenolic and flavonoid content. Soong and Barlow (2004) underscored the significant ascorbic acid equivalent antioxidant effects and 27.7 & 0.9 gallic acid equivalent phenolic contents found in fresh jackfruit seeds and flesh. These compounds are estimated to contribute substantially, accounting for approximately 70% of the overall antioxidant activity.

In summary, jackfruit emerges as a noteworthy source of phytonutrients, particularly carotenoids that exhibit potent antioxidant effects, thereby conferring protective benefits against oxidative stress and its associated detrimental consequences on biological systems.

2A.18.2 Anti-inflammatory and anticancer properties

Several investigations have highlighted the anti-inflammatory attributes of bioactive compounds extracted from *A. heterophyllus* fruits (Fang *et al.*, 2008). The flavonoids identified in jackfruit have demonstrated the ability to impede the release of inflammatory mediators which is indicative of their potential anti-inflammatory effects (Wei *et al.*, 2005).

In addition to their well-documented anti-inflammatory attributes, the bioactive compounds found in jackfruit including lignans, isoflavones and saponins contribute to a diverse array of health advantages. These encompass not only anti-inflammatory effects but also extend to anticancer, antihypertensive, antiulcer and anti-ageing properties. The presence of these phytonutrients in jackfruit suggests potential benefits in neutralizing cancer-causing free radicals within the body, thereby impeding the degeneration of cells that may lead to various degenerative diseases. Particularly noteworthy is their implication in exhibiting chemoprotective properties, suggesting a capacity to potentially mitigate the mutagenicity of aflatoxin B1 (AFB1) and curtail the proliferation of cancer cells (Baliga *et al.*, 2011; Ruiz-Montañez *et al.*, 2014; Priya Devi *et al.*, 2014). This multifaceted array of health-promoting attributes underscores the potential of jackfruit-derived compounds in fostering overall well-being.

2A.18.3 Antifungal properties

Jackfruit demonstrates noteworthy antifungal properties attributed to specific compounds within its composition. One such compound 'jackin' ~ a chitin-binding lectin has been identified for its ability to impede the growth of select fungi including *Fusarium moniliforme* and *Saccharomyces cerevisiae* (Trindade *et al.*, 2006). The fruit displays hemagglutination activity against human and rabbit erythrocytes as reported by Trindade *et al.* (2006).

2A.18.4 Antibacterial and nematicidal activities

The methanolic extracts derived from diverse components of the jackfruit plant encompassing the stem, root, bark, heartwood, leaves, fruits and seeds showcase an extensive range of antibacterial activities against various bacterial strains. Shoots of jackfruit plant demonstrate notable nematicidal activity, effectively combatting a spectrum of nematodes including *Rotylenchulus reniformis*, *Tylenchorhynchus brassicae*, *Tylenchus filifofmis* and *Meloidogyne incognita* (Khan *et al.*, 2003; Sharma and Trivedi, 1995). This multifaceted antimicrobial and nematicidal potential underscores the pharmacological significance of different parts of the jackfruit plant, presenting promising avenues for further exploration in the realms of medicine and agriculture.

2A.18.5 Cardiovascular health

An elevated blood LDL: HDL ratio is recognized as a substantial risk factor for the development of coronary heart diseases. However, jackfruit with its notable antioxidant properties presents a potential mitigating influence on this risk. These antioxidant properties are attributed to a diverse array of phytonutrients such as carotenoids found within the fruit.

Scientific evidence supports the assertion that the antioxidants in jackfruit including carotenoids play a crucial role in counteracting oxidative stress.

Oxidative stress is a known contributor to atherosclerosis and inflammatory modifications within arterial walls, both of which are implicated in the pathogenesis of cardiovascular diseases. Esmaillzadeh and Azadbakht (2008); Heinecke (2006); Fan and Watanabe (2003); Baliga *et al.* (2011) and Ranasinghe *et al.* (2019) have explored and documented the potential cardiovascular benefits associated with consumption of jackfruit.

In essence, the incorporation of jackfruit into the diet could be a strategic dietary choice to harness its antioxidant prowess, offering protection against oxidative stress-induced atherosclerosis and inflammatory changes in arterial walls thus contributing to a more favourable LDL:HDL ratio and potentially lowering the risk of coronary heart diseases.

2A.18.6 Blood pressure regulation

Jackfruit is endowed with a myriad of functional compounds that exhibit the potential to mitigate various health conditions including but not limited to high blood pressure, heart diseases, strokes and bone loss. Its remarkable properties extend to the improvement of muscle and nerve function as well as reduction of homocysteine levels in the bloodstream.

One notable attribute of jackfruit is its richness in potassium ~ a vital mineral that plays a pivotal role in the regulation of blood pressure and acts as a countermeasure against the adverse effects of sodium. The abundant potassium content in jackfruit not only contributes to the enhancement of muscle and nerve function but also serves as a protective factor against the onset of heart diseases, strokes and bone loss (Swami *et al.*, 2012; Priya Devi *et al.*, 2014).

Moreover, the presence of vitamin B6 in jackfruit serves as an additional health-promoting factor. Vitamin B6 aids in the reduction of homocysteine levels thereby lowering the risk of heart disease (Fernando *et al.*, 1991). The multifaceted nutritional composition of jackfruit positions it as a valuable dietary inclusion with potential preventive and therapeutic implications for various health concerns.

2A.18.7 Melanin biosynthesis inhibition

The extract derived from jackfruit wood has garnered recognition for its efficacy in inhibiting melanin biosynthesis. Specifically, prenylated flavones-based polyphenols isolated from jackfruit wood have exhibited notable capabilities in impeding melanin biosynthesis within B16 melanoma cells. These polyphenols demonstrate this inhibitory effect while maintaining a minimal or negligible level of cytotoxicity (Arung *et al.*, 2007).

2A.18.8 Glucose tolerance

The aqueous extracts obtained from jackfruit leaves have demonstrated notable enhancements in glucose tolerance among both normoglycemic individuals and those with diabetes (Fernando *et al.*, 1991).

2A.18.9 Strengthens immune system

Jackfruit stands out as a notable source of vitamin C ~ a vital nutrient renowned for its role in fortifying the body against both viral and bacterial infections. This essential vitamin serves as a cornerstone in supporting the immune system by enhancing the functionality of white blood cells ~ the frontline defenders against invading pathogens. In just one cup of jackfruit, the body receives a substantial and commendable supply of this potent antioxidant, contributing to overall immune resilience and well-being (Priya Devi *et al.*, 2014).

2A.18.10 Skin health and eye care

Jackfruit stands out as a nutritional powerhouse, offering a rich source of essential nutrients that contribute to overall skin health. Abundant in vitamin C, the fruit becomes a natural ally in the battle against aging and sun damage. Vitamin C is renowned for its role in promoting collagen synthesis ~ a key factor in maintaining skin firmness and strength (Lavker, 1991; Babitha *et al.*, 2004).

Moreover, jackfruit brings another vital component to the table in the form of beta-carotene. This compound is not only crucial for maintaining skin health but also plays a pivotal role in supporting optimal vision and bolstering the immune system. Delving into the specifics, the presence of lutein and zeaxanthin carotenoids in jackfruit adds an extra layer of health benefits particularly for the eyes. These carotenoids predominantly located in the retina are believed to serve as guardians against harmful high-energy light waves including ultraviolet rays. Thus, by incorporating jackfruit into one's diet, individuals can harness its diverse array of nutrients to nourish and protect the skin, support vision and to fortify the immune system.

2A.18.11 Boosts energy

Jackfruit is renowned for its energy-boosting properties attributed to the presence of simple sugars including fructose and sucrose. This unique composition makes it a go-to choice for a quick and sustained energy release. Despite being an energy-rich fruit, one of the remarkable health benefits of jackfruit is its absence of saturated fatty oils and cholesterol. This not only makes it a wholesome addition to diet but also positions jackfruit as a health-conscious option for those seeking both vitality and nutritional well-being. So, indulging in the delectable taste of jackfruit not only satisfies ones tastebuds but also provides a guilt-free source of energy for a vibrant and healthy lifestyle (Priya Devi *et al.,* 2014).

2A.18.12 Asthma control

The root of the jackfruit has demonstrated potential benefits for individuals dealing with asthma. Extracts obtained from boiling jackfruit roots have shown promise in effectively managing asthma symptoms (Priya Devi *et al.,* 2014).

2A.18.13 Metabolic benefits

Niacin, a key component found in jackfruit plays a pivotal role in supporting essential physiological functions such as energy metabolism, nerve function and hormone synthesis. Moreover, the dietary fibre inherent in jackfruit prevents constipation and serves as an effective bulk laxative contributing to the well-being of the colon mucous membrane by driving away the carcinogenic chemicals from the large intestine and facilitating seamless bowel movements (Morton, 1987; Priya Devi *et al.,* 2014). This nutritional synergy highlights jackfruit as not only a flavourful tropical fruit but also a source of beneficial compounds crucial for maintaining various aspects of overall health.

2A.18.14 Bone strengthening, preventing anaemia and maintaining healthy thyroid

Jackfruit is not only renowned for its delectable taste but also for its nutritional richness, particularly in essential minerals. Among these, magnesium stands out playing a pivotal role in facilitating calcium absorption, thereby contributing to the fortification of bones. The presence of iron in jackfruit serves as a preventive measure against anemia and supports optimal blood circulation. Inclusion of copper in this tropical fruit is notable for its involvement in thyroid gland metabolism highlighting the diverse range of health benefits that jackfruit offers through its mineral content (Singh *et al.,* 1991).

2A.19 Medicinal Use

Compounds such as morin, dihydromorin, cynomacurin, artocarpin, isoartocarpin, cyloartocarpin, artocarpesin, oxydihydroartocarpesin, artocarpetin, norartocarpetin, cycloartinone, betulinic acid, artocarpanone and heterophylol identified in jackfruit have a rich tradition of therapeutic applications. These compounds have been historically recognized for their effectiveness in addressing a diverse range of health concerns including fever, boils, wounds, skin diseases, convulsions, diuretic properties, constipation, ophthalmic disorders and even snake bites (Rama Rao and Venkataraman, 1973; Om Prakash *et al.,* 2009). The incorporation of these bioactive compounds from jackfruit into traditional medicine highlights their potential in various health remedies.

Chinese have long regarded jackfruit pulp and seeds as more than just a culinary delight as these are considered a tonic with cooling and nutritious properties. Interestingly, these versatile fruits are believed to be particularly effective in counteracting the effects of alcohol on the human system. The seeds in particular are prized for their starch which is administered to alleviate biliousness while the roasted seeds are esteemed as an aphrodisiac. Beyond the realm of culinary applications, various parts of jackfruit tree have found their place in traditional medicine. Ash derived from burning jackfruit leaves along with corn and coconut shells is utilized either independently or in combination with coconut oil to facilitate the healing of ulcers. Moreover, the dried latex extracted from the fruit yields artostenone ~ a compound that can be converted into artosterone which showcases notable androgenic properties (Morton, 1987).

Medicinal applications extend further to the treatment of ailments such as abscesses, snakebites and glandular swellings. A concoction of the latex mixed with vinegar is employed to promote the healing process. The root of the jackfruit tree is recognized as a remedy for skin diseases and asthma with an extract from the root prescribed in cases of fever and diarrhoea. Bark is skillfully transformed into poultices while heated leaves are strategically placed on wounds to aid in the recovery process (Morton, 1987).

Jackfruit tree exhibits a sedative property within its wood and the pith of the tree is reputed to have potential to induce abortion. The rich tapestry of applications attributed to various parts of jackfruit tree underscores its significance not only as a source of sustenance but also as a venerable component of traditional Chinese medicine.

2A.20 Ayurvedic description

- **Sanskrit name:** Panasa, Atibrhatphala
- **Properties:**
  - **Rasa:** Madhura, kasaya (unripe fruit)
  - **Guna:** Snigdha, guru, picchila
  - **Virya:** Sita; **Vipaka:** Madhura
- **Actions**:
  - Vatapittahara, kaphavardhaka, balaya, sukraprada, tarpana, durjara
  - Seeds- Vrsya, baddhavitkara, mutrala
- **Therapeutic uses:** Klibata, durbalya, raktapitta, varna

(Om Prakash *et al.*, 2009).

2A.21 Aroma

A fully mature and unopened jackfruit is recognized for releasing a uniquely robust, sweet and fruity aroma which some may find displeasing (Hargreaves and Hargreaves, 1964; Morton, 1987). The interior of the fruit is often characterized by a fragrance reminiscent of pineapple and banana (Morton, 1987). Following the roasting process, the seeds of the jackfruit present a noteworthy commercial alternative to the aroma of chocolate (Spada *et al.,* 2017).

2A.22 Uses

Jackfruit emerges as a versatile species demonstrating a multifaceted utility encompassing fodder, timber, fuel, medicinal and industrial applications (Kishore, 2018). This botanical marvel not only presents a spectrum of practical uses but also boasts nutritional richness within its fruits and nuts. Abundant in protein, dietary fibre and essential vitamins such as A, B & C, the jackfruit offers a commendable array of minerals as well. In addition to its nutritional value, jackfruit is appreciated for its meaty texture, making it a popular meat substitute in vegetarian and vegan dishes. It is also a good source of energy and can be incorporated into various savoury and sweet recipes.

2A.22.1 Culinary use

Jackfruit undergoes a remarkable culinary transformation depending on its ripeness. When tender, raw or semi-ripe it is savoured as a vegetable making its way into an array of dishes such as curries, stews, dry stir-fries, biryani and even as a meat substitute in vegetarian and vegan recipes. The culinary possibilities extend further as it takes on various forms ~ deep-fried as delectable koftas and fritters, dried and crisped into chips or pickled for a tangy twist (Lobo, 2019).

Flakes and seeds are processed into various products considering their nutritional properties (Swami *et al.,* 2012). The nutritious jackfruit seeds can be prepared by boiling or roasting and enjoyed as to those of chestnuts. These can be incorporated into flour for baking adding a unique and nutritious element to various recipes and dishes (Om Prakash *et al.,* 2009). Sun-dried seeds of jackfruit become integral components in curries and simple roasted ones to create delightful snacks. The leaves of the jackfruit tree contribute to the culinary tapestry as they serve as vessels for steaming food. Even the core of the fruit is not overlooked and is sometimes prepared in savory dishes. As the jackfruit matures and ripens, it metamorphoses into a sweet delight that finds its way into an entirely different category of culinary creations. Ripe jackfruit is often eaten fresh or utilized in the creation of sweets, syrups, jams and chutneys ~ adding a natural sweetness and distinct flavour profile to these confections. The culinary creativity surrounding jackfruit has transcended traditional boundaries as evidenced by innovative concoctions such as the jackfruit cocktail. In Goa, pioneers like Hansel Vaz of Cazulo Premium Feni and bartender Rohan Barbosa have ventured into uncharted territory experimenting with jackfruit-infused creations. Castaad Ponos, for instance, employs a syrup derived from jackfruit mixed with cashew feni, crafting a uniquely tantalizing sour flavour (Lobo, 2019).

Westerners typically prefer jackfruit in its full-grown but unripe stage as it lacks objectionable odour and surpasses cooked green breadfruit and plantain in taste. To prepare, the fruit is cut into large chunks, boiled in lightly salted water until tender and then the delicious flesh including seeds is served as a vegetable. The only challenge is dealing with the gummy latex which can be minimized by rubbing hands and knives with salad oil. Unripe fruit can be canned in brine, with curry or dried for long-term storage. In Thailand, dried unripe jackfruit cross sections are available in native markets. Young tender fruits lend themselves well to pickling with or without spices (Morton, 1987).

If allowed to ripen, jackfruit bulbs and seeds can be extracted outdoors because of its stains and smell which attract flies and other creatures. Enjoy the ripe bulbs raw or cooked (with coconut milk or otherwise) or transform them into icecream, chutney, jam, jelly, paste, leather or papad. Canning in syrup made with sugar or honey along with added citric acid is an option with preference given to crisp types for this process. The resulting canned product often referred to as 'vegetable meat', is more visually appealing than fresh pulp. Mechanically pulping ripe bulbs produce jackfruit nectar or concentrates and synthetic flavouring such as ethyl and n-butyl esters of 4-hydroxybutyric acid at 120ppm and 100ppm, respectively significantly enhances the flavour of both canned fruit and nectar (Morton, 1987).

Boiling jackfruit bulbs in milk creates a delightful orange-coloured custard upon cooling. In India, a patented method allows ripe bulbs to be dried, fried in oil and salted which resemble the potato chips. Brazil marketed candied jackfruit pulp in boxes as early as 1917. CFTRI, Mysuru has developed advanced techniques for preserving and candying jackfruit pulp. Slicing ripe bulbs, packing them in syrup with added citric acid and freezing them ensures excellent colour, flavour and texture retention for up to one year. Canned jackfruit maintains quality for 63 weeks at room temperature (23.89°-26.67°C) with just a 3% loss of B-carotene. The frozen canned pulp remains well-preserved for two years (Morton, 1987).

In Malaya, small jackfruits embracing the scent of ripeness are halved, deseeded, chilled and presented filled with ice cream (Morton, 1987). The mature bulbs after fermentation and distillation yield a potent liquor (Morton, 1987). The versatile seeds appealing to various palates can be boiled, roasted or preserved in syrup akin to chestnuts. Canning in brine, curry and tomato sauce is successful and these are frequently featured in curried dishes. Roasted and dried seeds are ground into a flour and blended with wheat flour for baking (Morton, 1987).

Localities with abundance of jackfruit can resourcefully utilise the inedible parts as rind can yield a satisfactory jelly with citric acid. Pectin extraction from the peel, undeveloped perianths and core or the inner rind is possible and the byproduct produces a syrup used in tobacco curing. Tender jackfruit leaves and young male flower clusters can be cooked and served as delectable vegetables (Morton, 1987).

2A.22.2 Other uses (Morton, 1987)

i) **Jackfruit in animal husbandry:** In certain regions, jackfruit serves as a nutritious feed for cattle. The trees are strategically planted in pastures providing a natural source of sustenance for animals through fallen fruits. Surplus jackfruit rind is recognized as valuable stock food.

ii) **Jackfruit leaves:** Young leaves of jackfruit tree are a favoured choice for livestock contributing to their nutrition and weight gain. In India, these leaves take on a dual role as food wrappers in cooking and are ingeniously fastened together for use as eco-friendly plates.

iii) **Latex:** Jackfruit latex often used as birdlime demonstrates versatility in non-agricultural domains. When heated, this transforms into a household cement ideal for repairing chinaware, earthenware and even caulking boats & buckets. While not a rubber substitute, its resin content holds potential for varnishes boasting bacteriolytic properties comparable to papaya latex.

iv) **Jackwood (the timber treasure):** Renowned as an important timber in Sri Lanka and India, jackwood stands out for its durability and resistance. Changing colour with age, it is termite-proof and finds application in construction, furniture, turnery and even musical instruments. The roots of old jackfruit trees are highly sought after for carving, and dried branches play a role in religious ceremonies.

v) **As a dye source:** From the sawdust and heartwood chips of jackwood a rich yellow dye is derived by boiling with alum. This dye is widely used for colouring silk and the cotton robes of Buddhist priests. In Indonesia, the wood's splinters are introduced into bamboo tubes collecting coconut toddy to impart a yellow tone to sugar.

vi) **Bark ~ a modest contributor:** While its bark contains a relatively low tannin percentage (3.3%), it is occasionally repurposed to make cordage or cloth showcasing the resourcefulness of every part of the tree.

2A.23 Jackfruit Latex

Jackfruit latex is a sticky gummy substance that is often referred to as the 'latex' or 'gummy latex' derived from latex vessels of jackfruit trees (*Artocarpus heterophyllus*).

Some key points about jackfruit latex:

i) **Extraction:** The latex is extracted by making incisions in the tree's bark or by cutting the fruit. When the fruit is cut latex oozes out from the cut surfaces.

ii) **Composition:** The latex contains a mixture of substances including proteins, enzymes and other compounds. One of the significant components is a group of proteins known as chitinases which are responsible for sticky nature of the latex.

iii) **Use in cooking:** In some traditional cuisines, jackfruit latex is used as a natural adhesive in cooking. It can be used to bind and thicken sauces, gravies or stews. The latex has a strong but distinct flavour so it is typically used in dishes where its taste won't overpower the overall flavour profile.

iv) **Potential allergen:** The latex of jackfruit can be a potential allergen to a few individuals. It may cause skin irritation or allergic reactions in some people. Handling the fruit or its latex may lead to skin itching or rashes in certain individuals.

v) **Processing:** In processed jackfruit products available commercially, efforts are often made to reduce the latex content. This is because the gummy texture may not be desirable in certain culinary applications.

vi) **Medical uses:** Traditionally, the latex has been used in some cultures for medicinal purposes. It has been applied to wounds, and its potential antimicrobial properties have been explored.

It is important to exercise caution when handling the latex, especially for individuals with latex allergies. While the gummy latex has culinary and potential medicinal uses, its distinctive characteristics should be taken into account in different applications.

2A.24 Toxicity

In India, the acceptance of jackfruit encounters resistance, often linked to the misconception that excessive consumption may lead to digestive issues. This resistance is often based on the belief that the raw and unripe fruit is astringent and difficult to digest. However, according to Burkill, a renowned authority on plants, the astringency and indigestibility are specific to the raw or unripe jackfruit. The ripe fruit possesses mild laxative properties so excessive consumption may result in diarrhoea.

Another noteworthy aspect is the presence of a potent trypsin inhibitor in raw jackfruit seeds, rendering them indigestible. However, this inhibitory element can be neutralized either by boiling or baking the seeds. This additional information sheds light on the specific conditions that affect the digestibility of different parts of jackfruit.

2A.25 Cultivation

As a crop, Jackfruit is easy to cultivate, grows well in neglected conditions, and gives abundant yield (Lobo, 2019).

2A.25.1 Climate

The cultivation of jackfruit is intricately tied to its climatic requirements which play a pivotal role in determining the success and quality of the harvest. Jackfruit thrives in regions characterized by equatorial to subtropical climates, ideally situated at elevations ranging from sea level to 1600m a.s.l. Warm, humid and frost-free climates with minimum temperatures of 16-22°C and mean temperatures of 25-30°C are desirable for good jackfruit bearing. Temperatures below 5°C severely damage trees and frost kills the developing shoots, fruit and sometimes even main branches (Bakewell-Stone, 2023). For proper cultivation jackfruit requires an average annual rainfall between 1000 to 2400mm (Fathin *et al.*, 2021). Jackfruit exhibits sensitivity to frost during its early stages of life and is intolerant to drought conditions. In instances where rainfall falls below the requisite levels, supplemental irrigation becomes imperative (Morton, 1987). Under more favourable environments such as areas with consistent light exposure and abundant rainfall, jackfruit trees have the remarkable ability to flower and bear fruit throughout the year (Li *et al.*, 2010; Nakintu *et al.*, 2019). However, in regions characterized by distinct dry and wet seasons, jackfruit tends to initiate its flowering phase exclusively during the wet season (Fathin *et al.*, 2021).

Cold, drought and flood limit the distribution of jackfruit to areas having greater than 1500mm rainfall evenly distributed throughout the year, without a prominent dry season. Trees do not perform well in exposed locations with drying winds. The trees have some salt tolerance but poor drought and flood tolerance.

A noteworthy consideration is that jackfruits cultivated above 1,200 meters are often suitable for cooking purposes. Morton's (1987) findings underscored this point providing valuable insight into the elevation-related nuances of jackfruit cultivation. In specific geographic locations, such as Kwangtung (China) the jackfruit trees have been observed to ascend to approximately 244 meters, emphasizing the adaptability of the tree to varied terrains.

2A.25.2 Soil

The successful cultivation of jackfruit hinges on providing an environment with specific soil characteristics conducive to its growth. Jackfruit trees exhibit robust development in deep and rich soil with a medium to open texture and with a pH range of 6-7.5 (Bakewell-Stone, 2023). These trees can thrive in environments with deep, gravelly or laterite soil. However, growth may occur but hindered significantly in shallow limestone conditions.

Within the nuanced interplay of soil and tree, regional nuances manifest in the characteristic stature of the jackfruit. In the Indian context, anecdotal insights propose that when rooted in sandy soil, the tree elegantly ascends, presenting a tall and slender profile. In contrast, when situated in stony terrain, tree tends to embrace a more condensed and robust structure. This underscores the tree's remarkable adaptability across varied soil compositions, albeit with discernible variations in growth patterns.

One critical caveat in jackfruit cultivation is the tree's intolerance to waterlogged conditions. The adage 'it cannot tolerate wet feet' succinctly encapsulates the vulnerability of jackfruit trees to excessive soil moisture. If the roots come into contact with water, it not only impedes fruit-bearing but may also jeopardize the overall health of the tree, potentially leading to its demise.

To optimize jackfruit cultivation, it is imperative to maintain well-draining soil conditions, ensuring that the roots remain free from prolonged exposure to water. This consideration is pivotal in fostering a conducive environment for the tree to bear fruit and thrive. Periodic soil testing and amendments can contribute to sustaining the health and productivity of jackfruit trees over the long term.

2A.25.3 Seasons

Sowing season

Jackfruit is a tropical fruit tree that is typically grown from seeds. The optimal season for sowing jackfruit seeds varies depending on the local climate, though generally it is recommended to sow the seeds during warm and rainy season. It is always a good idea to observe local agricultural practices, *via-a-vis* planting time in specific locations. Here are some guidelines based on common practices:

i) **Tropical regions:** In tropical regions where jackfruit is native or well-adapted such as parts of SE Asia and India, the best time for its seed sowing is during rainy season, *i.e.* from June to August. The combination of warmth and moisture promotes germination and initial seedling growth.

ii) **Subtropical regions:** In subtropical regions with a milder climate, it is still advisable to sow seeds during warmer months. Spring to early summer is often a suitable time, taking advantage of increasing temperatures and longer daylight hours.

iii) **Container planting:** In case of container raising one can opt to sow or plant any time of the year as one can make adjustments by controlling environmental conditions accordingly. However, it is still recommended to provide warmth and adequate moisture for successful germination and early growth.

iv) **Year-round planting in equatorial regions:** In equatorial regions where there is little variation in temperature and consistent rainfall throughout the year, jackfruit seeds can be sown almost any time. However, it is essential to ensure that the soil remains consistently moist during initial stages of germination.

Planting season

Transplanting jackfruit seedlings is often done after the rainy season or during the warm season. This allows the young plants to establish themselves with ample sunlight and warmth. Jackfruit trees prefer well-drained soil so planting seedlings in well-prepared soil after the rainy season helps avoid waterlogging issues. In regions where there is a risk of frost, it is crucial to time the planting of jackfruit seedlings to avoid exposure to frost conditions.

Harvest season

The harvesting season for jackfruit exhibits variability across different regions and climates, reflecting the adaptability of this tropical fruit to diverse environmental conditions. In Asia, the ripening period typically spans from March to June, April to September or June to August depending on specific climatic regions. There are instances of off-season crops occurring between September and December with sporadic fruiting at other times of the year.

Specifically in the West Indies, a notable concentration of ripening jackfruits is reported in June. Meanwhile, in the state of Florida, the harvest season is observed during late summer and fall, aligning with the climatic conditions of the region (Morton, 1987).

This temporal diversity in jackfruit ripening underscores the fruit's ability to adapt to a range of climates. It also provides opportunities for extended availability of jackfruit, contributing to its significance as a valuable crop in various parts of the world.

2A.25.4 Propagation

Propagation of jackfruit typically occurs through seeds with a limited shelf life of no more than a month before planting. The seed is generally sown fresh but where short-term storage is necessary it should not dry out. The germination process spanning 3 to 8 weeks can be expedited by soaking the seeds in water for 24 hours. Alternatively, a 10% solution of gibberellic acid has been found to yield 100% germination. Sowing seeds can be done either *in situ* or in a nursery and transplantation is recommended when the seedlings have no more than four leaves. Transplanting becomes challenging with advanced seedlings due to their long and delicate taproot (Morton, 1987).

Efforts to propagate jackfruit through budding and grafting have historically encountered challenges though modified Forkert method of budding emerges as a promising option. This technique allows for the use of either jackfruit or chempedak seedlings as rootstocks on which grafting process can be undertaken at any time of the year (Morton, 1987).

Softwood grafting employing the cleft method on 2-month-old seedlings with scions aged 3-4 months give 70-80% success. Given the low viability of jackfruit seeds it is imperative to sow them promptly after extraction to raise viable rootstocks. 'Rudrakshi' and *A. hirsuta* are utilized as alternative rootstocks in certain cases (Priya Devi *et al.*, 2014).

Inarching is also practiced but it shares the challenge of ensuring successful transplanting post-separation from the scion parent. To address this, air-layers produced with growth-promoting hormones are being distributed in India which consistently yield early-bearing fruits of known quality. In Florida, the successful rooting of cuttings from young wood under mist conditions has been reported. Calcutta University has achieved successful rootings using forced and etiolated shoots treated with IBA, preferably at a concentration of 5,000mg/l and kept under mist (Morton, 1987).

For commercial purposes, inarching on 10-month-old jack seedlings is employed to produce vegetative progenies. Grafts subjected to this method start bearing fruits within 4½-5 years ~ a notable advancement compared to seedlings which typically take 7-8 years to bear fruits under conventional cultivation practices (Priya Devi *et al.*, 2014).

Moreover, ongoing experiments in tissue culture have been conducted at IIHR, Bengaluru suggesting a continued interest in advancing jackfruit propagation techniques. This comprehensive approach to propagation encompasses various methods each with its own set of advantages and challenges providing horticulturists with multiple options for cultivating jackfruit (Morton, 1987).

2A.25.5 Cultivation practices

Optimal cultivation practices play a crucial role in fostering the growth and productivity of jackfruit trees.

2A.25.5.1 Planting and initial care

Digging pits with precision each measuring one cubic metre serves as the initial step in the cultivation process. These pits strategically spaced at intervals of 6-8 meters become reservoirs for a meticulously concocted mixture of topsoil and 10kg FYM. The opportune period for planting seeds, seedlings or grafts within these pits spans from June to September. Ensuring the robust establishment of the planted material involves more than just the act of planting. To prevent lodging and the subsequent risk of breakage at graft joints, proper staking is imperative. Post-planting care becomes an integral component of successful cultivation. Thatching or the provision of shade using coconut leaves coupled with comprehensive mulching of basins after the monsoon season concludes, emerges as a crucial practice. These measures as outlined by Priya Devi *et al.* (2014) safeguard against potential mortality of seedlings or grafts.

In the context of plantation settings the strategic spacing of trees becomes a pivotal consideration in maximizing growth potential. Morton (1987) recommends a distance of 9-12 metres between trees for optimal results. Young plantings demand extra attention against sunscald, *vis-a-vis* guarding against the interference of grazing animals such as hares and deer. Presence of rats introduces another challenge, requiring the implementation of protective measures to safeguard seeds in the field as suggested by Morton (1987). This comprehensive approach ensures not only the establishment of trees but also their sustained growth and resilience in the face of potential challenges.

A notable horticultural technique detailed in Morton's seminal work of 1987 referencing Firminger's account involves nurturing seedlings within bamboo tubes measuring approximately 90cm-1.2m in height. This method entails bending the pliable stem of the seedling skillfully coiling it beneath the soil and leaving only the tip exposed. Remarkably, after five years the resulting plant is known to yield substantial and exquisite fruits while maintaining a distinctive spiral growth pattern beneath the surface of earth.

In Travancore an alternative method involves burying the whole fruit. Multiple seedlings emerging from this process are bound together with straw so all the seedlings gradually fusing into one entity that begins bearing fruit in 6-7 years. Though typical gestation period for seedlings ranges 4-14 years certain precocious cultivars may commence bearing in as little as 30 to 42 months (Morton, 1987).

With a potential lifespan of up to 100 years, productivity tends to decline with age. In Thailand, a recommended practice involves planting alternate rows every 10 years, facilitating the routine removal of 20-year-old trees and their replacement with a new generation. Despite the tree's longevity, attention to fertilizer requirements is crucial. In India, severe symptoms of Mn deficiency have been observed underscoring the importance of comprehensive nutrient management in jackfruit cultivation (Morton, 1987).

Cultivating jackfruit involves a series of post-harvest practices aimed at maximizing the yield of this tropical fruit. One essential technique is the strategic pruning of fruiting twigs which can be selectively cut back to the trunk or branches to stimulate flowering in the subsequent season. In the Cachar district of Asom a distinct method involves the use of a hatchet to slash the tree promoting the emergence of shoots from the wounds. This practice is believed to enhance production of female flowers. A periodic practice in the region involves the regular lopping of branches every 3-4 years to sustain and optimize fruitfulness (Morton, 1987).

Diverse approaches to post-harvest management exist as evidenced by researches conducted at the University of Kalyani (West Bengal). Contrary to certain traditional methods, the studies revealed that neither scoring nor pruning of shoots has a significant impact on increasing fruit set. Surprisingly, the results indicated that ringing ~ a technique involving the removal of a ring of bark around the tree enhances fruit set exclusively in the first year with production showing a decline in the subsequent year (Morton, 1987).

The cultivation of jackfruit involves a nuanced understanding of regional practices and scientific findings where traditional methods such as slashing with a hatchet may coexist with more modern approaches like ringing all in the pursuit of optimizing fruit production over multiple seasons (Morton, 1987).

2A.25.5.2 Seedling treatment

To enhance shoot growth in one-month-old seedlings, a method involving application of 25-200ppm GA solution has proven effective. This treatment not only stimulates shoot development but also promotes root growth when applied through sprays and paste methods (Morton, 1987).

2A.25.5.3 Training

To cultivate jackfruit trees successfully it is recommended to train them into a single-stem structure. Early side-branches should carefully be pruned to encourage development of uniform and smooth trunk for a height of 1.5-2 meters. Once this height is achieved side branches should be allowed to emerge. It is crucial to keep the trunk free from vegetative growth as the flower buds essential for fruiting appear directly on the trunk (Priya Devi *et al.*, 2014).

2A.25.5 .4 Manuring

The fertilizer requirements for jackfruit cultivation depend on various factors such as soil conditions, age of the tree and local climate. However, here are some general guidelines for fertilizing jackfruit trees:

i) **NPK fertilizer:** Jackfruit trees benefit from a balanced fertilizer containing N, P & K. A fertilizer with an NPK ratio of 8-4-2 or 6-6-6 is commonly recommended. Nitrogen is crucial for vegetative growth, phosphorus supports root development and flowering while potassium aids in overall tree health *cum* fruit development.

ii) **Micronutrients:** In addition to primary nutrients (NPK), jackfruit trees require various micronutrients for optimal growth. These include iron, zinc, manganese, copper, boron and molybdenum. It is recommended to conduct soil tests to determine the specific micronutrient needs of the soil.

iii) **Organic matter:** Incorporating organic matter into the soil is essential for improving soil structure, water retention and nutrient availability. Well-rotted compost or manure can be applied around the base of the tree but care should be taken not to place it directly against the trunk.

iv) **Application timing:** Fertilizers should be applied during the growing season typically in the spring and early summer. It is advisable to divide the total annual fertilizer requirement into multiple applications to avoid over-fertilization.

v) **Mulching:** Mulching around the base of the jackfruit tree helps conserve soil moisture, suppressing weeds and gradually releasing nutrients as the organic mulch decomposes. This can be particularly beneficial in areas with fast-draining soils.

vi) **Foliar feeding:** Some growers also use foliar sprays containing micronutrients to supplement the nutritional needs of the tree. Foliar feeding involves spraying a nutrient solution directly onto the leaves so allowing for rapid nutrient absorption.

vii) **Adjustments based on soil tests:** Conducting regular soil tests is recommended to assess nutrient levels and pH. Based on the results, adjustments can be made to the fertilizer regimen to address specific deficiencies or imbalances.

Priya Devi et al. (2014) recommends application of nutrients as given below:

Manure/nutrients	*One year after planting(kg/plant)*	*Annual increase (kg/plant)*	*5th year onwards (kg/plant)*
FYM	10 kg	10 kg	50 kg
N	0.15	0.15	0.75
P	0.08	0.08	0.40
K	0.10	0.10	0.50

Fertilizers are typically administered during rainy season. In instances where irrigation is accessible, it is advisable to divide the fertilizer application into two doses allowing for application twice a year, specifically in June–July and

September–October. To optimize their effectiveness, manures and fertilizers can be strategically applied within a circular trench positioned approximately 50-60cm away from the base of the trunk. This method enhances nutrient absorption and promotes the overall health and growth of the plant (Priya Devi *et al.*, 2014).

It is important to note that fertilizer requirements can vary depending on factors such as soil type, climate and the age of the tree. Consultations with local agricultural extension services can provide more specific recommendations tailored to your particular growing conditions.

2A.25.5.5 Irrigation

Jackfruit trees generally thrive in tropical and subtropical climates where these can receive a consistent water supply. Adequate irrigation is crucial especially during critical stages of growth, flowering and fruit development. Specific irrigation requirements for jackfruit can vary based on factors such as soil type, climate and tree age. Here are some general guidelines:

i) **Establishment phase:** During first year of planting young trees require regular watering to help establish a strong root system. Watering frequency may vary based on soil and climatic conditions but in general young trees benefit from consistent moisture. It should be thorough and deep watering rather than frequent shallow watering.

ii) **Flowering and fruit development:** Jackfruit trees need ample water during flowering and fruiting to support the development of the large fruits. Maintaining a regular watering schedule ensures consistently moist soil though not waterlogged. Drip irrigation or soaker hoses can be effective in providing a slow *cum* steady water supply directly to the root zone.

iii) **Mature trees:** Once jackfruit trees are established these generally tolerate drier conditions better than during the establishment phase. However, during prolonged dry periods, mature trees still benefit from supplemental watering to ensure over all tree health and healthy fruit development.

iv) **Soil type:** Soil type influences the irrigation needs of jackfruit trees. Well-draining soils are important to prevent waterlogging which leads to root rot. Sandy soils may require more frequent watering while clayey soils may retain moisture for a more extended period.

v) **Rainfall considerations:** In regions with distinct wet and dry seasons, adjustments to the irrigation schedule may be necessary. During dry season or periods of insufficient rainfall, supplemental irrigation becomes especially important.

vi) **Mulching:** Applying a 10-15cm thick layer of organic mulch around the base of the tree helps retain soil moisture, reduction in weed competition and regulation of soil temperature.

It's important to note that while jackfruit trees require regular watering, they are also sensitive to waterlogged conditions. Balancing soil moisture is crucial to prevent both drought stress and waterlogged roots. Observing the specific conditions of the planting site and adjusting irrigation practices accordingly will contribute to the health and productivity of jackfruit trees.

2A.25.5.6 Intercultivation

In the initial pre-bearing stage of jackfruit trees it is conducive to cultivate pulses as an intercrop. Spreading of dry leaves beneath the trees serves the dual purpose of acting as a natural mulch as well as aiding in the conservation of moisture. As the tree matures and the trunk undergoes growth, active buds of the female inflorescence emerge not only from the trunk but also from the primary scaffold branches. This distinctive bearing habit, characterized by the development of flowers and fruits directly from the trunk and branches is referred to as the 'cauliflorous' pattern (Priya Devi *et al.*, 2014).

2A.25.6 Harvest

Jackfruit typically begins to bear fruits around 7-8th year under natural conditions. However, grafted plants exhibit an earlier onset of yield usually starting from 4th to 5th year. Singapore jack stands out with even seedlings showcasing fruit-bearing capabilities as early as the 3rd year. Regular fruiting season spans from March to June while at elevated regions the harvest may extend up to September. Certain genotypes in plains demonstrate the ability to bear an off-season crop during October-December. This variability in fruiting periods allows for a more extended availability of jackfruit catering to different climatic conditions and elevations (Priya Devi *et al.*, 2014).

The harvesting of jackfruit is a crucial process that requires careful consideration of various maturity indicators. Fruits reach maturity usually 3 to 8 months after flowering. In certain regions such as Jamaica, a strategic 'X' incision at the apex of the fruit is employed to expedite ripening and enhancing flavour (Morton, 1987).

Maturity signs for jackfruit are multi-faceted and involve a comprehensive assessment of several indices:

i) **Maturation timeline:** The fruit generally matures within 12-16 weeks after flower anthesis.

ii) **Colour transformation:** The fruit undergoes a noticeable change in colour transitioning from green to green-yellowish hues.

iii) **Acoustic cues:** A mature jackfruit emits a dull hollow sound when tapped by the finger.

iv) **Spinel development:** The spinel of the fruit becomes well developed and exhibits widened spacing.

v) **Leaf transformation:** The last leaf of the peduncle turns yellow.

vi) **Aroma presence:** An aromatic odour becomes evident as the fruit ripens.

vii) **Skin characteristics:** The number of spikes on the outer skin decreases and the remaining spikes become flatter.

viii) **Size changes:** Both fruit length and girth increase as a sign of maturity.

For optimal harvesting, it is recommended to employ sharp-bladed equipment to cut the fruit from the stalk. In cases where fruit is positioned high on the tree a sack should be secured around the fruit with a rope. Subsequently, the stalk can be carefully severed allowing the fruit to be gently lowered to the ground.

Accurate determination of maturity coupled with precise timing for harvesting and adherence to correct harvesting practices ensures minimal loss of fruits. This comprehensive approach not only maximizes the yield but also contributes to the overall quality of the harvested jackfruit.

2A.25.7 Yield

The jackfruit yield in India is considered substantial with an optimal harvest averaging around 150 large fruits per tree annually. Remarkably, certain trees exhibit an even more prolific yield boasting up to 250 fruits each year. Fully mature jackfruit trees have been known to achieve an exceptional production rate of 500 medium or smaller fruits annually. This variation in yield underscores the diverse potential of jackfruit cultivation, contributing to its significance in the agricultural landscape of the region (Morton, 1987). The fruit weight varies from 10 to 30kg (Priya Devi *et al.*, 2014).

2A.25.8 Storage

The storage of jackfruit poses a challenge due to its tendency to rapidly turn brown and undergo deterioration once it reaches the ripened stage. To address this concern extensive cold storage trials have been conducted, revealing a promising solution for prolonging the shelf life of ripe jackfruits. Ripe jackfruits can effectively be stored for an extended duration of 3-6 weeks. The optimal storage conditions involve maintaining a controlled temperature range of 11.11°-12.78°C coupled with a specific relative humidity level ranging from 85-

95%. These carefully regulated environmental parameters play a crucial role in mitigating the natural tendency of jackfruits to deteriorate quickly, allowing for a significantly extended storage period without compromising the fruit quality (Morton, 1987).

This refined approach to jackfruit storage not only enhances the preservation of its freshness but also provides valuable insights into the intricacies of temperature and humidity control for optimal fruit storage. It stands as a practical solution for farmers, distributors and retailers seeking to maximize the market availability of ripe jackfruits while minimizing post-ripening losses.

2A.26 Nutrient Defeciencies

Nutrient deficiencies in jackfruit can lead to various symptoms affecting the overall health and productivity of the tree. Here are some common nutrient deficiencies in jackfruit and their associated symptoms:

i) **N-deficiency:** Yellowing of older leaves starting from the tips and spreading towards the base, overall stunted growth and reduced fruit development. Application of nitrogen-rich fertilizers, organic matter or compost is advised. Regular fertilization practices should be implemented.

ii) **P-deficiency:** Dark green leaves with a purplish tint, poor root development and delayed flowering *cum* fruiting. In this case, phosphorus-containing fertilizers should be used. In the soil, organic matter should be enhanced and root health should be improved.

iii) **K-deficiency:** Marginal leaf chlorosis (yellowing) and necrosis (death) starts from leaf edges, fruit size *cum* quality is reduced and susceptibility to pests and diseases is increased. Potassium rich fertilizers should be applied, *vis-à-vis* proper soil pH should be maintained along with proper drainage facilities.

iv) **Mg-deficiency:** Interveinal chlorosis on older leaves is observed, leaf margins may become necrotic and leaf curling may also be observed. Magnesium-containing fertilizers or amendments such as epsom salts should be applied along with adjustment of proper soil pH.

v) **Ca-deficiency:** New leaves exhibit distorted growth along with tip burn, occurrence of blossom end rot is observed and fruits are of poor quality. Calcium-containing fertilizers, lime or gypsum should be applied and proper soil pH should be maintained.

vi) **Cu-deficiency:** In its deficiency, wilting, dieback of shoot tips and chlorosis in young leaves occur. Copper-containing fungicides or fertilizers should be applied apart from ensuring proper soil pH.

vii) **S-deficiency:** Due to its deficiency uniform yellowing of young leaves, stunted growth, delayed flowering and reduced fruit yield occur.

viii) **Fe-deficiency:** This causes overall growth retardation along with reduced leaf size and interveinal chlorosis in young leaves while veins remain green. Application of iron chelates or foliar sprays containing iron may improve plant health. However, soil drainage should also be improved.

ix) **Zn-deficiency:** Its deficiency causes leaf distortion *cum* mottling, shortened internodes, small & narrow leaves and reduced fruit development. It should be ensured that soil pH is proper. Its deficiency can be improved by applying zinc-containing fertilizers or foliar sprays.

x) **Mn-deficiency:** Its deficiency causes interveinal chlorosis similar to iron deficiency, leaf mottling and poor fruit development. Application of Mn-containing fertilizers or foliar sprays should be followed and proper pH of the soil should be maintained.

xi) **B-deficiency:** Its deficiency causes terminal bud dieback, distorted or misshapen leaves and cracking or occurrence of corky tissues in fruits. Application of boron-containing fertilizers should be used but not in excess otherwise B-toxicity may occur.

It is important to mention that visual symptoms alone may not always accurately diagnose nutrient deficiencies as symptoms can sometimes overlap or be influenced by other factors such as pH imbalances, diseases or environmental stress. It is advisable to conduct a soil test to determine nutrient levels accurately. Based on the test results one can use various nutrients to address specific deficiencies.

2A.27 Physiological Disorders

Physiological disorders in jackfruit can arise from non-pathogenic factors often related to environmental conditions, cultural practices or due to genetic factors. Here are some common physiological disorders observed in jackfruit:

- **Fruit cracking:** It occurs due to rapid changes in humidity and moisture levels during fruit development. It is symptomised due to vertical or radical cracks on fruit surface. At the stage of fruit development there should be consistent watering to the plants.
- **Blossom drop:** It occurs due to stress at flowering mainly due to adverse weather conditions such as high temperatures, *vis-a-vis* insufficient pollination. In this case premature blossom dropping occurs before fruit formation. It can be prevented by ensuring proper pollination and by maintaining optimum growing condition.

- **Fruit drop:** It is premature dropping of immature fruits due to nutrient imbalances, nutrient deficiencies and environmental stresses. To get rid of this problem the trees should be managed properly through nutrient adjustment, pest control and timely irrigation.
- **Fruit blackening:** In this case fruit skin starts blackening or darkening. It occurs due to high humidity or excess moisture during fruit developmental stage. This can be managed through proper tree spacing, improving ventilation and by providing controlled irrigation.
- **Sunburn:** This is discoloured or burnt areas on fruit surface occurring due to excessive exposure of fruits to direct sunlight especially during hot and dry periods. At developing stage the fruits are shaded or protected with reflective materials.
- **Internal browning:** Browning of internal fruit tissues occurs due to Ca-deficiency or irregular watering which can be managed through proper soil management by applying calcium and proper irrigation.
- **Uneven ripening:** Irregular ripening in a fruit can be managed through ensuring uniform growing conditions and harvesting of fruits at proper maturity stage.
- **Corky fruit tissue:** Formation of corky or woody tissues in the fruits occurs due to boron deficiency therefore boron should be applied in the soil.
- To effectively manage physiological disorders in jackfruit, it is essential to maintain optimal growing conditions, implement proper cultural practices and addressing any nutritional deficiency promptly. Regular monitoring and timely intervention can help mitigate the impact of these disorders.

2A.28 Pests and diseases

2A.28.1 Insect-pests

Jackfruit ~ a fruit of considerable economic importance in India and South China confronts formidable challenges posed by a variety of insect-pests that can significantly impede its growth and overall yield. In India, noteworthy among these insect-pests are the shoot-borer caterpillar (*Diaphania caesalis*), mealybugs (*Nipaecoccus viridis*, *Pseudococcus corymbatus* & *Ferrisia virgata*), the spittle bug (*Cosmoscarta relata*) and jack scale (*Ceroplastes rubina*). The most destructive and widespread threats come from bark borers, particularly *Indarbela tetraonis* and *Batocera rufomaculata*. Other significant pests attacking this tree are stem and fruit borer (*Margaronia caecalis*) and the brown bud-weevil (*Ochyromera artocarpio*).

In the southern regions of China, the larvae of longicorn beetles such as *Apriona germarri*, *Pterolophia discalis*, *Xenolea tomenlosa asiatica* and *Olenecamptus bilobus* pose a serious threat by causing substantial damage to the fruit stem. The impact of leaf webbers' caterpillars (*Perina nuda* & *Diaphania bivitralis*) is relatively minor as are infestations by aphids such as *Greenidea artocarpi* & *Toxoptera aurantii*, *vis-a-vis* thrips such as *Pseudodendrothrips dwivarna* (Morton, 1987). The intricate interplay between jackfruit and these pests underscores the intricate balance that must be maintained for successful cultivation. It necessitates a comprehensive understanding of the various insect species, their life cycles and the adoption of effective pest management strategies to ensure the sustainable growth and productivity of jackfruit crops.

Adopting integrated pest management (IPM) practices including cultural, biological and chemical control methods is crucial for effectively managing these pests and preserving the health and productivity of jackfruit trees. Regular monitoring and timely interventions are key components of a successful pest management strategy as furnished below (Butani, 1978; Morton, 1987; Vikaspedia, 2023),

2A.28.1.1 Shoot-borer caterpillar (*Diaphania caesalis*)

This borer undergoes a comprehensive life cycle that includes distinct stages such as egg, larva, pupa and adult. The intricacies of this life cycle are particularly pronounced during flowering and fruiting periods where females exhibit increased egg-laying activity and accelerated development. Visually identifiable by its reddish-brown hue adorned with conspicuous black spots, the caterpillar's destructive tendencies manifest as it bores into delicate shoots and emerging fruits, occasionally resulting into significant damage. The adult stage of this pest is characterized by a pale-brown hue adorned with a plethora of dark brown spots and orange stripes delicately outlined in dark brown embellishing each wing. This borer can be controlled by cutting and destroying the affected shoots. Its natural enemies such as *Bracon* sp., *Chelonus* sp. and *Systasis* sp. can be introduced on trees to parasitize this pest. Spraying of insecticides such as 0.1% carbaryl or 0.03% phosphamidon or dusting with 0.2% BHC/DDT will control this borer.

2A.28.1.2 Mealybugs (*Nipaecoccus viridis*, *Pseudococcus corymbatus*, *Ferrisia virgata*)

Mealybugs undergo a comprehensive life cycle encompassing distinct stages of egg, nymph and adulthood often accompanied by secretion of honeydew. The reproductive process involves females laying eggs directly on the host in a fluted ovisac attached to the adult female's body. Inseminated eggs give rise to hermaphrodites while uninseminated eggs result only in males. The initial nymph stage (crawlers) is mobile and settles on plants initiating sap extraction and the formation of colonies. Interestingly, females function as hermaphrodites and frequently self-inseminate while adult males engage in mating. The adult bugs are coated in a whitish powder and these along with nymphs colonise under the bark, young shoots, in the panicles and on emerging shoot tips which causes floral drop. These also excrete honeydew which causes development of sooty mould. Their infestation causes yellowing, wilting and sooty mould infection. By adopting stringent plant hygiene practices, use of *neem* oil and insecticidal soap solutions, *vis-a-vis* introduction of natural enemies such as *Menochilus sexmaculatus, Rodolia fumida* and *Cryptolaemus montrozieri* will minimise their population drastically. Spraying of 0.05% dichlorvos (DDVP), fenitrothion or carbophenothion will not only control mealy bugs but also the scale insects.

2A.28.1.3 Spittlebug (*Cosmoscarta relata*)

Pine spittlebug follows a distinctive life cycle that spans from overwintering as eggs to maturing into winged adults. During winter, the eggs are nestled within dead twigs or lodged in slits carved into the bark of living stems. With onset of summer, *i.e.* May the eggs hatch giving rise to young nymphs that gravitate towards the new supple growth. While feeding, these nymphs generate protective spittle masses, serving as a shield against predators, parasites and arid conditions. As the spittle descends onto lower branches it may attract black sooty mould which can cover the needles. The nymphs exhibit a colouration shift from yellowish with black markings in their early stages to a deeper brown as they mature. These nymphs attain maturity by July and by undergoing moulting these transform into winged adults. Their infestation causes stunted growth. To get rid of this pest, the spindle masses should be removed, base around the tree should regularly be dug up and watered through summer until winter, alkathene banding around the tree base should be done by March end to discourage climbing of nymphs up the tree and its predators or parasitoids should be introduced on such trees to predate over these bugs. Targeted insecticides such as 0.04% diazinon or monocrotophos during its peak infestation should be sprayed or soil application of 5% aldrin, chlordane, heptachlor dust (300-500g/tree) will be quite effective if applied just before start of winter.

2A.28.1.4 Jack scale (*Ceroplastes rubina*)

Scale is a plant pest with a life cycle encompassing distinct stages, *viz.* egg, nymph and adult typically characterized by the presence of a protective scale covering. Identifying this infestation is crucial as it manifests through the presence of scales on both leaves and stems, leading to adverse effects such as yellowing and weakening of the host plant. Heavy infested areas are drastically pruned. Parasitoids can be introduced to feed on these scales. Horticultural oil can be applied to eliminate its population. The affected organs can be sprayed with 0.05% dichlorvos (DDVP), fenitrithion or carbophenothion.

2A.28.1.5 Bark borers (*Indarbela tetraonis, Batocera rufomaculata*)

Bark borers exhibit a distinctive life cycle characterized by their impact on host trees. Its females lay about 2000 eggs in clusters of 15-20 on host tree bark. These eggs hatch in 8-10 days featuring dark brown heads and dirty brown bodies. These larvae reach full size by December and continue feeding until April when these pupate, each pupa measuring 1.6-2cm long and being reddish-brown, stout and equipped with two rows of spines on each abdominal segment. Pupal period lasts for 21-31 days. Upon adulthood, these pests present as pale brown with dark brown heads and thoraxes, while their abdomens and forewings exhibit a pale reddish-brown hue adorned with numerous dark brown bands. The larvae's mode of infestation involves boring into the bark, causing considerable damage during their developmental stages. Their feeding is recognizable by entry and exit holes on the bark, sawdust-like frass and weakened branches. Newly hatched larvae initially nibble on the tree's bark, subsequently boring into the same after 2-3 days, disrupting the sap flow and resulting in poor growth and reduced fruiting. Silk webs composed of excreta and chewed wood particles are commonly observed at the junction of branches on infested trees. Infested branches should be removed, and after cleaning of the affected trunk or branches cotton wool swab soaked in a potent fumigant such as chloroform, carbon bisulphide or petral or kerosene oil or a mixture of ethyl glycol and kerosene (3:1) is quite effective. However, after inserting the fumigant, the holes should be mud-sealed.

2A.28.1.6 Stem and fruit borer (*Margaronia caecalis*)

Stem *cum* fruit borer exhibits a distinctive and potentially destructive lifecycle that poses a significant threat to trees as well as to fruits. This species is characterized by the female's unique behaviour of cutting into tree bark to lay up to 200 brownish-white eggs individually, each measuring 6.2mm. After hatching the tiny larvae start boring into the trunk and branches where these feed for up to two years making about 2-3cm wide tunnels which interfere with sap flow so foliage and fruit production is badly affected. The adult beetles come out through a zigzag short tunnel leading to circular exit holes. Adults can survive for up to eight months. Its presence is felt through entry holes, frass (excrement) and damage to stems and fruits. Prompt pruning of infested branches, use of parasitoids such as *Bracon brevicornis* and *Brachymeria euploeae* along with predators such as *Chrysoperla zastrowisillemi,* ladybird beetle, reduviid bugs, spiders, fire ants, robber flies, black drongos (king crows), common mynhas, big-eyed bugs (*Geocoris* sp.), earwigs, ground beetles, pentatomid bugs (*Eocanthecona furcellata*) and preying mantises play crucial role in controlling population of the borer and maintaining ecological balance.

2A.28.1.7 Brown bud-weevil (*Ochyromera artocarpi*) and leaf-eating weevil (*Onychocnemis careyae* & *Teluropa ballardi*)

The brown bud-weevil poses a significant threat to jackfruit cultivation due to its life cycle and feeding habits. This weevil follows a well-defined reproductive and developmental pattern that warrants careful attention for effective control measures. The life cycle of this pest begins with the deposition of smooth, pearly white and oblong-oval eggs measuring some 0.42mm long × 0.28mm wide in bud tissues which hatch in 3-4 days. Emerging pale-white larvae with pale-brown head are initially legless that grow afterwards 0.56mm long. Over the next 12 to 15 days the larvae bore through the tissue in various directions growing to 4-9mm and of whitish colour. Its head is light brown but with darker tone at the frontal region. Its head is quite narrower than the thorax. Pupal stage lasts 5 to 6 days. Adult weevil is a small and active greyish-brown insect measuring 3.5mm in length and 1mm in breadth, possessing a distinctive appearance. Its body is thinly clothed with fine setiform golden scales and set with suberect setae while the elytra bears numerous irregular and ill-defined small bare spots. The adult weevils are often found in groups feeding on inflorescence tissues.

The small whitish grubs of brown bud-weevil infests by boring tender floral buds and fruits, distorts the growth and causes reduced flowering so dropping of damaged organs occurs prematurely. Adult weevils further exacerbate the issue by nibbling on leaves. To get rid of their damage, infested buds should be pruned and destroyed, *vis-a-vis* contact insecticides should be sprayed at 0.2%.

2A.28.1.8 Longicorn beetles and larvae (*Apriona germarri*, *Pterolophia discalis*, *Xenolea tomenlosa asiatica*, *Olenecamptus bilobus*)

Longicorn beetles and their larvae are known for their destructive impact on fruit-bearing plants. The larvae of these beetles bore into fruit stems exhibiting entry and exit holes so inflicting significant damage during crucial stages of plant development as is noticeable through stem wilting and premature fruit drop. Removing affected plant parts helps prevent further damage and limits the reproductive success of the pests. Parasitic wasps or predatory beetles should be introduced on infested trees to control this pest. Spraying with 0.1% carbaryl will control these beetles and their larvae.

2A.28.1.9 Leaf webbers (*Perina nuda*, *Diaphania bivitralis*)

Leaf webber females lay minute white eggs measuring some 0.4-0.6mm in width and 0.8mm in length, ranging from spherical to flattened and these eggs turn to the yellow hue in 24 hours. The eggs are laid in the groups of 2-7 on buds, flowers and other quick growing parts of host plants. A female lays some 300-400 eggs. In about four days these eggs hatch and in a fortnight these larvae attain all the five instars, finally with 2.5cm length. Larvae engage in web-spinning activities on leaves during their feeding phase. As larvae approach pupation, a noticeable transformation occurs with the colour changing to a dark copper shade. Typically pupation occurs under a folded leaf, often utilizing dead or dry material and the stage lasts some 8-9 days. The elongated pupa is about 1.3cm in length and 4mm in width displaying a colour range from a light to dark brown and tapering to points at both the ends. Moth emerges 3-5 hours after sunset, followed by peak flight at midnight. For their control, webs and caterpillars are removed manually, proper sanitation is maintained, young developing fruits are covered with perforated alkathese bags, natural enemies are introduced and 0.1% Carbaryl or 0.05% Phosalone are sprayed.

2A.28.1.10 Aphids (*Greenidea artocarpi*, *Toxoptera aurantii*) and thrips (*Pseudodendrothrips dwivarna*)

Aphids, specifically *Greenidea artocarpi* and *Toxoptera aurantii*, along with thrips *Pseudodendrothrips dwivarna* collectively exhibit a simple metamorphic life cycle comprising of white distinct eggs deposited along the leaf veins, 4 nymphal stages (instars) in 8-9 days and adult stages. Aphids being soft-bodied pear-shaped small creatures equipped with cornicles (wax-secreting tubes) projecting from the 5th or 6th abdominal segment. The wingless female aphids exhibit varying shades of yellowish green, gray green or olive green coated with a white waxy bloom. The winged female adults showcase a dusky green abdomen distinguished by dark lateral stripes separating body segments and dusky wing veins. Male aphids display an olive-green to brown colouration. Generally, aphid attacks occur during 2nd and 3rd weeks of December, persisting until March.

The deleterious effects of aphid and thrips infestations manifest through sap-sucking activities resulting in the yellowing, curling and distortion of leaves. Both nymphs and adults extract sap from leaves, buds and newly emerging fruits, *vis-a-vis* inducing leaf curling. Its infestation may cause stunted growth *cum* plant withering, development of sooty mould, stunted growth and finally withering and death of the plant. For effective control, a multi-faceted approach is recommended, encompassing insecticidal soaps, neem oil, natural predators and meticulous plant hygiene practices. Spraying with 2% *mahua* oil, 0.03% Dimethoate or 0.05% Monocrotophos will kill these pests. Sprayed fruits should be used only after two weeks of spraying.

2A.28.1.11 Castor capsule borer (*Dichocrocis punctiferalis*)

Castor capsule borer is a significant pest of castor plants though feeds on fruits also of citrus, guava, jackfruit, mango, mulberry, peach, pear, plum, pomegranate, tamarind as well as ginger. Female moths lay rough-textured oval and pink eggs measuring 0.5mm in diameter primarily on tender parts of host plant. The dark pinkish-brown caterpillars having spiny warts covering the abdomen reach to a length of 2.5-3.5cm. The caterpillars burrow into buds and young fruits eating inside and plugging the bored holes with excreta. The moths feature a brownish-yellow colouration with numerous black dots on all the four wings. These complete their life cycle in 4-5 weeks (eggs for 6-7 days, larvae 12-16 days and pupae 7-10 days) during summer though a little longer in winter. To mitigate the infestation, affected buds and fruits should be sheared off and destroyed. Natural enemies such as *Bracon brevicornis* and *Brachymeria euploeae* as well as predators such as *Chrysoperla zastrowisillemi, Geocoris* sp. (big-eyed bugs), *Eocanthecona furcellata* (pentatomid bugs), ladybird beetles, reduviid bugs, earwigs, ground beetles, preying mantises, fire ants, robber flies, spiders, common mynhas, black drongos (king crows), *etc.* may be introduced.

2A.28.2 Diseases

Significant to jackfruit cultivation are pink disease [*Pelliculana* (*Corticium*) *salmonicolour*], stem rot, fruit rot and male inflorescence rot all attributed to the pathogenic influence of *Rhizopus artocarpi*. Leafspot diseases are *Phomopsis artocarpina, Colletotrichum lagenarium, Septoria artocarpi* and certain others of minor importance. Jackfruits are also affected with gray blight (*Pestalotia elasticola*), charcoal rot (*Ustilana zonata*), collar rot (*Rosellinia arcuata*) and rust (*Uredo artocarpi*) but to certain specific regions (Morton, 1987).

2A.28.2.1 Soft rot or fruit rot

Soft rot (fruit rot) poses a significant threat to jackfruit cultivation. This disease targets primarily male inflorescences and young fruits leading to a considerable reduction in fruit maturity. Female inflorescences and matured fruits are generally spared from the disease. Recognizing the disease symptoms, understanding survival *cum* spread mechanisms and identifying favourable conditions for its development are crucial for effective management. It is often caused by various fungi and bacteria that infect the fruit leading to its decay. The specific pathogens responsible for fruit rot in jackfruit can though vary but most common ones include *Colletotrichum gloeosporioides* (anthracnose), *Fusarium* spp., *Rhizopus artocarpi, Erwinia* spp. and *Botryobasidium salmonicolor*. In this case a substantial number of affected fruits fall off the tree. Initially the pathogen appears as grayish growth but afterwards turning into dense black growth and causing the inflorescence or fruit to rot and falling, especially during warm, humid and rainy seasons. To manage this problem, there should be proper sanitation in the orchard, provision should be made to permit proper air circulation inside the trees by giving proper spacing, through regular pruning followed by 0.3% application of copper oxychloride after rainy season and reducing humidity. Carbendazim/thiophanate methyl/Bitertanol/Prochloraz at 0.1% against anthracnose should be applied during rainy season. Moreover, proactive approaches should also be followed by using resistant varieties.

2A.28.2.2 Dieback

Jackfruit dieback is a serious concern for cultivators, characterized by distinctive symptoms that manifest in the plant's bark, twigs and leaves. Dieback disease in jackfruit is typically caused by a combination of factors including various fungi and environmental stressors as over 80% RH and 25-31°C prevailing atmospheric temperatures are more conducive in the spread of this disease. One of the common pathogens associated with dieback symptoms is the fungus *Phytophthora*. Several *Phytophthora* species such as *P. palmivora* have been reported to be involved in causing dieback in jackfruit. *Phytophthora* species are known as water moulds soil-borne pathogens causing root rot and dieback. These pathogens thrive in waterlogged soil conditions causing symptoms such as wilting, branch dieback, gummosis and overall decline in tree health. Poor soil drainage, improper cultural practices and environmental stresses contribute to the development and severity of the disease. Soil management, proper cultural practices and fungicidal treatments are only the way out of this dieback. Proper spacing *cum* air circulation, adequate nutrition, needbased irrigation, prompt pruning of infected parts and proper disposal & isolation of infected plants are necessary. Copper fungicides should be pasted on cut ends and 0.3% copper oxychloride should be sprayed after the rain is over.

2A.28.2.3 Leaf spot

Jackfruit is susceptible to various diseases among which leaf spot caused by *Phyllostricta artocarpina* is of significant concern. Occurrence of light brown with double margin dark brick-red spots on leaves cause complete defoliation on the trees. On maturation, these spots transform into a greyish appearance adorned with small, dark-coloured and pin-headed fruiting bodies of the causative fungus. Though initially this fungus targets primarily young and rapidly expanding leaves but overwinters on old fallen leaves for further infections. A temperature around 25°C and RH of 95-87% are quite conducive in its spread. This problem is mitigated through combined use of cultural, preventive and chemical uses. There should be proper aeration among and inside the trees, infected leaves should be removed and burnt and only balanced fertilization should be used. Disease resistant varieties if available should be used. Good sanitation practices should be adopted. Kavach or mancozeb at 0.2% spraying on clear days during rainy season should be used.

2A.28.2.4 Rust

Rust disease surviving through teliospores (thick-walled resting spores) poses a significant threat to jackfruit planting particularly during late summer and these spores survive on fallen leaves or on soil surface though spread through wind-borne uredospores. The severity of the rust can lead to rapid and recurrent defoliation so impacting overall tree growth and yield. Early summer defoliation prompts new growth that becomes vulnerable to damage from early frosts though fall defoliation induces early dormancy which protects trees from potential frost damage. Temperature range of 25.5-30.5°C coupled with 86-92% RH is conducive to its high intensity. Initially it invades upper leaf surface as small yellowish spots which later on evolve into reddish-brown lesions though retaining the smooth texture. Later on the lesions on lower leaf surface take on a reddish-brown hue with slightly raised blister-like appearance. Heavily infected leaves often turn yellow to brown, especially around the edges and drop prematurely so impacting tree health and productivity. To manage this disease, there should be proper disposal of infected leaves, *vis-a-vis* proper orchard sanitation. Proper monitoring of the disease from the beginning allows for prompt intervention. If available, only resistant varieties should be used. Spraying with 0.2% mancozeb or 0.1% Bayleton, Baycor, ditenaconazole or hexaconazole will control jackfruit rust.

2A.28.2.5 Pink disease

Pink Disease of jackfruit is caused by *Botryobasidium salmonicolor* fungus as is diagnosed by a distinctive pinkish powdery coating on the stems reflecting the extensive conidial production of the fungus. In its infection, young branches loose their leaves, followed by occurrence of die back. Lower shaded side of the stems exhibits thin pink inclusions. To get rid of this problem spacing of trees should be quite proper to allow proper ventilation, *vis-a-vis* only at proper time watering should be done. Affected plant parts are pruned immediately and the pruned material is disposed off properly & pasted with copper fungicides such as blue copper, Blitox and Fytolan. If available, tolerant or resistant variety should be planted,

2A.29 Pest Prevention and Control

At quite young stage the fruits may be paper-bagged to protect them from pests and diseases. As per Burkill, bagging encourages ants to swarm over the fruit to safeguard it from its enemies (Morton, 1987; Vikaspedia, 2023).

2A.29.1 Stage-wise crop IPM

Management	Activity
Pre-planting	
	Common cultural practices • Deep ploughing of orchard immediately after harvest to expose eggs and pupae of mealy bugs. • Heavy irrigation of orchard in October also helps in destruction of eggs of mealy bugs. • After removing the webs of bark eating caterpillars and all the borer holes except the fresh ones should be plugged with mud plastering.
Nutrients	• Dig pits of 1 m×1m×1m. • Fill up the pits with top soil mixed with 10kg FYM and 1kg of neem cake per pit
Weeds	• Deep summer ploughing should be undertaken to destroy stubbles and rhizomes of perennial weeds. • Ploughing the field before planting to destroy existing weeds in the field.
Planting	
	Common mechanical practices • Remove new sprouts emerging from rootstock at frequent intervals. • Shift the grafts frequently from one place to another to prevent them from striking roots into the ground.
Nutrients	• Transplant the saplings/planting materials in pits already prepared and filled with manures.
Weeds	• Before transplanting, pits and surrounding areas should be kept free from weeds
Pests, soil-borne pathogens	**Cultural control** • Nursery beds should be raised and fumigated with 4% formalin.

Management	Activity
Vegetative (1-5 years onward)	
	Common cultural practices • Collect and destroy tree debris affected with diseases and pests • Enhance parasitic activity by avoiding chemical spray when only 1-2 larval parasitoids are observed • Apply timely irrigation, manuring cum fertilization as per recommended dose, proper drainage, weeding, mulching, intercultural operations, etc. **Common mechanical practices** • After 25cm mud plastering, wide 400 gauge alkathene should be fastened to the tree trunk above the ground level to prevent migration of freshly hatched first instar mealy bug nymphs in December-January. • The affected leaves and young shoots may be pruned and then these may be destroyed along with the pest. This practice will help in bringing down the pest population. • Prune and burn affected twigs for the control of scales and mealy bugs. Sticky band should be fixed on main trunk to prevent movement of ants. • The affected parts should be nipped off and destroyed. **Common biological practices** • Conserve natural enemies through ecological engineering • Augmentative release of natural enemies. • Ladybird beetle, *Adalia* sp., *Synharmonia* sp., *Exochomus* sp., & *Stethorus* sp. @ 30-50 adults/infested tree should be introduced. • Lacewing and *Syrphus* sp. @ 10-20 first instar larvae should be introduced/tree.
Nutrients	The manures and fertilizers should be applied in two splits during May-June and September–October on the basis of soil test report or in general as per schedule

Management	Activity
Weeds	• Keep the pits and surrounding areas weed-free by hand. • Use straw or plastic mulch to manage the weeds and conserve soil moisture in the tree basins. • Grow annual or perennial recommended intercrops like corn, vegetables (okra, brinal, chilli, tomato, etc.), citrus, banana, pineapple, etc. or cover crops like pulses (green gram, Bengal Gram, etc.) and groundnut. • Whenever intercrops are not grown between the rows, slashing and mowing of weed may be adopted.
Shoot and fruit borer	**Cultural control** • Attacked shoots should be clipped off and destroyed. • Clean hole and pour kerosene/petrol/crude oil or formalin into holes of stem borer and subsequently closing entrance of tunnel by plugging with cotton wool and mud paste. • Use light trap@1/acre **Physical control** • To protect them from egg laying, fruit may be covered with polythene bags and the affected parts removed and destroyed. **Botanical control** • Spraying neem oil may be recommended.
Spittle bugs	**Cultural control** • Keep orchard clean and healthy. • Cut dried branches **Physical control** • Light accessible spittlebug infestations can be removed by hand or by a strong water spray. • Cut dried branches **Biological control** • Pipunculid fly, *Verrallia virginica* causes 50-60% parasitism of adult spittlebugs.

Management	Activity
Mealybug	**Cultural control** • Flooding of orchard with water in October kills the eggs. • Ploughing of orchard in November. • Raking of soil around tree trunk to expose the eggs to natural enemies and sun, cum manual removal of weeds • Fastening of 25cm wide 400 gauge alkathene sheet/grease band, and then after mud plastering of trunk at 30cm above the ground in mid- December. • In July–August, scale infested fallen leaves should be burnt **Biological control** • Raking of soil around tree trunk to expose eggs to sun and natural enemies, removal of weeds and releasing 10-15 parasitoid grubs • Releasing 10-15 grubs of cocinellid predator, *C. montrozieri* per tree.
Bud weevil	**Physical control** • Remove the infested shoots, flower buds and fruits to check infestation. **Biological control** • Pipunculid fly, *Verrallia virginica* causes 50-60% parasitism of adult spittle bugs.
Bark eating caterpillar	**Cultural control** • Remove and destroy dead and severely affected branches of the tree • Remove alternate host, silk cotton and other hosts
Aphid	**Physical control** • Collect and destroy the damaged plant parts along with nymphs and adults **Biological control** • Release coccinellid predators

Management	Activity
Pink waxy scale	**Cultural control** • Prune heavily infested plant parts to open the tree canopy and destroy them immediately. • Prune infested parts (branches and twigs) preferably during summer. • These should be placed in a pit constructed on one corner of the orchard. Allow branches and twigs to dry until the parasites escape. • Burn the remaining debris. • Removal of attendant ants may permit natural enemies to control the insect.
Leaf webber	**Physical control** • Mechanical clipping and burning of affected shoots **Biological control** • Release of pupal parasitoid, *Tetrastichus howardi* @50,000/hectare. • Release of egg parasitoid, *Trichogramma chilonis* @ 5cc/hectare
Thrips	**Cultural control** • Spraying strong jet of water to dislodge and wash out the pest
Castor capsule borer	**Cultural control** • Prune heavily infested plant parts to open the tree canopy and destroy them immediately **Biological control** • The natural enemies, *Hexamermis* spp and *Apanteles taragammae* have been found to be potential bio-control agents of the pest.

Management	Activity
Dieback	**Cultural control** • Every care should be taken to prevent introduction of disease in newly planted orchards. **Mechanical control** • Any infected portion should immediately be pruned, followed by spraying/pasting of copper oxychloride or pasting with cow dung at the cut ends. • Pruning should be done in such a way that some healthy portion is also removed to ensure complete eradication of pathogen.
Leaf spot	**Cultural control** • Affected branches should be pruned
Rust	**Cultural control** • Affected branches should be pruned
Anthracnose	Follow common cultural, mechanical and biological practices **Cultural control** • Diseased leaves, twigs as well as gall-midge infected leaves and fruits should be collected and burnt. • Covering the fruits on tree 15 days prior to harvest with newspapers or brown paper bags.
Pink disease	**Cultural control** • Pruning of infected branches
Reproductive*	
Nutrients	• Micronutrient deficiency, if any, should be corrected by application of particular nutrients.
Weeds	• Use straw or plastic mulch to manage the weeds and conserve the soil moisture in the tree basins. • Through hand weeding, keep the pits and surrounding areas weed-free. • Whenever intercrop(s) not grown between the rows of trees, slashing and mowing of weed may be adopted.

Management	Activity
Anthracnose, dieback, pink disease	Same as vegetative stage
Soft fruit rot	**Physical control** • Remove affected fruits.
Maturation *	
Stem borer, shoot *cum* fruit borer, pink waxy scale, thrips, aphid, castor capsule borer	Same as vegetative stage
Post harvest	
Anthracnose	**Physical control** • HWT at 520C for 4-5 minutes • Fruits should be sprayed with bioinoculant's mixture.

Source: Vikaspedia (2023)

References

1. Arung, E.T., K. Shimizu and R. Kondo, 2007. Structure-activity relationship of prenyl-substituted polyphenols from *Artocarpus heterophyllus* as inhibitors of melanin biosynthesis in cultured melanoma cells. *Chem. & Biodiv.*, **4**(9): 2166–71.

2. Arung, E.T., K. Shimizu, H. Tanaka and R. Kondo, 2010. 3-Prenyl luteolin, a new prenylated flavone with melanin biosynthesis inhibitory activity from wood of *Artocarpus heterophyllus. Fitoterapia*, **81**: 640-43.

3. Babitha, S., C. Sandhya and A. Pandey, 2004. Natural food colourants. *Appl. Bot. Abstr.*, **23**:258–66.

4. Bakewell-Stone, P. 2023. *Artocarpus heterophyllus* (jackfruit). CABI - Digital Library, 11 January 2023 https://www.cabidigitallibrary.org/doi/10.1079/cabicompendium.1832. Accessed on: 14 Jan 2024.

5. Balamaze, J., J.H. Muyonga and Y.B. Byaruhanga, 2019. Production and utilization of jackfruit (*Artocarpus heterophyllus*) in Uganda. *Afr. J. Food Agric. Nutr. Dev.*, **19**(2): 14289-302.

6. Baliga, M.S., A.R. Shivashankara, R. Haniadka, J. Dsouza and H.P. Bhat, 2011. Phytochemistry, nutritional and pharmacological properties of Artocarpus heterophyllus Lam. (jackfruit): A review. Food Res. intern., 44(9): 1800–1811.

7. Barik, B.R., T. Bhaumik, A.K. Dey and A.B. Kundu, 1994. Triterpenoids from *Artocarpus heterophyllus. Phytochem.*, **35**: 1001-4.

8. Barik, B.R., T. Bhaumik, A.K. Kundu and A.B. Kundu, 1997. Triterpenoids of *Artocarpus heterophyllus. J. Ind. Chem. Soc.*, **74**: 163-4.

9. Barua, A.G. and B.R. Boruah, 2004. Minerals and functional groups present in the jackfruit seed: a spectroscopic investigation. *J. Food Sci. Nutr.*, **55**(6): 479- 83.

10. Bruno, J. 2024. Jackfruit varieties: 30+ types of jackfruit & how to identify them. Retrieved on 14 January, 2024. https://thrivecuisine.com/jackfruit/jackfruit-varieties/.

11. Butani, D.K. 1978. Pests and diseases of jackfruit in India and their control . *CIRAD Journ.*, **33**(5): 351-7.

12. Chandrika, U.G., E.R. Jansz and N.D. Warnasuriya, 2004. Analysis of carotenoids in ripe jackfruit (*Artocarpus heterophyllus*) kernel and study of their bioconversion in rats. *J. Sci. Food & Agric.*, **85**(2): 186–90.

13. Chawdhary, F.A. aqnd M.A. Raman, 1997. Distribution of free sugars and fatty acids in jackfruit. *Food Chem.*, **60**(1): 25-8.

14. Choudury, A.G., S.O. Bhutia, M.A. Hasan and B.C. Das, 2017. Fruiting behaviour of different jackfruit genotypes. *Trends in Biosci*, **10**(16): 2847-8.

15. Chung, M., C. Lu, P. Huang and C. Lin, 1995. Prenyl flavonoids of *Artocarpus heterophyllus. Phytochem.*, **40**(4): 1279-82.

16. Coyne, T., T.I. Ibiebele, P.D. Baade, A. Dobson, C. McClintock, S. Dunn, D. Leonard and J. Shaw, 2005. Diabetes mellitus and serum carotenoids: findings of a population-based study in Queensland, Australia. *Amer. J. Clinic. Nutr.*, **82**(3): 685–93.

17. Dayal, R. and T.R. Seshadri, 1974. Colourless compounds of the roots of *Artocarpus heterophyllus*. Isolation of new compound artoflavone. *Indian J. Chem.*, **12**: 895-6.

18. Esmaillzadeh, A. and L. Azadbakht, 2008. Food intake patterns may explain the high prevalence of cardiovascular risk factors among Iranian women. *J. Nutr.*, **138**(8): 1469–75.

19. de Faria, A.F., V.V. de Rosso and A.Z. Mercadante, 2009. Carotenoid composition of jackfruit (Artocarpus heterophyllus), determined by HPLC-PDA-MS/MS. Pl. Foods Human Nutr., 64(2): 108–15.

20. en.wikipedia.org, 2024a. *Artocarpus integer*. Accessed on 26 January 2024. https://en.wikipedia.org/wiki/Artocarpus_integer.

21. en.wikipedia.org, 2024b. Jackfruit. Accessed on 10 January, 2024. https://en.wikipedia.org/wiki/Jackfruit.

22. Fan, J. and T. Watanabe, 2003. Inflammatory reactions in the pathogenesis of atherosclerosis. *J. Atheroscl. & Thrombosis*, **10**(2): 63–71.

23. Fang, S.-C., C.-L. Hsu and G.-C. Yen, 2008. Anti-inflammatory effects of phenolic compounds isolated from the fruits of *Artocarpus heterophyllus. J. agric. & Food Chem.*, **56**(12): 4463–8.

24. Fathin, A.N., D. Astuti, W.W. Winarni and Y.W.N. Ratnaningrum, 2021. Flowering and fruiting phenology of jackfruit (*Artocarpus heterophyllus* Lam.) from Sumatra landraces in *ex situ* conservation area in Karangmojo, Yogyakarta. *IOP Conf. Series: Earth & environ. Sci.*, **914**: 012052 IOP Publishing doi:10.1088/1755-1315/914/1/012052. Via https://iopscience.iop.org/article/10.1088/1755-1315/914/1/012052.

25. Fernando, M.R., S.M.D.N. Wickramasinghe, M.I. Thabrew, P.L. Ariyananda and E.H. Karunanayake, 1991. Effect of *Artocarpus heterophyllus* and *Asteracanthus longifolia* on glucose tolerance in normal human subjects and in maturity-onset diabetic patients. *J. Ethnopharmac.*, **31**(3): 277–82.

26. FTG, 2024. Jackfruit Cultivar Descriptions. Virtual Herbarium, Fairchild Tropical Garden Herbarium. Accessed on 14 January, 2024. http://www.virtualherbarium.org/tropicalfruit/jackfruit-cultivars.html.

27. Fu, Y.H., J.M. Guo, Y.T. Xie, X.M. Yu, Q.T. Su, L. Qiang, L.Y. Kong and Y.P. Liu, 2020. Prenylated chromones from the fruits of *Artocarpus heterophyllus* and their potential anti-HIV-1 activities. *J. Agric. Food Chem.*, **68**(7): 2024-30.

28. Gardner, E.M., R.J. Gagne, P.E. Kendra, W.S. Montgomery, R.A. Raguso, T.T. McNeil and N.J.C. Zerega, 2018. A flower in fruit's clothing: pollination of jackfruit by a new species of gall midge *Clinodiplosis ultracrepidata*. *Int. J. Pl. Sci.*, **179**(5): 350-67.

29. Halliwell, B. 1997. Antioxidants and human diseases: A general introduction. *Nutr. Rev.*, **55**(1): 44–9.

30. Hargreaves, D. and B. Hargreaves, 1964. Tropical Trees of Hawaii. Hargreaves, Kailua, Hawaii, 30p. ISBN 9780910690027.

31. Heinecke, J.W. 2006. Chemical knockout of C-reactive protein in cardiovascular disease. *Nat. Chem. Biol.*, **2**(6): 300–301.

32. Jagadeesh, S.L., B.S. Reddy, N. Basavaraj, G.S.K. Swamy, K. Gorbal, L. Hegde, G.S.V. Raghavan and S.T. Kajjidoni, 2007. Inter tree variability for fruit quality in jackfruit selections of Western Ghats of India. *Sci. Hort.*, **112:** 382-7.

33. Jagtap, U.B., S.N. Panaskar and V.A. Bapat, 2010. Evaluation of antioxidant capacity and phenol content in jackfruit (*Artocarpus heterophyllus* Lam.) fruit pulp. *Pl. Foods for Human Nutr.*. **65**(2): 99–104.

34. Jha, P.K., R.S. Choudhary and S.K. Choudhary, 1997. Studies of medicinal plants of Palamau (Bihar)-IInd Part. *Biojournal*, **9**: 21-38.

35. Kamaluddin, M., M. Ali and M.K. Bhuiyan, 1997. Effect of auxin on rooting of cuttings and growth of stecklings of jackfruit (*Artocarpus heterophyllus* Lam.). *Chittagong Univ. Stud. Sci.*, **20**(1): 71-5.

36. Khan, M.R., A.D. Omoloso and M. Kihara, 2003. Antibacterial activity of *Artocarpus heterophyllus*. *Fitoterapia*, **74**(5): 501–5.

37. Khan, R., N. Zerega, S. Hossain and M.I. Zuberi, 2010. Jackfruit (*Artocarpus heterophyllus* Lam.) diversity in Bangladesh: land use and artificial selection. *Econ. Bot.*, **64:** 124-36.

38. Kishore, K. 2018. Phenological growth stages of jackfruit (*Artocarpus heterophyllus*) according to the extended BBCH scale. *Ann. Appl. Biol.*, **172** (2018): 366-74.

39. Krinsky, N.I., J.T. Landrum and R.A. Bone, 2003. Biologic mechanisms of the protective role of lutein and zeaxanthin in the eye. *Ann. Rev. Nutr.*, **23**:171–201.

40. Kumar, G.S., P.S. Appukttan and D.K. Basu, 1982. D-Galactose – specific lectin from jack fruit seed. *J. Biosci.*, **4**: 257-61.

41. kvk, 2024. Jackfruit. https://kvk.icar.gov.in/API/Content/PPupload/k0358_122.pdf Accessed on 14 January, 2024.

42. Lavker, R.M. 1991. Cutaneous aging: chronologic versus photoaging. *Photodamage* (ed. Gilchrest, B.A). Blackwell Science, Cambridge, pp. 123–35.

43. Li, Y.Z., M.A.O. Qi, F. Feng and C.H. Ye, 2010. Genetic diversity within a jackfruit (*Artocarpus heterophyllus* Lam.) germplasm collection in China using AFLP markers. *Agric. Sci.* in China, **9**(9): 1263-70.

44. Lin, C.-N. and C.-M. Lu, 1993. Heterophylol, a phenolic compound with novel skeleton from *Artocarpus heterophyllus. Tetrahedron Let.*, **34**(17): 8249-50.

45. Lin, C.-N., C.-M. Lu and P.-L. Huang, 1995. Flavonoids from *Artocarpus heterophyllus. Phytochem.*, **39**(6): 1447-51.

46. Lina, D.P. and C.M. Protacio, 2015. Phenology of jackfruit (*Artocarpus heterophyllus* Lam.) and effects of paclobutrazol on off-season fruit production. *Acta Hort.*, No. 1088, pp. 495-500.

47. Liu, Y.P., X.M. Yu, W. Zhang, T. Wang, B. Jiang, H.X. Tang, Q.T. Su and Y.H. Fu, 2020. Prenylated chromones and flavonoids from *Artocarpus heterophyllus* with their potential antiproliferative and anti-inflammatory activities. *Bioorg. Chem.*, **101**:104030.

48. Lobo, Joanna, 2019. Jackfruit: The King of India. https://tenderly.medium.com/jackfruit-india-eee207285d87.

49. Lu, C.-M. and C.-N. Lin, 1993. Two 2′, 4′, 6′-trioxygenated flavanones from *Artocarpus heterophyllus. Nat. Prod. Res. Cent.*, **33**(4): 909-911.

50. Morton, Julia F. 1987. Jackfruit. *Fruits of Warm Climates.* Center for New Crops & Plant Products, Purdue University Department of Horticulture and Landscape Architecture, West Lafayette, Indiana, USA, pp. 58–64.

51. Nakintu, J., A.E. Olet, M. Andama and B.J. Lejju, 2019. Ethno-varieties and distribution of jackfruit tree (*Artocarpus heterophyllus* Lam.) in Uganda: implications for trade, food security and germplasm conservation. *East Afric. J. Sci., Tech. Innov.*, **1**(1): 27-51.

52. Om Prakash, R. Kumar, A. Mishra and R. Gupta, 2009. *Artocarpus heterophyllus* (Jackfruit): An Overview. *Phcog. Rev.*, **3**(6): 353-8.

53. Ong, B.T., S.A.H. Nazimah, C.P. Tan, H. Mirhosseini, A. Osman, D. Hashim and G. Rusul, 2008. Analysis of volatile compounds in five jackfruit (*Artocarpus heterophyllus* L.) cultivars using solid-phase microextraction (SPME) and gas chromatography-time-of-flight mass spectrometry (GC-TOFMS). *J. Food Comp. & Anal.*, **21** (5): 416–22.

54. Pavanasasivam, G., M. Uvais and S. Sultanbawa, 1973. Cycloartenyl acetate, cycloartenol and cycloartenone in the bark of *Artocarpus* species. *Phytochem.*, **12**(11): 2725-6.

55. Pereira-da-Slva, G., A.N. Moreno, F. Marques, C. Oliver, M.C. Jamur, A. Panunto-Castelo and M.C. Roque-Barreira, 2006. Neutrophil activation induced by the lectin KM+ involves binding to CXCR2. *Biochim. Biophys. Acta*, **1**: 86-94.

56. Perkin, A.V. and C. Cope, 1895. The constituents of *Artocarpus integrifolia. J. Chem. Soc.*, **67**: 937-44.

57. POWO, 2023. *Artocarpus heterophyllus* Lam. Plants of the World Online. Royal Botanic Gardens, Kew Science. Retrieved on 9 January, 2024. (https://powo.science.kew.org/taxon/urn:lsid:ipni.org:names:850389-1). First published in *Encycl.* 3: **209** (1789).

58. Pradeepkumar, T., B. Jyothibhaskar, Suma and K.N. Satheesan, 2008. Management of Horticultural Crops. *Horticultural Science Series*. Vol. **11** (ed. Peter, K. V. Peter), New India Pub. House, New Delhi, 81p. ISBN 978-81-89422-49-3.

59. Priya Devi, S., S. Talaulikar, M.J. Gupta, M. Thangam and N.P. Singh, 2014. A Guide on Jack Fruit - Cultivation and Value Addition. Technical Bulletin No. 41, ICAR (RC), Goa, PP. 1-11. https://ccari.icar.gov.in/Technical%20Bulletin%20No.%2041.pdf.

60. Pushpakumara, D.K.N.G. 2006. Floral and fruit morphology and phenology of *Artocarpus heterophyllus* (Moraceae). *Sri Lankan J. agric. Sci.*, **43**: 83-106.

61. Rama Rao, A.V. and M.V. Venkataraman, 1973. Colouring matter of the *A. hetetophyllus. Indian J. Chem.*, **11**: 298–9.

62. Ranasinghe, R.A.S.N., S.D.T. Maduwanthi and R.A.U.J. Marapana, 2019. Nutritional and Health Benefits of Jackfruit (*Artocarpus heterophyllus* Lam.): A Review. *Intern. J. Food Sci.*, **43**: 271-83. https://doi.org/10.1155/2019/4327183.

63. Ruiz-Montañez, G., A. Burgos-Hernández, M. Calderón-Santoyo, C.M. López-Saiz., C.A. Velázquez-Contreras, A. Navarro-Ocaña. And J.A. Ragazzo-Sánchez, 2014. Screening antimutagenic and antiproliferative properties of extracts isolated from jackfruit pulp (*Artocarpus heterophyllus* Lam). *Food Chem.*, **175**:409–416.

64. Sawe, B.E. 2017. World Leaders In Jackfruit Production. WorldAtlas. Retrieved 17 January 2024. < https://www.worldatlas.com/articles/world-leaders-in-jackfruit-production.html >.

65. Senavirathna, H.M.T.N., L.T. Ranaweera, M.M.A.W.P. Mudannayake, P.W.I. Nawanjana, W.M.D.A. Wijesundara, H.S.M. Jayaranthne, M.A.P. Ratnasuriya, C.K. Weebadde and S.D.S.S. Sooriyapathirana, 2020. Assessment of the taxonomic status of the members of genus *Artocarpus* (Moraceae) in Sri Lanka. *Genet. Res. & Crop Evol.*, **67**(5): 1163-79.

66. Septama, A.W., I. Jantan, P. Panichayupakaranant, M.F.F.M. Aluwi and E.P. Rahmi, 2020. Immunosuppressive and antibacterial activities of dihydromorin and norartocarpetin isolated from *Artocarpus heterophyllus* heartwoods. *Asian Pacific J. Trop. Biomed.*, **10**: 361-8.

67. Shanmugapriya, K., P.S. Saravana, H. Payal, S. Peer Mohammed and W. Binnie, 2011. Antioxidant activity, total phenolic and flavonoid contents of *Artocarpus heterophyllus* and Manilkara zapota seeds and its reduction potential. *Intern. J. Pharm. & Pharmaceut. Sci.*, **3**(5): 256–60.

68. Sharma, W. and P.C. Trivedi, 1995. Nematocidal and nematostostatic response of aqueous extract of certain plants of semi-arid niche. *Curr. Nematol.*, **6**: 45–53.

69. Shinomiya, K., M. Aida, Y. Hano and T. Nomura, 2004. A diels-alder-type adducts from *Artocarpus heterophyllus. Phytochem.*, **40**(4): 1317-9.

70. Silver, Mark, 2014. Here's The Scoop On Jackfruit, A Ginormous Fruit To Feed The World. *NPR*. Retrieved 10 January, 2024. https://www.npr.org/sections/thesalt/2014/05/01/308708000/heres-the-scoop-on-jackfruit-a-ginormous-fruit-to-feed-the-world.

71. Sindhu, A.S. 2012. Jackfruit Improvement in the Asia-Pacific Region – A Status Report. Asia-Pacific Association of Agricultural Research Institutions (APAARI), Bangkok, Thailand. Via https://iopscience.iop.org/article/10.1088/1755-1315/914/1/012052/pdf.

72. Singh, A., S. Kumar and I.S. Singh, 1991. Functional properties of jackfruit seed flour. *Lebensm-Will u Technol.*, **24**: 373–4.

73. Soong, Y.-Y. and P.J. Barlow, 2004. Antioxidant activity and phenolic content of selected fruit seeds. *Food Chem.*, **88**(3): 411–7.

74. Spada, F.P., L.M. Zerbeto, G.B.C. Ragazi, É.M.R. Gutierrez, M.C. Souza, J.K. Parker, S.G. Canniatti-Brazaca, 2017. Optimization of Postharvest Conditions To Produce Chocolate Aroma from Jackfruit Seeds (PDF). *J. agric. & Food Chem.*, **65**(6): 1196–1208.

75. Stahl, W. and H. Sies, 2005. Bioactivity and protective effects of natural carotenoids. *Biochimica et Biophysica Acta (BBA) – Molec. Basis Dis.*, **1740**(2): 101–7. doi: 10.1016/j.bbadis.2004.12.006.

76. Stanner, S.A. J., Hughes, C.N.M. Kelly and J. Buttriss, 2004. A review of the epidemiological evidence for the antioxidant hypothesis. *Pub. Health Nutr.*, **7**(3): 407–22.

77. Swami, S.B., N.J. Thakor, P.M. Haldankar and S.B. Kalse, 2012. Jackfruit and its many functional components as related to human health: a review. *Compreh. Rev. in Food Sci. & Food Saf.*, **11:** 565-76.

78. Trindade, M.B., J.L.S. Lopes, A. Soares-Costa, A.C. Monteiro-Moreira, R.A. Moreira, M.L.V. Oliva and L.M. Beltramini, 2006. Structural characterization of novel chitin-binding lectins from the genus *Artocarpus* and their antifungal activity. *Biochimica et Biophysica Acta—Proteins and Proteomics*, **1764**(1): 146–52.

79. USDA, 2018. Jackfruit, raw. FoodData Central. Released in April 2018. Accessed on 11 January 2024. https://fdc.nal.usda.gov/fdc-app.html; https://fdc.nal.usda.gov/fdc-app.html#/food-details/174687/nutrients.

80. Venkataraman, K. 2001. Wood of phenolics in the chemotaxonomy of the Moraceae. *Phytochem.*, **11**(5): 1571-86.

81. Vikaspedia, 2023. Jackfruit Crop: Stagewise IPM. Accessed on 9 December, 2023. https://vikaspedia.in/agriculture/crop-production/integrated-pest-managment/ipm-for-fruit-crops/ipm-strategies-for-jackfruit/jackfruit-crop-stage-wise-ipm.

82. Wang, X.L., X.X.T. Shen, S.Q. Wang and X.N. Wang, 2017. New phenolic compounds from the leaves of *Artocarpus heterophyllus. Chin. Chem. Lett.*, **28**: 37-40.

83. Wei, B.-L., J.-R. Weng, P.-H. Chiu, C.-F. Hung, J.-P. Wang and C.-N. Lin, 2005. Antiinflammatory flavonoids from *Artocarpus heterophyllus* and *Artocarpus communis. J. agric. & Food Chem.*, **53**(10):3867–71.

84. WFO, 2024. *Artocarpus heterophyllus* Lam. Pub. on the Internet;http://www.worldfloraonline.org/taxon/wfo-0000550491. Accessed on: 14 Jan 2024.

85. www.foreststreesagroforestry.org, 2024. *Artocarpus heterophyllus* L. Research Program on Forests, Trees and Agroforestry, CGIAR. https://www.foreststreesagroforestry.org/tree/artocarpus-heterophyllus-l/. Accessed on 17 January, 2024.

86. Xu, S. and R.M. Touyz, 2006. Reactive oxygen species and vascular remodelling in hypertension: Still alive. *Canad. J. Cardiol.*, **22**(11): 947–52.

87. Yuan, W.J., J.B. Yuan, J.B. Peng, Y.Q. Ding, J.X. Zhu and G. Ren, 2017. Flavonoids from the roots of *Artocarpus heterophyllus. Fitoterapia*, **117**: 133-7.

88. Zheng, Z.P., S. Chen, S. Wang, X.C. Wang, K.W. Cheng, J.J. Wu, D. Yang and M. Wang, 2009. Chemical components and tyrosinase inhibitors from the twigs of *Artocarpus heterophyllus. J. agric. Food Chem.*, **57**(15): 6649-55.

89. Zheng, Z.P., K.W. Cheng, J.K. To, H. Li and M. Wang, 2008. Isolation of tyrosinase inhibitors from *Artocarpus heterophyllus* and use of its extract as anti-browning agent. *Mol. Nutr. Food Res.*, **52**: 1530-8.

90. Zheng, Z.P., Y. Xu, C. Qin, S. Zhang and X. Gu, 2014. Characterization of antiproliferative activity constituents from *Artocarpus heterophyllus. J. Agric. Food Chem.*, **62**: 5519-27.

2B

Chempedak [*Artocarpus integer* (Thunb.) Merr.]

Accepted intraspecifics: *Artocarpus integer* var. *integer* and *A. integer* var. *silvestris* Corner.

Synonyms: *Artocarpus integrifolius* L.f., nom. superfl.; *A.i.* var. *hirsutus* Stokes, not validly publ.; *Radermachia integra* Thunb.; *Sitodium cauliflorum* Gaertn., nom. superfl.; *S. macrocarpon* Thunb., nom. superfl.; *Saccus integer* (Thunb.) Kuntze (POWO, 2024); *Artocarpus champeden* (Thunb.) Merr.; and *A. polyphema* Pers. (en.wikipedia.org, 2024).

Other common names: In Malaya: Chempedak, Cempedak, Sempedak, Temedak; in Thailand: Cham-pa-da; in Indonesia: Tjampedak; and in the Philippines: Lemasa. The Chempedak's wild form in Malaya is known as bangkong or baroh (Morton, 1987).

2B.1 Introduction, Origin and Cultivation

Chempedak is a native fruit that thrives in the wild across Malaya, flourishing at elevations up to 1,300 metres. This tropical gem is not only indigenous but also extensively cultivated throughout Malaysia, signifying its importance as a regional crop (Morton, 1987). Beyond its Malaysian roots, Chempedak has found its way to other parts of SE Asia, being introduced into India (Assam), Laos, and Vietnam (POWO, 2024). Recognized as a vital agricultural asset in Malaysia, it has also gained popularity in S. Thailand and various regions of Indonesia, presenting promising potential for further expansion (Wang *et al.*, 2018).

2B.2 Cultivation Conditions

Chempedak trees, typically planted in non-eroded and well-drained soils exhibit a remarkable resilience to temporary flooding. Thriving in a temperature range of 13–47°C and requiring an annual rainfall between 1,250–2,500mm, these trees showcase their adaptability to diverse environmental conditions.

2B.2.1 Propagation and growth

Preferred by many enthusiasts over its relative ~ the jackfruit, Chempedak can be propagated either from seeds or by budding onto self-seeded jackfruit or other *Artocarpus* species (Wang *et al.*, 2018; Morton, 1987). The trees demonstrate their fruit-bearing potential at 3–6 years for seed-planted trees and 2–4 years for clonal ones. Blossoms adorn the trees from February to April and again from August to October in southern Malaysia with a notable deviation observed in W. Java where flowering predominantly occurs in July and August. The transition from flowering to ripe fruit typically spans 2–4 months (en.wikipedia.org, 2024).

2B.2.2 Harvesting and maturity indicators

Harvesting proves to be a critical juncture in ensuring the quality of Chempedak fruit. Recognizing maturity involves a multi-faceted approach including tapping the fruit to make a dull hollow sound, monitoring skin colour changing from green to a yellower hue and noting the development of a durian-like odour. The spines on the fruit's skin become flattened as it matures. Harvesting before natural falling of fruits is crucial to prevent damage, ensuring an extended shelf life and averting premature ripening.

2B.2.3 Post-harvest handling

Upon harvesting, the fruit exudes a latex. The harvested fruit is left to drain the latex in the field before being transported from the orchard. Despite its delectable appeal, the fruit has a relatively short shelf life of 2–3 days.

2B.3 Description

The Chempedak is a distinctive deciduous tree exhibiting a majestic stature and achieving heights of approximately 18m in cultivation and soaring to impressive heights ranging between 30 to 45.5 metres in its untamed habitat. One of its defining characteristics is the presence of long, stiff and brown hairs on young branchlets, leaves, buds and peduncles ~ a feature that sets it distinctly apart from its close relative ~ the jackfruit. The leaves of the Chempedak initially manifest often in 3-lobed formations, gradually maturing into obovate, oblong or elliptical shapes spanning a length of 15-28cm.

Male flower spikes, merely 5cm in length lead to development of cylindrical or irregular fruits not exceeding dimensions of 35.5cm in length and 15cm in thickness. When fully ripe, the chempedak fruit showcases a vibrant mustard-yellow to golden-brown hue with a reticulated and warty exterior. Renowned for its distinctive olfactory profile, it holds the reputation of possessing the strongest and richest smell of any fruit in creation (Morton, 1987).

Distinguishing itself from the jackfruit, chempedak boasts a thin rind facilitating the ease of extraction of seeds and surrounding pulp by cutting open the base and pulling on the fruit stalk. The deep-yellow pulp described as tender, slimy, juicy and sweet exhibits variations in the wild form, featuring a thinner, subacid and odourless composition (Morton, 1987).

Within the thick layer surrounding the large seeds, fleshy and edible arils are found, presenting a spectrum of culinary possibilities. These arils with a colour palette ranging from yellowish-white to orange deliver a sweet and fragrant taste coupled with a soft, slippery and slightly fibrous texture on the tongue. Ripe chempedak fruit emits a pungent aroma often likened to the harsh and penetrating smell of durian (Wang *et al.*, 2018).

2B.4 Uses

Chempedak prized for its delectable pulpy flesh, characterized by a vibrant yellow/orange hue and a high content of beta-carotene, stands out as a sought-after tropical fruit. Its distinct sweet flavour reminiscent of a blend between durian and mango contributes to its popularity. The fruit is a culinary delight often enjoyed alongside rice, with its roasted seeds serving as a unique addition to various dishes.

In regions where chempedak is cultivated the large fruits are commonly opened, sliced into pieces and sold. The seeds possessing a taste akin to water chestnuts can be prepared through frying, boiling or grilling then peeled and enjoyed with a sprinkle of salt. Furthermore, akin to young jackfruit, the young chempedak fruit can be utilized as a vegetable. After peeling, slicing and boiling it is often seasoned or incorporated into other dishes such as curries.

In South and East Kalimantan (Indonesia), a distinctive culinary tradition involves fermenting the inner skin of chempedak using lactic acid bacteria which is referred to as 'dami' or 'mandai'. This fermented product is typically consumed after frying, adding a unique dimension to the culinary repertoire.

Beyond its culinary applications, chempedak extends its utility to other domains. The wood of the chempedak tree is recognized for its strength and durability ~ making it an ideal material for building homes, furnishings and boats. The wood yields a yellow dye while the bark is rich in tannin. This fibrous bark finds use in the creation of ropes. The versatility of the chempedak tree is further emphasized by its ability to produce yellow dye not only from the wood but also from other parts of the tree (Morton, 1987; Wang *et al.*, 2018; en.wikipedia.org, 2024).

References

1. en.wikipedia.org, 2024. Jackfruit. Accessed on 10 January, 2024. https://en.wikipedia.org/wiki/Jackfruit.
2. Morton, Julia F. 1987. Jackfruit. *Fruits of Warm Climates.* Center for New Crops & Plant Products, Purdue University Department of Horticulture and Landscape Architecture, West Lafayette, Indiana, USA, pp. 58–64.
3. POWO, 2024a. *Artocarpus integer* Merr. Plants of the World Online. Royal Botanic Gardens, Kew Science. Retrieved on 9 January, 2024. (https://powo.science.kew.org/taxon/urn:lsid:ipni.org:names:582622-1). First published in *Interpr. Herb. Amboin.*: **190** (1917).
4. Wang, M.M., E.M. Gardner, R.C. Chung, M.Y. Chew, A.R. Milan, J.T. Pereira and N.J. Zerega, 2018. Origin and diversity of an underutilized fruit tree crop, chempedak (*Artocarpus integer*, Moraceae). *Amer. J. Bot.*, **105**(5): 898–914.

2C

Lakoocha (*Artocarpus lakoocha* Buch.-Ham.)

Synonyms: *Artocarpus cumingianus* var. *stenophyllus* Diels; *A. ficifolius* W.T.Wang; *A. lakoocha* Roxb.; *A. yunnanensis* H.H.Hu; *Prainea rumphiana* Becc.; and *Saccus lakoocha* (Roxb.) Kuntze (POWO, 2024).

Common names: In India: Monkey jack, Monkey fruit or Lakuchi; in Malaya: Tampang and various other native names; and in Thailand: Lokhat (Morton, 1987).

2C.1 Introduction

Having its roots in the humid sub-Himalayan region of India at altitudes reaching up to 1,200 meters, the lakoocha extends its presence to the regions of Malaya and Sri Lanka (Morton, 1987). This species exhibits a widespread distribution encompassing the entire Indian subcontinent and SE Asia (Oudhia and Paull, 2008). Flourishing in tropical climates, monkey jack trees are prominently situated in both evergreen and deciduous forests showcasing their adaptability to a diverse range of soil types. Renowned for this remarkable adaptability, the monkey jack tree stands as a testament to its ability to thrive in varying environmental conditions across its expansive geographic range.

2C.2 Description

Lakoocha, a deciduous tree standing 6-9 metres in height, presents itself with expansive leathery leaves exhibiting a downy texture on their undersides.

An intriguing botanical feature of this tree is the coexistence of both male and female flowers on the same tree. The male flowers showcase a vibrant orange-yellow hue while the female counterparts display a striking reddish colouration.

The tree bears large spiky fruits characterized by their nearly round or irregular shape, measuring 5-12.5 centimeters in width. These fruits boast a velvety dull-yellow exterior, subtly tinged with pink, and encase a delectable sweet-sour pulp. During the unripe stage, the fruit exhibits a green colouration, transforming into a vivid yellow as it reaches maturity. The focal edible components of lakoocha fruit are the fleshy arils renowned for their consumption either in their raw state or following culinary preparations. This remarkable tree with its distinctive foliage, diverse floral manifestations and enticingly textured fruits provides a source of both visual and gastronomic delight (Morton, 1987).

2C.3 Uses

Often grown for shade or its fruits, seedlings come into production in 5 years. The tree is valued for its wood, *vis-à-vis* its fruit as it is edible and is believed to have medicinal values. In NE Thailand, the wood is used to make pong lang ~ a local traditional instrument (Cruz-Garcia and Price, 2011). Wood known as lakuch is heavier than jackfruit wood ~ akin to teak it is durable outdoors and underwater but lacks polishing. The bark of the *Artocarpus lakoocha* tree has been traditionally used for producing a natural dye. Dye obtained from the bark is yellowish and has been utilized for colouring fabrics (Morton, 1987).

The fleshy arils of lakoocha are renowned for their distinctive starchy and fibrous texture. While the pulp is occasionally enjoyed in its raw state, it predominantly finds its place in culinary creations such as curries, chutneys and various savory dishes. The seeds can be subjected to roasting for consumption. The acid and astringent male flower spike of lakoocha tree is adeptly transformed into pickles (Morton, 1987).

Beyond its culinary utility, the lakoocha tree serves a multitude of purposes. Bark contains 8.5% tannin and it is not merely relegated to structural applications but also chewed reminiscent of betelnut. This versatile bark contributes to endeavours such as construction, boat building, furniture crafting and cabinetwork. Furthermore, the lakoocha tree yields valuable fibre for cordage while both its seeds and milky latex possess purgative properties. The bark esteemed for its medicinal attributes is applied to address various skin ailments and the fruit is believed to function as a liver tonic (Morton, 1987).

Delving into the pharmacological realm, the heartwood of *A. lakoocha* harbours oxyresveratrol ~ a notable stilbenoid (Maneechai *et al.*, 2009). This compound is not only intrinsic to the tree but also plays a pivotal role in formulation of the traditional drug 'Puag-Haad'. This drug obtained as a light brown powder from the aqueous extract of *A. lakoocha* wood chips through a meticulous process involving boiling, slow evaporation and subsequent cooling, has demonstrated efficacy against the intestinal fluke *Haplorchis taichui* (Wongsawad *et al.*, 2005). Furthermore, 'Puag-Haad' has exhibited efficacy in combating taeniasis (Charoenlarp *et al.*, 1989). The multifaceted utility of *A. lakoocha* extends from culinary practices to traditional medicine showcasing its significance in various domains.

References

1. Charoenlarp, P., P. Radomyos and D. Bunnag, 1989. The optimum dose of Puag-Haad in the treatment of taeniasis. *J. Med. Assoc. Thailand = Chotmaihet Thangphaet*, **72**(2): 71–3.

2. Cruz-Garcia, G.S. and L.L. Price, 2011. Ethnobotanical investigation of wild food plants used by rice farmers in Kalasin, Northeast Thailand. *J. Ethnobiol. & Ethnomed.*, **7**: 33.

3. Maneechai, S., K. Likhitwitayawuid, B. Sritularak, C. Palanuvej, N. Ruangrungsi and P. Sirisa-Ard, 2009. Quantitative analysis of oxyresveratrol content in *Artocarpus lakoocha* and 'Puag-Haad'. *Med. Princ. & Pract.*, **18**(3): 223–7.

4. Morton, Julia F. 1987. Jackfruit. *Fruits of Warm Climates*. Center for New Crops & Plant Products, Purdue University Department of Horticulture and Landscape Architecture, West Lafayette, Indiana, USA, pp. 58–64.

5. Oudhia, P. and R.E. Paull, 2008. Monkey Jack *Artocarpus lakoocha* Roxb., Moraceae pp. 485-7. Encyclopedia of Fruit and Nuts (eds J. Janick and R. E. Paull). CABI, Wallingford, United Kingdom.

6. POWO, 2024. *Artocarpus lacucha* Buch.-Ham. Plants of the World Online. Royal Botanic Gardens, Kew Science. Retrieved on 9 January, 2024. (https://powo.science.kew.org/taxon/urn:lsid:ipni.org:names:927744-1). First published in Mem. Wern. Nat. Hist. Soc. **5**: 333 (1826).

7. Wongsawad, C., P. Wongsawad, P. Luangphai and K. Kumchoo, 2005. *In vitro* effects of aqueous extract from *Artocarpus lakoocha* Roxb on tegumental surface of *Haplorchis taichui* (PDF). *The Southeast Asian J. Trop. Med. & Public Health*, **36**(Suppl 4): 112–6.

Kwai Muk (*Artocarpus parvus* Gagnep.)

Synonyms: *Artocarpus lingnanensis* Merr., *A. nitidus* ssp. *lingnanensis* (Merr.) F.M.Jarrett and *A. sampor* Gagnep (POWO, 2024).

2D.1 Introduction

Originating in Kwangtung (China), the Kwai Muk though sparingly introduced beyond its native region has left a notable mark through experimental planting efforts. Noteworthy instances include its successful cultivation in Florida in 1927 where it not only survived but thrived even in Puerto Rico by 1929. Its natural habitat in China extends to an altitude of 152 metres.

Despite its resilience, young Kwai Muk trees exhibit sensitivity to brief drops in temperature, particularly in the range of -2.22 to -1.11°C. However, the hardiness of mature trees is evident in their ability to endure even lower temperatures, withstanding as much as -3.89 to -3.33°C in Florida homesteads. Nevertheless, the species has succumbed to extreme cold, with temperatures reaching -6.67°C leading to fatalities in central Florida (Morton, 1987).

2D.2 Description and Uses

Kwai Muk, an elegant and gradually developing ornamental tree, stands tall at 6-15 metres. It possesses a slender erect stature and is uniquely characterized by its dark-red rough bark. Tree's distinctive features include milky latex and evergreen leaves measuring 5-12.5cm in length. Noteworthy is the presence of both tiny male and female flowers tinged in yellow on the same tree. Female flowers coalesce into globular heads reaching up to 1cm in length (Morton, 1987).

Kwai Muk fruits are more or less oblate and irregular spanning a diameter of 2.5-5cm. Encased in a velvety, brownish, thin and tender skin, the unripe fruits are abundant with latex. Upon ripening, the fruits transform into a vibrant orange-red or red hue revealing soft and agreeably subacid to acid-flavoured pulp. Edible pulp may either be seedless or may contain 1 to 7 small pale seeds that can be enjoyed in its raw form, preserved in syrup or dried. The ripening period spans from August to October in Florida (Morton, 1987; Gardner and Zerega, 2020).

References

1. Gardner, E.M.and N.J.C. Zerega, 2020. Taxonomic updates to *Artocarpus* subgenus *Peudojaca* (Moraceae) with a particular focus on the taxa in Singapore (PDF). *Gardeners' Bull. Singapore*, **72**(2): 173-213.

2. Morton, Julia F. 1987. Jackfruit. *Fruits of Warm Climates*. Center for New Crops & Plant Products, Purdue University Department of Horticulture and Landscape Architecture, West Lafayette, Indiana, USA, pp. 58–64.

3. POWO, 2024. *Artocarpus hirsutus* Lam. Plants of the World Online. Royal Botanic Gardens, Kew Science. Retrieved on 9 January, 2024. (https://powo.science.kew.org/taxon/urn:lsid:ipni.org:names:850390-1). First published in Encycl. 3: 211 (1789).

Wild Jack/Hirsute Artocarpus (*Artocarpus hirsutus* Lam.)

Synonyms: *Artocarpus pubescens* Willd and *Saccus hirsutus* (Lam.) Kuntze (POWO, 2024).

Vernacular names: In **Tamil:** Aiyinipila, Anjili; in **Malayalam:** Anjili, Ayani; and in **Kannada:** Hebbalasu (biotik.org, 2024), Aini and Aini-maram (USDA, 2024).

2E.1 Introduction

Artocarpus hirsutus is endemic to the Western Ghats and found in the South, Central and Maharashtra Sahyadris regions. Ecologically, these trees thrive as canopy species in disturbed evergreen forests, specifically on elevations up to 900 meters (biotik.org, 2024). Despite its distinctive characteristics, this species remains relatively overlooked, largely attributed to its exclusive distribution in the Western Ghats and the surrounding forests. The limited attention it has garnered has resulted in its status as an understudied botanical entity (Matthew *et al.*, 2006).

2E.2 Botany

Artocarpus hirsutus is a majestic tree species that can grow up to 35 metres tall. The bark of these trees is initially dark grey, exhibiting a smooth texture in their early stages and later developing a scaly or flaky characteristic accompanied by lenticels. The blaze of the trunk is distinguished by a cream hue. *A. hirsutus* features young branchlets that are terete, showcasing a tawny hirsute appearance and adorned with annular scars. Striking white latex produced by this species is profuse.

Its leaves are simple, alternate and spirally clustered at the ends of twigs. Stipules being lanceolate and tawny hirsute can grow up to 2.5cm long, proving to be caducous and leaving behind annular scars. The petioles measuring 1.3-3cm in length and displaying a stout, subterete or planoconvex form are covered in hirsute texture. The lamina ranging from 10 to 30.5cm in length and 5 to 14cm in width typically assumes an ovate to broadly elliptic shape. The apex is subacute or shortly acuminate while the base is rounded or subacute. The margin of the leaves is either entire or undulate and occasionally lobed when young. The coriaceous texture of the leaves dries to a brown colour, initially densely hirsute beneath when young but becoming glabrous except on the midrib and nerves. The midrib is flat above accompanied by 10-12 pairs of ascending secondary nerves with broadly reticulo-percurrent tertiary nerves.

In terms of reproduction this bears unisexual flowers with male flowers arranged in axillary spikes that are pendulous and reach a length of up to 14cm. Female flowers also appear in axillary spikes assuming an ovoid shape. Its fruit is a syncarp ~ known as sorosis and is subglobose or ellipsoid in form with long echinate processes. When ripe the fruit takes on an orange hue where inside there may be numerous ovoid seeds (biotik.org, 2024).

2E.3 Uses

Its fruits and seeds serve as a rich source of energy with constituents comparable or even superior to those found in other two *Artocarpus* species. This intriguing aspect warrants further exploration and research (Solanki *et al.*, 2020). The ripe fruit of *A. hirsutus* is eaten after removing spiny outer skin. The structure of the fruit is similar to that of the much larger jackfruit. The seeds are also edible so usually fried as a snack.

Beyond its utilitarian applications various components of 'wild jack' including leaves, seeds and bark have been integrated into traditional medicinal practices (Matthew *et al.*, 2006; Jagtap *et al.*, 2010; Solanki *et al.*, 2020; Buddhisuharto *et al.*, 2021; Meenu *et al.*, 2021). Its role in traditional medicine underscores its cultural and medicinal importance ~ adding another layer to its multifaceted significance.

In its indigenous habitat within the Western Ghats, wild jack assumes a pivotal role as a multipurpose plant, exhibiting both economic and ecological significance. Particularly noteworthy is its provision of long-lasting, high-quality and pathogen-resistant timber which has found extensive application in the construction of houses, boats and other durable structures (Matthew *et al.*, 2006;

Xavier *et al.*, 2014; Meenu *et al.*, 2021). Wild jack holds cultural significance in Kerala where it serves a unique purpose. In the local tradition it is employed in the construction of 'Palliyodam' ~ a traditional snake boat utilized during the annual water processions of 'Uthrattathi Jalamela' and 'Valla Sadhya' at the Aranmula Parthasarathy Temple. These deeply rooted events in the region's religious and cultural practices showcase its versatile use beyond its botanical attributes. The wood from this tree not only contributes to the ecological diversity but also plays a vital role in sustaining and enriching the cultural heritage of Kerala (TNM, 2021).

Despite its invaluable contributions and potential, the wild jack faces a critical challenge. With a maturity period of approximately 20 years for optimal timber harvesting coupled with its restricted distribution, the current population of the species is unable to meet the escalating demand in the timber market. This vulnerability places it at risk of population decline, emphasizing the urgency for conservation efforts and sustainable management practices to ensure its continued existence and contributions to both ecological balance and human well-being.

References

1. biotik.org, 2024. *Artocarpus hirsutus* – MORACEAE. Accessed on 27 January 2024. http://www.biotik.org/india/species/a/artohirs/artohirs_en.html.
2. Buddhisuharto, A.K., H. Pramastya and I. Fidrianny, 2021. An Updated Review of Phytochemical Compounds and Pharmacology Activities of *Artocarpus* genus. *Biointerface Res. Appl. Chem.*, **11**(6): 14898–905.
3. Jagtap, U.B., S.N. Panaskar and V.A. Bapat, 2010. Evaluation of antioxidant capacity and phenol content in jackfruit (*Artocarpus heterophyllus* Lam.) fruit pulp. *Pl. Foods for Human Nutr.*. **65**(2): 99–104.
4. Matthew, S.P., A. Mohandas, S.M. Shareef and G.M. Nair, 2006. Biocultural diversity of the endemic 'wild jack tree' on the Malabar coast of South India. *Ethnobot. Res. & Applic.*, **4**: 025-040.
5. Meenu, M.T., G. Kaul, M. Shukla, K.V. Radhakrishnan and S. Chopra, 2021. Cudraflavone C from *Artocarpus hirsutus* as a promising inhibitor of pathogenic, multidrug-resistant *S. aureus* persisters and biofilms: a new insight into a rational explanation of traditional wisdom. *J. Nat. Prod.*, **84**(10): 2700-2708.
6. POWO, 2024. *Artocarpus hirsutus* Lam. Plants of the World Online. Royal Botanic Gardens, Kew Science. Retrieved on 9 January, 2024. (https://powo.science.kew.org/taxon/urn:lsid:ipni.org:names:850390-1). First published in Encycl. 3: 211 (1789).

7. Solanki, S., R. Bhardwaj, R. Vasudeva, S. Chourey and S. Archak, 2020. Biochemical composition of pulp and seed of Wild Jack (*Artocarpus hirsutus* Lam.) fruit. *Pl. Foods & Hum. Nutr.*, **75**(4): 659-60.

8. TNM, 2021. Explained: What is a Palliyodam, and why a Kerala actor was arrested for photoshoot on it. *Thenewsminute*. Retrieved 27 January 2024.

9. USDA, 2024. Agricultural Research Service, National Plant Germplasm System. 2024. Germplasm Resources Information Network (GRIN Taxonomy). National Germplasm Resources Laboratory, Beltsville, Maryland. URL: https://npgsweb.ars-grin.gov/gringlobal/taxon/taxonomydetail?id=317887. Accessed 26 January 2024.

10. Xavier, T.F., M. Kannan, L. Lija, A. Auxillia, A.K.F. Rose and S.S. Kumar, 2014. Ethnobotanical study of Kani tribes in Thodu hills of Kerala, South India. *J. Ethnopharmacol.*, **152**(1): 78–90.

Breadfruit [*Artocarpus altilis* (Parkinson) Fosberg.]

Synonyms: *Artocarpus altilis* var. *non-seminifer* (Duss), *A. communis* J.R.Forst. & G.Forst., *A. incisifolius* Stokes, *A. incisifolius* var. *apyrenus* Stokes, *A. incisus* (Thunb.) L.f., *A. incisus* var. *laevis* (Hassk.) Miq., *A. incisus* var. *non-seminifer* Duss, *A. laevis* Hassk., *A. rima* Blanco, *Radermachia incisa* Thunb., *Saccus communis* (J.R.Forst. & G.Forst.) Kuntze, *S. laevis* (Hassk.) Kuntze and *Sitodium altile* Parkinson (POWO, 2024).

Common names: Árbol de pan, fruta de pan, pan, panapen (Spanish); arbre à pain, fruit à pain (French); beta (Vanuatu); bia, bulo, nimbalu (Solomon Islands); blèfoutou, yovotévi (Bénin); breadfruit (English); brotfruchtbaum (German); broodvrucht, broodboom (Dutch); cow, panbwa, pain bois, prutapan and fruta de pan (Caribbean); fruta pao, pao de massa (Portuguese); kapiak (Papua New Guinea); kuru (Cook Islands); lemai, lemae (Guam, Mariana Islands); mazapan (Guatemala, Honduras); meduu (Palau); mei, mai (Federated States of Micronesia, Kiribati, Marshall Islands, Marquesas, Tonga, Tuvalu); mos (Kosrae); rata del (Sri Lanka); rimas (Philippines); shelisheli (Tanzania); sukun (Indonesia, Malaysia); ulu (Hawaiʻi, Samoa, Rotuma, Tuvalu); uru (Society Islands); uto, kulu and buco (Fiji) (ntbg.org, 2024).

Indian names: In Hindi: Bakri-chajhar; in Marathi: Nirphanas, Vilayati phanas; in Tamil: Rottipala, Irppla, Kadapila; in Kannada: Jeegujje, Deevi halasu,

Rotti halasum, Gujju halasu, Seeme halasu; In Konkani: Jivi kadgi; in Malayalam: Kadachakka, Kadapilavu, Seemaplavu; in Sanskrit: Nagadamani (flowersofindia.net, 2024).

3.1 Introduction

Artocarpus altilis

By Kowloonese - English Wikipedia (http://en.wikipedia.org/wiki/Image:Breadfruit.JPG), CC BY-SA 3.0, https://commons.wikimedia.org/w/index.php?curid=130671

Renowned as one of the eminent food producers in its domain, the breadfruit, enriched with a history romanticized and dramatized stands as a widely recognized entity if not by direct experience then certainly by its resounding name. The breadfruit tree is known for its adaptability to different climates. It stands out as a highly versatile crop with its fruit being palatable and suitable for consumption at every stage of maturity. The seeds too are edible when cooked. Esteemed for its starchy and nutritious qualities breadfruit serves as a staple food in numerous tropical regions. With a rich history spanning over 3,000 years it has been a crucial staple crop and a key element in traditional agroforestry systems in the Pacific (ntbg.org, 2024).

The tree has historical and cultural significance in many societies where it is cultivated. This botanical marvel, belonging to the illustrious mulberry family, Moraceae, carries with it a rich tapestry of nomenclature and cultural significance. Universally referred to as 'breadfruit' its influence transcends linguistic boundaries. In the enchanting cadence of Spanish it takes on the guise of 'fruta de pan' (fruit) or 'arbor de pan'/'arbor del pan' (tree), alternatively known as 'pan de pobre'. French language lends its elegance so christen it 'fruit à pain' (seedless), 'châtaignier' (with seeds), and 'arbre à pain' (tree). Meanwhile, in the Portuguese tongue, it assumes the identity of 'fruta pão' or 'pão de massa'. In the Dutch vernacular, it reveals itself as 'broodvrucht' (fruit) and 'broodboom' (tree). Within the vibrant tapestry of Venezuela, myriad names reflect its diverse roles and attributes. From 'pan de ano' and 'pan de todo el año' to 'pan de palo' and 'pan de name' the breadfruit is woven into the culinary and cultural fabric of the region. Similarly, Guatemala and Honduras celebrate its essence as 'mazapan' (seedless) and 'castana' (seeded one). In Peru, it takes on the name 'marure', while in Yucatan it is known as 'castano de Malabar' (with seeds). Puerto Rico embraces it as 'panapen' (seedless) and 'pana de pepitas' (seeded) (Morton, 1987). Far beyond the shores of the Americas, in the lush landscapes of Malaya and Java, the breadfruit goes by 'suku' or 'sukun' (seedless), and 'kulur', 'kelur', or 'kulor' (seeded one). In the vibrant tapestry of Thailand, it resonates as 'sa-ke' while the Philippines affectionately refer it as 'rimas' (seedless). Across the Pacific waves in

Hawaii, it proudly carries the name 'ulu'. The intricate diversity of names reflects not only linguistic nuances but also the myriad ways in which the breadfruit has embedded itself into the culinary and cultural identities of regions across the globe. Beyond its nomenclature, the breadfruit has also found a place in the lexicon of sustenance, showcasing versatility in its applications.

It is important to note that breadfruit is not a monolith, with distinctions emerging based on the presence or absence of seeds. The type with seeds occasionally bears the moniker 'breadnut' ~ a title more aptly reserved for *Brosimum alicastrum* Swartz. This related species, indigenous to Yucatan, Central America and neighboring areas boasts edible seeds with its Spanish name being 'ramon'. The seeds, leaves and twigs of this tree are esteemed for their value in livestock feed ~ adding another layer to the intricate narrative of botanical diversity and utility.

3.2 Etymology

Nomenclature of breadfruit expressed in its scientific or Latin appellation finds its etymological roots in the Greek language wherein '*artos*' signifies bread and '*karpos*' the fruit. Complementing this linguistic amalgamation, the epithet '*altilis*' is bestowed upon the breadfruit. The word '*altilis*' is derived from the Latin '*altus*' meaning fat or nourishing (ntbg.org, 2024). The choice of 'altilis' is not arbitrary as it encapsulates the essence of this prolific tree as a source of rich sustenance. The term '*altilis*' further traces its origin to the Latin word '*utile*' meaning useful thereby enshrining the breadfruit tree as an entity revered for its practical utility. This nomenclatural interplay thus memorializes the breadfruit tree as a utilitarian species bestowing upon it the dual distinction of being both nutritionally enriching and purposefully advantageous.

The colloquial name 'breadfruit' itself is a testament to the sensory experience associated with its fruits. When subjected to the transformative process of baking or roasting, these fruits emanate a gratifying and aromatic fragrance. Moreover, the resulting texture is starchy and dense reminiscent of traditional staples such as bread or tuber crops like potatoes, yams or sweet potatoes.

In essence, the name 'breadfruit' not only encapsulates the linguistic homage to its bread-like and fruitful qualities but also serves as a sensory prelude to the culinary delights and nutritional bounty that this remarkable tree graciously bestows upon those who partake in its harvest.

3.3 Origin and Distribution

Artocarpus altilis is believed to be a domesticated descendant of *A. camansi* originating in New Guinea ~ the Maluku Islands, and the Philippines (Matisoo-

Smith, 2015; Morton, 1987). Believed to originate in a vast expanse ranging from New Guinea through the Indo-Malayan Archipelago to Western Micronesia, breadfruit is a botanical treasure with a rich history of dissemination and cultivation. The journey of this starchy fruit through time and across diverse regions has shaped its global presence and significance (Morton, 1987). Tree is grown on most Pacific Islands but with exception of New Zealand and Easter Island. Now it is cultivated throughout the tropics (ntbg.org, 2024).

Introduction of breadfruit to the Pacific region is credited to migrating Polynesians who as per Hawaiian tradition are believed to have transported it from the Samoan island of Upalu to Oahu around the 12th Century A.D. Initial European encounter with breadfruit occurred in the Marquesas in 1595, followed by Tahiti in 1606. By the end of 18th century, English explorers were extolling its virtues, prompting British West Indies plantation owners to appeal to King George III for importation of seedless breadfruit trees during periods of famine in Jamaica between 1780 and 1786 (Morton, 1987). In 1772, French navigator Sonnerat reportedly acquired seeded breadfruit in the Philippines, introducing it to the French West Indies. Both seedless and seeded breadfruit plants made their way to Jamaica with at least two seeded breadfruit plants documented on the island in 1784. Captain Bligh's infamous 1787 voyage to Tahiti resulted in the loss of his cargo of 1,015 potted breadfruit plants but his 1791 journey successfully delivered various types totalling 2,126 plants to Jamaica. Seedless variety flourished on the island and propagated to other Caribbean islands, as well as the lowlands of Central and northern South America (Morton, 1987).

Polynesian breadfruit varieties spread across the Caribbean and extended to C&S America, Africa, India, SE Asia, Madagascar, the Maldives, the Seychelles, Indonesia, Sri Lanka, northern Australia, and south Florida. Presently, breadfruit is cultivated in nearly 90 countries (ntbg.org, 2024). While Jamaica remains a leading producer of seedless breadfruit, the distribution of seeded and seedless varieties varies across regions. In Haiti, the seeded type is more prevalent. In New Guinea, only the seeded type is cultivated for consumption (Morton, 1987).

Historical connections are drawn between the seeded breadfruit carried by Spaniards from the Philippines to Mexico and Central America where it is a common food for domestic swine on the Pacific coast. On the Atlantic Coast, seedless varieties are popular among people of African origin. Yucatan utilizes the breadfruit tree for shade and in Colombia it is a common food in certain regions but primarily used as livestock feed in other regions (Morton, 1987).

In Guyana and Trinidad, both seeded and seedless breadfruits are appreciated due to the significant Asian population. However, in some Caribbean areas breadfruit is considered a food for the poor which is reserved for emergencies.

Recently, it has gained attention from gourmets with some islands exporting small shipments to the United States, Canada and Europe for specialized ethnic markets (Morton, 1987).

While breadfruit faces challenges from competing food sources in certain regions, efforts are on by USDA to introduce breadfruit plants from the Canal Zone. Panama in 1906 contributed to its presence in Florida. Despite limitations in outdoor cultivation due to mild winters, some seedless breadfruit trees have thrived in Florida, particularly in Key West and Vaca Key. Greenhouses and conservatories in establishments like the Rare Plant House of Fairchild Tropical Garden and the indoor garden of the Jamaica Inn (Key Biscayne) have also played a role in preserving these plants. Journey of breadfruit from its ancestral roots to its diverse global adaptations showcases its resilience and the cultural and agricultural tapestry it weaves across continents (Morton, 1987).

Tree showing form of the buttress root

By Ping an Chang - Own work, CC BY-SA 4.0, https://commons.wikimedia.org/w/index.php?curid=123935047

3.4 Botany

3.4.1 Tree structure

The breadfruit is a striking, large (15-26m high), single-trunked evergreen and rapidly growing tree with spreading evergreen canopy. It typically features a clear trunk extending to 4-6m before branching out, with a width ranging from 0.6-1.8m. Its wood is gold in colour but when cut soon it turns darker when comes in contact with air. Milky gummy latex can be seen in all its parts even in unripe fruits (Ragone, 2006; Sikarwar *et al.,* 2014; Morton, 1987).

3.4.2 Root system

The root system of the breadfruit tree is typically well-developed and buttressed at the base providing stability for the fast-growing tree (Morton, 1987).

3.4.3 Foliage

The tree showcases numerous spreading branches varying in thickness and length. Some branches are thick with lateral foliage-bearing branchlets while others are long and slender with foliage clustered only at their tips. The leaves, evergreen or deciduous depending on climate are alternate, broadly obovate to broadly ovate, 15-90cm long and 20-50cm wide, almost entire with only slight lobing to deeply pinnately lobes having sinuses up to 2/3 or more of the distance from margin to midrib with up to six pairs of lobes and a large apical tip. Blade is generally smooth, glossy and dark green on the upper surface, featuring conspicuous green or yellow-green veins and few to many white to reddish-white hairs on the midrib and veins while the underside is dull yellowish and coated with minute stiff hairs. Leaves on new shoots and root suckers are generally larger and more hirsute than those on mature branches. Size is variable depending on the variety (Ragone, 2006; Morton, 1987).

3.4.4 Flowers

Male inflorescence

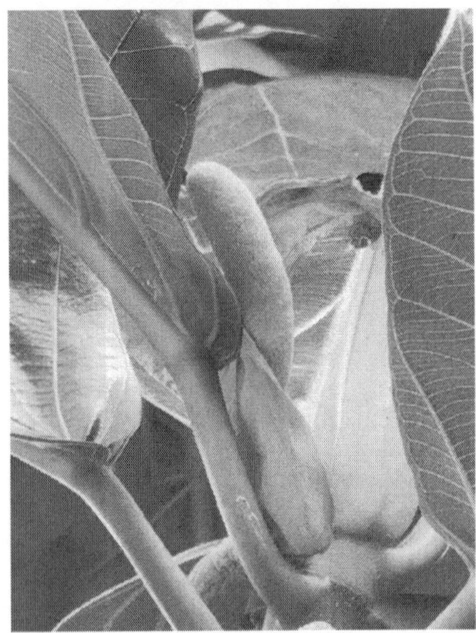

By Ping an Chang - Own work, CC BY-SA 4.0, https://commons.wikimedia.org/w/index.php?curid=103498957

Female inflorescence

By Ping an Chang - Own work, CC BY-SA 4.0, https://commons.wikimedia.org/w/index.php?curid=103498958

Breadfruit tree is monoecious in nature with separate male and female inflorescences coexisting on the same tree where male inflorescence appears first. The tree bears numerous tiny flowers with club-shaped male flowers densely set on a drooping spike, 12.5-30cm or even up to 45cm long and 2.5-3.75cm or sometimes even up to 5cm thick, initially yellowish and later turning brown. Thousands of tiny flowers with two anthers are attached to a central spongy core (Ragone, 2006; Morton, 1987).

Female inflorescences consist of 1500–2000 reduced flowers attached to a spongy core. The female flowers are massed in a green prickly head, 6.35cm long and 3.8cm across. The flowers fuse together and develop into the fleshy edible portion of the fruit. It is cross-pollinated but pollination is not required for the fruit to form (Ragone, 2006; Morton, 1987). It has diploid ($2n=2x=56$) and triploid ($2n=3x=84$) chromosomes (Jarrett, 1959; Barrau, 1976).

3.4.5 Fruits and seeds

Breadfruit whole, sliced lengthwise and in cross-section

By US Pacific Basin Agricultural Research Center - US Pacific Basin Agricultural Research Center (http://pbarc.ars.usda.gov/pages/research/tpgrmu/Hart-49.jpg; http://pbarc.ars.usda.gov/pages/research/tpgrmu/HART49.shtml; http://pbarc.ars.usda.gov/pages/research/tpgrmu/artocarpus.shtml), Public Domain, https://commons.wikimedia.org/w/index.php?curid=131476

Fruits are variable in shape, size and surface texture. There are both seed-bearing and seedless varieties with the latter being more widely cultivated. The compound syncarp fruits of breadfruit tree is oblong, cylindrical, ovoid, rounded or pear-shaped measuring 9-45cm in length and 5-30cm in diameter that may weigh 0.25 to 6kg each. Thin rind is patterned with irregular 4- to 7-sided faces with variations in texture from smooth to slightly bumpy or spiny and the colouration being light green, yellowish-green or yellow/yellow-brown when mature though one unusual variety ('Afara' from the Society Islands) has pinkish or orange-brown skin. Mature skin typically bears dried latex exudations which stains its surface. In its unripe state the fruit is firm with a white, starchy and somewhat fibrous interior. Upon reaching full ripeness the fruit softens revealing a creamy or yellow pasty interior that is sweetly fragrant. Fruits are either singularly borne or clustered in groups of 2 to 3 at the branch tips. The fruit stalk (pedicel) is usually 2.5-12.5cm long. Two strap-shaped reflexed stigmas protrude from centre of the disk and often leave a small distinctive scar when they blacken and wither. Fruits are typically mature and ready to harvest for consumption as a starchy staple in 15–19 weeks. The flesh of ripe fruit can be eaten raw but quite rarely in the Pacific (Ragone, 2006; Sikarwar *et al.*, 2014; Morton, 1987).

This is of two types ~ the normal wild type with seeds though with little pulp and the more widely grown cultivated seedless type. Occasionally, seedless cultivars possess fully developed seeds (Ragone, 2006; Sikarwar *et al.*, 2014; Morton, 1987). Variations with entire leaves as well as both the seeds and edible pulp have been classified separately but may integrate with highly variable *A. altilis* species. At the core of seedless fruits lies a cylindrical or oblong structure often adorned with hairs that carries brown, flat and underdeveloped seeds measuring approximately 3mm in length. India's breadfruits are seedless while in the Philippines a spiky and seeded relative called 'Kamansi' is cultivated.

Seeds are thin-walled, subglobose or obovoid, irregularly compressed, rounded at one end, pointed at the other, about 2cm long, 1–2cm thick, dull-brown with darker stripes and embedded in the pulp. The outer seed coat is usually shiny dark brown with a light brown inner seed coat. Seeds have little or no endosperm or dormancy as these germinate immediately and are unable to withstand desiccation. Seeds are distributed by flying foxes where these occur, however, these are rarely used for propagation (Ragone, 2006; Morton, 1987).

3.4.6 Rooting habit

The breadfruit plant, specifically varieties such as the hybrid *A. altilis* × *A. mariannensis* manifests a robust root system that extends and flourishes both at the surface and just below the soil. Certain cultivars showcase the emergence of significant buttress roots contributing to the overall structural integrity and support of the plant (Ragone, 2006).

3.4.7 Botanical classification

Dr. Fosberg has meticulously classified certain botanical forms characterized by complete leaves, seeds and edible pulp as a distinct species such as *A. mariannensis* Trecul. However, despite this distinct classification, these variations frequently intermingle with *A. altilis*, prompting some botanists to regard them as integral components of the exceptionally diverse *A. altilis* species (Morton, 1987).

3.5 Growth and Development

3.5.1 Growth rate

Under favourable conditions breadfruit demonstrates rapid growth averaging 0.5–1.5 metres annually, attaining the diameter of almost 1m within its initial 10–12 years of life. Though small branches may experience dieback at the tip post-fruiting, the tree consistently exhibits the ability to generate new shoots and branches throughout its entire lifespan (Ragone, 2006). Upon experiencing damage or cutting of its roots, breadfruit responds by producing numerous root shoots. Furthermore, the tree promptly regenerates new shoots and branches in the aftermath of wind damage or intentional topping to facilitate harvesting. Notably resilient, even large trees exceeding 1 metre in diameter can recover and bear fruits within as little as 2 years following extensive pruning.

3.5.2 Flowering and fruiting

Breadfruit exhibits a seasonal bearing pattern with majority of varieties yielding one or two crops annually. The primary harvest usually takes place amid the warm and rainy summer months, succeeded by a smaller secondary crop approximately 3-4 months later. Trees derived from seed commence flowering and bearing fruit within a span of 6-10 years duration though sometimes even earlier. Vegetatively propagated trees initiate fruiting in a more accelerated timeframe, *i.e.* from 3 to 6 years (Ragone, 2006).

3.6 Classification

3.6.1 Breadfruit diversity in the South Pacific

Throughout the Pacific breadfruit exhibits great morphological variability ranging from true seedless varieties to those with several small aborted seeds or with one to a few viable seeds to varieties with numerous viable seeds. Seeded types are most common in the SW Pacific. Seedless varieties are most common in Micronesia and the eastern islands of Polynesia. All of the breadfruit varieties elsewhere in the tropics are seedless (Ragone, 2006).

Delving into the rich history of breadfruit cultivation in the South Pacific, this comprehensive analysis merges insights from a 1921 unpublished report covering 200 cultivars in the Marquesas and the 1966 survey published by the South Pacific Commission. An additional contribution from Koroieveibau provides a key to the eight classes ~ each defined by leaf form and fruit characteristics. The report unveils a trove of information describing 166 named sorts across Tonga, Niue, Western and American Samoa, Papua New Guinea, New Hebrides and Rotuma. Fiji with its own distinct breadfruit landscape boasts 70 named varieties intricately categorized into eight classes based on leaf form (Morton, 1987).

3.6.2 Varietal classes and excellent selections

The ensuing list provides a succinct overview of those entries deemed to be of exceptional quality categorized as 'very good'. It is noteworthy that certain varietal names are documented in multiple classifications, adding a layer of complexity to the assessment process. This underscores the nuanced nature of the evaluations and the versatility of certain entries as they transcend conventional categorizations (Morton, 1987).

i) **Class I: Leaf entire or with one or two (occasionally three) lobes**

a. 'Koqo': Round, 4-10cm wide; seedless; slow deterioration.

b. 'Tamaikora': Gourd-shaped, 11.5cm long, 7.5cm wide; many seeds; raw consumption; highly perishable.

ii) **Class II: Leaf dissected at apex**

a. 'Temaipo': Round, about 9cm long; seedless; raw consumption; oblong form with many seeds.

iii) **Class III: Leaf moderately but deeply dissected at apex**

a. 'Uto Kuro': Round, 12.5cm long; slow deterioration.

iv) **Class IV: Leaf moderately deeply dissected on upper half**

a. 'Samoa' ('Kasa Balavu'): Round, 10-15cm long; sparse to many seeds.

b. 'Uto Yalewa': Oblong, up to 20cm long, 15cm wide; seedless.

c. 'Kulu Dina': Oblong, up to 40cm long, 33cm wide; seedless; no need to peel after cooking; year-round fruiting.

d. 'Sogasoga': Oblong, up to 23cm long, 16.5cm wide; seedless.

e. 'Uto Dina': Oblong, up to 15cm long, 7.5-9cm wide; seedless; no need to peel after cooking; tree height 18-21m.

f. 'Buco Ni Viti': Oblong, 28-35.5cm long, 15-18cm wide; seedless; outstanding cultivar.

g. 'Tamaikora': Oblong, 18-23cm long, 12.5-16.5cm wide; sparse seeds; pulp eaten raw; tree height 23-26m; two crops per year.

h. 'Kulu Mabomabo': Oval, 15-20cm long, 10-14cm wide; seedless.

v) **Class V: Leaf moderately deeply dissected; shape of leaf base variable**

a. 'Uto Dina': Round, 11.5-12.5cm wide; seedless; highly recommended; tree height 7.5-9m.

b. 'Balekana Ni Samoa': Round, 10-12.5cm long; sparse seeds; best Samoan variety; oval form seedless but deteriorates quickly.

c. 'Balekana Ni Vita': Round, 9-10cm long; seedless; slow deterioration.

d. 'Balekana Dina': Oval, 15-20cm long, 7.5-12.5cm wide; sparse seeds; excellent when boiled.

e. 'Tabukiraro': Round, 20cm long; seedless; skin sometimes eaten after cooking.

f. 'Sici Ni Samoa': Oval, 12.5-15cm long, 7.5-9cm wide; seedless; highly recommended Samoan variety.

g. 'Uto Me': Oval, 12.5-17cm long, 11.5cm wide; many seeded; slow deterioration.

h. 'Uto Wa': Oval, 15-19cm long, 12.5-14cm wide; most recommended variety.

i. 'Kulu Vawiri': Oval, 22-30cm long, 20-22cm wide; especially good when boiled.

vi) Class VI: Leaf deeply dissected

a. 'Kulu Dina': Round, 7.5-10cm long; seedless; no need to peel after boiling; highly recommended.

b. 'Balekana': Oval, 10cm long, 7.5cm wide; top-quality; tree height 21-24m.

c. 'Balekana Ni Samoa': Round, 7.5cm long; sparse seeds; best of Samoan varieties.

d. 'Balekana Ni Viti': Oblong, 12.5-15cm long, 7.5-10cm wide; seedless; best native-type variety.

e. 'Uto Dina' ('Kasa Leka'): Round, 10cm long; seedless.

f. 'Uto Matala': Round, 7.5-10cm long; excellent when boiled; three crops per year.

vii) Class VII: Leaf deeply dissected; apex pointed

a. 'Balekana Ni Samoa': Round, 12.5-14cm long; sparse seeds; best of Samoan varieties.

b. 'Kulu Dina' ('Kasa Balavu'): Oval, 15-18cm long, 10-12.5cm wide; seedless.

c. 'Uto Dina' (Large): Oval, 20-22cm long, 10-18cm wide; seedless; also a form with moderately dissected leaves.

d. 'Bokasi': Round, 10cm long, 7.5cm wide.

viii) Class VIII: Leaf deeply dissected, wide spaces between lobes

a. 'Savisavi Ni Samoa': Oval, 10-12.5cm long, 7.5-9cm wide; ranks with the best Samoan varieties.

b. 'Savisavi Ni Viti': Oblong, 16-20cm long, 10-15cm wide; seedless; especially good when boiled.

c. 'Savisavi': Round, 7.5-9cm wide; especially good when boiled.

d. 'Balawa Ni Viti': Oval, 15-18cm long, 9-10cm wide; seedless.

e. 'Uto Kasekasei': Round, 10-12.5cm long; sparse seeds.

f. 'Via Loa': Oblong, 15-18cm long, 10-12.5cm wide; seedless; slow deterioration.

Pacific Islands has published a detailed information on 52 cultivars. Further enriching this knowledge, G.P. Wilder's book '**The Breadfruit of Tahiti**', offers intricate insights complete with close-up, black-and-white photographic illustrations of foliage and fruit for 30 named varieties. One particular cultivar 'Aata' stands out for its oblong fruit deemed of suboptimal quality for human consumption until more preferable breadfruits are scarce. Nonetheless, it plays a crucial role as feed for pigs and horses, showcasing its agricultural significance and multifaceted utility.

Several standout varieties each possessing unique attributes have been documented, showcasing the diversity within the breadfruit spectrum:

- **'Aravei'**: Notable for its ellipsoidal large fruit with a rough, yellowish-green rind and a light-yellow deliciously flavoured pulp after minimal cooking. Tree is known for its heavy fruiting.
- **'Havana'**: Characterized by oval-round fruit with a spiny yellowish-green rind and a golden-yellow moist pulp with an excellent sweet flavour. Fruit is highly perishable, requiring consumption within two days.
- **'Maohi'**: A round fruit with bright yellow-green rind having red-brown patches and smooth-cooking creamy pulp. Despite slow cooking, it boasts excellent flavour and is a prolific bearer.
- **'Paea'**: Ellipsoidal and very large with a yellowish-green spiny rind, pulp being bright-yellow and moist, core being oval and thick with abortive seeds. Reserved for chiefs in the past, it requires one hour of roasting over an open fire.
- **'Pei'**: Broad-ellipsoidal, large fruit with light-green and relatively smooth rind. Its pulp is light-yellow, flaky and aromatic that cooks quickly. Its fruits can be stored for up to one year in the ground.
- **'Pucro'**: Spherical or elongated large fruit with a rough, spiny and yellow-green rind, *vis-a-vis* smooth light-yellow pulp. Highly esteemed and is a best breadfruit.
- **'Rare'**: Broad-ovoid fruit with a bright-green, rough and spiny rind, and with a deep-cream tone, fine-grained and sweet-flavoured pulp. Recognized in three distinct forms. It must be cooked for about an hour.
- **'Tatara'**: Quite large broad-ellipsoid fruit with a spiny rind and smooth-cooking light-yellow pulp. Esteemed for its flavour, it grows in a specific coastal valley with heavy rainfall.

- **'Vai** Paere': Obovoid fruit with yellow-green rind having red-brown splotches, and an excellent-flavoured slightly acidic light-yellow pulp. The tree is tall and bears fruit in clusters.

Beyond these varieties, 'Ponape' and 'Truk' boast at least 50 distinct cultivars with Samoa's 'Maopo' recognized as one of the best due to its almost entire or shallowly lobed leaves. 'Puou' and 'Ulu Ea' varieties known for deeply cut leaves and rounded fruits have been favoured since early times.

Over the past three decades there has been a renewed focus on increasing food supplies in tropical regions by cultivating selected varieties of seedless breadfruit. Comparative trials were initiated in 1958 collecting appealing varieties from the South Pacific and transferring them to Western Samoa, Tahiti and Fiji. Two years later, efforts were made to introduce Polynesian varieties into Micronesia, distributing propagating material of 36 Micronesian types to various regions, highlighting ongoing endeavours to enhance breadfruit cultivation.

3.7 Nutritional Value

3.7.1 Fruit

Nutritional value per 100g of breadfruit, raw (USDA, 2018a)			
Energy	***431 kJ (103 kcal)***	**Minerals**	**Quantity (%DV†)**
Carbohydrates	27.12 g	Calcium	17 mg (2%)
Sugars	11 g	Iron	0.54 mg (4%)
Dietary fibre	4.9 g	Magnesium	25 mg (7%)
Fat	0.23 g	Manganese	0.06 mg (3%)
Protein	1.07 g	Phosphorus	30 mg (4%)
Vitamins	**Quantity (%DV†)**	Potassium	490 mg (10%)
Vitamin A equiv. lutein zeaxanthin	22 μg	Sodium	2 mg (0%)
Thiamine (B1)	0.11 mg (10%)	Zinc	0.12 mg (1%)
Riboflavin (B2)	0.03 mg (3%)	**Other constituents**	**Quantity**
Niacin (B3)	0.9 mg (6%)	Water	70.65 g
Pantothenic acid (B5)	0.457 mg (9%)	†Percentages are roughly approximated using US recommendations for adults.	
Vitamin B6	0.1 mg (8%)		
Folate (B9)	14 μg (4%)		
Choline	9.8 mg (2%)		
Vitamin C	29 mg (35%)		
Vitamin E	0.1 mg (1%)		
Vitamin K	0.5 μg (0%)		

3.7.1.2 Seed

Seeds, breadfruit (100g)				
Name	**Boiled**	**Raw**	**Roasted**	**Unit**
Water	59.3	56.3	49.7	g
Energy	168	191	207	kcal
Energy	703	799	866	kJ
Protein	5.3	7.4	6.2	g
Total lipid (fat)	2.3	5.59	2.7	g
Ash	1.1	1.5	1.3	g
Carbohydrate, by difference	32	29.2	40.1	g
Fiber, total dietary	4.8	5.2	6	g
Calcium, Ca	61	36	86	mg
Iron, Fe	0.6	3.67	0.9	mg
Magnesium, Mg	50	54	62	mg
Phosphorus, P	124	175	175	mg
Potassium, K	875	941	1080	mg
Sodium, Na	23	25	28	mg
Zinc, Zn	0.83	0.9	1.03	mg
Copper, Cu	1.07	1.15	1.32	mg
Manganese, Mn	0.132	0.142	0.163	mg
Vitamin C, total ascorbic acid	6.1	6.6	7.6	mg
Thiamin	0.29	0.482	0.41	mg
Riboflavin	0.17	0.301	0.24	mg
Niacin	5.3	0.438	7.4	mg
Pantothenic acid	0.816	0.877	1.01	mg
Vitamin B-6	0.298	0.32	0.42	mg
Folate, total	49	53	59	µg
Folate, food	49	53	59	µg
Folate, DFE	49	53	59	µg
Vitamin A, RAE	12	13	15	µg
Vitamin A, IU	238	256	294	IU
Vitamin D (D2 + D3)	0	0	0	µg
Fatty acids, total saturated	0.621	1.51	0.729	g
SFA 16:0	0.411	0.999	0.483	g
SFA 18:0	0.21	0.51	0.246	g
Fatty acids, total monounsaturated	0.293	0.712	0.344	g

Seeds, breadfruit (100g)				
Name	**Boiled**	**Raw**	**Roasted**	**Unit**
MUFA 16:1	0.012	0.03	0.015	g
MUFA 18:1	0.281	0.682	0.329	g
Fatty acids, total polyunsaturated	1.22	2.98	1.44	g
PUFA 18:2	0.942	2.29	1.11	g
PUFA 18:3	0.283	0.687	0.332	g
Tryptophan	0.088	0.123	0.103	g
Threonine	0.276	0.385	0.323	g
Isoleucine	0.317	0.443	0.371	g
Leucine	0.403	0.563	0.472	g
Lysine	0.408	0.57	0.477	g
Methionine	0.069	0.096	0.081	g
Cystine	0.083	0.116	0.097	g
Phenylalanine	0.571	0.797	0.668	g
Tyrosine	0.39	0.544	0.456	g
Valine	0.383	0.535	0.448	g
Arginine	0.354	0.494	0.414	g
Histidine	0.148	0.207	0.173	g
Alanine	0.241	0.336	0.282	g
Aspartic acid	0.585	0.817	0.684	g
Glutamic acid	0.742	1.04	0.868	g
Glycine	0.333	0.465	0.39	g
Proline	0.264	0.369	0.309	g
Serine	0.355	0.496	0.415	g
Source: USDA, 2018b,c,d				

Food value per 100g of edible portion (Mortonj, 1987)					
	Fruit (underripe, raw)	**Ripe (cooked)**	**Seeds (fresh)**	**Seeds (roasted)**	**Seeds (dried)**
Calories	105-109				
Moisture	62.7-89.16 g	67.8 g	35.08-56.80 g	43.80 g	
Protein	1.3-2.24 g	1.34 g	5.25-13.3 g	7.72 g	13.8-19.96 g
Fat	0.1-0.86 g	0.31 g	2.59-5.59 g	3.30 g	5.1-12.79 g
Carbohydrates	21.5 29.49 g	27.82 g	30.83-44.03 g	41.61 g	15.95 g
Fiber	1.08 2.1 g	1.5 g	1.34-2.14g	1.67 g	3.0-3.87 g
Ash	0.56-1.2 g	1.23 g	1.50-5.58 g	1.90 g	3.42-3.5 g

Food value per 100g of edible portion (Mortonj, 1987)					
	Fruit (underripe, raw)	**Ripe (cooked)**	**Seeds (fresh)**	**Seeds (roasted)**	**Seeds (dried)**
Calcium	0.05 mg	0.022 g	0.11 mg	40 mg	0.12 mg
Phosphorus	0.04 mg	0.062mg	0.35 mg	178 mg	0.37 mg
Iron	0.61-2.4 mg		3.78 mg	2.66 mg	
Carotene	0.004 mg (35-40 I.U.)				
Thiamine	0.08-0.085 mg		0.25 mg	0.32 mg	180 mcg
Riboflavin	0.033-0.07 mg		0.10 mg	0.10 mg	84 mcg
Niacin	0.506 0.92 mg		3.54 mg	2.94 mg	2.6 mg
Ascorbic Acid	15 33 mg		13.70 mg	14 mg	
Amino Acids	[N = 16 p. 100])				
Arginine	4.9		0.66		
Cystine	-		0.62		
Histidine	1.6		0.91		
Isoleucine	6.7		2.41		
Leucine	7.4		2.60		
Lysine	5.8				
Methionine	1.2		3.17		
Phenylalanine	8.3		1.05		
Threonine	6.8		0.78		
Tryptophan	7.0				
Valine	7.8				
Aspartic Acid	10.8				
Glutamic Acid	11.3		0.98		
Alanine	3.9		1.53		
Glycine	7.2		0.95		
Proline	6.5		0.72		
Serine	5.7		2.08		
Tyrosine			1.45		

The above mentioned nutritional value of breadfruit has been extensively studied through a composite of analyses conducted in various regions including Central America, Mexico, Colombia, Africa and India. This comprehensive examination provides a nuanced understanding of the diverse nutritional profile of breadfruit, shedding light on its potential as a staple food source.

The breadfruit is said to harbour a duo of enzymes, *viz.* papayotin and artocarpine according to available reports. These enzymes ~ pivotal components of the breadfruit's biochemical profile contribute to its unique nutritional composition and potential health benefits. Papayotin, recognized for its proteolytic activity plays a crucial role in breaking down proteins within the fruit. On the other hand, artocarpine with its distinctive chemical properties adds to the overall physiological characteristics of the breadfruit. Together, these enzymes not only enhance the palatability of the fruit but also offer potential therapeutic implications making the breadfruit a fascinating subject for further research and exploration in the realms of nutrition and medicine.

Upon closer examination of aforementioned data a discernible pattern emerges revealing that the seedless variety of breadfruit exhibits a relatively low protein content while its seeded counterpart boasts a considerably higher protein concentration. Consequently, it becomes evident that the seeded breadfruit holds superior nutritional value as a food source.

Delving into specific nutritional profiles, breadfruit flour showcases a composition of 4.05% protein, 76.70% carbohydrates and an energy content of 331 calories per 100g. In stark contrast, cassava flour ~ a pertinent alternative, manifests a protein content of 1.16%, a higher carbohydrate concentration at 83.83% and a slightly elevated caloric content of 347 calories per 100 grammes.

Expanding the understanding of these nutritional nuances, it is essential to underscore the significance of protein as a vital macronutrient for human health. Proteins play a crucial role in various physiological functions including muscle development, immune system maintenance and enzyme synthesis. Therefore, the discernable disparity in protein content between the seedless and seeded breadfruit varieties bears substantial implications for their respective nutritional contributions to human diet. Furthermore, the carbohydrate composition of these flour variants merits attention as carbohydrates serve as the primary source of energy. The juxtaposition of breadfruit and cassava flour underscores their divergent carbohydrate profiles presenting consumers with nuanced choices based on their dietary preferences and nutritional requirements (Morton, 1987).

In conclusion, beyond the immediate comparison of protein, carbohydrate and caloric content, a comprehensive understanding of the nutritional nuances between seedless and seeded breadfruit, as well as their alternatives like cassava flour empowers individuals to make informed dietary decisions aligned with their health and nutritional goals.

3.8 Phytochemicals

Artocarpus genus showcases a remarkable capacity for synthesizing a diverse spectrum of secondary metabolites, particularly abundant in phenylpropanoids such as flavonoids and flavones. Moreover, phenolic compounds including flavonoids, stilbenoids and arylbenzofurons are prevalent within this genus. A thorough investigation of *Artocarpus altilis* has unravelled the identification of over 130 compounds distributed across various organs of the tree with more than 70 arising from the phenylpropanoid pathway (Handa *et al.,* 2008).

Furthermore, the chemical constituents of *Artocarpus altilis* encompass morin, moracin, dihydromorin, cynomacurin, oxydihydroartocarpesin, artocarpetin, norartocarpetin (20, 40, 5, 7 tetrahydroxyflavone), cycloartinone, cyclogeracommunin, cyclocommunol, cycloartenyl acetate, β-sitosterol, sitosterol-β-D-glucopyranoside, ursolic acid, betullic acid acetate, artocarpanone, cudraflavone, artoflavon, orartocarpanone (20, 40, 5, 7 tetrahydroxyflavanone), oxyresveratrol and artoindonesianin F(40-[3-methyl-1(E)-butenyl]-(E)-2,30,4,50-stilbenetetrol) (Solanki and Nagori, 2012). Extending the roster of compounds isolated from different plant parts of *Artocarpus altilis,* notable additions include cycloartocarpin from the root stem, cycloartobiloxanthone from the root bark, cudraflavone B and C from the root, morusin from the root bark and an array of compounds such as broussochalcone A, kazinol A, broussoaurone A, cycloartocarpin A, cycloheterophyllin, broussoflavonol F, ursolic acid, cycloartenol, β-sitosterol, artocarpetin, norartocarpetin and certain others (Sikarwar *et al.,* 2014; Lewis, 1961).

Moreover, the leaf extracts of *Artocarpus altilis* obtained through methanolic ethyl acetate and petroleum ether methods comprise steroids, phytosterols, gums and resins. The leaf extract also features 72.5% amino acids, 68.2% fatty acids and 81.4% carbohydrates. The starch content in breadfruit is reported at 15.52g/100g fresh weight. Essential amino acids like cystine, arginine, histidine, leucine, lysine, methionine, threonine, tryptophan along with sucrose, fatty acids and ellagic acid constitute essential constituents of *Artocarpus altilis.* The mineral composition includes calcium, iron and sodium (Huie, 2002; Sikarwar *et al.,* 2014).

Specific to *Artocarpus* species, isoprenylated flavonoids with an isoprenyl side chain at C-3 and 2', 4'-dioxygenation or 2', 4', 5' trioxigenation patterns in ring B of the flavone skeleton are identified. The plant synthesizes prenylflavonoids and geranyl flavonoids (Delazar *et al.,* 2012).

Furthermore, breadfruit is distinguished by the presence of geranyl dihydrochalcones including 1-(2,4-dihydroxyphenyl)-3-[3,4-dihydro-3,8-dihydroxy-2-methyl-2-(4-methyl-3-pentenyl)-2H-1-benzopyran-5-yl]-1-propanone, 1-(2,4-dihydroxyphenyl)-3-[8-hydroxy-2-methyl-2-(3,4-epoxy-4-methyl-1-pentenyl)-2H-1-benzopyran-5-yl]-1-propanone (51-(2,4 dihydroxyphenyl)-3-{4-hydroxy-6,6,9-trimethyl-6a,7,8,10a-tetrahydro-6H-dibenzo[b,d]pyran-5-yl}-1-propanone,), 2-[6-hydroxy-3,7-dimethylocta-2(E),7-dienyl]-2',3,4,4'-tetrahydroxy dihydrochalcone, and 1-(2,4-dihydroxyphenyl)-3-

[8-hydroxy-2-methyl-2-(4-hydroxy-4-methyl-2-pentenyl)-2H-1-benzopyran-5-yl]-1-propanone (Raymond and Scott, 2003; Sikarwar *et al.*, 2014).

The phytochemical profile of *Artocarpus altilis* is rich and varied encompassing a multitude of compounds with potential therapeutic benefits. It is essential to note that further research is warranted to fully elucidate and confirm the specific uses and effects of these phytochemicals.

3.9 Beneficial Effects of Breadfruit on Human Health

In the Pacific and Caribbean regions, various plant parts of breadfruit are extensively utilized for their medicinal and therapeutic properties. Latex, leaf tips, inner bark and other components contribute to a range of traditional treatments (Ragone, 2006).

- Breadfruit latex is applied topically through massage, particularly for addressing issues like broken bones and sprains. It is strategically bandaged on the spine to alleviate symptoms of sciatica. Widely recognized for its efficacy, the latex is a common remedy for treating skin ailments and fungal diseases including conditions like 'thrush'.
- Diluted latex is ingested internally to combat gastrointestinal issues such as diarrhoea, stomachache and dysentery.
- Crushed leaves are specifically employed in the treatment of 'thrush'.
- For ear infections or sore eyes, the sap derived from the crushed stems or leaves is utilized.
- Plant roots possess astringent properties and are administered as a purgative. The macerated root is employed as a poultice for addressing various skin ailments.
- Bark is also enlisted in traditional medicine, particularly in the treatment of headaches across multiple islands.
- In the West Indies, a unique medicinal preparation involves brewing tea from the yellowing leaves of the breadfruit plant. This tea is believed to have beneficial effects on high blood pressure and asthma. It is perceived as having a potential role in controlling diabetes.
- Fruit itself with its yellow-green, soft, sweet and creamy flesh serves not only as a nutritional powerhouse rich in minerals, vitamins, antioxidants and other nutrients but also harbours medicinally significant compounds.

Despite the nutritional attributes and historical usage in traditional medicine, there is a notable scarcity of comprehensive studies demonstrating various biological effects of *Artocarpus altilis*. Existing studies often fail to elucidate the precise mode of action of the medicinal compounds present in the plant. Additional compounds found in breadfruit further contribute to its pharmacological potential underscoring the need for further research to unveil the full spectrum of its therapeutic benefits.

3.10 Uses

Similar to bananas & plantains, breadfruit offers a versatile culinary experience. The fruit can be enjoyed either ripe as a sweet fruit or underripe as a starchy vegetable. The culinary practices vary across regions, showcasing the adaptability of breadfruit in different cuisines (Morton, 1987). Unripe breadfruits exhibit a resilient green outer skin accompanied by a starchy, bitter and firm white interior. Upon reaching maturity, these fruits undergo a transformative process featuring a transition to a lustrous golden-brown exterior and a shift in the inner composition to a yellow custard-like mildly sweet pulp. When fully ripe, breadfruits emanate a subtle aroma and boast a soft yielding texture. The optimal ripeness of a breadfruit is contingent upon its intended culinary purpose. For endeavours such as desserts, purees and puddings selecting a fully ripe fruit is advised capitalizing on its sweet and custardy attributes. Conversely, when embarking on the creation of savoury cooked dishes, a firm green fruit is the preferred choice which ensures a more robust and less sweet flavour profile (theindianvegan.blogspot.com, 2012).

For underripe breadfruit, traditional Pacific Island cooking involves roasting it in an underground oven on pre-heated rocks sometimes cored and stuffed with coconut before roasting. Malayans fry firm-ripe breadfruit slices in syrup or palm sugar until crisp and brown while Filipinos enjoy it cooked with coconut and sugar. Fully ripe breadfruit being sweeter can be baked whole with a little water. Some cooks remove the stem and core and add butter and sugar in the cavity for serving. Alternatively, it can be halved or quartered and steamed for 1-2 hours, seasoned in the same manner as baked fruits. In Hawaii, underripe breadfruits are diced, boiled and served with butter and sugar, salt and pepper or diced and cooked with other vegetables, bacon and milk as chowder. In the Bahamas, soup is prepared by boiling underripe chunks of breadfruit until the liquid thickens, followed by addition of cooked salted pork, chopped onion, white pepper, salt, milk and butter and then finished with a touch of sherry. The versatility of breadfruit extends to desserts. The pulp scraped from soft and ripe breadfruits can be combined with coconut milk, salt and sugar then baked to create a pudding. A more elaborate dessert involves mashed ripe breadfruit with butter, beaten eggs, sugar, nutmeg, cinnamon, rosewater and a dash of sherry or brandy, followed by boiling to perfection.

Beyond direct consumption, breadfruit undergoes various preservation and fermentation processes in different regions. In Micronesia, the peel is scraped off or peeled with a knife then soaked in the sea for about 2 hours while being beaten or trampled. In Polynesia and Micronesia, fruits are baked in a native oven and left to ferment over weeks. In some Pacific Islands, breadfruit is dried and in the New Hebrides, peeled breadfruits are wrapped in leaves and placed in piles of stones on open beaches for fermentation. In Samoa, seeded breadfruits are fermented in a pit lined with banana and *Heliconia* leaves. 'Poi' ~ a traditional dish involves fermenting breadfruit in stone pits, preserving it for several years. Modern poi is made by boiling firm-ripe whole fruits, followed by processing, straining and thinning with water before consumption.

Breadfruit has also been explored for its potential as a substitute for wheat flour. In Jamaica, breadfruit flour is boiled, sweetened and eaten as porridge for breakfast ~ offering a nutritious alternative. Various parts of breadfruit such as fallen male floral spikes and seeds are used in culinary practices. The seeds can be boiled, steamed, roasted or made into a puree in different parts of the world. Furthermore, experiments by USDA have shown that breadfruit can be dehydrated commercially through tunnel drying or freeze-drying where its waste serves as a highly-digestible stock feed.

3.10.1 Other uses

Various parts of breadfruit tree serve multiple purposes showcasing its versatility in traditional practices across different cultures (Morton, 1987) as defined below:

Leaves: Its leaves play a crucial role in livestock nutrition. In India, these form a significant part of diet for cattle and goats while in Guam these are consumed by cattle, horses and pigs. Young trees face a threat from horses as they tend to graze on its bark which necessitates protective measures for new plantings.

Latex: Latex extracted from breadfruit has a diverse range of applications. In the past, it was employed as birdlime to catch birds when applied to the tips of posts. Early Hawaiians used it to remove the gummy substance from birds' feet for ceremonial cloaks, utilizing oil from the candlenut or sugarcane juice. Boiled with coconut oil, the latex serves as an effective caulking material for boats. When mixed with coloured earth, it is utilized as a paint for boat exteriors.

Wood: Breadfruit tree wood which is light in weight exhibits strength, elasticity and resistance to termites (except for drywood termites). In Samoa, it is a standard material for house-posts and rounded roof-ends of native houses. Wood of specific varieties such as 'Aveloloa' and 'Puou' is preferred for various construction purposes. In places like Guam and Puerto Rico, breadfruit wood is utilized for interior partitions. Its lightness also makes it a sought-after material for crafting surfboards. Traditional Hawaiian drums are crafted from sections of breadfruit trunks, contributing to the cultural significance of the tree.

Fibre: Extracting of bark fibres is though challenging but results in highly durable material. Traditionally, Malaysians fashion clothing from its fibre while tape cloth is derived from the inner bark of young trees and branches. In the Philippines, breadfruit fibre is transformed into harnesses for water buffaloes, highlighting its diverse applications.

Flowers: Male breadfruit floral spikes when blended with fibre of mulberry paper were historically used to create elegant loin cloths. Moreover, when thoroughly dried, these flower spikes serve as effective tinder for various purposes.

Medicinal Uses: Beyond its utility in material and nutrition, the breadfruit tree also boasts medicinal properties. In Trinidad and the Bahamas, a decoction of breadfruit leaves is believed to lower blood pressure and alleviate asthma. Crushed leaves are applied to treat thrush and the leaf juice is used as ear drops. Burnt leaf ashes are employed for skin infections and a powder of roasted leaves is used as a remedy for enlarged spleen. Various parts of the tree including the crushed fruit and toasted flowers find applications in traditional healing practices for ailments ranging from tumours to dental issues and sciatica. Diluted latex is even taken internally to counteract diarrhoea. This multifaceted medicinal use underscores the holistic significance of the breadfruit tree in indigenous healthcare practices.

3.11 Slicing or Break-Opening Breadfruit

The meticulous preparation of breadfruit involves a series of steps to ensure a seamless and enjoyable culinary experience. To begin with, the hands and knife are first meticulously coated with oil forming a protective barrier against the sticky latex that envelops the fruit. This precautionary measure is crucial in preventing the latex from adhering to the blade, facilitating a smoother and more efficient cutting process. With this precaution in place, the breadfruit's stem is perfectly removed and the fruit is artfully inverted to allow the latex to drain. Any residual latex that may cling to surfaces is then swiftly wiped away using an oiled cloth, maintaining cleanliness and ease of handling.

When dealing with ripe breadfruit, a delicate approach is adopted. The fruit is carefully halved and its doughy flesh is skillfully scooped out. However, unripe breadfruit demands a more intricate preparation process. The outer skin is peeled away ~ a task akin to peeling a potato in its manageability. The peeled fruit is then halved, further divided into quarters and then a sharp knife is employed to excise the firm core from each wedge. The peeled and de-cored wedges are then meticulously cut into the desired-sized pieces. For those destined to become breadfruit chips a fine and thin precision is applied while those intended for stews, bakes or salads are crafted into pieces reminiscent of potato chunks.

Breadfruit being versatile in its culinary applications can be cooked akin to a pumpkin or squash. A spiky and seeded relative named 'Kamansi' cultivated in the Philippines adds a distinct variation to the breadfruit family. Careful consideration and craftsmanship invested in each step of the preparation process underscore the culinary artistry associated with this unique and globally cherished fruit (theindianvegan.blogspot.com, 2012).

3.12 Flavour

The flavour profile of breadfruit is intricately tied to its ripeness. At full maturity, the taste is akin to unadorned bland pudding ~ its doughy and bready nuances might not resonate with every palate. Nevertheless, much like tofu (soya paneer) its culinary potential blossoms when paired with specific ingredients and adept cooking techniques.

Raw and unripe breadfruit characterized by excessive starch content is unsuitable for direct consumption and necessitates cooking to unleash its gastronomic charm (theindianvegan.blogspot.com, 2012). When subjected to transformative embrace of baking or roasting, the fruit undergoes a metamorphosis adopting a starchy texture and emanating an enticing fragrance reminiscent of freshly baked bread (ntbg.org, 2024). Delving further into the edible realm of breadfruit, its seeds akin to those of jackfruit present a delectable versatility. Boiling, deep-frying or roasting are viable methods for their preparation each imparting a distinct character. Cooked breadfruit seeds boast a delightful amalgamation of nutty sweetness, crispiness and crunchiness reminiscent of the nuanced flavour profile of chestnuts (theindianvegan.blogspot.com, 2012).

3.13 Toxicity

Breadfruit is a tropical tree fruit which is renowned for its versatile culinary applications but it harbours certain toxic elements that necessitate careful handling and preparation. Raw consumption of most breadfruit varieties poses a purgative risk so prompting the need for specific precautions. One commonly adopted measure involves boiling certain varieties twice with the water discarded during the process. This meticulous approach is implemented to diminish the potential unpleasant effects associated with raw consumption. However, it is crucial to note that not all varieties necessitate such extensive preparation as there exist named cultivars that can be safely consumed without cooking (Morton, 1987).

3.14 Breadfruit in India

Potential for thriving breadfruit cultivation along India's warm coastline is vast yet its current distribution is confined primarily to Karnataka, Kerala and Lakshadweep (including islands of Agatti, Amini, Kalpeni, Kavaratti and Minicoy) with sporadic plantations existing in Tamil Nadu. Despite this, the prominence of breadfruit in Indian cuisine is noteworthy, with Kerala standing out as the only state to feature it prominently (Rao and Kumar, 2023).

Somewhat limited prevalence of breadfruit may be attributed to the challenges associated with its cultivation. A single tree has capacity to yield an impressive 200 fruits per season but the intricate nature of its cultivation process acts as a deterrent. The need for fresh seeds to be planted within a few weeks coupled with the susceptibility of ripe fruits to fungal infections adds complexity to its growth. Furthermore, high perishability of breadfruit poses challenges for growers looking to transport and sell the fruit in distant markets. In regions where breadfruit is cultivated such as coastal areas of Karnataka, Kerala and Tamil Nadu as well as the Lakshadweep Islands, the likelihood of locals obtaining the fruit is relatively high. However, the prevalence of breadfruit in larger metropolitan areas is limited largely due to insufficient consumer demand. This has resulted in a scenario where large retail outlets seldom stock breadfruit, leaving the responsibility of distribution to smaller and mid-sized vendors who frequent wholesale markets. The flowering of breadfruit trees in March marks the beginning of fruiting season which extends from June to July in India. Despite its seasonal availability, the challenges associated with cultivation coupled with consumer preferences contribute to the intricate dynamics surrounding the popularity and distribution of breadfruit in the country (theindianvegan.blogspot.com, 2012).

3.15 Cultivation

3.15.1 Climate adaptability

Known for its tropical origin, breadfruit exhibits a delicate nature compared to robust tree-like mangoes. Despite its versatility in adapting to different ecological conditions, breadfruit thrives in climates characterized by summer rains. Ideal conditions for optimal growth include tropical lowlands situated below 650 metres elevations with an annual rainfall of 1500-3000mm. The tree flourishes well under temperatures between 21–32°C with a relative humidity range of 70 to 80%. Though its preferred conditions are within these parameters, breadfruit can still be found thriving at elevations up to 1550 metres, enduring temperatures between 15–40°C. It can endure drought only for a brief period though prolonged dry spells can lead to premature fruit drop (Ragone, 2006).

In southern India, breadfruit is cultivated both at sea level and on humid slopes up to an altitude of 1,065 metres. Even in dry regions, it is successfully grown in thickets where irrigation is feasible. Marquesas Islands characterized by an 'equatorial dry climate', rely on breadfruit as a staple crop despite experiencing limited rainfall (1000-1500mm) and frequent droughts. Similarly, in Central America, breadfruit cultivation is constrained at altitudes below 600-650 metres (Morton, 1987).

3.15.2 Tolerance and environmental resilience

Breadfruit thrives in full sun forming canopy in mixed agroforests. While young trees prefer 20–50% shade, they can adapt to full sun conditions. The tree can regenerate from the roots after a small fire but the trunk and branches are not fire-tolerant. Frost can cause significant damage leading to the loss of fruits and leaves, with potential branch die-back. The tree demonstrates limited tolerance to waterlogged soils, enduring such conditions only briefly. While it can withstand some salt spray for short durations, extended exposure results in yellowing of leaves and leaf fall. Heavy winds especially with laden fruit load can cause branch breakage but the tree exhibits rapid regeneration of new shoots and branches (Ragone, 2006).

3.15.3 Soil preferences

Breadfruit thrives in deep, fertile and well-drained soils with a pH range of 6.1-7.4. It shows adaptability to various soil textures including sandy, sandy loams, loams and clayey sandy loams. Certain varieties have even adapted to shallow sandy soils found in coral atolls. Notably, seedless breadfruit performs well in sandy coral soils while seeded types naturally grow on coraline limestone islands in Micronesia. Adaptability varies among strains, emphasizing the importance of matching each variety with its optimal climatic and soil conditions. Specific varieties such as the Tahitian 'Manitarvaka' and the Gilbert Island's 'Mai-Tarika' exhibit resistance to drought and tolerance to salt, respectively (Ragone, 2006; Morton, 1987).

3.15.4 Seasonal patterns

Male flowers and fruits emerge predominantly at the distal ends of branches with initial appearance of male flowers preceding the onset of fruiting. The fruiting phase generally aligns with the moist, rainy summer months, however, certain varieties may exhibit a secondary, albeit smaller, fruiting episode approximately five months later. The continuous production of new leaves persists throughout the year marked by a substantial flush following a period of dormancy subsequent to the conclusion of the fruiting season.

In South Seas, breadfruit trees exhibit continuous fruiting with two or three main fruiting periods throughout the year. Seasonal variations are observed in different regions. For instance, in the Caroline and Gilbert Islands, the primary ripening season is from May to July or September. In the Society Islands and New Hebrides, fruits ripen from November to April, with a secondary crop in July and August. Breadfruits are commonly available in Hawaiian markets from July to February. Flowering in northern India commences in March with fruits ready for harvest approximately three months later. Seasonal patterns also depend on the variety with some bearing fruits once a year while others such as seedless varieties from Ponape yield 2-3 times annually. In the Bahamas, breadfruit availability is mainly from June to November though some of the fruits may mature at other times of the year. The adaptability and continuous fruiting nature make it a valuable and resilient crop in various tropical climates (Morton, 1987).

3.15.5 Propagation

Breadfruit offers multiple avenues for propagation, allowing for the cultivation of diverse varieties. The plant can be readily propagated through several techniques, showcasing its adaptability and accessibility. One of the primary methods involves utilizing root shoots or root cuttings, providing a straightforward and efficient means of reproducing breadfruit. This approach is applicable to both seeded and seedless varieties, underscoring its versatility. Air-layering branches represent another viable technique, contributing to successful propagation efforts. For those seeking to preserve specific characteristics or traits, grafting techniques offer a controlled method of reproduction. Breadfruit can be grafted using various techniques allowing for preservation and transmission of desired qualities across generations

When considering the propagation of seeded breadfruit, seed is a potential option. However, it is crucial to note that seeds need to be planted promptly as these lose viability within a few weeks. Despite this, the use of seeds is relatively uncommon due to the challenge of maintaining true-to-type characteristics in the resulting plants. On the contrary, seedless varieties known for their desirable attributes necessitate vegetative propagation methods. Root shoots or root cuttings emerge as the preferred means for both seeded and seedless varieties, ensuring the faithful reproduction of desired qualities (Ragone, 2006).

3.15.5.1 Propagation by root shoots and root cuttings

Propagation of seedless breadfruit primarily involves the utilization of root shoots and root cuttings. This process crucial for maintaining and expanding breadfruit orchards is characterized by deliberate interventions and strategic timing. Propagation is commonly achieved through transplantation of naturally occurring suckers that emerge from roots of breadfruit tree. To enhance the generation of these suckers, deliberate actions such as uncovering and injuring the roots can be employed. Furthermore, prudent pruning of the parent tree proves effective in increasing the overall number of suckers. For optimal results, root pruning of each sucker multiple times over an extended period before extraction significantly contributes to its successful establishment upon transplantation (Ragone, 2006; Morton, 1987).

The timing of collecting root shoots and root cuttings is crucial for success of propagation process. Recommended time for collection is after the fruiting season has concluded and the tree has entered an active vegetative stage which is marked by production of new leaves. This phase typically aligns with conclusion of the dry season and the onset of rainy season. Notably, this specific timeframe is chosen as it corresponds to the peak of carbohydrate storage within the roots. Carbohydrate levels are at their highest during this transitional period, ensuring that the collected root shoots and cuttings possess optimal energy reserves for successful establishment and growth (Ragone, 2006). This strategic approach to timing aligns with the natural growth cycles of the breadfruit tree, enhancing the likelihood of successful propagation and subsequent orchard development.

3.15.5.1.1 Root shoots

The propagation of breadfruit through root shoots remains a traditional practice in the Pacific with certain varieties such as 'Puou' exhibiting prolific production of shoots. Optimal harvesting of root shoots involves selecting healthy and robust shoots that have attained a minimum height of 20–25cm, displaying a woody stem and lobed leaves. Shoots measuring up to 1m in height are deemed suitable for propagation.

To harvest root shoots, a meticulous process is employed. The shoot is carefully removed by cutting the attached root approximately 10–15cm on either side of the shoot. A sharp machete or clippers are utilized to sever the shoot from the parent tree so that efficient wound healing is necessitated. Given the inherent challenge of minimizing damage to root system of shoots during removal, it is imperative to nurture root shoots under nursery conditions before subsequent outplanting. The direct transplantation of shoots to a new location yields a relatively low success rate of approximately 25%.

Post-harvest leaves are systematically trimmed and tip of shoot is precisely cut at a 45^0 angle situated at a height ranging from 15 to 30cm. This practice not only enhances the manageability of the root shoot but also anticipates the natural tendency of the apical shoot to undergo dieback. Subsequently, shoots are planted in well-drained organic media within a 10-20cm pot. Maintaining an environment with up to 60% shade and ensuring consistent moisture without allowing saturation is crucial during initial growth phase. Misting is recommended so that shoot is never permitted to dry out.

Depending on the size of the shoot, a nursery period of 3-6 months is recommended before outplanting. The success of this propagation method is contingent upon factors such as rooting efficiency and shoot growth, with reported rates ranging from 50 to 90% (Ragone, 2006). This comprehensive approach to breadfruit propagation underscores the importance of meticulous techniques and nursery conditions in ensuring the successful establishment of root shoots for subsequent cultivation.

3.15.5.1.2 Root cuttings

Propagation of breadfruit through root cuttings is a viable method particularly when root shoots of a desired variety are not readily available. To initiate mass propagation, it is crucial to gather roots from robust and healthy trees, focusing on those growing just beneath the soil surface to ensure optimal success. Careful excavation is paramount, avoiding surface roots that are prone to drying out and exhibit a lower success rate. Ideal roots for cuttings display small rounded bumps on the surface which are adventitious buds poised to develop into new shoots. Roots within diameter range of 1.5–6cm can be utilized with recommended range for optimal results being 3–4cm. Large roots exceeding 6cm should be avoided as their removal could harm the tree by damaging the root system and impeding the healing process. Precision in root cutting is crucial and a sharp machete or clippers is employed to sever the root and facilitate efficient wound healing. Remaining attached root often undergoes the development of a root shoot at the cut end (Ragone, 2006; Morton, 1987).

The segmented roots typically cut into 15–25cm sections undergo a thorough cleaning process ~ removing soil and discarding any damaged or misshapen portion. A preventive treatment with fungicide is advisable to ward off pathogens causing root rots. Optionally, dipping the ends into a potassium permanganate solution helps coagulate the latex. Although hormone treatment is not mandatory though standard hormone mixes can be employed based on the manufacturer's recommendations.

Propagation beds, flats or individual pots serve as suitable environments for placing the root cuttings with recommended spacing of 10–15cm within rows and 15–20cm between rows in beds or flats. Use of well-drained potting media or clean and washed silica sand is recommended with caution against bleached sand due to its saline and alkaline nature. Cuttings are positioned either horizontally covered with media or at an angle avoiding an upright orientation, with a small upper portion of the root exposed.

Maintaining a shaded environment (up to 60% shade) and ensuring consistent moisture without over-saturation preferably through misting is essential for successful root cutting development. Preventing roots from drying out is imperative. Rooting success ranges from 75 to 85%, and shoots emerge from adventitious buds after 3–4 weeks. Once shoots reach a height of 20–25cm and develop their own root systems (typically within 4–6 months), careful uprooting and transplantation into 20–30cm pots with well-drained medium is performed. Fertilizer application such as a balanced 8–8–8 mix should be administered sparingly, adhering to less than half the manufacturer's recommendations. The transplanted plants thrive in partial shade requiring a weed-free environment and there these reach the size of 0.6–1.6m within 6–9 months.

3.15.5.2 Propagation by air layering

Air layering is a highly effective method for propagating trees and when applied to the cultivation of certain plants such as during commencement of rainy season, it can yield optimum results. This technique is particularly advantageous when the tree is actively engaged in a vegetative phase, marked by the emergence of new shoots and leaves, and notably before the onset of fruit development.

Selecting newly formed shoots is paramount with a discerning eye toward avoiding branches that have previously borne flowers or fruits. To initiate air layering branches measuring 2–4cm in diameter are carefully prepared, ideally by removing a 3–5cm wide strip of bark encircling the branch. Precision is crucial and the use of a sharp knife is recommended to ensure that the wood is not inadvertently cut. While rooting hormone is not mandatory but its application should strictly be as per manufacturer's guidelines.

The designated area is then enveloped with moistened sphagnum moss or another organic medium. To secure this arrangement a covering of plastic, aluminum foil, burlap or a copra bag is securely tied around the branch. It is noteworthy that not all the air-layers will successfully root but in some instances a ring of hardened callus forms along the cut end. The inherent brittleness of the branches susceptible to snapping in high winds can be mitigated by reinforcing them with bamboo splints placed over the wrapped air-layer.

Over a span of 2–4 months, the air-layered branch undergoes a transformative process ~ developing new roots that extend through the surrounding medium. Subsequent to this period, the air-layer is separated by making a cut directly below the roots. The excised air-layer is then transplanted into a 10-20cm pot containing a well-drained medium. This stage typically lasts about a year allowing the plant to establish a robust root system. It is worth noting that the tip of the branch may experience dieback and the success of the air-layering process is contingent upon the emergence of a new shoot from lower buds on the branch.

Attention to these details ensures a successful and viable propagation method through air layering (Ragone, 2006).

3.15.5.3 Propagation by stem cuttings

The propagation of *Artocarpus altilis* through stem cuttings represents a sophisticated technique aimed at achieving swift and efficient reproduction. This method focuses on utilizing stem cuttings derived from root shoots. In regions like Puerto Rico the process involves transplanting these cuttings into specialized plastic bags that house a carefully curated blend of soil, peat and sand.

The early stages of cultivation involve subjecting the cuttings to a conducive environment. Initially, these are placed under a misting system for optimal humidity control. Following this stage, the cuttings are strategically positioned under 65% shade to ensure a gradual acclimatization. During this crucial period, the plants receive a regimen of liquid fertilizer complemented by regular and attentive waterings.

As the root system undergoes robust development, transition of the plants to a phase of exposure to full sunlight is necessary. This pivotal stage sets the groundwork for the eventual transplantation of the well-established plants into the designated field. Meticulous combination of controlled environments, nutrient supplementation and attentive care at each stage of the propagation process contributes to the successful establishment of breadfruit through stem cuttings (Morton, 1987).

3.15.5.4 Grafting or budding

The art of horticultural propagation in the context of breadfruit cultivation in India involves the meticulous techniques of grafting or budding. It has been documented that the grafting or budding of breadfruit scions onto the seedlings of wild jackfruit trees has proven to be a successful and viable method. This practice underscores the adaptability and compatibility between breadfruit and wild jackfruit, harnessing the potential for enhanced cultivation and utilization of breadfruit through strategic propagation methods (Morton, 1987).

3.15.6 Full sun hardening in nursery

Optimal growth conditions for young breadfruit plants are typically achieved in partial shade during their nursery phase. While full sun hardening may not be imperative, circumstances may arise where plants are intended for transplantation into sun-exposed locations. In such instances, a gradual acclimatization process to full sun conditions in the nursery becomes essential to fortify the plants against the specific environmental demands of their intended site. This acclimatization process is recommended to commence at approximately 2 months of age.

During the hardening phase, meticulous care should be taken to ensure that the young plants are shielded from the detrimental effects of drying out and exposure to strong winds. The vulnerability of young plants to these stressors underscores the importance of a carefully managed transition to full-sun conditions. This nuanced approach not only enhances the adaptability of the plants but also plays a crucial role in their overall resilience to the site conditions they will encounter upon transplantation. As custodians of the nursery environment, practitioners should prioritize the well-being of the young breadfruit plants by implementing measures that safeguard against dehydration and wind-induced stress during full sun hardening process (Ragone, 2006).

3.15.7 Outplanting

Outplanting ~ a critical stage in the cultivation of breadfruit trees is executed when the plants attain the desired size ensuring optimal conditions for successful establishment. Given the expansive surface area of the leaves, a prudent approach involves reducing leaf size to mitigate transpiration. This is achieved by carefully removing 1/2 to 2/3rd of the lower leaves while preserving the crucial growing point where new leaves develop. During transport, it is imperative to shield the plants from wind and excessive heat to safeguard their integrity.

The transplantation process involves planting young breadfruit trees in meticulously prepared and enriched holes measuring 40cm deep and 90cm wide. These holes prior to planting undergo a preparatory phase where trash is burned to sterilize the soil and insecticide is integrated to shield the roots and shoots from potential grubs. To enhance nutrient availability, a small amount of slow-release fertilizer such as 8–8–8 can be introduced to the bottom of the hole before covering it with soil. Careful extraction of the plant from its container is recommended to prevent damage to the delicate root system. The tree is then positioned in the hole, ensuring that the soil level does not surpass the level of the plant in the pot. Top-dressing with compost and thorough watering completes the process, yielding an impressive success rate approaching 100%.

Trees should be planted 12–14m apart under orchard conditions though Caribbean trees are often spaced 8–12m apart in plantations ~ typically totalling around 85 trees per hectare. Breadfruit trees propagated from root suckers exhibit early productivity bearing fruit in 5 years and sustaining productivity for up to 50 years. Pruning practices, targeting branches that have borne fruit and would naturally die back, are advocated by some growers. This approach stimulates new shoots and manages height of trees for convenient harvesting.

Optimal conditions for young plants involve partial shade and the planting aligning with onset of rainy season. In instances of dry weather, irrigation during the initial 1–3 months of establishment is recommended. Once established, breadfruit trees showcase resilience during a dry season of 3–4 months though these thrive in consistently moist conditions. Employing mulch around young plants proves beneficial which ensures retention of soil moisture and a sustained nutrient supply. This practice concurrently aids in weed control around the root system. However, caution is warranted when considering herbicides as direct contact with surface roots or the young trunk can inflict damage.

The vulnerability of young trees to browsing animals such as cattle, goats, horses and pigs necessitates protective measures to safeguard the bark and tender shoots from feeding. These comprehensive practices collectively contribute to the successful outplanting and establishment of breadfruit trees promoting robust growth and long-term productivity (Morton, 1987; Ragone, 2006).

3.15.8 Manuring

Nutrient management, particularly in the form of standardized mixtures of N, P and K plays a crucial role in the cultivation of these trees. Seasonal applications of NPK ensure that the trees receive a balanced supply of essential nutrients. As the trees mature into bearing age, an additional enhancement of 2kg of superphosphate per tree per year should be given. This strategic application is designed to not only bolster the overall health of the trees but also to specifically augment the size and quality of the fruits yielded. The incorporation of superphosphate is a targeted approach aimed at optimizing fruiting capacity and enhancing overall productivity of the orchard (Morton, 1987).

3.15.9 Harvest

Breadfruits are picked when maturity is indicated by the appearance of small drops of latex on the surface. Harvesters climb the trees and break the fruit stalk with a forked stick to fall the fruit down. Though this may cause some bruising or splitting even then it is better than catching the fruits by hand as broken pedicel leaks much latex. They are packed in cartons in which they are separated individually by dividers (Morton, 1987).

3.15.10 Yield

Productivity levels of breadfruit exhibit regional variations, with the South Pacific, southern India and the West Indies offering distinct yield patterns. In the South Pacific these resilient trees showcase an annual yield ranging from 50 to 150 fruits. Similarly, in southern India, the standard production rate ranges 150 to 200 fruits per tree annually. Productivity levels fluctuate in response to the contrasting environmental conditions prevalent in wet and dry areas. In West Indies, a more conservative estimate suggests an average of 25 fruits per tree.

However, studies conducted in Barbados have unveiled a promising potential ~ projecting yields within the range of 16 to 32 tonnes per hectare. Despite optimistic forecasts predicting even higher yields, experts exercise caution and express skepticism regarding these projections as potentially unrealistic dividers.

These disparities in breadfruit yield across diverse geographical regions underscore the intricate interplay of environmental factors and agricultural practices that influence the productivity. Further research and ongoing assessments are essential to refine our understanding of the factors impacting breadfruit yield and to develop sustainable practices for maximizing its production (Morton, 1987).

3.15.11 Storage

Storage practices for breadfruits in Jamaica involve strategic methods to preserve their quality and extend shelf life. Surplus breadfruits are commonly submerged in water until needed, providing a protective environment. For fully ripe fruits that have naturally fallen from the tree, an alternative storage approach involves wrapping them in polyethylene or placing them inside polyethylene bags. Under these conditions, these ripe fruits can be stored for up to 10 days at a temperature of 12°C. However, it is crucial to note that exposing fruit to lower temperatures may result in chilling injury.

When harvesting slightly unripe fruits by hand before they fall, these can also be stored for a longer duration ~ up to 15 days by using the same polyethylene wrapping method at the specified temperature. The thickness of the polyethylene material is a critical factor in ensuring effective storage. Bags with 38- or 50-micrometer thickness are deemed beneficial whereas those with 25-micrometer thickness are not recommended for optimal preservation.

In the context of exportation, some Jamaican exporters employ a unique preservation technique. They partially roast whole breadfruits to coagulate the latex within, allowing them to cool before shipping. This specialized approach is particularly applied when transporting the fruits by sea to destinations such as New York and Europe (Morton, 1987).

These storage practices underscore the importance of careful handling and controlled environmental conditions in maximizing the post-harvest life of breadfruits, ensuring their availability and quality for both local consumption and international markets.

3.16 Nutrient defeciencies

Same as in Jackfruit.

3.17 Physiological disorders

Physiological disorders in plants arise due to internal factors, environmental conditions or cultural practices. While both jackfruit and breadfruit belong to the same genus, they are distinct species with their own set of characteristics. Here are some differences in the physiological disorders observed in jackfruit and breadfruit:

i) **Blossom end rot:** This malady may occur only in breadfruit but not in jackfruit, especially when there are fluctuations in calcium availability. It may result in dark and leathery patches on the fruit.

ii) **Fruit splitting:** Some jackfruit varieties are prone to fruit splitting which can occur due to irregular watering or excessive rainfall though this malady is quite less common in breadfruit.

iii) **Leaf curling:** Leaf curling in jackfruit may occur due to environmental stress, pests or diseases but in breadfruit it may be associated with factors such as water or nutrient deficiencies.

iv) **Leaf drop:** Excessive leaf drop can occur in both the plants due to environmental stress, inadequate watering or nutrient deficiencies.

v) **Sunburn:** Jackfruit may be affected when exposed to sudden intense sunlight after a prolonged period of shade but in case of breadfruit it may occur in young plants after sudden opening of canopy.

vi) **Wind damage:** Both the trees are susceptible to wind damage which may lead to broken branches or uprooting, especially during storms or high winds.

vii) **Physiological leaf spotting:** It is quite uncommon in jackfruit though some varieties may exhibit sometimes but in breadfruit it occurs due to certain environmental stressors.

viii) **Genetic variability:** Different varieties within each species may exhibit varying degrees of susceptibility to physiological disorders.

Understanding the specific conditions and responses of jackfruit and breadfruit is essential for effective management and care of these fruit trees. Site-specific factors and individual species characteristics should be considered when addressing physiological disorders in orchards.

3.18 Diseases

In southern India, the fruits on the tree are susceptible to soft rot ~ a fungal disease that can be effectively managed with two sprays of Bordeaux mixture applied one month apart. In Trinidad, young breadfruit trees have faced mortality due to a disease caused by *Rosellinia* sp. In the Pacific Islands, dieback is believed to be associated with *Fusarium* sp., and cases of root rot are suspected to involve *Pythium* sp. A mysterious ailment known as 'Pingalap disease' decimated thousands of trees from 1957 to 1960 in the Gilbert and Ellice Islands, the Caroline Islands, Marshalls and Mariannas. This condition manifests as wilting foliage followed by branch dieback, occasionally affecting the entire tree and all these may lead to root death. *Phytophthora palmivora* attacks the fruit on the island of Truk though stem-end rot is attributed to *Phomopsis*, *Dothiorella* and *Phylospora* (Morton, 1987).

Uredo artocarpi, initially identified on *Artocarpus* spp. in India and the Philippines though prevalent on islands in the South Pacific, made its first appearance in Hawai′i in 1991 (Gardner, 1991). Among the documented diseases, anthracnose caused by *Colletotrichum gloeosporioides* is reported (Ragone, 2006; Stice *et al.*, 2007). *Diplodia* collar rot (Sangchote *et al.*, 2003); fruit, stem and root rot due to *Phytophthora* spp. (Cerqueira *et al.*, 2006); pink disease due to *Erthricium salmonicolor* (Sangchote *et al.*, 2003); *Rhizopus* rot (Sangchote *et al.*, 2003; Ragone and Paull, 2008); and 'Pingelap disease' initially thought to cause an epidemic decline in the Pacific but later linked to environmental factors (Zaiger and Zentmyer, 1966) are major diseases of breadfruit. Coates-Beckford and Pereira (1992) noted the prevalence of *Pseudomonas* sp. and *Fusarium* sp. on both declining and non-declining breadfruit trees in Jamaica. Brown set/root rot caused by *Phellinus noxius* (Ragone and Paull, 2008; Stice *et al.*, 2007) and fruit rot (*Phytophthora palmivora*) (Redfern, 2010) further contribute to the challenges faced by breadfruit cultivation, particularly in tropical areas.

Diseases affecting breadfruit can be broadly classified into three main categories: leaf, fruit and stem/root diseases. Each category poses distinct challenges to tree vitality and overall productivity. Understanding these diseases is crucial for effective management strategies (Lincoln *et al.*, 2021).

1. **Leaf diseases**
    a. **Leaf spots:** Leaf spots are common with pathogens like *Corynespora cassiicola* causing irregular spots with undulate borders, often leading to shot holes and defoliation. *Alternaria* sp., *Fusarium* sp., *Lasiodiploidia* sp., *Macrophoma* sp., *Phoma* sp. and *Phyllosticta* sp. are associated with different types of dark-brown spots.

    b. **Leaf anthracnose:** Spores spread by rain or wind infecting leaves, fruits and stems. Leaf anthracnose is caused by *Colletotrichum gloeosporioides* which results in reddish-brown spots that evolve into dark-brown lesions with reddish-brown margins. Severe cases can lead to premature defoliation and reduced canopy thus impacting fruit production. Pruning the infected parts, application of copper-based fungicides and following proper hygiene overcome this problem.

c. **Rust:** This rust is caused by *Uredo artocarpi* which is characterized by russeting on both leaf surfaces. It can be kept under check through pruning the infected leaves and subsequent application of 1% Bordeaux mixture or 0.4% COC during wet seasons.

d. **Algae leaf spots:** This is caused by *Cephaleuros virescens* which shows as orange velvety spots on upper leaf surface. It is controlled through pruning the infected leaves followed by spraying of 0.4% Cuman-L, 0.2% Baycor/Folicur or 0.1% Nativo/Baan/Beam fungicides during wet seasons.

2. **Fruit Diseases**

   a. **Fruit spots:** Fruit spot is primarily caused by *Fusarium* and *Phoma* spp. It is manifested as light to dark-brown fruit spots. Generally the spots are superficial though excessive spotting can affect fruit quality. It can be controlled through 0.1% soil drenching of Bavistin/Bengard. Use of *Trichoderma viridae* + *Paecilomyces lilanicus* alongwith neem cake is quite useful.

   b. **Fruit anthracnose:** It is caused by *Colletotrichum gloeosporioides.* Its spores spread by rain or wind infecting leaves, stems and fruits, and on fruits it causes anthracnose leading to large dark-brown spots that harden the flesh and may deform the fruit. Spraying with 0.1% methyl thiophanate/Score/Prochloraz or 0.2% mancozeb or Kavach will control this problem.

   c. **Soft rot:** *Rhizopus* and *Geotrichum* cause soft watery rot, impacting fruit ripening on the tree and post-harvest during storage or shipment. The treatment described under 'Fruit anthracnose' will control this disease. Moreover, through proper irrigation management also this problem may be minimised.

   d. **Fruit rot:** *Phytophthora palmivora* and *P. tropicalis* cause fruit rot with characteristic white bands, leading to severe consequences if left unmanaged. Through spraying with 0.2% zineb or 0.6% Tricyclazole this problem can be solved.

3. **Stem and root diseases**

   a. **Seedling dieback:** *Sclerotium rolfsii* and *Rosellinia* sp. contribute to seedling dieback characterized by yellowing, necrosis, wilting and branch dieback. Spraying cum drenching with 0.2% captan/sprint/ thiride/subeej may control this problem.

b. **Phytophthora root rot:** It is caused by *Phytophthora* spp., a soil-borne pathogen where roots start decaying. Symptoms include yellowing, root rot and wilting. By improving drainage, avoiding waterlogged conditions and by use of resistant rootstocks one can get rid of this problem. Spraying with 1% Bordeaux mixture or 0.2% mancozeb this problem may be solved.

c. **Crown and root rot:** *Phellinus noxius* induces brown root rot, causing tree decline and death. Other causal agents include *Lasiodiplodia theobromae*, *Rhizoctonia solani*, *Pythium* sp., *Fomes* sp. and *Polyporus zonalis*. By minimising soil wetness and proper drainage are essential steps in minimising this problem. Soil should be treated with *neem* cake + *Trichoderma harzianum* + *Pseudomonas florescence*. Moreover, 1% Bordeaux mixture or 0.2% Aliette/Ridomil MZ/Matco spraying is quite effective.

d. **Stem rot:** Fusarium spp. and Phoma sp. infect small branches and stems, disrupting water and nutrient translocation, leading to browning and dieback. Infected parts should be sheared off and destroyed. The soil should be treated with neem cake + *Trichoderma viridae* + *Pseudomonas florescence*. Soil may also be drenched with 0.2% Bavistin/Derosal/Bengard along with liming of the pits.

Furthermore, the susceptibility of breadfruit to pest and disease issues is exacerbated by the lack of genetic diversity ~ a common challenge faced by crops in Oceania. The absence of genetic variability often leads to the buildup of pathogen populations, contributing to the deterioration of agronomic performance (Lebot, 1992).

3.19 Insect-pests and nematodes

Breadfruit is though generally resilient to diseases and pests but may encounter challenges with mealybugs which are of a local concern. Ripe fruits on the tree and fallen ones are susceptible to infestations by fruit flies. Other potential pests of breadfruit include mites, aphids, mealybugs, scale insects and nematodes. In West Indies breadfruit trees can attract soft-scales and mealybugs, while ants may infest branches that experience dieback after fruiting. These pests can cause damage to the foliage, sap and overall health of breadfruit trees.

1. **Breadfruit weevil (*Oryctes rhinoceros*)**: Its adult female lays eggs in decaying organic matter, larvae start feeding on roots and pupate in the soil. Its attack causes stunted growth, yellowing and wilting of infested plants. To control its infestation, pheromone traps should be used, drenching of soil with entomopathogenic nematodes is suggested and on the trees parasitoid wasps (*Tetrastichus* spp.) should be introduced.

Basins of the affected trees are drenched with Chlorpyrifos 20EC at 0.2% twice in a year.

2. **Scale insects:** Its crawlers settle on leaves and form protective shells. Its infestation causes leaf yellowing, sticky honeydew and sooty mould. Affected parts should be pruned, 0.2% Dimethoate 30EC or Chlorpyrifos 20EC should be sprayed. Predators such as ladybugs should also be introduced on the trees.
3. **Aphids:** *Greenidia artocarpi* and *Toxoptera aurantii* aphids feed in colonies on tender parts of plants (twigs, laves and buds) so plants get distorted and devitalised. Strong water spray will kill these aphids. Introduction of natural enemies, especially lacewings and parasitoid wasps may prove effective. Spraying of 0.2% Dimethoate 30EC or 0.03% Imidacloprid 17.8SL will control these pests but its fruits should be consumed only after a fortnight of spraying.
4. **Soil-borne nematodes:** These nematodes feed on the roots which causes stunted growth, root-galling and plant wilting. Solarization, crop rotation, use of any resistant variety and application of biofumigants may control nematodes. Carbofuran @1g/m^2 followed by irrigation will kill these nematodes.

Other related breadfruit species

The 'breadfruit complex' comprises three interrelated species ~ *Artocarpus altilis* (Parkinson) Fosberg, *A. camansi* Blanco and *A. mariannensis* Trécul. This botanical trio presents a versatile and invaluable resource with both ~ the fruits and seeds. These trees not only contribute to the nutritional landscape but also embody multifunctionality as these offer numerous benefits to the environment. These multipurpose trees mature readily to bear fruit in 3-5 years of planting and continue fruiting for many decades. Apart from yielding edible fruits, these present a repertoire of resources that encompass medicinal components, fabric, adhesive substances, insect repellents, animal fodder and construction materials, *etc.* (ntbg.org, 2024).

3.20 Breadnut or seeded breadfruit (*Artocarpus camansi* Blanco)

Synonyms: *Artocarpus incisifolius* var. *seminifer* Stokes, *A. incisus* var. *seminifer* (Stokes) Duss, *A.a.* var. *seminifer* (Stokes) Fournet, *A. incisus* var. *muricatus* Becc., *A. leeuwenii* Diels, and *A. papuanus* Diels ~ nom. illeg. (POWO, 2023a).

Common names: Breadnut (English); Castaña (Spanish); Chataigne (Caribbean); Chataignier (French); Dulugian, Kamansi, Kolo, Pakau, Ugod

(Philippines); Kapiak (New Guinea); Kos del (Sri Lanka); Mei kakano (Marquesas); Pana de pepitas (Puerto Rico); and Kelur, Kulor, Kulur, Kuror, Timbul (Indonesia, Malaysia) (ntbg.org, 2024).

3.20.1 Introduction, origin and distribution

Breadnut is indigenous to the lush landscapes of New Guinea with potential occurrences in the Moluccas (Indonesia) and the Philippines (POWO, 2023a). Its propagation primarily relies on seeds, marking a distinctive characteristic of this species. In its native habitat of New Guinea, the breadnut thrives in alluvial forests within lowland regions where it is intricately intertwined with the local ecosystem. Dispersal of breadnut seeds is facilitated by avian (birds) and chiropteran species (fruit bats) which after feeding the fruits disperse a sizeable number of seeds. Moreover, cultivation of breadnut is not limited only to the wild but it also finds a place in home gardens contributing to its broader distribution in New Guinea. In the Philippines, it is typically cultivated as a backyard tree, showcasing its adaptability to various cultivation practices.

Despite its native origins, breadnut has ventured beyond its original habitat and has been introduced to several Pacific regions such as New Caledonia, Pohnpei (Federated States of Micronesia), the Marquesas, Tahiti, Palau and Hawaii. Introduction to these areas is attributed to immigrants from the Philippines in recent years, underscoring the human-mediated spread of this species. While breadnut remains relatively uncommon in Oceania, it boasts a rich history of cultivation and utilization in other tropical regions. Its presence can be traced back to the late 1700s when French played a pivotal role in disseminating it across tropical regions. Consequently, it has established itself as a widespread species in the Caribbean, C&S America, SE Asia, and West Africa.

Beyond its geographical distribution, breadnut holds significance as the progenitor of both seeded and seedless breadfruit (*Artocarpus altilis*). This distinction highlights breadnut as a separate species, despite its occasional classification as a form of seeded breadfruit. Understanding its evolutionary role adds depth to the appreciation of breadnut's contribution to the diversity within the *Artocarpus* genus (ntbg.org, 2024).

The breadnut's journey from its native New Guinea to diverse regions across the globe exemplifies the intricate interplay between natural dispersal mechanisms, human cultivation practices and historical influences all making it a noteworthy species with both ecological and cultural importance.

3.20.2 Description

The majestic tree of *Artocarpus camansi* attains an impressive height, reaching up to 35 meters. It is characterized by development of sturdy buttresses at the base of its trunk. The species stands apart from its botanical relatives, *Artocarpus altilis* or *A. mariannensis*, through its distinctive characteristics. One notable feature distinguishing this species is its impressive canopy which surpasses that of *Artocarpus altilis* or *A. mariannensis*. The arboreal wonder showcases generously sized leaves measuring 40-60cm in length and 25–45cm in width, displaying moderate dissection with 4-6 pairs of lobes (Zerega *et al.*, 2005).

This monoecious plant exhibits male and female flowers at the tips of its branches. The 2-anthered club-shaped male flowers are grouped together in a 3cm diameter and 25–35cm long inflorescence, showcasing thousands of individual flowers (Ragone, 2006). The breadnut's distinctive oblong fruit adorned with spikes undergoes a subtle transformation from a dull green to a greenish-brown hue upon ripening (ntbg.org, 2024). These fruits measure 16–20cm in length and 8–15cm in width, weighing approximately 800g. A mature tree can yield an impressive 600–800 fruits per year (Ragone, 2006; Zerega *et al.*, 2005). When ripe, the interior of the fruit is off white with a sweet taste and aroma. The fruit bears number of seeds per fruit ranging from 12 to 150 where an individual seed weigh around 7–10g. Dissemination of seeds is primarily facilitated by flying foxes and other mammals (Ragone, 2006).

Distinguishing *Artocarpus camansi* from its close relative *Artocarpus altilis*, involves noting its spinier fruits with numerous seeds. Meanwhile, differentiation from *Artocarpus mariannensis* can be achieved by observing the dark green elongated fruits with darker yellow flesh as well as the entire or shallowly lobed leaves of the latter (Ragone, 2006).

As indicated by its nomenclature, *Artocarpus camansi* is primarily cultivated for its abundant and sizeable seeds often colloquially referred to as 'nuts' (ntbg.org, 2024). Beyond its botanical significance, the species holds cultural and economic value, making it a subject of interest for researchers, horticulturists and enthusiasts.

4.20.3 Uses

4.20.3.1 Culinary uses

Elsewhere the breadnut fruits may also be known as 'kluweh'~ similar to breadfruit though smaller and are a good source of carbohydrates, fibre and essential nutrients. These provide energy and contribute to a balanced diet in regions where these are a dietary staple. The fruit can be roasted, boiled or baked and its starchy flesh is consumed as a staple food in many tropical regions.

Most significant use of breadnut tree is for its seeds that can be roasted, boiled or ground into flour for use in a variety of culinary applications through baking and cooking. In addition to their culinary significance, the seeds are known for their nutritional richness so making them a valuable dietary component. It adds a unique flavour and nutritional content to recipes. By skilfully thin slicing of entire immature fruits and seeds and then subjecting to a boiling process find their way into a variety of culinary delights including soups, stews and salads. Reminiscent of chestnuts in both ~ texture and flavour the seeds contribute a unique gastronomic element to a range of dishes.

3.20.3.2 Other uses

Besides its role in gastronomy, the tree holds ecological importance contributing to the biodiversity of its habitat. As an evergreen presence in the landscape, it provides shelter and sustenance to various fauna, adding to the intricate tapestry of the natural environment in which it thrives.

i) **Animal feed:** In some regions the by-products of processing breadnut such as the outer husk or pulp are used as animal feed. This helps in reducing waste and maximizing the utility of the tree.

ii) **Agroforestry:** Breadnut trees are sometimes integrated into agroforestry systems. These can provide shade to other crops, contribute to soil fertility and serve as a valuable component in sustainable farming practices.

iii) **Cultural and social significance:** Breadnut trees hold cultural importance in many communities where they are grown. They are often part of traditional rituals and the fruit plays a role in local cuisines, contributing to the cultural identity of the region.

iv) **Biodiversity conservation:** Breadnut trees as native species contribute to the overall biodiversity of tropical ecosystems. Conservation efforts may involve the preservation and sustainable management of breadnut populations to maintain biodiversity.

v) **Erosion control:** The extensive root system of breadnut trees helps in preventing soil erosion. Planting them in certain areas can contribute to soil conservation efforts.

vi) **Timber and wood products:** The wood from breadnut trees is used for construction and making various wooden products. The timber is valued for its strength and durability.

vii) **Traditional medicine:** In traditional medicine, various parts of the breadnut tree including the leaves, bark and roots are believed to have medicinal properties. Extracts or concoctions from these parts are used to treat ailments though scientific validation of these uses may vary.

These varied uses highlight the versatility of the breadnut tree and its importance in supporting livelihoods, nutrition and cultural practices in tropical regions.

3.21 Marianas Breadfruit or Seeded Breadfruit (*Artocarpus mariannensis* Trécul)

Synonym: *Saccus mariannensis* (Trécul) Kuntze (POWO, 2023b).

Other common names: Chebiei, Ebiei, Meduuliou, Mai (Palau); Dugdug, Dokdok (Mariana Islands); Maiyah (Puluwat, Yap); Marianas breadfruit, Seeded breadfruit (English); Mei chocho (Chuuk); Mei kole (Pohnpei); Mejwaan (Marshall Islands); Mos en Kosrae (Kosrae); Te mai (Kiribati) (ntbg.org, 2024); *etc.*

3.21.1 Introduction, origin and distribution

The indigenous breadfruit relative *Artocarpus mariannensis* thrives in the pristine ecosystems of Marianas to Caroline Islands (Palau) (POWO, 2023b) establishing its habitat in the dynamic interplay of limestone and ravine forests. Its resilient presence extends from the coastal zones to the lower mountain slopes, offering a testament to its adaptability. Nature's intricate choreography sees the dissemination of this invaluable species carried out by fruit bats ~ their delicate wings aiding the dispersion of seeds across its native range.

Unfortunately, the ecological harmony that once characterized the wild populations of *Artocarpus mariannensis* is now under siege. Typhoon onslaughts have taken their toll wreaking havoc on the delicate ecosystems. The encroachment of feral deer enticed by the delectable offerings of the trees adds another layer of threat contributing to the alarming decline in wild populations. A poignant backdrop to this decline is the disconcerting disappearance of fruit bats leaving the symbiotic relationship between these creatures and the breadfruit relative in jeopardy.

Remarkably, *Artocarpus mariannensis* has not succumbed to the challenges in isolation. It has engaged in a natural dance of genetic exchange with *A. altilis* resulting in a plethora of interspecific hybrid varieties. Whether adorned with seeds or rendered seedless these hybrids are collectively recognized as 'breadfruit'. This natural hybridization adds an intriguing dimension to the conservation narrative as it underscores the resilience and adaptability of these botanical wonders.

In the culinary landscape, *Artocarpus mariannensis* and its hybrid progeny hold a position of eminence. Serving as a major staple food the tree is cultivated with reverence and ubiquity across a vast expanse spanning Palau, the Mariana Islands, Federated States of Micronesia, Kiribati, the Republic of the Marshall Islands, Tokelau, Tuvalu, Nauru and Banaba Island. These regions bear witness to the enduring legacy of a tree that not only sustains but also shapes cultural and dietary practices.

Interestingly, the embrace of *Artocarpus mariannensis* extends beyond its native territories. While its stronghold remains in the mentioned regions, a few outliers including Hawaii and Rabi Island in Fiji have welcomed the presence of these trees. The latter introduced in the 1940s from Banaba further attests to the far-reaching impact of this breadfruit-relative.

The hybrid varieties with their fortitude against sandy soils and saline conditions present an opportunity for broader cultivation. Atoll countries and coastal areas in the tropics grappling with challenging soil and climatic conditions may find solace in these hybrids. The adaptability of these varieties could potentially offer a sustainable solution to food security challenges in these regions expanding the legacy of *Artocarpus mariannensis* beyond its native realms (ntbg.org, 2024).

3.21.2 Description

The majestic trees of *Artocarpus mariannensis* along with their diverse hybrid counterparts stand as botanical giants attaining the impressive heights of 20 metres or more. Distinguished by their sheer magnitude these trees surpass their botanical relatives ~ the breadfruit and breadnut, boasting large trunks that can measure up to 2m in width at their base. Robust buttresses extend gracefully from the trunk, providing structural support to these arboreal wonders while their canopies unfurl in full rounded splendour.

In addition to their imposing stature these trees exhibit a remarkable resilience to salinity which render these hardy in coastal environments. The leaves of *Artocarpus mariannensis* and its hybrids are typically entire or shallowly dissected, featuring 1-3 lobes on the upper third of the blade. This unique foliage adds to the distinctive allure of these botanical specimens.

The fruit, a testament to nature's artistry is a small wonder with a weight ranging between 250 to 500g. Cloaked in dark green skin even when ripe, the fruit boasts a pebbly texture that engages the senses. Upon ripening, the flesh transforms into a luscious deep yellow emitting a sweet aroma that beckons with temptation. Unlike its denser counterparts in the breadfruit family, *Artocarpus mariannensis* features flesh that is not as solid, interspersed with few to many rounded, dark brown and shiny seeds. Notably, this species relies on seed propagation for its continuation.

Venturing into the realm of hybrid varieties in Micronesia, a kaleidoscope of diversity unfolds. These hybrids characterized by their distinctive fruit and leaf forms present a fascinating tapestry of horticultural innovation. Ranging from seeded to seedless, these varieties exhibit considerable variability. Rough-skinned or pebbly fruit showcases a spectrum of hues from light to dark green with a glossy finish that adds an element of sophistication. The flesh not as dense as that of seedless Polynesian breadfruit varieties offers a delightful palate of creamy white to yellow tones.

Seeded hybrid varieties with their lumpy asymmetrical fruits measuring 12-30cm entail of its local adaptation, each variety unique to its specific geographical enclave. Some outliers defy convention, presenting narrow and elongated fruits of up to 45cm. Among the myriad hybrids, 'Meinpadahk' stands out as a seedless gem, widely distributed across high islands and coral atolls, adding a touch of continuity to the diverse landscapes of Micronesia.

The botanical narrative of *Artocarpus mariannensis* and its hybrids with nuanced intricacies of their form, flavour and adaptability entails a testament to richness of nature's creations (ntbg.org, 2024).

3.21.3 Uses

Artocarpus mariannensis holds significance in various cultural, culinary and ecological aspects. Its primary and traditional use is as a food source. When ripe the fruit has deep yellow flesh with a sweet aroma and flavour. It is consumed in various culinary preparations. The fruit can be prepared in numerous ways including boiling, roasting, frying or even in traditional dishes. It can be used in both sweet and savory recipes.

References

1. Barrau, J. 1976. Breadfruit and relatives. In: Evolution of Crop Plants (ed. Simmonds, N.W.), pp. 201–202. Longman, London, UK.
2. Cerqueira, A.O., E.D.M. Luz and J.T.D. Souza, 2006. First record of *Phytophthora tropicalis* causing leaf blight and fruit rot on breadfruit in Brazil. *Plant Path.*, **55**: 296.
3. Coates-Beckford, P.L. and M.J. Pereira, 1992. Survey of root-inhabiting microorganisms on declining and nondeclining breadfruit (*Artocarpus altilis*) in Jamaica. *Nematropica*, **22**: 55- 63.
4. Delazar, A., L. Nahar, S. Hamedeyazdan and S.D. Sarker, 2012. Microwave assisted extraction in natural products isolation. *US Nat. Libr. Medic. Nat. Inst. Health*, **2**(1): 27-30.
5. flowersofindia.net, 2024. Breadfruit. *Flowers of India*. Accessed on 25 January, 2024. http://www.flowersofindia.net/catalog/slides/Breadfruit.html.
6. Gardner, D.E. 1991. Occurrence of breadfruit rust, caused by *Uredo artocarpi*, in Hawai'i. *Pl. Dis.*, **75**: 968.

7. Handa, S.S., S.P.S. Khanuja and Gennaro Longo, 2008. Extraction technologies for medicinal and aromatic plants. *Intern. Cent. Sci. High Technol.*, [Online]. Available from: http://agritech.tnau.ac.in/horticulture/extraction_techniques%20_medicinal_plants.pdf [cited in Sikarwar *et al.* (2014)].

8. Huie, C.W. 2002. A review of modern sample-preparation techniques for the extraction and analysis of medicinal plants. *Anal. Bioanal. Chem.*, **373**: 23–30 [Online]. Available from http://www.ncbi.nlm.nih.gov/pubmed/12012169 [cited in Sikarwar *et al.* (2014)].

9. Jarrett, F.M. 1959. Studies in *Artocarpus* and allied genera. III. A revision of *Artocarpus* subgenus *Artocarpus*. *J. Arnold Arboretum*, **40**: 113–55; 298–326.

10. Lebot, V. 1992. Genetic vulnerability of Oceania's traditional crops. *Experim. Agric.*, **28**: 309-23.

11. Lewis, W.K. 1961. The principle of counter-current extraction. *J. Ind. Engin. Chem.*, **8**(9): 825–33.

12. Lincoln, N., D. Autufuga, J. Uchida, T. Redfern, A. Jones and D. Ragone, 2021. Breadfruit and breadfruit diseases in Hawai'i history, identification and management. *Fruit, Nut and Beverage Crops*, FN-63. College of Tropical Agriculture and Human Resources, University of Hawai'i at Manoa. *Extens. Publ.*, **808:** 956-7036. Rev. by CTAHR Fac. https://www.ctahr.hawaii.edu/oc/freepubs/pdf/FN-63.pdf.

13. Matisoo-Smith, Elizabeth, A. 2015. Tracking Austronesian expansion into the Pacific via the paper mulberry plant. *Proc. Natl. Acad. Sci.*, **112**(44): 13432–3. Bibcode:2015PNAS..11213432M. doi:10.1073/pnas.1518576112. PMC 4640783. PMID 26499243.

14. Morton, Julia, F. 1987. Breadfruit. *Fruits of Warm Climates*, pp. 50-58. West Lafayette, Indiana, USA: Center for New Crops & Plant Products, Department of Horticulture and Landscape Architecture, Purdue Univ., USA. ISBN 0-9610184-1-0.

15. ntbg.org, 2024. Breadfruit Species. The Breadfruit Institute of the National Tropical Botanical Garden. https://ntbg.org/breadfruit/.

16. POWO, 2024. *Artocarpus altilis* (Parkinson) Fosberg. Plants of the World Online. Royal Botanic Gardens, Kew Science. Retrieved on 22 January, 2024. (https://powo.science.kew.org/taxon/urn:lsid:ipni.org:names:771522-1). First published in *J. Washington Acad. Sci.* 31: **95** (1941).

17. Ragone, D. 2006. *Artocarpus altilis* (breadfruit), ver. 2.1. Species Profiles for Pacific Island Agroforestry (ed. Elevitch, C.R.), pp. 85-100. Permanent

Agriculture Resources (PAR), 816p. Holualoa, Hawai'i. <http://www.traditionaltree.org>.

18. Ragone, D. and R.E. Paull, 2008. *Artocarpus altilis*: Breadfruit. The Encyclopedia of Fruit and Nuts (eds Janick, J. and R.E. Paull), pp. 476-79. CABI North American Office, Cambridge, MA, 976p.

19. Rao, K.S. and D. Kumar, 2023. *Artocarpus incisus* (Thunb.) L.f. India Flora Online.Indian Institute of Science. https://indiaflora-ces.iisc.ac.in/herbsheet.php?id=1103&cat=13 Accessed on 25 January 2024.

20. Raymond, P. and W. Scott, 2003. Principles and Practice of Chromatography. *Chrom-Ed Book Series*, 5967 [Online] Available from: http://faculty.ksu.edu.sa/MFarouk/Mfarouk/PRINCIPLES%20of%20Chromatography.pdf [cited in Sikarwar *et al.* (2014)].

21. Redfern, T. 2010. Etiological study of breadfruit diseases in Hawai⊠i. Masters of Science Thesis. University of Hawai⊠i at Mānoa, 107p.

22. Sangchote, S., J.G. Wright and G.I. Johnson, 2003. Diseases of breadfruit, jackfruit and related crops. In: Diseases of Tropical Fruit Crops (ed. Ploetz, R.C.), pp. 135-44. CABI Publishing, Wallingford, UK, 544p.

23. Sikarwar, M.S., B.J. Hui, K. Subramaniam, B.D. Valeisamy, L.K. Yean and K. Balaji, 2014. A Review on *Artocarpus altilis* (Parkinson) Fosberg (breadfruit). *J. Appl. Pharmaceut. Sci.*, **4**(8): 91-7.

24. Solanki, Renu and B.P. Nagori, 2012. New method for extracting phytoconstituents from plants. *Intern J. Biomed. & Adv. Res.* [Online], **03**(10): 770-73. Available from: www.ssjournals.com.

25. Stice, K.N., A.M. McGregor and S.N. Kumar, 2007. A review of fresh breadfruit exports from Fiji. 1st Intern. Symp. Breadfruit Res. & Dev. *Acta Hort.*, No. 757, pp. 243-9.

26. theindianvegan.blogspot.com, 2012. All About Breadfruit. The Earth of India. Tuesday, September 25, 2012. Retrieved on 25 January, 2024. http://theindianvegan.blogspot.com/2012/09/all-about-breadfruit-in-india.html

27. USDA, 2018a. Breadfruit, raw. FoodData Central. Released in April 2018. Accessed on 11 January 2024. https://fdc.nal.usda.gov/fdc-app.html#/food-details/171714/nutrients.

28. USDA, 2018b. Seeds, breadfruit seeds, boiled. FoodData Central. Released in April 2018. Accessed on 11 January 2024. https://fdc.nal.usda.gov/fdc-app.html#/food-details/170145/nutrients.

29. USDA, 2018c. Seeds, breadfruit seeds, raw. FoodData Central. Released in April 2018. Accessed on 11 January 2024. https://fdc.nal.usda.gov/fdc-app.html#/food-details/170144/nutrients.

30. USDA, 2018d. Seeds, breadfruit seeds, roasted. FoodData Central. Released in April 2018. Accessed on 11 January 2024. https://fdc.nal.usda.gov/fdc-app.html; https://fdc.nal.usda.gov/fdc-app.html#/food-details/170595/nutrients.

31. Zaiger, D. and G.A. Zentmeyer, 1966. A new lethal disease of breadfruit in the Pacific islands. *Pl. Dis. Rep.*, **50**: 892-6.

32. Zerega, N.J.C., D. Ragone and T.J. Motley, 2005. Systematics and Species Limits of Breadfruit (*Artocarpus*, Moraceae). *System. Bot.*, **30**(3): 603–15. doi:10.1600/0363644054782134. S2CID 85907225.

Drumstick tree/Moringa (*Moringa oleifera* Lam.)

Family: Moringaceae.

Synonyms: *Guilandina moringa* L., Hyp*eranthera moringa* (L.) Vahl, and *Moringa pterygosperma* Gaertn. nom. illeg. **Chromosome number**: 2n = 28.

Common names: 'Drumstick tree' (GRIN, 2023), 'horseradish tree' (GRIN, 2023) due to its roots tasting as horseradish, 'malunggay' as known in maritime or archipelagic areas in Asia (Serafico *et al,* 2017), 'miracle tree/tree of life' owing to its excellent health, nutritional and environmental effects (Rajamanickam and Arokiamary, 2022), 'benzolive tree' and 'ben oil tree' (due to oil being derived from its seeds); in Hindi: Sajana (Saha *et al*., 2012)/Sahgan/Sahjan and Seeng.

4.1 Introduction

Moringa is a monogeneric genus encompassing 13 distinct species thriving in tropical and subtropical regions of Africa and Asia. Exhibiting rapid growth, *Moringa* species flourish in diverse environments, varying in size from diminutive herbs to towering trees. Word *Moringa* originates from regional names for the plant ~ 'moringo' in Malabar, 'murunga' in Sinhalese, 'murungai'/'murunggai' in Tamil and 'muringa' in Malayalam. These names collectively refer to the characteristic twisted pods of the young fruit (Olson, 2010; Saha *et al*., 2012). The specific epithet *'oleifera'* originates from the Latin '*oleum*' meaning 'oil' and 'ferre' or 'fero' meaning 'to bear'.

Moringa oleifera ~ the most extensively cultivated species within its genus is native to the sub-Himalayan regions of NW India (Mark Olson, 2010; Leone *et al.*, 2016) which has achieved widespread distribution in tropical and subtropical regions, thriving at altitudes of up to 2000m (Leone *et al.*, 2015a). Ancient Indian literature makes mention of *Moringa* as an interesting plant due to its widespread use in agriculture, medicine and industry (Saha *et al.*, 2012). A comprehensive examination integrating both morphological traits and DNA analysis reveals that *M. oleifera* is most closely affiliated with *M. concanensis* Nimmo (indigenous to northern India) (Leone *et al.*, 2015a). The common ancestor of these species diverged from the lineage of *M. peregrina* (Forssk.) Fiori (Mark Olson, 2002) which is native to the Arabian Peninsula, Horn of Africa and southern Sinai of Egypt (Dadamouny *et al.*, 2016).

Moringa is now grown worldwide in the tropics and sub-tropics for its tender pods, leaves and flowers. Widely utilised in the S&SE Asia, it holds significant importance in various applications. Moringa pod is quite popular vegetable in south Indian cuisine, being valued for its unique flavour, aroma and taste. Currently, *Moringa oleifera* is predominantly distributed in the Middle East and various African and Asian countries. However, its remarkable adaptability has led to its expansion into additional regions, particularly in tropical and subtropical areas that are susceptible to drought (Leone *et al.*, 2016).

Globally the second fastest-growing plant *Moringa* can achieve an impressive growth rate of up to 5cm per day. Remarkably, a lone moringa tree can mature rapidly, providing enough sustenance to feed an entire family within a year. Unlike other superfood greens, moringa tree continues to produce long after the first harvest, making it a long-lasting source of nutrition. Moringa is considered as a good quality meal, *vis-a-vis* has garnered significant recognition for its nutraceutical benefits, establishing itself as an essential plant for health maintenance. The plant boasts a myriad of nutritional and medicinal properties demonstrating immense potential in addressing malnutrition, *vis-a-vis* preventing and treating various ailments.

4.2 Production area

In S&SE Asia, moringa is cultivated as live fences in backyard gardens and is usually sold in the neighbourhood markets. It is frequently planted for its edible leaves in Indonesia and the Philippines. The World Vegetable Centre in Taiwan also actively cultivates moringa. Countries in C. America and the Caribbean, northern South America, Africa, S&SE Asia and several Oceanian nations are home to wild or cultivated moringa (Radovich, 2011). Large-scale cultivation of *Moringa oleifera* in East Africa has experienced significant growth in the past four years. The primary source of seeds for planting in Kenya, Tanzania and more recently Uganda is predominantly Mbololo (Kenya). These seed repositories have been initiated and sustained by local farmers. However, there is a notable lack of information regarding the genetic quality of these seed sources (Ramesh *et al.*, 2014).

In India, *Moringa* is widely cultivated across the subcontinent where it is cherished for its tender pods, leaves and flowers. Its pods renowned for their uniquely inviting flavour have become a staple in South Indian cuisine. India stands as the predominant global producer of *Moringa*, annually yielding 2.2 million tonnes of tender fruits across 43,600 hectares. This translates to an impressive productivity rate of approximately 51t/ha. Its primary cultivation in India is concentrated in the southern states of Tamil Nadu, Karnataka, Kerala, Andhra Pradesh and Telangana. This versatile tree has found a home in the daily lives of over two million households in South India serving as a convenient backyard resource. Andhra Pradesh takes the lead among states in both cultivation area (15,665 hectares) and production, closely followed by Tamil Nadu (13,042 hectares) and Karnataka (10,280 hectares). Other states collectively contribute to an additional 4,613 hectares (MIR, 2023).

Noteworthy is Tamil Nadu ~ a pioneering state in *Moringa* cultivation showcasing diverse genotypes sourced from varied geographical areas including introductions from Sri Lanka. The widespread cultivation of *Moringa* across these states highlights its integral role in Indian agriculture and culinary traditions (Rajangam *et al.*, 2001). Internationally, India serves as the primary supplier of *Moringa*, meeting approximately 80% of the global demand. The cultivation landscape in India predominantly features large plantations employing conventional methods. This approach enables the market to offer *Moringa* at competitive *cum* cost-effective prices (MIR, 2023).

4.3 Botany

4.3.1 Tree and leaves

Moringa oleifera characterized by its rapid growth, resilience to drought and distinctive appearance is a multi-branched, somewhat contorted and softwood tree adorned with a corky whitish-gray bark. When wounded, the plant releases a milky sap which adds to its intriguing features. The young shoots exhibit purplish or greenish-white hairy bark and despite being deciduous the tree tends to retain some of its leaves throughout the year. Overall structure includes an open crown adorned with delicate drooping branches where foliage consists of feathery clusters of green to dark green tripinnate compound leaves where elliptical leaflets measure 1-2cm in length.

Height of tree is subject to various influences such as environmental factors, soil composition and cultivation methods. Generally recognized for their swift growth, usually these trees can attain a height of 3-3.7m in their inaugural year under favourable conditions. In optimum settings, these trees can eventually soar to impressive heights of 6-9 metres or in some exceptional cases even up to 12m with a trunk diameter of 50cm. However, typically performed annual regular pruning can maintain a convenient height of 1–2 metres as is common practice in commercial growing areas. This strategic trimming fosters bushier growth thereby ensuring the pods and leaves remaining easily accessible within arm's reach.

Moringa foliage

By Obsidian Soul - Own work, CC0, https://commons.wikimedia.org/w/index.php?curid=58536866

4.3.2 Floral biology and pollination

Moringa Tree Flowers

By Venkatx5 - Own work, CC BY-SA 3.0, https://commons.wikimedia.org/w/index.php?curid=20134455

Moringa flowers

By Harvey McDaniel from Naalehu, HI - Flickr - image description page, CC BY 2.0, https://commons.wikimedia.org/w/index.php? curid=346366

Flowering typically commences within the initial six months post-planting. In regions characterized by seasonal coolness, blossoming occurs singularly each year, typically in late spring and early summer (April-June in the Northern Hemisphere and October-December in the Southern Hemisphere). Conversely, areas with consistent seasonal temperatures and ample rainfall may witness flowering twice a year or even persistently throughout entire year. Its inflorescences manifest as lax panicles extending 10-25cm in length, gracefully drooping from the axils (Leone *et al.*, 2016). Fragrant bisexual zygomorphic flowers some 12mm long range in colour from white to cream with a hint of pink in some varieties. These consist of 5 pale green sepals, 5 uneven and thinly veined white petals, 5 stamens with anthers and 5 staminoid stamens (Heuze *et al.*, 2016). Some varieties of flowers exhibit a delicate pink tinge. The flowers undergo extensive cross-pollination ~ a process facilitated by delayed stigma receptivity. Successful pollination necessitates the involvement of a substantial number of insects (Bhattacharya and Mandal, 2004). It is important to mention that appearance of its flowers may exhibit slight variation depending on the specific variety and growing conditions. These flowers typically emerge on slender and hairy stalks arranged in spreading or drooping clusters with a length ranging 10-25cm.

Moringa trees are pollinated primarily by insects especially bees, butterflies and certain other flying insects that transfer pollen from the male floral parts (anthers) to the female parts (stigma) within the same or different flowers on the tree. This pollen-transfer is essential for production of seeds. The stigma becomes receptive one day before flower opening and continues with maximum receptivity (88-96%) in 'PKM1' and 'PKM2', respectively, based on pollen adherence, and 72-84% by controlled pollination on the day of opening, then a sudden decline in receptivity thereafter (Kanthaswamy, 2006). The tree is usually self-pollinating as it can produce seeds without involvement of the pollen of other *Moringa* trees. However, cross-pollination can occur naturally with the help of insects carrying pollen from one tree to another. For commercial cultivation or to enhance the yield, some growers might manually assist in pollination process by transferring pollen between flowers using a small brush or similar tool to ensure better fruit set and seed production.

Its fruits are slender, pendant and tri-lobed capsule commonly referred to as a pod or drumstick. Measuring 20-60cm in length, the capsules undergoes ripening approximately three months after flowering. When immature the pods boast a vibrant green hue with certain varieties showcasing hints of reddish colouration. At maturity the pods turn to a brown and dry state, ultimately splitting open into three longitudinal parts. Each pod containing usually 12-35 sizeable globular seeds that measure around 1cm in diameter. These seeds are characterized by three angles and three wings, with the wings extending from the base to the apex, measuring 2–2.5cm in length and 4–7mm in width (Ramachandran *et al.*, 1980). The seeds are enveloped in semi-permeable hulls, typically displaying shades ranging from brown to black though instances of white appearance may occur in kernels with lower viability. The viable seeds have ability to germinate within a two-week timeframe (Radovich, 2011). A single tree can produce anywhere from 15,000 to 25,000 seeds, each weighing about 0.3 grammes (Foidl *et al.*, 2001). The kernel bears majority of the weight accounting for 70-75% (Ramachandran *et al.*, 1980).

Moringa pods in Panchkhal (Nepal)

By Krish Dulal - Own work, CC BY-SA 3.0, https://commons.wikimedia.org/w/index.php? curid=26452783

Moringa tree and dry seed pods

By Prof.Chen Hualin - Own work, CC BY-SA 4.0, https://commons.wikimedia.org/w/index.php?curid=42392849

4.3.3 Root system

Its robust root system plays a pivotal role in fostering growth of the plant and ensuring its survival. Characterized by a taproot system, the primary root extends vertically downward giving rise to lateral roots. This dual-root structure is instrumental in providing stability as the taproot penetrates deep into the soil while the lateral roots facilitate absorption of water and nutrients from surrounding soil.

Seedlings exhibit formation of a white, swollen and tuberous taproot characterized by a distinct pungent odour and sparse lateral roots. As these mature into trees from seeds, a robust taproot develops accompanied by a broad system of thick lateral tuberous roots that spread widely. The development of taproot ~ the primary root that grows vertically is more common in moringa plants grown from seeds. Seeds are the natural means of propagation for moringa and the plants grown from seeds often develop a taproot as part of their early root system.

Formation of taproots in cutting-propagated moringa plants can be influenced by various factors including the specific propagation method used and the environmental conditions. When moringa is propagated through cuttings it tends to develop a fibrous root system rather than a taproot (Lahjie and Siebert, 1987). Moringa cuttings, especially those taken from mature trees typically produce a network of fibrous roots that spread horizontally in the soil. The cuttings may not give rise to a taproot readily as compared to plants grown from seeds, root structure and its success can be influenced by factors such as age of cutting, the rooting medium, humidity and temperature. However, moringa is known for its adaptability and hardiness so cuttings can still result in healthy and vigorous plants with a fibrous root system.

4.4 Growth and Development

Plants grown from seeds can experience accelerated growth under optimal conditions. Certain early flowering varieties often termed 'annual types' such as 'PKM-1' and 'PKM-2' (developed by TNAU) have meticulously been chosen for their capacity to produce vegetable pods efficiently within a year facilitating timely harvesting and subsequent replanting. In contrast to alternative types, these early flowering variants can reach market-mature pod-stages in just six months ~ a significant reduction from the typical year-long duration for other varieties. In general moringa demonstrates resilience to similar climatic conditions. Following coppicing, the branches undergo rapid regrowth, and immature pods can be harvested within a six-month timeframe.

Moringa is characterized by its free-flowering nature with blossoms typically appearing 4–12 months after planting depending on the specific type. Some selections even showcase flowering within a remarkable 4–5 months after planting (Radovich, 2011). Early flowering varieties exhibit pod production in 6-7 months while other variants necessitate more than a year to reach the same stage.

4.5 life-Cycle Duration

Though *Moringa oleifera* is a perennial herbaceous tree known for its long lifespan ~ often surviving for several years ~ more than a decade under favourable conditions but HCRI, TNAU, Periyakulam (TN) has introduced seed-propagated improved annual varieties. Usually it behaves as annual in colder climates or areas with distinct seasons where it cannot survive frost or freezing temperatures and there it acts as an annual, reseeding itself yearly or requiring replanting each year during March due to frost damage or inability to survive the winter. Understanding the specific species and local climatic conditions is crucial when determining whether a particular moringa plant will behave as a perennial or an annual. In frost-fee warmer climates, *Moringa oleifera* is more likely to exhibit perennial characteristics. *Moringa concanensis* is often considered as an annual or short-lived perennial which usually completes its life cycle, *i.e.* seed germination, growth, flowering, and seed production within one year though where these are unable to survive the rigours of winter this behaves as an annual.

4.5.1 Perennial type

Perennial types have been in cultivation for a very long time. In India these types are typically propagated from cuttings. Nevertheless, the cultivation of perennial types is hindered by numerous production constraints including a relatively prolonged pre-fruit bearing period, unavailability of planting materials such as stem cuttings, the need for a higher number of rainy days in regions with water scarcity and susceptibility to pests and diseases. These may not be suited for areas with short growing seasons or where is shortage of water. Notable moringa varieties encompass 'Chavakacheri', 'Chemmurungai', 'Jaffna-type', 'Kattumurungai', 'Kodikkalmurungai', 'Moolanur', 'Valayapatti', 'Palmurungai', 'Punamurungai' and 'Palamedu' (sustainablebioresources.com, 2023).

4.5.2 Annual type

Cultivation of *Moringa oleifera* in the country has undergone a significant transformation following introduction of seed-propagated moringa varieties by HCRI, TNAU, Periyakulam (TN). Development of two enhanced annual varieties, *viz.* 'Periyakulam-1' (PKM-1) and 'Periyakulam-2' (PKM-2) is a result of recent advancements in plant breeding research. These varieties have demonstrated substantial success in commercial cultivation across both traditional and non-traditional farming areas (sustainablebioresources.com, 2023).

'Periyakulam-1' and 'Periyakulam-2' being seed-grown annual moringa cultivars have effectively replaced perennial moringa in the southern states of India. Their rapid adoption is attributed to their rapid maturation, higher yields and greater adaptability to diverse soil and climatic conditions. In consequence, these now constitute a significant proportion of moringa production particularly in marginal and small farm holdings. Disadvantages of annual types compared with perennial types may include shorter lifespans, requirements for more frequent replanting and reduced genetic diversity (sustainablebioresources.com, 2023).

4.6 Classification

Moringa cultivars can be classified based on several criteria including their purpose, growth habits and regional adaptations as furnished below:

4.6.1 Classification based on growth habit

- **Dwarf/semi-dwarf cultivars:** Compact varieties suitable for smaller spaces or container gardening which are characterized by their reduced height and sometimes bushier growth.
- **Tall cultivars:** Varieties that grow to substantial heights, *i.e.* 10m or little more which are suitable for larger agricultural spaces or as tree plantation settings.

Various compact varieties of *Moringa oleifera* have been cultivated in India with limited information available on the origins of some strains. Dwarfs derived from the annual determinant height 'PKM-1' and 'PKM-2' varieties are widely accessible. These dwarf varieties are particularly suitable for individuals with limited outdoor space or those desiring to grow these in containers so making them adaptable for indoor cultivation in cold climates or temporary indoor relocation during winter in temperate climates.

True dwarf varieties like 'Bhagya KDM1', 'Dhanaraj (S 6/4)', 'Periyakulam 1 (PKM1) Dwarf', 'Periyakulam 2 Dwarf' (PKM2) and 'Periyakulam 2 Extra Dwarf hybrid' (PKM2) ~ mimic standard-sized counterparts but growing only 1–2.5m tall compared to 10–12m potential height of untrimmed standard forms. Semi-dwarf options with intermediate heights such as 'Andipatty', 'Moolanu', 'Multiplex' and 'Shobhanjana' are also available. Regular trimming is crucial for both dwarf and semi-dwarf types where side branches are encouraged to optimise leaf production. Initiation of trimming is recommended when tree reaches approximately 1m in height. Repeated pruning of trees approximately 50% of their height encourages lateral growth and a more balanced structure. Untrimmed trees tend to grow rapidly reaching their maximum height with leaves and pods concentrated only on the top.

Dwarf trees offer an additional benefit as they are less prone to root binding when cultivated in containers compared to standard varieties. A closer inspection of the root system of container-grown 'PKM-2' extra dwarf trees unveiled the emergence of several tubers accompanied by small roots. Notably, there were no signs of root circling or any other indications suggesting that the tree was becoming root bound (sustainablebioresources.com, 2023).

4.6.2 Classification based on purpose

- **Nutritional cultivars:** Specifically cultivated for their high nutritional values in terms of leaves, pods and seeds.
- **Oilseed cultivars:** Primarily grown for oil extraction from seeds due to their high oil content.
- **Fodder cultivars:** Mainly used for livestock feed due to their leaf and stem properties.
- **Dual-purpose cultivars:** Varieties that serve multiple purposes offering both high nutritional value and good oil or fodder production.

4.6.3 Classification based on regional adaptation

- **Local or regional cultivars:** Cultivars adapted to specific climate and soil conditions of a particular region are developed through traditional farming practices.

- **Improved or hybrid cultivars:** Cultivars bred through selection or hybridization to enhance specific traits such as yield, disease resistance or adaptability to various environments.

4.6.4 Ayurvedic classification

Three varieties of *Moringa* referred in Ayurvedic textbooks are 'Shyama' being most common black variety, 'Shweta' ~ the white variety and 'Rakta' ('Madhu Shigru') ~ red variety (Agarwal, 2023).

4.6.5 Species-specific classification

- ***Moringa oleifera* cultivars:** Most common and widely cultivated for its edible leaves, pods and seeds and for its medicinal properties.
- **Other species:** There are certain other species in *Moringa* genus such as *M. stenopetala* which might have specific cultivars or varieties with distinct properties or adaptations.

Having knowledge about the varieties the growers may choose cultivars based on their specific needs such as for nutritional supplementation, commercial oil production, livestock fodder or those specific to environmental conditions and space-specific ones.

4.7 Breeding

In breeding, we enhance various useful aspects such as yield, nutritional content, disease resistance and adaptability to different environments in the offsprings as elaborated here under:

1. **Trait selection:** Breeders identify and select plants with desirable traits such as high yield, drought resistance, better nutritional content and disease resistance.
2. **Crossbreeding:** Controlled crosses are made between parent plants with complementary traits to combine their beneficial characteristics in the offspring.
3. **Hybridization:** Hybrid varieties are developed by crossing genetically distinct parent plants. These hybrids often display improved vigour, yield and other desired characteristics.
4. **Genetic diversity:** To broaden the genetic base, breeders try to incorporate genetic material from different moringa varieties or related species. This helps introducing new traits and enhancing overall diversity.
5. **Selective breeding:** Through successive generations breeders select and propagate plants that exhibit desired traits more consistently. This process helps fixing and stabilising the desired characteristics.

6. **Molecular breeding:** Modern techniques like molecular markers and genetic analysis aid in identifying specific genes associated with desired traits. This targeted approach can accelerate the breeding process.
7. **Field testing:** Newly developed varieties undergo extensive field trials to evaluate their performance under various environmental conditions, ensuring they meet the desired standards before widespread adoption.

Breeding programmes aim to create the varieties that are well-adapted to different climates, have increased yield with higher nutritional content, improved disease resistance and other desirable traits. Such efforts contribute to the development of superior cultivars that can benefit farmers and communities by providing more resilient and productive crops.

In India, rich diversity of wild moringa types provides an excellent foundation for breeding programmes. However, in regions where moringa has been introduced, the variety among cultivated types is notably narrower. Nevertheless, locally adapted wild moringa types can still be found across most regions.

Given the diverse uses of moringa, breeding goals differ significantly between aiming for an annual or perennial plant. In India, where moringa is primarily cultivated annually, enhancing fruit yield stability stands as a crucial breeding objective. Conversely, in less favourable environments, perennial cultivation offers substantial advantages, notably in erosion control. Pakistan has conducted tests on moringa varieties to assess the nutritional composition of their leaves across different locations (Iqbal and Bhanger, 2006). India emphasizes on selection of varieties with a higher pod count and focuses on developing dwarf or semi-dwarf varieties. Meanwhile, breeders in Tanzania prioritize elevating the oil contents in their selections (Wikipedia, 2023).

4.8 Varieties Grown in India

Moringa demonstrates significant diversity in various growth attributes such as growth rate, branching pattern, time to flowering, leaflet dimensions and pod length. The selection of preferred varieties for leaf production typically favours those characterized by large and dark green leaves. Desirable pod traits include length, tenderness and a lack of bitterness. Among the recognized varieties, the early flowering types 'PKM-1' and 'PKM-2' stand out as widely acknowledged, alongside others like 'Bombay', 'Chavakacheri', 'Chemmurungai' and 'Jaffna' (Radovich, 2011).

India in particular boasts a plethora of commercially cultivable *Moringa* varieties numbering at least 20. TNAU has achieved success in developing two high-yielding annual seed-propagated types 'PKM-1' (Selection) and 'PKM-2' (MP 31×MP 28). Introduction of 'Dhanraj', a dwarf type has revolutionized the moringa cultivation in the country. Another noteworthy release is the drumstick variety 'Anupama' by KAU, Kerala. Certain high-yielding land races such as 'Jaffna Melanor', 'Saragva', 'Chavakacheri', 'Chemmurungai', 'Pal Murungai', 'Puna Murungai', 'Kodaikal Murungai' and 'Saragvi' are cultivated in Tamil Nadu as well as in Andhra Pradesh and these are often identified by local farmers in specific regions. Public sector contributions include varieties like 'KM 1', 'GKVK 1', 'GKVK 2', 'GKVK 3', 'KDM 1', 'Konkan Ruchira' and 'Rohit 1' as documented by Saha *et al.* (2012).

4.8.1 Perennial types

A detailed overview of various moringa cultivars each possessing unique characteristics and adaptability to different regions is given below. This information is compiled from reliable sources such as TNAU, Ecoursesonline (2023), sustainablebioresources.com and certain others.

i) **Jaffna Type (Yazhpanam):** This Sri Lankan cultivar is believed to be one of the earliest varieties introduced which is well adapted to the coastal tracts of Tamil Nadu, Kerala and Karnataka, having 60-90cm long elongated pods with soft flesh and excellent taste and its peak season being March to June. Though in the second year it yields about 40 pods but third year 500-600 pods per tree though a mature tree of 40 years yields about 1,500 pods. It is almost resistant to insect-pests and diseases. Its seed oil analysis was performed by Bhatnagar and Krishna (2013).

ii) **Chavakacheri Moringa:** It is an ecotype of 'Jaffna Moringa' which is grown in Jaffna and Tamil Nadu, yielding 500-600 pods per tree per year having 90-120cm of pod length.

iii) **Chemmurungai:** This medium-sized tree is a regional ecotype of 'Jaffna Moringa' which has year-round blooming and fruiting with vivid red pod tips.

iv) **Palmurungai (Pal Moringa):** It is preferred for its 400-500 pods per year having good taste with thick pulp.

v) **Punamurungai:** It is grown in the home gardens of Tirunelveli and Kanyakumari districts, yielding thick and tasty pulp though there are a few references expressing its bitterness too.

vi) **Kodikkal/Kodikal Murungai:** This short-statured tree is grown in betel vine gardens to shade and train the vines in Tiruchirapalli district, bearing small leaves and shorter (20-25cm) pods with thicker flesh.

vii) **Andipatty:** It is a semi-dwarf type tree (up to 7.5m) growing quite fast in poor soils. It is resistant to insect-pests and diseases, vis-a-vis poor soils. It produces some 25-35 branches, good quality leaves and long pods with 22-25 seed count per pod.

viii) **Bhagya KDM-1 (2011):** It was developed at Kittur Rani Channamma College of Horticulture, Arabhavi (Paramaddi, Karnataka) in 2011 from var. 'Periyakulam 3' *i.e.* 'PKM 3'. This fast-growing dwarf to semi-dwarf variety flowering after 100-110 days has about 15 years of lifespan. Its height is maintained at 2-4m through pruning and the pods can be harvested after 160-180 days after transplanting. In the 1st year a tree produces 350-400 pods but in 2nd year 800-1000 with an average yield of 42-50 tonnes per hectare. Its seeds yield 38% crude oil with 30-33% net oil content (Hiraman, 2020).

ix) **Kattumurungai (Kadumurungai or Kadu):** It is a wild ecotype of *Moringa concan*ensis collected from forests in Tamil Nadu. Its pods are small and of inferior type.

x) **Mol'e:** It is a non-GMO (genetically modified organism) developed cultivar by Advanced Biofuel Center (Maharshi, 2020) having 5-8 years of life span. It is recommended for intensive cultivation for leaf production. It attains the height of 1-2m. Per hectare some 30,000-167,000 plants can be accommodated and for first cutting it requires 50 days from planting. Some 5-7 cuttings can be taken annually, yielding some 90-500 tonnes of wet biomass (dry weight being some 6% of wet weight). The protein content in fresh leaves is about 10%, in dry leaves some 22% and the dry leaf powder some 29%.

xi) **MOMAX3 © (Advanced Biofuel Center):** It is a proprietary variety for seed oil production which was released in 2016. It is a non-GMO proprietary variety for seed oil production (Maharshi, 2020). Its trees have life span of 15-20 years with uniform growth and maturity. It provides harvest of pods twice a year. It withstands adverse weather conditions and is free from insect-pests and diseases. In the first year a tree produces 350-500 pods, a pod containing 16 seeds where per 100 seed-weight comes to 30.2g. Its highest yield is 5-9 tonnes of seeds per hectare per year with 40% oil content ~ world's highest oil yielding variety. It withstands adverse weather conditions.

xii) **Moolanur Moringa:** It is a perennial semi-dwarf ecotype growing up to 4.5m tall and bearing 30-53cm long pods. It is cultivated in Moolanur, Karur and Dharapuram regions of Tamil Nadu. On an average a tree yields 500-600 pods per year.

xiii) **MS Series ('MS01' and 'MS02', Ancient Greenfield Pvt. Ltd):** It is a perennial type with high productivity.

xiv) **Palamedu Moringa:** Its average yield is 100 pods per tree with pod length of 60cm, each pod weighing 95-100 grammes. It yields 37+1% crude oil with 30-33% oil content as per the test of International Seed Testing Association methods.

xv) **Sarpan SD2:** This medium-sized perennial variety was developed by Dr. Nijagunadev Gaddagimath of Dharwad's Sarpan Agri-Horticultural Research Center and Sarpan Hybrid Seeds requiring pruning every alternate year at 1.5-1.8m to manage the canopy and for ease of pod harvesting. It produces fleshy, pulpier and soft-seeded 300-500 pods per year with each pod measuring 37.5-45cm. It provides two flushes annually, one in December-March and another in June-July.

xvi) **Rohit 1:** This variety was developed by a farmer Balasaheb Marale at Sinnar in Nashik district where trees remain productive up to 10 years. It starts producing first dark green pods some 45-60cm long 4 to 6 months after planting. The pods are quite tasty and pulpy. A 5-year old tree produces 585-650 drumsticks per tree annually. In welldrained irrigated soils the pod yield is 17-30 tonnes per hectare.

xvii) **Valayapatti Moringa:** It is grown in Usilampatti and Andipatti regions of Tamil Nadu. It yields 1000-1200 pods per tree per year, each pod weighing about 120g with an average length of 65cm.

1.1.2 Improved seed-propagated annual types

i) **Periyakulam 1 (PKM-1):** This dwarf variety was released in 1989 by HCRI, Periyakulam (T.N.). This grows 4-6m in height in a year producing 6-12 primary branches. Its pods are ready for harvesting in 7-8 months of planting with peak of harvesting being from March to August. It can be grown as intercrop in coconut plantings during early period. In its own field chilli, onion and groundnut can be intercropped. High density planting at 1.5×1m spacing with two plants/hill and with a plant population of 13,333/ha along with application of 135:23:45g NPK/pit through drip increases the yield. In this, phosphorus and muriate orf potash should be applied basally in the soil, and nitrogen in the form of urea through drip. After every year the trees have to be cut back to 1m from ground level. Ratoon crops can be maintained for 3-4 years. It thrives well when grown in containers, avoiding the issue of becoming root-bound. Its 40cm long pinnate leaves have small leaflets where upper side is dark green though lower pale green and these can be used as nutritive greens. It starts flowering in 5-6 months of sowing bearing 12-150 flower per truss but some 60-75cm long 2-4 fleshy and tasty pods develop per cluster having girth of 6-6.3cm, each pod weighing some 150g containing 70% flesh. Pods attain edible maturity 65 days after flowering. Average yield is 200-250 pods/tree/year and estimated yield is around 52t/ha. Its ratoon crop can be maintained for 3-4 years.

ii) **Periyakulam 2 (PKM-2):** This is a hybrid derivative of a cross between 'MP 31' ('Eppothum Vendran Local') and 'MP 28' ('Arasaradi Local'), released by HCRI, Periyakulam (T.N.). It is an extra dwarf quick-growing variety as an improvement over 'PKM-1' with some 12 lateral branches and is desirable for more leaf production at lower height for ease of harvesting. It comes to bearing 6 months after planting. Its clustered flowers bear 3-4 pods per cluster, each pod measuring some 1.26m with 8.3cm girth and 280g pod weight having 70% flesh but is less seeded. A tree produces about 220 pods per tree per year so yielding about 98-138t/ha of pods. Pinching of main shoots on 80th day of sowing will register highest fruit yield. Its ratoon crops can be harvested up to three years. It is suitable for growing in Tamil Nadu and is adapted to most soil types varying from sandy loam to clay loam with good drainage. The closest spacing of 1.2×1.2m is ideal to obtain highest yield of 138t/ha. Pinching of main shoots on 80th day after sowing will also help register the highest yield of fruits. It is suitable for various cropping systems. Ratoon crop can also be taken up for three years.

iii) **Dhanaraj (S 6/4):** It was developed by TNAU which is an annual dwarf variety with a canopy height of 4-6m. It is early in flowering, i.e. 7-8 months of planting yielding 250-300 pods per plant per year with per hectare total of 90-100 tonnes of green pods which are of good cooking quality.

iv) **Dhanraj (a Karnataka variety):** This variety was developed at Rani Chennamma College of Horticulture, Arabhavi. The trees grow to a height of 4–6 metres and come to flower in 90–100 days after planting with first harvesting at 160-170 days after planting. Each tree produces an average of 200-225 pods per year.

v) **Kudumianmalai 1 (KM-1):** It was developed through pureline selection. After each harvest, the plants can be ratooned for 2-3 years by cutting the trunk at a height of 1m. Fresh planting can be taken after 3 years. It is a shrubby dwarf plant which starts bearing pods after six months of planting. Its pods are 25-30cm long and thick with an average yield of 400-500 pods per tree per year.

vi) **Multiplex:** It is a dwarf hybrid of 'PKM1' and 'PKM2' attaining the height of 4-6m. It has wide adaptability. This produces pods twice a year with an average yield of 200-350 pods per tree.

vii) **GKVK-1, GKVK-2 and GKVK-3:** These three improved dwarf varieties were released by UAS, Bangalore. These grow to a height of 1.5m and are suitable for high density planting. 'GKVK-1' produces 35cm long pods with average yield of 250-300 pods, 'GKVK-2' produces 300-400 pods/year, and 'GKVK-3' 250-300 fruits/plant/year where pods are triangular having blackish green colour.

viii) **Konkan Ruchira:** This variety was developed by the public sector.

1.1.3 Some other popular varieties

i) **Amar 32** (Amar Seeds Pvt. Ltd in Pune area of the western Indian state of Maharashtra): It is a high yielding rapid maturing variety ready to harvest within 5-6 months of planting.

ii) **Coimbatore-1:** This variety is widely available in India and considered superior for quality drumstick production. This continues yielding fruits for 8-10 years. It produces 45-60cm long pods twice a year.

iii) **Coimbatore-2:** This Tamilian variety is more popular than 'Coimbatoire-1', yielding 25-35cm long and bulky pods with per tree production being 500-900 pods. Economic tree life is 3-5 years.

iv) **MX3 (Advanced Biofuel Centre, India):** This is a high yielding variety suitable for seed oil extraction. The seeds have oil content reportedly in the range of 35-40%. As per ISTA, the seeds have a crude oil content of 37+1 % and an oil content of 35–38%.

v) **Oddanchathiran (ODC):** This renowned drumstick variety extensively cultivated throughout India is result of collaborative efforts between the public and private sectors. Named after Oddanchatram ~ a region in the Dindigul district of Tamil Nadu, this variety stands out for its efficient use of water and fertilizer, impressive pod yield, delightful taste, prolonged shelf life and widespread consumer preference. 'ODC' can be cultivated using standard spacing of 3m row to row and 2.4m plant to plant or with more intensive spacing of 2×1.5m. The trees exhibit rapid growth and initiate profuse branching just three months after planting. Edible-sized pods are ready for harvest in 65 days after flowering. Anticipated yield includes 200-300 pods per plant thus contributing a total annual yield of approximately 60-70 tonnes per hectare. Ratooned trees can be sustained for 10-15 years with an annual cutting back at 1m from ground level (Thampalayagowder, 2017).

vi) **ODC3:** The variety known as 'ODC 3' stems from a meticulous 6-generation process of continuous selfing in tree selection which culminated its release in 2017. The resulting trees boast delectable fleshy fruits. Their flowering commences within 3-4 months of sowing with harvestable pods emerging at the 6-month mark. These trees exhibit rapid growth reaching 2.4-3m annually with 6-10 primary branches. Clusters of 50–200 flowers adorn the trees though per cluster this yields only 1-5 pods with 60-75cm length and 6.25cm girth and with a pod weight of 70–80g which contains some 80% flesh. Pod yield per tree is around 300 which comes to 60-75t/year. In comparison with PKM varieties, ODC 3 stands out as a superior choice for various reasons (OMAR, 2021). The drumstick pods exhibit a smoother surface, uniform fleshy appearance, greater girth and with enhanced taste, aroma and keeping quality. ODC 3 also boasts approximately twice the oil yield. Furthermore, it demonstrates water efficiency, heightened drought tolerance and reduced susceptibility to pests during dry conditions. Its adaptability to diverse soil types and resilience in the face of weather and climatic fluctuations make it a highly favourable option for cultivation.

vii) **PAVM:** The PAVM variety originating from the agricultural innovation of a farmer in the Dindigul district (Tamil Nadu), has gained widespread adoption. This high-yielding drumstick variant begins producing pods as early as the 5th or 6th month after planting. It is propagated through cuttings and air layering (Prabu, 2009).

viii) **'Bombay' and 'Madurai':** As mentioned by Radovich (2011), these varieties are of uncertain origin.

ix) **Anupama** (KAU, Vellanikara).

x) Durga is a regional landrace identified by farmers and others.

xi) Saragvais: It is a regional landrace identified by farmers and others.

xii) Shobhanjana: A dwarf variety of undetermined origin.

xiii) Yalpanamis: A regional landrace identified by farmers and others.

xiv) Yazhpanamis: A regional landrace identified by farmers and others.

1.1.4 Varieties from other regions

i) **Mbololo:** The Kenya Forestry Research Institute (KFRI) in Nairobi has cultivated a distinct variety that stands out as a robust pod producer. Thriving without need for irrigation and displaying resilience against wind, this variety holds significant agricultural merit. Moreover, the seeds of this cultivar yield oil with impressive qualities—it boasts high resistance to oxidation and a favourable monounsaturated to saturated fatty acids ratio. This unique combination positions it as a potentially ideal substitute for olive oil in dietary applications (Tasakinis *et al.*, 1999).

ii) **Malawi:** A Malawi-developed variety with heightened oil content has undergone characterization, revealing identified antioxidants in its seed oil (Lalas and Tsaknis, 2002; Tsaknis et al., 2002).

iii) **Congo-Brazzaville:** This name was applied to trees growing in neighborhood gardens around the Marien Ngouabi University Campus in Brazzaville (Congo). The oil derived from moringa seeds demonstrated favourable physicochemical properties. This oil holds promise for applications in both edible and industrial contexts (Nzikou et al., 2009).

iv) **STX-1 and STX-2:** These cultivars were developed by Toni Ramirez of Lorado, Texas.

   a. **STX-1** exhibits remarkable cold-hardiness, along with impressive heat and drought tolerance, making it an ideal choice for diverse climatic conditions. Renowned for its accelerated growth and high productivity, this strain boasts an impressive lifespan with some trees thriving for over 25 years in their designated growing areas. Unlike its counterpart STX-2, STX-1 accommodates increased soil moisture and can endure colder winters. Distinguished by its focus on leaf production rather than flowering and pod development, STX-1 allocates its energy strategically. While flowering may commence after 2-3 years, pod production remains relatively low but the pods are stout measuring about 50cm.

   b. **STX-2** is an enhanced non-GMO landrace variety derived through meticulous pureline selection from the PKM-1 strain. This demonstrates a little lower cold-hardiness than STX-1 though compensates with comparable heat and drought tolerance, rapid growth and exceptional productivity. This initiates flowering within 90 days. This excels in leaf and pod production with substantial pods measuring 60-75cm long. Sheer abundance of pods may lead to branches sagging under their weight, a testament to STX-2's prolific nature. The grower firmly asserts that STX-2 surpasses the esteemed qualities and productivity of the PKM-1 variety.

4.9 Nutritional value

Moringa boasts a wealth of nutrients across all its plant components. Moringa has been employed to ward off and shield against various physiological disorders.

Its nutritional profile is impressive, featuring seven times more vitamin C than oranges, four times more calcium than milk, four times more vitamin A than carrots, twice the protein content of milk, three times more potassium than bananas, and a substantial amount of iron content (Singh, 2010).

It offers a myriad of health benefits through its flowers, seeds, pods, leaves, gum and bark. Rich in essential nutrients, these components play a vital role in addressing various health concerns. Moringa has been recognized for its potential to alleviate vitamin and mineral deficiencies, regulating blood glucose levels and contributing to a healthy cardiovascular system. It supports the anti-inflammatory mechanisms in the body, enhances the immune system and acts as a potent antioxidant ~ neutralizing free radicals. It has shown promise in combating malnutrition, providing essential support to lactating mothers, alleviating general weakness and offering relief from conditions such as depression, menopause, osteoporosis and arthritis. The by-products of the drumstick tree including greens, flowers and fresh pods are packed with antioxidants, anti-inflammatory agents, vitamins, nutrients and minerals that meet the recommended dietary allowance (RDA). It is a rich source of vitamins A, C, D & E along with minerals such as calcium, phosphorus, potassium, copper, iron and magnesium (Janick and Paull Robert, 2008). Its composition also includes alpha-linolenic acid, amino acids and plant-based omega-3 which further enhance its nutraceutical properties (MIR, 2023).

Extensive research conducted at AVRDC revealed that its leaves are abundant in nutrients and antioxidants. Boiled leaves or using leaf powder results in at least three times more bioavailable iron, *vis-a-vis* increased antioxidant activities compared to raw leaves. Experiments at University of Baroda (India) underscore the benefits of cooking its leaves with oil. This cooking method not only helps in retaining ß-carotene but also promotes its conversion into vitamin A within the human body (Saha *et al.*, 2012).

Moringa leaflets boast exceptional nutritional value, offering a multitude of health benefits. Even in their early stages, these plants can be utilised as a tender vegetable. The leaves stand out as an excellent source of vitamin A and when consumed in their raw form it provides a significant dose of vitamin C. These serve as a rich reservoir of vitamin B and rank among the finest plant sources of essential minerals. Its Ca and Fe contents are remarkably high. These leaves prove to be an outstanding source of protein while maintaining minimal levels of fat and carbohydrates. What sets them apart further is their unmatched contribution to sulphur-containing amino acids such as methionine and cystine ~ addressing a common deficiency in many diets (Ramesh *et al.*, 2014).

4.9.1 Nutrient content of fresh *Moringa oleifera* leaves

Nutrient content of 100 grammes fresh, raw *Moringa oleifera* leaves			
Amount in 100 grammes (about 5 cups)			
References	**USDA (2013)**	**Goplan *et al.* (1999)**	**Stadlmayr *et al.* (2012)**
Nutrient	**USDA National Nutrient Database**	**Nutritive Value of Indian Foods**	**FAO West African Food Composition Table**
Energy (Kcal)	64	92	86
Moisture (mg)	78.66	75.9	76.5 +/- 1.9
Protein (g)	9.4	6.7	8.3 +/- 0.7
Carbohydrates (g)	8.28	12.5	9.6
Fiber crude (g)[A]	2	0.9	2.0
Fiber, total dietary (g)[A]	(5.3 – 7.3**)	Na	Na
Fat (g)	1.4	1.7	1.2 +/- 0.5
Ca (mg)	185	440	434 +/- 181
P (mg)	112	70	90 – 112
Na (mg)	9	Na	4 – 9
K (mg)	337	Na	337 – 470
Mg (mg)	147	Na	70 +/- 67
Mn (mg)	0.36	Na	Na
Fe (mg)	4	0.85	6.1 +/- 4.0[D]
Zn (mg)	0.6	Na	0.6 – 1.1
Cu (mg)	0.15	Na	0.11 – 0.21
Thiamin (B_1) (mg)	0.257	0.06[C]	0.23 +/- 0.02
Riboflavin (B_2) (mg)	0.66	0.05	0.73 +/- 0.49
Niacin (B_3) (mg)	2.22	0.8	2.7 +/- 0.05
Pantothenic acid (B_5) mg	0.125	Na	Na
Vitamin B_6 (mg)	1.200	Na	1.2 – 1.2
Folate (B_9) (µg)	40	Na	40 – 370
Vitamin A equiv. (µg RAE[B]	378	1640	738
Vitamin C (mg)	51.7	220	164 +/- 79
Vitamin E (mg)	Na	Na	3.07

[A]Crude fibre measures significantly underestimate dietary fibre for humans.

[B]Estimated from µg ß-carotene using 12 µg ß-carotene=1 µg RAE. This likely underestimates actual vitamin A activity slightly as other carotenes contribute to vitamin A activity, though to a lesser extent than ß-carotene.

[C]This value may be an error which occurred when the value was copied forward from older references (McBurney *et al.*, 2004).

[D]source indicates data being of poor quality

4.9.2 Nutrient content of dried *Moringa oleifera* leaves

Nutrient content (mean + std dev) of dried *M. oleifera* leaves			
Nutrient	**Amount in 100g (300ml or 1.25 cups)**	**Nutrient**	**Amount in 100g (300ml or 1.25 cups)**
Energy (Kcal, MJ)	304 ± 87 kcal, 1.3 ± 0.36 MJ	Mg (mg)	473 ± 429.4
Moisture (mg)	7.4 ± 2.89	Fe (mg)	32.5 ± 10.78
Protein (g)	24 ± 5.8	Zn (mg)	2.4 ± 1.12
Carbohydrates (g)	36 ± 9.2	Cu (mg)	0.9 ± 0.48
Fat (g)	6 ± 2.5	Thiamin (mg)	2.6
Fibre, crude (g)[A]	9 ± 7.45	Riboflavin (mg)	1.29 – 20.5
Fibre, total dietary (g)[A]	20.6 – 28.6	Niacin (mg)	8.2
Oxalate (g)	2.6 ±1.25	Vitamin B-6 (mg)	2.4
Tannins (g)	1.2 – 1.4	Folate (µg)	540
Ca (mg)	1897 ± 748.4	Vitamin A (µg RAE)[B]	3639 ± 1979.8
P (mg)	297 ± 149.0	Vitamin C (mg)	172 ± 37.7
Na (mg)	220 ± 180.0	Vitamin E (mg)	56 – 113
K (mg)	1467 ± 636.7		

[A]Crude fibre measures significantly underestimate dietary fibre for humans. The total dietary fibre value provided is based on the fibre content of the dried leaves.

[B]Estimated from µg beta-carotene using 12 µg beta-carotene=1 µg RAE, this likely underestimates actual vitamin A activity slightly as other carotenes contribute to vitamin A activity, though to a lesser extent than beta-carotene.

4.9.3 Nutrient content of raw pods of *Moringa oleifera*

Nutritional value per 100g			
Energy	***37 kcal***	**Minerals**	**Quantity**
Carbohydrates	8.53g	Calcium	30mg
Dietary fiber	3.2g	Iron	0.36mg
Fat	0.2g	Magnesium	45mg
Protein	2.1g	Manganese	0.259mg
Vitamins	**Quantity**	Phosphorus	50mg
Vitamin A equiv.	4µg	Potassium	461mg
Thiamine (B1)	0.053mg	Sodium	42mg
Riboflavin (B2)	0.074mg	Zinc	0.45mg
Niacin (B3)	0.620mg	**Other constituents**	**Quantity**
Pantothenic acid (B5)	0.794mg	Water	88.20 g
Vitamin B6	0.12mg		
Folate (B9)	44µg		
Vitamin C	141 mg	**Source: USDA (2013)**	

4.10 Phytochemicals

Phytochemicals are compounds synthesized by plants. However, the term is commonly used to describe specific chemicals that can influence factors such as health, flavour, texture and smell or colour in plants. These compounds are not essential nutrients for humans but their presence can contribute to various aspects of plant characteristics. Moringa is widely recognized for its substantial nutritional content and serves as a reservoir of diverse phytochemicals. Phytochemicals play a pivotal role in contributing to their health-promoting attributes. In particular, this plant family boasts a wealth of compounds featuring the simple sugar ~ rhamnose. It is abundant in a distinctive category of compounds known as glucosinolates and isothiocyanates (Bennett *et. al.*, 2003). Noteworthy components found in moringa preparations reported for their hypotensive, anticancer and antibacterial properties encompass 4-(4'-O-acetyl-α-L-rhamnopyranosyloxy) benzyl isothiocyanate, 4-(α-L-rhamnopyranosyloxy) benzyl isothiocyanate, niazimicin, pterygospermin, benzyl isothiocyanate, and 4-(α-L-rhamnopyranosyloxy) benzyl glucosinolate. While these compounds are relatively exclusive to the moringa family, it also boasts a plethora of vitamins, minerals and other well-recognized phytochemicals such as carotenoids, including α-carotene or pro-vitamin A (TNAU, 2023).

Moringa is recognized as a vibrant and affordable source of phytochemicals having potential applications in medicines, functional food preparations, water purification and biodiesel production. Main phytochemicals isolated from different parts of moringa are (TNAU , 2023):

i) **Roots:** 4-(α-L-rhamnopyranosyloxy) – benzyl glucosinolate and benzyl glucosinolate

ii) **Stem:** 4-hydroxymellein, vanillin, β-sitosterone, octacosanic acid and β-sitosterol

iii) **Bark:** 4-(α-L-rhamnopyranosyloxy)-benzyl glucosinolate

iv) **Whole gum exudates:** L -arabinose, D-galactose, D-glucuronic acid, L-rhamnose, D-mannose, D-xylose and leuco anthocyanin

v) **Leaves:** Glycoside niazirin, niazirinin and three mustard oil glycosides, 4-[4'-O-acetyl-α-L-rhamnosyloxy) benzyl] isothiocyanate, niaziminin A and B

vi) **Mature flowers:** D-mannose, D-glucose, protein, ascorbic acid, polysaccharide

vii) **Whole pods:** Nitriles, isothiocyanate, thiocarbanates, 0-[2'-hydroxy-3'-(2"-heptenyloxy)]- propyl undecanoate, 0-ethyl-4-[(α-1-rhamnosyloxy)-benzyl] carbamate, methyl-p-hydroxybenzoate and β-sitosterol

viii) Mature seeds: Crude protein, crude fat, carbohydrate, methionine, cysteine, 4-(α-L- rhamnopyranosyloxy)-benzyl glucosinolate, benzyl glucosinolate, moringyne, mono-palmitic and di-oleic triglyceride

ix) Seed oil: Vitamin A, betacarotene, precursor of Vitamin A

4.10.1 Compounds extracted from *Moringa oleifera* leaves and seeds

The leaves of *Moringa oleifera* have been acknowledged to contain numerous bioactive compounds encompassing vitamins, carotenoids, polyphenols (including phenolic acids and flavonoids), alkaloids, glucosinolates, isothiocyanates, fatty acids, amino acids, tannins, saponins, steroids, terpenes, oxalates and phytates. It is imperative to note that the concentrations of these compounds in *Moringa oleifera* leaves and seeds exhibited significant inter-study variations. This variability can be attributed to diverse environmental conditions, harvesting seasons and genetic factors influencing the plant, drying methods employed, leaf maturity stages, extraction techniques and the distinct sensitivities of analytical methods utilised. Awareness of these influencing factors is crucial for a comprehensive understanding of the composition of *Moringa oleifera* and its potential health benefits.

Compounds	Molecular formula	Quantity		Method	Reference
		Leaves	Seeds		
Fatty acid					
Caproic acid	$C_6H_{12}O_2$	0.1%	-	GC-MS	Lee *et al.*, 2017
Capric acid	$C_{10}H_{20}O_2$	0.1%	-	GC-MS	Lee *et al.*, 2017
Lauric acid	$C_{12}H_{24}O_2$	0.72%	0.03%	GC-MS	Lee *et al.*, 2017
Myristic acid	$C_{14}H_{28}O_2$	1.13%	0.16%	GC-MS	Lee *et al.*, 2017
Palmitic acid	$C_{16}H_{32}O_2$	23.65%	7.92%	GC-MS	Lee *et al.*, 2017
		-	Madurai: 6.17% [a]	GC-MS	Chelliah *et al.*, 2017
		1.08% [a]	-	GC-MS	Chuang *et al.*, 2007
		0.15%	-	GC-MS	Zhao and Zhang, 2013
		2.5703% [a]	-	GC-MS	Bhalla *et al.*, 2021
Palmitoleic acid	$C_{16}H_{30}O_2$	-	1.49%	GC-MS	Lee *et al.*, 2017
Heptadecanoic acid	$C_{17}H_{34}O_2$	0.46%	0.32%	GC-MS	
Stearic acid	$C_{18}H_{36}O_2$	1.2051% [a]	-	GC-MS	Bhalla *et al.*, 2021
		3.93%	7.10%	GC-MS	Lee *et al.*, 2017
Oleic acid	$C_{18}H_{34}O_2$	5.92%	74.50%	GC-MS	
		-	Madurai: 5.12% [a]	GC-MS	Chelliah *et al.*, 2017
Linoleic acid	$C_{18}H_{32}O_2$	6.84%	0.62%	GC-MS	Lee *et al.*, 2017

Compounds	Molecular formula	Quantity		Method	Reference
		Leaves	Seeds		
Arachidic acid	$C_{20}H_{40}O_2$	0.72%	4.28%	GC-MS	Lee *et al.*, 2017
Cis-11-Eicosenoic acid	$C_{20}H_{38}O_2$	54.44%	-	GC-MS	Lee *et al.*, 2017
Linolenic acid	$C_{18}H_{30}O_2$	-	2.50%	GC-MS	Lee *et al.*, 2017
Cis-11, 14-Eicosadienoic acid	$C_{20}H_{36}O_2$	-	0.48%	GC-MS	Lee *et al.*, 2017
Cis-5, 8, 11, 14, 17-Eicosapentaenoic acid.	$C_{21}H_{32}O_2$	1.99%	0.60%	GC-MS	Lee *et al.*, 2017
6-octadecenoic acid	$C_{18}H_{34}O_2$	-	Madurai: 52.24% [a]	GC-MS	Chelliah *et al.*, 2017
Ethyl oleate	$C_{20}H_{38}O_2$	-	Chennai: 4.33% [a]	GC-MS	Chelliah *et al.*, 2017
Fatty acid derivatives - carboxylic acid amide derivatives of fatty acids					
13-docosenamide, (Z)-	$C_{22}H_{43}NO$	-	Chennai: 13.62% [a]	GC-MS	Chelliah *et al.*, 2017
Propionamide	C_3H_7NO	0.7335% [a]	-	GC-MS	Bhalla *et al.*, 2021
		-	Chennai: 4.48% [a]	GC-MS	Chelliah *et al.*, 2017
Amino acid					
Lysine - Essential amino acid	$C_6H_{14}N_2O_2$	Madurai: 4.10 g/16 gN Chennai: 3.30 g/16 gN	Madurai: 4.20 g/16 gN Chennai: 3.70 g/16 gN	HPLC	Chelliah *et al.*, 2017
		129.4 mg/100g dw	-	HPLC	Lalas *et al.*, 2017
Histidine - Essential amino acid	$C_6H_9N_3O_2$	Madurai: 2.50 g/16 gN Chennai: 3.12 g/16 gN	Madurai: 6.30 g/16 gN Chennai: 5.70 g/16 gN	HPLC	Chelliah *et al.*, 2017
		73.9 mg/100g dw	-	HPLC	Lalas *et al.*, 2017
Valine - Essential amino acid	$C_5H_{11}NO_2$	Madurai: 4.55 g/16 gN Chennai: 4.75 g/16 gN	Madurai: 3.36 g/16 gN Chennai: 2.37 g/16 gN	HPLC	Chelliah *et al.*, 2017
		111.5 mg/100g dw	-	HPLC	Lalas *et al.*, 2017

Compounds	Molecular formula	Quantity		Method	Reference
		Leaves	Seeds		
Leucine - Essential amino acid	$C_6H_{13}NO_2$	Madurai: 4.80 g/16 gN Chennai: 4.40 g/16 gN	Madurai: 4.80 g/16 gN Chennai: 4.70 g/16 gN	HPLC	Chelliah *et al.*, 2017
		211.5 mg/100g dw	-	HPLC	Lalas *et al.*, 2017
Isoleucine - Essential amino acid	$C_6H_{13}NO_2$	Madurai: 5.90 g/16 gN Chennai: 9.10 g/16 gN	Madurai: 8.30 g/16 gN Chennai: 7.70 g/16 gN	HPLC	Chelliah *et al.*, 2017
		81.3 mg/100g dw	-	HPLC	Lalas *et al.*, 2017
Threonine - Essential amino acid	$C_4H_9NO_3$	Madurai: 4.10 g/16 gN Chennai: 4.50 g/16 gN	Madurai: 3.20 g/16 gN Chennai: 3.80 g/16 gN	HPLC	Chelliah *et al.*, 2017
		91.8 mg/100g dw	-	HPLC	Lalas *et al.*, 2017
Alanine - non-essential amino acid	$C_3H_7NO_2$	Madurai: 3.23 g/16 gN Chennai: 3.90 g/16 gN	Madurai: 4.29 g/16 gN Chennai: 3.55 g/16 gN	HPLC	Chelliah *et al.*, 2017
		121.2 mg/100 g dw	-	HPLC	Lalas *et al.*, 2017
Aspartic acid - non-essential amino acid	$C_4H_7NO_4$	Madurai: 6.44 g/16 gN Chennai: 6.86 g/16 gN	Madurai: 6.02 g/16 gN Chennai: 6.37 g/16 gN	HPLC	Chelliah *et al.*, 2017
		195.8 mg/100 g dw	-	HPLC	Lalas *et al.*, 2017
Serine - non-essential amino acid	$C_3H_7NO_3$	Madurai: 4.25 g/16 gN Chennai: 4.66 g/16 gN	Madurai: 4.22 g/16 gN Chennai: 4.11 g/16 gN	HPLC	Chelliah *et al.*, 2017
		90.2 mg/100 g dw	-	HPLC	Lalas *et al.*, 2017

Compounds	Molecular formula	Quantity		Method	Reference
		Leaves	Seeds		
Proline - non-essential amino acid	$C_5H_9NO_2$	128.9 mg/100 g dw	-	HPLC	Lalas *et al.*, 2017
Glutamic acid - non-essential amino acid	$C_5H_9NO_4$	Madurai: 15.86 g/16 gN Chennai: 15.33 g/16 gN	Madurai: 14.23 g/16 gN Chennai: 14.74 g/16 gN	HPLC	Chelliah *et al.*, 2017
		268.7 mg/100 g dw	-	HPLC	Lalas *et al.*, 2017
Glycine - non-essential amino acid	$C_2H_5NO_2$	Madurai: 5.13 g/16 gN Chennai: 5.11 g/16 gN	Madurai: 5.00 g/16 gN Chennai: 4.70 g/16 gN	HPLC	Chelliah *et al.*, 2017
		102.4 mg/100 g dw	-	HPLC	Lalas *et al.*, 2017
Arginine - semi-essential amino acid	$C_6H_{14}N_4O_2$	Madurai: 4.30 g/16 gN Chennai: 3.70 g/16 gN	Madurai: 9.44 g/16 gN Chennai: 11.00 g/16 gN	HPLC	Chelliah *et al.*, 2017
		125.4 mg/100 g dw	-	HPLC	Lalas *et al.*, 2017
Cysteine - semi-essential amino acid	$C_3H_7NO_2S$	29.6 mg/100 g dw	-	HPLC	Lalas *et al.*, 2017
Tyrosine - non-essential amino acid	$C_9H_{11}NO_3$	79.3 mg/100 g dw	-	HPLC	Lalas *et al.*, 2017
Methionine - essential amino acid	$C_5H_{11}NO_2S$	Madurai: 1.90 g/16 gN Chennai: 3.40 g/16 gN	Madurai: 3.50 g/16 gN Chennai: 3.90 g/16 gN	HPLC	Chelliah *et al.*, 2017
		221.4 mg/100 g dw	-	HPLC	Lalas *et al.*, 2017
Phenylalanine - essential amino acid	$C_9H_{11}NO_2$	Madurai: 4.20 g/16 gN Chennai: 4.60 g/16 gN	Madurai: 3.50 g/16 gN Chennai: 3.90 g/16 gN	HPLC	Chelliah *et al.*, 2017
		117.2 mg/100 g	-	HPLC	Lalas *et al.*, 2017

Compounds	Molecular formula	Quantity		Method	Reference
		Leaves	Seeds		
Alkanes and their derivatives					
1,30-Triacontanediol	$C_{30}H_{62}O_2$	3.06%	-	GC-MS	Zhao and Zhang, 2013
		Madurai: 14.98% [a]	-	GC-MS	Chelliah *et al.*, 2017
Octacosane	$C_{28}H_{58}$	Madurai: 8.57% [a]	-	GC-MS	Chelliah *et al.*, 2017
Z-14-nonacosane	$C_{29}H_{58}$	Madurai: 8.3% [a]	-	GC-MS	Chelliah *et al.*, 2017
Nonacosane	$C_{29}H_{60}$	Chennai: 15.55% [a]	-	GC-MS	Chelliah *et al.*, 2017
		18,65%	-	GC-MS	Zhao and Zhang, 2013
Nonadecane	$C_{19}H_{40}$	0.22%	-	GC-MS	Zhao and Zhang, 2013
Eicosane	$C_{20}H_{42}$	0.31%	-	GC-MS	Zhao and Zhang, 2013
Heptacosane	$C_{27}H_{56}$	7.45%	-	GC-MS	Zhao and Zhang, 2013
Tetracosane	$C_{24}H_{50}$	1.45% [a]	-	GC-MS	Chuang *et al.*, 2007
Pentacosane	$C_{25}H_{52}$	17.41% [a]	-	GC-MS	Chuang *et al.*, 2007
		2.14%	-	GC-MS	Chuang *et al.*, 2007
Hexacosane	$C_{26}H_{54}$	11.20% [a]	-	GC-MS	Chuang *et al.*, 2007
Docosane	$C_{22}H_{46}$	0.28% [a]	-	GC-MS	Chuang *et al.*, 2007
Undecane	$C_{11}H_{24}$	0.12% [a]	-	GC-MS	Chuang *et al.*, 2007
Tridecane	$C_{13}H_{28}$	0.16% [a]	-	GC-MS	Chuang *et al.*, 2007
Hepane,2-methyl-	C_8H_{18}	0.68%	-	GC-MS	Zhao and Zhang, 2013
Heptane,3-methyl-	C_8H_{18}	0.50%	-	GC-MS	Zhao and Zhang, 2013
Cyclohexane,1,3-dimethyl-,cis-	C_8H_{16}	0.78%	-	GC-MS	Zhao and Zhang, 2013
Octane	C_8H_{18}	1.75%	-	GC-MS	Zhao and Zhang, 2013
Cyclohexane,1,4-dimethyl-,cis-	C_8H_{16}	0.14%	-	GC-MS	Zhao and Zhang, 2013

Compounds	Molecular formula	Quantity		Method	Reference
		Leaves	Seeds		
Octane,2-methyl-	C_9H_{20}	0.17%	-	GC-MS	Zhao and Zhang, 2013
Cyclohexane,ethyl-	C_8H_{16}	0.43%	-	GC-MS	Zhao and Zhang, 2013
Nonane	C_9H_{20}	0.14%	-	GC-MS	Zhao and Zhang, 2013
Cyclic siloxanes					
2,2-dimethyl-1-oxa-2-silacyclotrid ecanone-13	$C_6H_{14}OSi$	Madurai: 8.28% [a]	-	GC-MS	Chelliah *et al.*, 2017
Phytosterols					
γ-Sitosterol	$C_{29}H_{50}O$	0,86%	-	GC-MS	Zhao and Zhang, 2013
		Chennai: 9.56% [a]	-	GC-MS	Chelliah *et al.*, 2017
Campesterol	$C_{28}H_{48}O$	Chennai: 5.00% [a]	Chennai: 1.21% [a]	GC-MS	Chelliah *et al.*, 2017
Glucosinolates - glucomoringin isomers					
Glucomoringin	$C_{21}H_{31}NO_{14}S_2$	16.14–17.3 µmol/g dry weight	-	UPLC	Lopez-Rodriguez *et al.*, 2020
		Domesticated: 75.29 µmol/g dry weight Wild: 18.00 µmol/g dry weight	-	HPLC	Lopez-Rodriguez *et al.*, 2020
		110–112 µmol/g dry weight	-	HPLC	Lopez-Rodriguez *et al.*, 2020
		21.84–22.56 mg/g dry weight	32.88 mg/g dry weight	HPLC	Lopez-Rodriguez *et al.*, 2020
		0.77 mg/g dry weight	86.19 mg/g dry weight	HPLC-ESI-MS/MS	Lopez-Rodriguez *et al.*, 2020
		33.9–59.4 mg/g dry weight	202 mg/g dry weight		Lopez-Rodriguez *et al.*, 2020
Glucomoringin Isomer I	$C_{21}H_{31}NO_{14}S_2$	7.15–8.92 µmol/g dry weight	-	UPLC	Lopez-Rodriguez *et al.*, 2020

Compounds	Molecular formula	Quantity		Method	Reference
		Leaves	Seeds		
Glucomoringin Isomer II	$C_{21}H_{31}NO_{14}S_2$	3.85–4.84 µmol/g dry weight	-	UPLC	Lopez-Rodriguez *et al.*, 2020
Glucomoringin Isomer III	$C_{21}H_{31}NO_{14}S_2$	18.71–22.58 µmol/g dry weight	-	UPLC	Lopez-Rodriguez *et al.*, 2020
Glucosinolates - Moringin and isomers of 4-O-(α-L acetylrhamnopyranosyloxy)-benzyl-isothiocyanate					
4-(α-L-rhamnopyranosyloxy)-benzyl-isothiocyanate (moringin)	$C_{14}H_{17}NO_5S$	22.56 mg/g [e,f]	-	HPLC-DAD-ESI-MS	Leone *et al.*, 2015a
		21.84 mg/g [g,f]	-		Leone *et al.*, 2015a
		33.9 mg/g [c]	-	LC-MS	Leone *et al.*, 2015a
		59.4 mg/g [d]	-		Leone *et al.*, 2015a
4-O-(α-Lacetylrhamnopyranosyloxy)-benzyl-isothiocyanate isomer 1	$C_{16}H_{19}NO_6S$	2.76 mg/g [e,f]	-	HPLC-DAD-ESI-MS	Leone *et al.*, 2015a
		2.16 mg/g [g,f]	-		Leone *et al.*, 2015a
		2.9 mg/g [c]	-	LC-MS	Leone *et al.*, 2015a
		5.0 mg/g [d]	-		Leone *et al.*, 2015a
4-O-(α-Lacetylrhamnopyranosyloxy)-benzyl-isothiocyanate isomer 2	$C_{16}H_{19}NO_6S$	1.80 mg/g [e,f]	-	HPLC-DAD-ESI-MS	Leone *et al.*, 2015a
		1.52 mg/g [g,f]	-		Leone *et al.*, 2015a
		1.2 mg/g [c]	-	LC-MS	Leone *et al.*, 2015a
		1.5 mg/g [d]	-		Leone *et al.*, 2015a
4-O-(α-Lacetylrhamnopyranosyloxy)-benzyl isomer 3	$C_{16}H_{19}NO_6S$	20.16 mg/g [e,f]	-	HPLC-DAD-ESI-MS	Leone *et al.*, 2015a
		12.76 mg/g [g,f]	-		Leone *et al.*, 2015a
		17.4 mg/g [c]	-	LC-MS	Leone *et al.*, 2015a
		50.2 mg/g [d]	-		Leone *et al.*, 2015a

Compounds	Molecular formula	Quantity		Method	Reference
		Leaves	Seeds		
Carotenoids					
β-carotene	$C_{40}H_{56}$	33.48 mg/100 g	-	HPLC	Leone *et al.*, 2015a
		27.22 mg/100 g	-	N/A	Leone *et al.*, 2015a
		36 mg/100 g	-	AOAC 2004	Leone *et al.*, 2015a
		39.6 mg/100 g	-	N/A	Leone *et al.*, 2015a
		37.8 mg/100 g	-	N/A	Leone *et al.*, 2015a
		17.62 mg/100 g	-	N/A	Leone *et al.*, 2015a
		20.44 mg/100 g	-	HPLC	Leone *et al.*, 2015a
		66 mg/100 g	-	HPLC	Leone *et al.*, 2015a
Lutein	$C_{40}H_{56}O_2$	35.05 mg/100 g	-	HPLC	Leone *et al.*, 2015a
		102 mg/100 g	-	HPLC	Leone *et al.*, 2015a
Polyphenol					
Phenolic acids					
Caffeic acid	$C_9H_8O_4$	0.409 mg/g	-	HPLC	Leone *et al.*, 2015a
		0.536 mg/g	-	HPLC	Leone *et al.*, 2015a
Chlorogenic acid	$C_{16}H_{18}O_9$	0.489 mg/g	-	HPLC and MS/MS	Leone *et al.*, 2015a
o-Coumaric acid	$C_9H_8O_3$	6.457 mg/g	-	HPLC	Leone *et al.*, 2015a
Ellagic acid	$C_{14}H_6O_8$	0.018 mg/g	-	HPLC and MS/MS	Leone *et al.*, 2015a
		0.189 mg/g	-	HPLC and MS/MS	Leone *et al.*, 2015a
Ferulic acid	$C_{10}H_{10}O_4$	0.078 mg/g	-	HPLC and MS/MS	Leone *et al.*, 2015a
		0.078 mg/g	-	HPLC	Leone *et al.*, 2015a
		0.128 mg/g	-	HPLC and MS/MS	Leone *et al.*, 2015a

Compounds	Molecular formula	Quantity		Method	Reference
		Leaves	Seeds		
Gallic acid	$C_7H_6O_5$	1.034 mg/g	-	HPLC and MS/MS	Leone *et al.*, 2015a
		1.034 mg/g	-	HPLC and MS/MS	Leone *et al.*, 2015a
Flavonoids (Polyphenols)					
Epicatechin	$C_{15}H_{14}O_6$	5.68 mg/g	-	HPLC	Leone *et al.*, 2015a
Isorhamnetin	$C_{16}H_{12}O_7$	0.118 mg/g	-	HPLC	Leone *et al.*, 2015a
		0.52 mg/g [e,f]	-	HPLC-DAD-ESI-MS	Leone *et al.*, 2015a
		0.72 mg/g [f,g]	-		Leone *et al.*, 2015a
Kaempferol	$C_{15}H_{10}O_6$	0.04 mg/g	-	HPLC	Leone *et al.*, 2015a
		2.360 mg/g	-	HPLC	Leone *et al.*, 2015a
		0.198 mg/g	-	HPLC and MS/MS	Leone *et al.*, 2015a
		1.412 mg/g	-	HPLC	Leone *et al.*, 2015a
		0.8 mg/g	-	LC/MS	Leone *et al.*, 2015a
		1.23 mg/g	-		Leone *et al.*, 2015a
		4.59 mg/g	-		Leone *et al.*, 2015a
		3.92 [e,f]	-	HPLC-DAD-ESI-MS	Leone *et al.*, 2015a
		2.16 mg/g [f,g]	-		Leone *et al.*, 2015a
		2.25 mg/g	-	HPLC-DAD	Leone *et al.*, 2015a
		1.75 mg/g	-		Leone *et al.*, 2015a
		1.05 mg/g	-		Leone *et al.*, 2015a
		2.9 mg/g [d]	-	LC/MS	Leone *et al.*, 2015a
		2.3 mg/g	-		Leone *et al.*, 2015a
		3.5 mg/g	-		Leone *et al.*, 2015a
		0.3 mg/g [c]	-		Leone *et al.*, 2015a
		0.16 mg/g [d]	-		Leone *et al.*, 2015a

Compounds	Molecular formula	Quantity		Method	Reference
		Leaves	Seeds		
Myricetin	$C_{15}H_{10}O_8$	5.804 mg/g	-	HPLC	Leone *et al.*, 2015a
Quercetin	$C_{15}H_{10}O_7$	0.281 mg/g	-	HPLC	Leone *et al.*, 2015a
		0.207 mg/g	-	HPLC and MS/MS	Leone *et al.*, 2015a
		0.207 mg/g	-	HPLC	Leone *et al.*, 2015a
		0.807 mg/g	-	HPLC and MS/MS	Leone *et al.*, 2015a
		3.529 mg/g	-	HPLC	Leone *et al.*, 2015a
		5.2 mg/g	-	LC/MS	Leone *et al.*, 2015a
		5.8 mg/g	-		Leone *et al.*, 2015a
		7.57 mg/g	-		Leone *et al.*, 2015a
		12.84 mg/g [e,f]	-	HPLC-DAD-ESI-MS	Leone *et al.*, 2015a
		16.64 mg/g [f,g]	-		Leone *et al.*, 2015a
		9.26 mg/g	-	HPLC-DAD	Leone *et al.*, 2015a
		6.34 mg/g	-		Leone *et al.*, 2015a
		7.70 mg/g	-		Leone *et al.*, 2015a
		5.47 mg/g [b]	-	LC-MS	Leone *et al.*, 2015a
		9.1 mg/g	-		Leone *et al.*, 2015a
		15.2 mg/g	-		Leone *et al.*, 2015a
		0.58 mg/g [c]	-		Leone *et al.*, 2015a
		0.46 mg/g [d]	-		Leone *et al.*, 2015a
Rutin	$C_{27}H_{30}O_{16}$	1.674 mg/g	-	HPLC	Leone *et al.*, 2015a

Compounds	Molecular formula	Quantity		Method	Reference
		Leaves	Seeds		
Miscellaneus compounds					
Toluene	C_7H_8	1.27%	-	GC-MS	Zhao and Zhang, 2013
		0.03% [a]	-	GC-MS	Chuang *et al.*, 2007
5-tert-Butyl-1,3-cyclopentadiene	C_9H_{14}	0.07% [a]	-	GC-MS	Chuang *et al.*, 2007
Benzaldehyde	C_6H_5CHO	0.55% [a]	-	GC-MS	Chuang *et al.*, 2007
5-Methyl-2-furaldehyde	$C_6H_6O_2$	0.27% [a]	-	GC-MS	Chuang *et al.*, 2007
Benzeneacetaldehyde	C_8H_8O	2.16% [a]	-	GC-MS	Chuang *et al.*, 2007
2-Ethyl-3,6-dimethylpyrazine	$C_8H_{12}N_2$	0.12% [a]	-	GC-MS	Chuang *et al.*, 2007
Isophoron	$C_9H_{14}O$	0.10% [a]	-	GC-MS	Chuang *et al.*, 2007
Benzylnitrile	C_8H_7N	1.10% [a]	-	GC-MS	Chuang *et al.*, 2007
2,6,6-Trimethyl-2-cyclohexane-1,4-dione	$C_9H_{12}O_2$	0.05% [a]	-	GC-MS	Chuang *et al.*, 2007
2,2,4-Trimethyl-pentadiol	$C_{12}H_{24}O_3$	0.09% [a]	-	GC-MS	Chuang *et al.*, 2007
2,3-Epoxycarane	$C_{10}H_{16}O$	0.16% [a]	-	GC-MS	Chuang *et al.*, 2007
p-Menth-1-en-8-ol	$C_{10}H_{18}O$	0.08% [a]	-	GC-MS	Chuang *et al.*, 2007
2,6,6-Trimethylcyclohexa-1,3-dienecarbaldehyde	C_9H_{14}	0.23% [a]	-	GC-MS	Chuang *et al.*, 2007
Indole	C_8H_7N	1.20% [a]	-	GC-MS	Chuang *et al.*, 2007
Ionone	$C_{13}H_{20}O$	0.13% [a]	-	GC-MS	Chuang *et al.*, 2007
		0.03% [a]	-	GC-MS	Chuang *et al.*, 2007
1,1,6-Trimethyl-1,2-dihydronaphthalene	$C_{13}H_{16}$	0.41% [a]	-	GC-MS	Chuang *et al.*, 2007
Ionene	$C_{13}H_{18}$	0.09% [a]	-	GC-MS	Chuang *et al.*, 2007
Damascenone	$C_{13}H_{18}O$	0.28%[a]	-	GC-MS	Chuang *et al.*, 2007
Ledene oxide	$C_{15}H_{24}O$	0.60% [a]	-	GC-MS	Chuang *et al.*, 2007
2-tert-Butyl-1,4-dimethoxybenzene	$C_{12}H_{18}O_2$	0.39% [a]	-	GC-MS	Chuang *et al.*, 2007

Compounds	Molecular formula	Quantity		Method	Reference
		Leaves	Seeds		
(E)-6,10-dimethylundeca-5,9-dien-2-one	$C_{13}H_{22}O$	0.26% [a]	-	GC-MS	Chuang *et al.*, 2007
4,6-Dimethyldodecane	$C_{14}H_{30}$	0.29% [a]	-	GC-MS	Chuang *et al.*, 2007
3,3,5,6-Tetramethyl-1-indanone	$C_{13}H_{16}O$	0.23% [a]	-	GC-MS	Chuang *et al.*, 2007
Dihydro-actiridioide	-	1.21% [a]	-	GC-MS	Chuang *et al.*, 2007
2,3,6-Trimethyl-naphthalene	$C_{13}H_{14}$	0.37% [a]	-	GC-MS	Chuang *et al.*, 2007
Megastigmatrienone	$C_{13}H_{18}O$	0.57% [a]	-	GC-MS	Chuang *et al.*, 2007
1-[2,3,6-Trimethyl-phenyl]-2-butanone	$C_{13}H_{18}O$	3.44% [a]	-	GC-MS	Chuang *et al.*, 2007
1-[2,3,6-Trimethyl-phenyl]-3-buten-2-one	$C_{13}H_{16}O$	0.75% [a]	-	GC-MS	Chuang *et al.*, 2007
Isolongifolene	$C_{15}H_{24}$	0.56% [a]	-	GC-MS	Chuang *et al.*, 2007
Hexahydrofarnesylactone	$C_{18}H_{36}O$	1.30% [a]	-	GC-MS	Chuang *et al.*, 2007
Farnesylacetone	$C_{18}H_{30}O$	0.08% [a]	-	GC-MS	Chuang *et al.*, 2007
Methyl palmitate	$C_{17}H_{34}O_2$	0.08% [a]	-	GC-MS	Chuang *et al.*, 2007
[6E,10E]-7,11,15-trimethyl-methylene-1,6,10,14-hexadeca-tetraene	$C_{20}H_{32}$	0.11% [a]	-	GC-MS	Chuang *et al.*, 2007
Phytol	$C_{20}H_{40}O$	0.9664% [a]	-	GC-MS	Bhalla *et al.*, 2021
		7.66% [a]	-	GC-MS	Chuang *et al.*, 2007
1-Docosene	$C_{22}H_{44}$	0.41% [a]	-	GC-MS	Chuang *et al.*, 2007
Pyridine	C_5H_5N	0.78%	-	GC-MS	Zhao and Zhang, 2013
Ethylbenzene	C_8H_{10}	0.83%	-	GC-MS	Zhao and Zhang, 2013
Benzene,1,3-dimethyl-	C_8H_{10}	4.63%	-	GC-MS	Zhao and Zhang, 2013
o-Xylene	C_8H_{10}	2.82%	-	GC-MS	Zhao and Zhang, 2013
Benzene,(1-methylethyl)-	C_9H_{12}	0.16%	-	GC-MS	Zhao and Zhang, 2013

Compounds	Molecular formula	Quantity		Method	Reference
		Leaves	Seeds		
Benzene,propyl-	C_9H_{12}	1.04%	-	GC-MS	Zhao and Zhang, 2013
Benzene,1-ethyl-3-methyl-	C_9H_{12}	5.95%	-	GC-MS	Zhao and Zhang, 2013
Benzene,1-ethyl-4-methyl-	C_9H_{12}	2.39%	-	GC-MS	Zhao and Zhang, 2013
Benzene,1,3,5-trimethyl-	C_9H_{12}	4.31%	-	GC-MS	Zhao and Zhang, 2013
Benzene,1-ethyl-2-methyl-	C_9H_{12}	2.55%	-	GC-MS	Zhao and Zhang, 2013
Benzene,1,2,4-trimethyl-	C_9H_{12}	16.96%	-	GC-MS	Zhao and Zhang, 2013
Benzene,1,2,3-trimethyl-	C_9H_{12}	4.14%	-	GC-MS	Zhao and Zhang, 2013
Indane	C_9H_{10}	0.39%	-	GC-MS	Zhao and Zhang, 2013
Benzene,1-ethyl-2,4-dimethyl-	$C_{10}H_{14}$	0.64%	-	GC-MS	Zhao and Zhang, 2013
Benzene,1-ethyl-2,3-dimethyl-	$C_{10}H_{14}$	0.99%	-	GC-MS	Zhao and Zhang, 2013
Benzene,1,2,3,5-tetramethyl-	$C_{10}H_{14}$	1.59%	-	GC-MS	Zhao and Zhang, 2013
Benzene,1,2,4,5-tetramethyl-	$C_{10}H_{14}$	2.12%	-	GC-MS	Zhao and Zhang, 2013
Benzene,1,2,3,4-tetramethyl-	$C_{10}H_{14}$	0.58%	-	GC-MS	Zhao and Zhang, 2013
Naphthalene	$C_{10}H_8$	1.21%	-	GC-MS	Zhao and Zhang, 2013
Dimethoate	$C_5H_{12}NO_3PS_2$	0.52%	-	GC-MS	Zhao and Zhang, 2013
γ-Tocopherol	$C_{28}H_{48}O_2$	0.75%	-	GC-MS	Zhao and Zhang, 2013
Sulfurous acid, hexyl pentadecyl ester	$C_{21}H_{44}O_3S$	0.34%	-	GC-MS	Zhao and Zhang, 2013
dl-α-Tocopherol	$C_{29}H_{50}O_2$	1.05%	-	GC-MS	Zhao and Zhang, 2013
β-Amyrin	$C_{30}H_{50}O$	4.60%	-	GC-MS	Zhao and Zhang, 2013
1,3-Dihydroxyacetone dimer	$C_6H_{12}O_6$	3.8551% [a]	-	GC-MS	Bhalla *et al.*, 2021
Acetic acid, [(amino carbonyl)amino]oxo-	$C_3H_4N_2O_4$	3.2396% [a]	-	GC-MS	Bhalla *et al.*, 2021
4(1H)-Pyrimidinone, 2,6-diamino-	$C_4H_6N_4O$	2.2433% [a]	-	GC-MS	Bhalla *et al.*, 2021

Compounds	Molecular formula	Quantity		Method	Reference
		Leaves	Seeds		
4H-Pyran-4-one, 2,3-dihydro-3,5-dihydroxy-6-methyl-	$C_6H_8O_4$	4.1801%	-	GC-MS	Bhalla *et al.*, 2021
		8.9858% [a]	-	GC-MS	Bhalla *et al.*, 2021
2-Hexynoic acid	$C_6H_8O_2$	1.1214% [a]	-	GC-MS	Bhalla *et al.*, 2021
Butanedioic acid, 2-hydroxy-2-methyl-, (S)-	$C_7H_{12}O_5$	3.1422% [a]	-	GC-MS	Bhalla *et al.*, 2021
3,3-Iminobispropylamine	$C_6H_{17}N_3$	1.9275% [a]	-	GC-MS	Bhalla *et al.*, 2021
1-Hexanamine	$C_6H_{15}N$	0.0774% [a]	-	GC-MS	Bhalla *et al.*, 2021
1,3-Dioxolan-2-one, 4,5-dimethyl-	$C_5H_8O_3$	6.1627% [a]	-	GC-MS	Bhalla *et al.*, 2021
2-Butenethioic acid, 3-(ethylthio)-, S-(1-methylethyl) ester	$C_9H_{16}OS_2$	1.3083% [a]	-	GC-MS	Bhalla *et al.*, 2021
Propanamide, N,N-dimethyl-	$C_5H_{11}NO$	3.0349% [a]	-	GC-MS	Bhalla *et al.*, 2021
2-Isopropoxyethyl propionate	$C_8H_{16}O_3$	16.8738% [a]	-	GC-MS	Bhalla *et al.*, 2021
D-Mannoheptulose	$C_7H_{14}O_7$	2.5622% [a]	-	GC-MS	Bhalla *et al.*, 2021
Azetidin-2-one 3,3-dimethyl-4-(1-aminoethyl)-	$C_7H_{14}N_2O$	4.6738% [a]	-	GC-MS	Bhalla *et al.*, 2021
Carbonic acid, butyl 2-pentyl ester	$C_{10}H_{20}O_3$	20.6431% [a]	-	GC-MS	Bhalla *et al.*, 2021
Tetra acetyl-d-xylonic nitrile	$C_{14}H_{17}NO_9$	5.0379% [a]	-	GC-MS	Bhalla *et al.*, 2021
α-D-Glucose	$C_6H_{12}O_6$	3.445% [a]	-	GC-MS	Bhalla *et al.*, 2021
1H-Cyclopenta[c]furan-3(3αH)-one, 6,6α-dihydro-1-(1,3-dioxolan-2-yl)-, (3αR,1-trans,6α-cis)-	$C_{10}H_{12}O_4$	1.2587% [a]	-	GC-MS	Bhalla *et al.*, 2021
3-[1-(4-Cyano-1,2,3,4-tetrahydronaphthyl)] propanenitrile	$C_{14}H_{14}N_2$	1.5253% [a]	-	GC-MS	Bhalla *et al.*, 2021
Quinolinium, 1-ethyl-, iodide	$C_{11}H_{12}IN$	1.4067% [a]	-	GC-MS	Bhalla *et al.*, 2021
N-Isopropyl-3-phenylpropanamide	$C_{12}H_{17}NO$	0.9462% [a]	-	GC-MS	Bhalla *et al.*, 2021
1,2-Ethanediamine, N-(2-aminoethyl)-	$C_4H_{13}N_3$	0.8805% [a]	-	GC-MS	Bhalla *et al.*, 2021

Compounds	Molecular formula	Quantity		Method	Reference
		Leaves	Seeds		
1,4-Benzenediol, 2-methyl-	$C_7H_8O_2$	4.3169% [a]	-	GC-MS	Bhalla *et al.*, 2021
Ethene, ethoxy-	C_4H_8O	0.5981% [a]	-	GC-MS	Bhalla *et al.*, 2021
Dihydroxyacetone	$C_3H_6O_3$	2.4651% [a]	-	GC-MS	Bhalla *et al.*, 2021
Glycerine	$C_3H_8O_3$	1.8656% [a]	-	GC-MS	Bhalla *et al.*, 2021
Erythritol	$C_4H_{10}O_4$	0.5327% [a]	-	GC-MS	Bhalla *et al.*, 2021
Monomethyl malonate	$C_4H_6O_4$	2.5684% [a]	-	GC-MS	Bhalla *et al.*, 2021
4,5-Diamino-6-hydroxypyrimidine	$C_8H_{12}N_8O_2$	0.6434% [a]	-	GC-MS	Bhalla *et al.*, 2021
Furan, 2,3-dihydro-4-methyl-	C_5H_8O	0.2105% [a]	-	GC-MS	Bhalla *et al.*, 2021
Catecholborane	$C_6H_5BO_2$	0.6806% [a]	-	GC-MS	Bhalla *et al.*, 2021
2-Fluoropyridine	C_5H_4FN	0.8121% [a]	-	GC-MS	Bhalla *et al.*, 2021
1,2,3-Propanetriol, 1-acetate	$C_5H_{10}O_4$	1.4375% [a]	-	GC-MS	Bhalla *et al.*, 2021
3,4-Furandiol, tetrahydro-, trans-	C_4H_8O	0.1746% [a]	-	GC-MS	Bhalla *et al.*, 2021
1-Nitro-β-d-arabino furanose, tetraacetate	$C_{13}H_{17}NO_{11}$	0.5114% [a]	-	GC-MS	Bhalla *et al.*, 2021
1,8-Diamino-3,6-dioxaoctane	$C_6H_{16}N_2O_2$	0.1172% [a]	-	GC-MS	Bhalla *et al.*, 2021
1,7-Diaminoheptane	$C_7H_{18}N_2$	1.7997% [a]	-	GC-MS	Bhalla *et al.*, 2021
N,N-Dimethylacetamide	C_4H_9ON	0.454% [a]	-	GC-MS	Bhalla *et al.*, 2021
2-Oxoglutaric acid	$C_5H_6O_5$	0.5466% [a]	-	GC-MS	Bhalla *et al.*, 2021
Oxazolidine, 2-ethyl-2-methyl-	$C_6H_{13}NO$	0.9008% [a]	-	GC-MS	Bhalla *et al.*, 2021
Heptanal	$C_7H_{14}O$	0.7112% [a]	-	GC-MS	Bhalla *et al.*, 2021
6-Methoxy-3-pyridazinethiol	$C_5H_6N_2OS$	0.4293% [a]	-	GC-MS	Bhalla *et al.*, 2021
3-Piperidinol	$C_5H_{11}NO$	0.5971% [a]	-	GC-MS	Bhalla *et al.*, 2021
1,3-Propanediol, 2-ethyl-2-(hydroxymethyl)-	$C_6H_{14}O_3$	21.1909% [a]	-	GC-MS	Bhalla *et al.*, 2021

Compounds	Molecular formula	Quantity		Method	Reference
		Leaves	Seeds		
Benzeneacetonitrile, 4-hydroxy-	C_8H_7NO	3.4763% [a]	-	GC-MS	Bhalla *et al.*, 2021
Benzenebutanal, γ, 4-dimethyl-	$C_{12}H_{16}O$	0.395% [a]	-	GC-MS	Bhalla *et al.*, 2021
2(4H)-Benzofuranone, 5,6,7,7α-tetrahydro-4,4,7α-trimethyl-	$C_{11}H_{16}O_2$	0.4491% [a]	-	GC-MS	Bhalla *et al.*, 2021
Ethanamine, N-ethyl-N-nitroso-	$C_4H_{10}N_2O$	5.2161% [a]	-	GC-MS	Bhalla *et al.*, 2021
Propanoic acid, 2-methyl-, octyl ester	$C_{12}H_{24}O$	15.0279% [a]	-	GC-MS	Bhalla *et al.*, 2021
3-Deoxy-d-mannoic lactone	$C_6H_{10}O_5$	3.2947% [a]	-	GC-MS	Bhalla *et al.*, 2021
d-Glycero-d-ido-heptose	$C_7H_{14}O_7$	0.3814% [a]	-	GC-MS	Bhalla *et al.*, 2021
D-erythro-Pentose, 2-deoxy-	$C_5H_{10}O$	0.3314% [a]	-	GC-MS	Bhalla *et al.*, 2021
N-Methoxy-1-ribofuranosyl-4-imidazolecarboxylic amide	$C_{10}H_{15}N_3O_6$	0.5345% [a]	-	GC-MS	Bhalla *et al.*, 2021
Formamide, N,N-dimethyl-	C_3H_7NO	0.5847% [a]	-	GC-MS	Bhalla *et al.*, 2021
d-Talonic acid lactone	$C_6H_{10}O_6$	0.5651% [a]	-	GC-MS	Bhalla *et al.*, 2021
Sorbitol	$C_6H_{14}O_6$	0.482% [a]	-	GC-MS	Bhalla *et al.*, 2021
Allo-Inositol	$C_6H_{12}O$	0.8749% [a]	-	GC-MS	Bhalla *et al.*, 2021
		0.5189% [a]	-	GC-MS	Bhalla *et al.*, 2021
		1.1254% [a]	-	GC-MS	Bhalla *et al.*, 2021
D-chiro-Inositol, 3-O-(2-amino-4-((carboxyiminomethyl) amino)- 2,3,4,6-tetradeoxy-α-D-arabino-hexopyranosyl)-	$C_{14}H_{25}N_3O_9$	1.595% [a]	-	GC-MS	Bhalla *et al.*, 2021
Muco-Inositol	$C_6H_{12}O_6$	2.0264% [a]	-	GC-MS	Bhalla *et al.*, 2021
Inositol	$C_6H_{12}O_6$	2.0545% [a]	-	GC-MS	Bhalla *et al.*, 2021
Cyclohexane, 1-methyl-4-(2-hydroxyethyl)-	$C_9H_{18}O_1$	0.4822% [a]	-	GC-MS	Bhalla *et al.*, 2021
Hexadecanoic acid, methyl ester	$C_{17}H_{34}O$	0.8519% [a]	-	GC-MS	Bhalla *et al.*, 2021

Compounds	Molecular formula	Quantity		Method	Reference
		Leaves	Seeds		
Phenol, 2-methyl-	C_7H_8O	0.3737% [a]	-	GC-MS	Bhalla *et al.*, 2021
(1S)-Propanol, (2S)-[(tert. butyloxycarbonyl)amino]-1-phenyl-	$C_{14}H_{21}NO_3$	0.967% [a]	-	GC-MS	Bhalla et al., 2021
9-Octadecenoic acid (Z)-, methyl ester	$C_{19}H_{36}O_2$	1.0307% [a]	-	GC-MS	Bhalla *et al.*, 2021
9,12,15-Octadecatrienoic acid, (Z,Z,Z)-	$C_{18}H_{30}O_2$	5.0063% [a]	-	GC-MS	Bhalla *et al.*, 2021
4-Allyl-3-(dimethylhydrazono)-2-methylhexane-2,5-diol	$C_{12}H_{24}N_2O_2$	0.4257% [a]	-	GC-MS	Bhalla *et al.*, 2021
Benzyl- β -d-glucoside	$C_{13}H_{16}O_7$	0.6056% [a]	-	GC-MS	Bhalla *et al.*, 2021
4,6-dimethyl-2-propyl-1,3,5-dithiazinane	$C_8H_{17}NS_2$	0.1698% [a]	-	GC-MS	Bhalla *et al.*, 2021
1,3-Benzenediol, 2-methyl-	$C_7H_8O_2$	0.545% [a]	-	GC-MS	Bhalla *et al.*, 2021
9-Octadecenamide, (Z)-	$C_{18}H_{35}NO$	1.4694% [a]	-	GC-MS	Bhalla *et al.*, 2021

Source: Anzano *et al.*, 2021

Abbreviations: a: calculated as peak area; b: mean value of different samples; c: mature/old leaves; d: tender/young leaves; e: vegetative plants; f: obtained considering a moisture of 75%; g: flowering plants.

AOAC 2004: Association of Official Agricultural Chemists 2004

GC-MS: Gas Chromatography-Mass Spectrometry

HPLC: High-Performance Liquid Chromatography (or High-Pressure Liquid Chromatography)

HPLC and MS/MS: High-Performance Liquid Chromatography and Mass Spectrometry

HPLC-DAD: High-Performance Liquid Chromatography with Diode Array Detection

HPLC-DAD-ESI-MS: High-Performance Liquid Chromatography–Diode Array Detection–Electro-Spray Ionization Mass Spectrometry

HPLC-ESI-MS/MS: High Pressure Liquid Chromatography Electrospray Ionisation tandem Mass Spectroscopy

LC-MS: Liquid Chromatography-Mass Spectrometry

UPLC: Ultra Performance Liquid Chromatography

4.10.2 Known potential applications

Ingredients	Key applications (if any readily available)
1,3-Dihydroxyacetone dimer	Synthesis of polymeric biomaterials (Zelikin and Putnam, 2005)
4(1H)-Pyrimidinone, 2,6-diamino-	Hydroxy and amino pyrimidines are of great interest in natural products and in the development of new drugs for various diverse areas like anti-cancer (Skoweranda *et al.*, 1990)
4H-Pyran-4-one, 2,3-dihydro-3,5-dihydroxy-6-methyl-	Strong anti-oxidant (Yu *et al.*, 2013; Čechovská *et al.*, 2011)
2-Hexynoic acid	As precursor for synthesizing various biological metabolites like leucotrienes, oxylipins (Starostin *et al.*, 2000)
Butanedioic acid, 2-hydroxy-2-methyl-, (S)-	One of the rarely identified components in plant extracts. Mobilizes the phosphates in soil for agricultural applications to enhance the P availability (Khorassani *et al.*, 2011)
3,3′-Iminobispropylamine	This is also among the rare component identified in plant having various roles in plant as well as medicinal antitumour activities (Rodriguez-Garay *et al.*, 1989; Sunkara *et al.*, 1988; Nishio *et al.*, 2019)
1,3-Dioxolan-2-one, 4,5-dimethyl-	Precursor for various drugs and cyclic carbonates having applications in electrochemical energies (Hagiyama *et al.*, 2008)
D-Mannoheptulose	D-Mannoheptulose widely studied for its activity against breast cancer and to suppress the D-glucose induced insulin release (Al-Ziaydi *et al.*, 2020; Courtois *et al.*, 2001)
Azetidin-2-one 3,3-dimethyl-4-(1-aminoethyl)-	Anti-inflammatory, ulcerogenic, and analgesic activities (Siddiqui *et al.*, 2010)
Tetra acetyl-d-xylonic nitrile	Anti-tumour and anti-oxidant (Imad *et al.*, 2015; Kanhar and Sahoo, 2018)
Alpha-D-Glucose	Source of energy, highly effective and used in food, medicine and its derivatives also have many medicinal uses (Shendurse and Khedkar, 2016; Cancelas *et al.*, 2000)
Propanamide	Propanamide derivatives studied for antimicrobial and antiviral efficacy (Ölgena *et al.*, 2008)
Dihydroxyacetone (DHA)	Dihydroxyacetone (DHA) is being used in sunless tanning type of products (Huang *et al.*, 2017)
Glycerin	Used as humectants as well as moisturizer having application in cosmetics and medicines (Sagiv *et al.*, 2001)
Erythritol	Antioxidant; Improving blood vessel function in people with type 2 diabetes (den Hartog *et al.*, 2010)

Ingredients	Key applications (if any readily available)
4,5-Diamino-6-hydroxypyrimidine	Analogues are used for various medicinal properties like antimicrobial (Abbas *et al.*, 2017)
4H-Pyran-4-one, 2,3-dihydro-3,5-dihydroxy-6-methyl-	Strong anti-oxidant (Yu *et al.*, 2013, Čechovská *et al.*, 2011)
Catecholborane	Versatile compound for various organic synthesis (Paquette *et al.*, 2009)
1,7-Diaminoheptane	Malaria treatment (Kaiser *et al.*, 2001)
N,N-Dimethylacetamide	Novel antiviral agent (He *et al.*, 2005)
2-Oxoglutaric acid	Biosynthesis of carotenoids in chloroplasts (Liu *et al.*, 2018)
3-Piperidinol	Anti-tuberculosis agent (Markad *et al.*, 2015)
2(4H)-Benzofuranone, 5,6,7,7a-tetrahydro-4,4,7a-trimethyl-	Antiarthritic activity (Rhew *et al.*, 2020)
Ethanamine, N-ethyl-N-nitroso-	Boosts immune system (Zaitseva *et al.*, 2018)
Propanoic acid, 2-methyl-, octyl ester	Anti-microbial properties (Hamedi *et al.*, 2019)
d-Glycero-d-*ido*-heptose	Inhibition of insulin secretion and Hexokinase (Scruel *et al.*, 1998)
Sorbitol	Helps to prevent hyperglycemia (Wick *et al.*, 1951) Protects and rejuvenates the skin from oxidative stress (Manca *et al.*, 2018)
Allo-Inositol	Reduces the inflammation (Bizzarri *et al.*, 2020)
D-chiro-Inositol, 3-O-(2-amino-4-((carboxyiminomethyl) amino)-2,3,4,6-tetradeoxy-.alpha.-D-*arabino*-hexopyranosyl)-	Type 2 diabetes treatment (Pintaudi *et al.*, 2016)
Allo-Inositol	Reduces the inflammation (Bizzarri *et al.*, 2020)
Scyllo-Inositol	Treats mild to moderate Aalzheimer's disease (Choi *et al.*, 2010)
Muco-Inositol	Oligomer of muco-inositol act as glycosidase inhibitors (Freeman and Hudlicky, 2004)
Allo-Inositol	Reduces the inflammation (Bizzarri *et al.*, 2020)
Inositol	Inositol and its derivative are known for treatment of polycystic ovary syndrome (PCOS) (Kamenov and Gateva, 2020; Mariano and Gianfranco, 2014)

Ingredients	Key applications (if any readily available)
Hexadecanoic acid, methyl ester	Anti-inflammatory (Saeed *et al.*, 2012)
n-Hexadecanoic acid (palmitic acid)	Anti-inflammatory (Aparna *et al.*, 2012)
(1S)-Propanol, (2S)-[(tert. butyloxycarbonyl)amino]-1-phenyl-	Human immunodeficiency virus -1 (HIV-1), protease inhibitors (Ghosh *et al.*, 2015)
9-Octadecenoic acid (Z)-, methyl ester	Lubricant (Faujdar and Singh, 2021)
Phytol	Controls ganoderma boninense (Plant disease) (Aziz *et al.*, 2019) Antioxidant, anti-inflammatory (Islam *et al.*, 2018)
9,12,15-Octadecatrienoic acid, (Z,Z,Z)-	Reduces complications in covid-19 patients. (Weill *et al.*, 2020) Neuroprotective properties. (Blondeau *et al.*, 2015)
Octadecanoic acid	Playing roles in food reward (Li *et al.*, 2020) Lowers high density lipoprotein (HDL) cholesterol (van Rooijen *et al.*, 2021)
9-Octadecenamide, (Z)-	Anti-depressive effects (Ge *et al.*, 2015)

Source: Anzano *et al.*, 2021

These compounds collectively contribute to various health benefits associated with moringa consumption. However, concentration of these compounds may vary depending on age of plant, growing conditions and the parts of the plant consumed (leaves, seeds, *etc.*). Though moringa has many potential health benefits but it is important to consult a healthcare professional before making significant dietary changes or using it as a supplement, especially when a person is already under medication.

4.10.3 Non-polar compounds

4.10.3.1 Fatty acids in leaves and seeds

Fatty acids are abundant in *Moringa oleifera* as analysed by Lee *et al.* (2017) of powdered *M. oleifera* leaves. The extraction process involved 70% methanol initially, followed by partitioning the resulting solution with hexane, dichloromethane, ethyl acetate, butanol and water. In the last stage, supercritical fluid extraction (SFE) was used. The predominant fatty acid identified in the extracts was cis-11-eicosenoic acid (gondoic acid), a monounsaturated acid commonly found in plant-derived oils. Subsequently, palmitic acid constituted 23.65% followed by linoleic acid and oleic acid at 6.84% and 5.92%, respectively.

Other fatty acids were present in quantities lower than 5%. In seed extracts, oleic acid dominated comprising of 74.5% of fatty acids, followed by palmitic acid (7.92%) and stearic acid (7.10%). Minor fatty acids were present in smaller quantities (Lee *et al.*, 2017).

These fatty acids play a pivotal role in metabolic processes leading to synthesis of omega-3 fatty acids crucial for regulating cellular development and the immune system (Lalas *et al.*, 2017). Lee *et al.* (2017) demonstrated that cis-11-eicosenoic acid, linoleic acid and oleic acid exhibit antimicrobial effects on *Staphylococcus aureus*, inhibiting biofilm formation by impeding cell growth, aligning with findings by Chelliah *et al.* (2017). Furthermore, the oily fraction of *M. oleifera* leaves holds significance in cosmetic applications serving as a fundamental component in skin and hair products (Chelliah *et al.*, 2017).

4.10.3.2 Phytosterol glycosides

Phytosterols are plant-derived compounds with a chemical structure similar to that of cholesterol. These compounds are known for their potential health benefits, particularly in helping to lower cholesterol levels. Within the spectrum of bioactive components present in moringa, phytosterols such as β-sitosterol, γ-sitosterol, campesterol and stigmasterol are identified. Nevertheless, the specific content of each phytosterol exhibits variability. The β-sitosterol was isolated from the leaves and seeds of *M. oleifera* (Maiyo *et al.*, 2016). The **γ-sitosterol ~ a structural isomer of β-sitosterol (structurally similar to cholesterol) was isolated from the leaves of some moringa varieties,** *e.g.* var. 'Madurai'. A sterol glycoside, β-sitosterol-3-*O*-β-*D*-galactopyranoside was isolated from a chloroform extract of *M. oleifera* stem bark (Bargah and Das, 2014).

4.10.3.3 Carotenoids

Two main carotenoids in moringa leaves are lutein and beta-carotene. Many fruits and vegetables have a green colouration due to presence of beta-carotene. Body can convert beta-carotene into active vitamin A which is essential for healthy skin, immunity and eyesight. Similar to beta-carotene, lutein is recognised for its function in maintaining eye health and for its antioxidant qualities. Lutein along with zeaxanthin is concentrated in the macula of the eye and is believed to help protect against age-related macular degeneration (AMD). The presence of chlorophyll which aids in photosynthesis together with carotenoids is shown by the green colour of moringa leaves. The amount of carotenoids in moringa varies as to those of other chemicals depending on the age of the plant, growth circumstances and section of the plant that is consumed. Furthermore, the bioavailability of carotenoids might be affected by processing or cooking method used for moringa.

4.10.3.4 Alkanes and their derivatives

Alkanes (saturated hydrocarbons) are compounds consisting of carbon and hydrogen atoms with single covalent bonds. They are hydrocarbons with general formula CnH_2n+2. In moringa, alkanes such as methane (CH_4), ethane (C_2H_6), propane (C_3H_8) and butanes (C_4H_{10}) might be present. These compounds are typically found in plant waxes and lipids. Alkyl groups such as methyl and ethyl are derived from alkanes by removing one hydrogen atom and may be part of various compounds in moringa. Alkane derivatives may include compounds like alkanols (alcohols), alkanoic acids and others. These derivatives often have functional groups attached to the alkane backbone.

Moringa being a plant with various bioactive components also contains alkanes and their derivatives though specific details may vary based on factors such as varieties, growing conditions and extraction methods. Using GC-MS method, the main alkanes and/or their derivatives in leaves of var. 'Madurai' were identified as 1, 30-triacontanediol (14.98%), octacosane (8.57%) and Z-14-nonacosane (8.3%) (Chelliah *et al.*, 2017) whereas nonacosane (15.55%) was identified in var. 'Chennai' though other alkanes and alkenes identified were pentacosane (2.14%-17.41%), hexacosane (11.20%), tetracosane (1.45%) (Chuang *et al.*, 2007), heptacosane (7.45%), eicosane (0.31%) and nonadecane (0.22%) (Zhao and Zhang, 2013).

4.10.3.5 Terpenes and terpenoids

While not alkanes in the strictest sense, the terpenes and terpenoids derived from isoprene units are organic compounds found in many plants including moringa. These may contain alkane-like structures within their molecules.

4.10.3.6 Non-polar compound analysis in varieties, various plant parts and oil

Within moringa extracts alkanes and their derivatives constitute a prominent category of molecules. Chelliah *et al.* (2017) made a comprehensive and extensive analysis in *Moringa oleifera* employing varied extractions with hexane, methanol and water. Subsequent analysis through GC-MS revealed several compounds with noteworthy concentrations in different varieties and plant parts. Var. 'Madurai' leaves presented predominant compounds as 1,30-triacontanediol (14.98%) and octacosane (8.57%) whereas nonacosane (15.55%) and γ-sitosterol (9.56%) prevailed in var. 'Chennai'. Seeds exhibited distinct compositions where var. 'Madurai' presented 6-octadecenoic acid (52.24%), n-hexadecanoic acid (palmitic acid) (6.17%) and oleic acid (5.12%) as major components. In contrast, var. 'Chennai' seeds contained (Z)-13-docosenamide (13.62%), propionamide (4.48%) and ethyl oleate (4.33%) (Chelliah *et al.*, 2017). Plants synthesize these substances to form an external waxy layer which is known as epicuticular wax which serves to safeguard leaves from surface wetting and water evaporation.

6-Octadecenoic acid or petroselinic acid and palmitic acid are prevalent fatty acids found in plant oils. γ-Sitosterol classified as a phytosterol represents an isomer of β-sitosterol with documented scientific evidence supporting its *in vivo* antidiabetic properties against type II diabetes (Balamurugan *et al.*, 2011, 2012). Similar compounds were identified by Zhao and Zhang (2013) through a Soxhlet extraction with hexane lasting 8 hours. Analysis of dried extract using GC-MS revealed nonacosane (18.65%), 1, 2, 4-trimethylbenzene (16.96%) and heptacosane (7.45%) as the most abundant compounds (Zhao and Zhang, 2013).

Essential oils represent intricate blends of numerous molecules, primarily volatile compounds falling within the classes of terpenes, alcohols, aldehydes, ketones, esters, phenols and others. Extraction methods such as hydrodistillation, solvent-free microwave extraction (SFME) or supercritical fluid extraction using supercritical carbon dioxide are commonly employed. Despite their widespread use in alternative medicine and aromatherapy, essential oils exhibit noteworthy properties. These demonstrate antibacterial activity against gram-positive species, potentially attributed to the disruption of external membranes by phenolic compounds. Their antioxidant activity is likely due to secondary metabolites with conjugated double bonds while anti-inflammatory effects result from a diverse array of compounds including antioxidants. Chuang *et al.* (2007) conducted a comprehensive analysis of essential oils extracted from *M. oleifera* leaves using hydrodistillation. The tested extract exhibited positive effects against fungal development, indicating potential antifungal activity by disrupting cellular membranes. GC-MS analysis of non-polar extracts revealed predominant compounds notably alkanes and alkenes such as hexacosane, pentacosane and tetracosane constituting almost 30% of total extracts. Phytol, a diterpene alcohol abundant in plants and a component of chlorophyll was identified (7.66% based on peak area). Phytol has applications as a food additive though researchers have suggested its potential in treating schistosomiasis ~ a parasitic disease (Dhifi *et al.*, 2016). Further examination unveiled a small quantity of terpenes and terpene derivatives with hexahydrofarnesylacetone being the most abundant (1.30%). Traces of linalool oxide, farnesylacetone, isolongifolene, α-ionene and α- and β-ionone were also present (Chuang *et al.*, 2007). This diverse composition highlights the complexity of essential oils and their potential applications in various fields including medicine and nutrition.

4.10.4 Polar compounds

4.10.4.1 Alkaloids

While moringa contains various phytochemicals, it is not particularly known for being rich in alkaloids. However, moringa does contain certain alkaloids in trace amounts. One of the alkaloids found in moringa is called moringine which has been studied for its potential medicinal properties including anti-inflammatory and antioxidant effects. Alkaloids are a group of naturally occurring chemical compounds that contain mostly basic nitrogen atoms. This nitrogen may occur in the form of a primary amine (RNH_2), a secondary amine (R_2NH) or a tertiary amine (R_3N). In addition to carbon, hydrogen and nitrogen most alkaloids contain oxygen (Cushnie *et al.*, 2014). The presence of these compounds has been confirmed in *Moringa oleifera* leaves (Sreelatha and Padma, 2009; Kasolo *et al.*, 2010). Several of these compounds such as N, α-l-rhamnopyranosyl vincosamide, 4-(α-l-rhamnopyranosyloxy) phenylacetonitrile (niazirin), pyrrolemarumine 4″-*O*-α-l-rhamnopyranoside, 4′-hydroxy phenylethanamide-α-l-rhamnopyranoside (marumoside A) and its 3-*O*-β-d-glucopyranosyl-derivative (marumoside B) and methyl 4-(α-l-rhamnopyranosyloxy)-benzylcarbamate have been isolated in *Moringa oleifera* leaves (Sahakitpichan *et al.*, 2011; Panda *et al.*, 2013). However, their amounts in the leaves are still unknown.

4.10.4.2 Amino acids

Moringa oleifera leaves are renowned for their exceptional protein content, making them a rich source of amino acids. These essential amino acids constitute more than 50% of the total amino acid content present in *M. oleifera* leaves. The abundance of amino acids, particularly highlighted by significant quantities of glutamic acid underscores the nutritional value of these leaves. Chelliah *et al.* (2017) while working on the var. 'Madurai' extracted 15.86g/16gN of glutamic acid while var. 'Chennai' displayed 15.33g/16gN. Lalas *et al.* (2017) reported 268.7mg/100g of glutamic acid. Methionine was found to be least abundant in var. 'Madurai' (1.9g/16gN) and histidine was least present in var. 'Chennai' (3.12g/16gN). Conversely, Lalas *et al.* (2017) reported cysteine as the least abundant amino acid in dried leaves with a concentration of 29.6mg/100g. In seeds, glutamic acid was the most prevalent amino acid (14.23g/16gN in var. 'Madurai', 14.74g/16gN in var. 'Chennai') while threonine and valine were the least abundant in 'Madurai' (3.2g/16gN) and 'Chennai' (2.37g/16gN), respectively (Chelliah *et al.*, 2017; Lalas *et al.*, 2017).

4.10.4.3 Glucosinolates and isothiocyanates

Glucosinolates in plants are a group of secondary metabolites. Structurally they are β-*S*-glucosides of thio-oxime-*O*-sulphates, synthesized from amino acids. Appreciable amounts of these compounds are found in *Moringa oleifera* leaves. In particular, around 116 and 63mg/g of DW in young and older leaves, respectively are reported (Bennett *et al.*, 2003; Forster *et al.*, 2015). These amounts are close to and in some case even larger than those found in many cruciferous vegetables such as broccoli, cabbage and radish (Ciska *et al.*, 2000).

Lopez-Rodriguez *et al.* (2020) conducted extractions using 70% and 80% methanol which revealed various glucomoringin isomers in *M. oleifera* leaves and seeds. These compounds of glucosinolate class play a crucial role in plant defense against disease infections. Aromatic glucosinolates, prevalent in *M. oleifera* leaf extracts are known precursors of stable isothiocyanates. Isothiocyanates from *M. oleifera* have exhibited noteworthy biological activities including hypoglycemic, antioxidant, anti-inflammatory, anticancer and chemopreventive effects. The stability of *M. oleifera* isothiocyanates, attributed to the presence of rhamnose, sets them apart from those in other crucifers (Lopez-Rodriguez *et al.*, 2020).

4-*O*-(α-L-rhamnopyranosyloxy)-benzyl glucosinolate has been identified as the dominant leaf glucosinolate of *M. oleifera* and is accompanied by lower levels of three isomeric 4-*O*-(α-L-acetylrhamnopyranosyloxy)-benzyl glucosinolates which reflect three positions of the acetyl group at the rhamnose moiety of the molecule (Bennett *et al.*, 2003; Forster *et al.*, 2015). Concentrations of these compounds seem to be affected by physiological stage of the plant and by the maturity stage of the leaves. Leone *et al.* (2015a) highlighted glucosinolates ~ specifically moringin and three isomers of 4-O-(α-L-acetylrhamnopyranosyloxy)-benzyl-isothiocyanate with concentrations ranging from 21.84mg/g to 59.4mg/g of dry weight. Concentrations of the three isomers of 4-*O*-(α-L-acetylrhamnopyranosyloxy)-benzyl glucosinolates range from 2.16 to 5mg/g of DW, 1.2 to 1.8mg/g of DW and 12.76 to 50.2mg/g of DW for isomer 1, 2 and 3, respectively (Bennett *et al.*, 2003; Amaglo *et al.*, 2010; Leone *et al.*, 2015a). Amaglo *et al.* (2010) reported presence of 4-hydroxybenzyl (sinalbin) with a concentration ranging between ND and 2.36mg/g of DW. Glucosinolates can be hydrolyzed by myrosinase to produce d-glucose and various other degradation products like isothiocyanates (Prakash and Gupta, 2012) which are also present in *Moringa oleifera* leaves (Waterman *et al.*, 2014; 2015). Both glucosinolates and isothiocyanates play an important role in health promoting and prevention of disease (Dinkova-Kostova and Kostov, 2012) and are associated with cancer chemoprotective, allelopathic and repellent/insecticidal activities in various studies (Vergara-Jimenez *et al.*, 2017).

4.10.4.4 Polyphenols

Polyphenols, a group of naturally occurring compounds found in plants have antioxidant properties and may offer various health benefits. *Moringa oleifera* dried leaves are a great source of polyphenols with concentrations ranging from 2090 to 12,200mg GAE (gallic acid equivalent) per 100g of DW (Bajpai *et al.*, 2007; Prakash *et al.*, 2007; Singh *et al.*, 2009; Sreelatha and Padma, 2009; Sultana *et al.*, 2009) or 1600 to 3400mg TAE (tannin acid equivalent) per 100g of DW (Makkar and Becker, 1996; Bhatta *et al.*, 2012). These amounts are greater than those found in fruits and vegetable (Brat *et al.*, 2006; Cieślik *et al.*, 2006; Fu *et al.*, 2011). Principal polyphenol compounds in *Moringa oleifera* leaves are flavonoids and phenolic acids.

4.10.4.4.1 Flavonoids

Moringa oleifera leaves are an interesting source of flavonoid compounds that are a sub-group of polyphenolic compounds having a benzo-γ-pyrone structure and are ubiquitous in plants as they are synthesized in response to microbial infections (Kumar and Pandey, 2013). Identified flavonoids include epicatechin, kaempferol, myricetin, quercetin and isorhamnetin. Epidemiological studies have consistently shown that high intake of flavonoids has protective effects against many infectious (bacterial and viral) and degenerative diseases such as cardiovascular diseases, cancer and other age-related diseases (Pandey and Rizvi, 2009; Kumar and Pandey, 2013). In particular, quercetin has been associated with various health benefits including vasorelaxation, positive effects on dyslipidemia & obesity, anti-inflammatory properties and antimicrobial activity. Kaempferol and myricetin likewise contribute to cardiovascular health and exhibit antioxidant, antiviral, anticarcinogenic, antiplatelet, hypoglycemic and cytoprotective properties.

Total flavonoid concentration in dried leaves ranges from 5.059 to 12.16mg/g of DW (Yang *et al.*, 2008; Leone *et al.*, 2015a) which is close to or larger than that in many fruits and vegetables normally consumed (Harnly *et al.*, 2006; Coppin *et al.*, 2013). Total concentration of flavonoids in freeze-dried leaves ranges from 21.0 to 61.62mgRE/g of DW (Siddhuraju and Becker, 2003; Zhang, *et al.*, 2011; Leone *et al.*, 2015a). In dried moringa leaves, myricetin concentration is approximately 5.804mg/g of DW while quercetin and kaempferol concentrations range from 0.207 to 7.57mg/g of DW and not in detectable amounts (ND) to 4.59mg/g of DW, respectively (Bajpai *et al.*, 2007; Prakash *et al.*, 2007; Sultana and Anwar, 2008; Yang *et al.*, 2008; Singh *et al.*, 2009; Coppin *et al.*, 2013; Leone *et al.*, 2015a). Higher amounts were found in freeze-dried leaves. In particular, quercetin and kaempferol concentrations range from 5.47 to 16.64mg/g and 1.5 to 3.5mg/g of DW, respectively (Bennett *et al.*, 2003; Siddhuraju and Becker, 2003; Amaglo *et al.*, 2010).

Isorhamnetin concentration in dried leaves is approximately 0.118mg/g of DW (Yang *et al.*, 2008; Leone *et al.*, 2015a) while in freeze-dried leaves its concentration was found up to 7 times larger ranging from 0.52 to 0.72mg/g of DW (Amaglo *et al.* 2010; Leone *et al.*, 2015a). Other flavonoids, such as luteolin, apigenin, daidzein and genistein were not found in detectable concentrations in *Moringa oleifera* leaves (Bajpai *et al.*, 2007; Yang *et al.*, 2008). However, there had been only a few such studies, therefore, further investigations are required.

4.10.4.4.2 Phenolic acids

Phenolic acids are a sub-group of phenolic compounds derived from hydroxybenzoic acid and hydroxycinnamic acid that are naturally present in plants and are known for their antioxidant, anti-inflammatory, antimutagenic and anticancer activities (Zhao and Moghadasian, 2008; Duthie and Wood, 2011; El-Seedi *et al.*, 2012; Verma *et al.*, 2013). Phenolic acids in moringa leaves were identified through HPLC and LC-MS analyses. Major compounds included caffeic acid, chlorogenic acid, o-coumaric acid, ellagic acid, ferulic acid and gallic acid with quantity ranging from 0.018mg/g (ellagic acid) to 6.457mg/g (coumaric acid) of dried leaves (Leone *et al.*, 2015a).

In dried leaves, gallic acid seems to be the most abundant with a concentration of about 1.034mg/g of DW (Prakash *et al.*, 2007; Singh *et al.*, 2009; Leone *et al.*, 2015a) though Bajpai *et al.* (2007) reported only in poorly detectable amounts. The concentration of chlorogenic and caffeic acids ranges from 0.018 to 0.489mg/g of DW and ND to 0.409mg/g of DW, respectively (Bajpai *et al.*, 2007; Prakash *et al.*, 2007; Singh *et al.*, 2009; Leone *et al.*, 2015a). Lower though appreciable concentrations of ellagic and ferulic acids were recorded. Their concentrations ranged from ND to 0.189mg/g and 0.078 to 0.128mg/g of DW, respectively (Bajpai *et al.*, 2007; Prakash *et al.*, 2007; Singh *et al.*, 2009; Leone *et al.*, 2015a). Some of these compounds were found more concentrated in freeze-dried leaves. Approximately 6.457mg/g of DW of o-coumaric acid and 0.536mg/g of DW of caffeic acid were found in freeze-dried leaves from Florida while p-coumaric, synaptic, gentistic and syringic acids were found in poorly detectable amounts (Zhang, *et al.*, 2011; Leone *et al.*, 2015a).

4.10.4.4.3 Tannins

Tannins are water-soluble phenolic compounds that bind and precipitate alkaloids, gelatin and other proteins. These exhibit various biological properties such as anti-cancer, antiatherosclerotic, anti-inflammatory, anti-hepatoxic, antibacterial and anti-HIV replication activity (Kancheva and Kasaikina, 2013). *Moringa oleifera* leaves are an appreciable source of tannins. Their concentrations range between 13.2 and 20.6g TAE/kg (Makkar and Becker, 1996; Bhatta *et al.*, 2012; Teixeira *et al.*, 2014) in dried leaves and 5-12g TAE/kg in freeze-dried leaves (Richter *et al.*, 2003; Makkar and Becker, 1997).

4.10.4.4.4 Oxalates and phytates

While moringa is praised for its rich nutrient profile, it also contains oxalates and phytates which are compounds that can have both positive and negative effects on health. Oxalates are salts or esters of oxalic acid containing polar covalent bonds. These compounds often form crystals in their solid state and are soluble in water, indicating their polar nature. Oxalates are naturally occurring compounds found in many plants. They can form crystals and in high concentrations contribute to the formation of kidney stones in some individuals. Moringa contains oxalates though the levels are generally considered moderate. For most people, consuming moringa as part of a balanced diet is unlikely to contribute significantly to oxalate-related health issues.

On the other hand phylates are salts of phytic acid that contain multiple phosphate groups which are polar due to the electronegativity difference between phosphorus and oxygen. Phytates are antioxidant compounds found in many plant foods. While they have health benefits, they can also bind to minerals reducing their absorption in the digestive tract. Moringa contains phytates which might affect the absorption of minerals such as calcium, iron and zinc. However, the impact of phytates in moringa on mineral absorption is generally not considered ~ a significant concern when moringa is consumed as part of a varied diet.

Oxalates content of dried leaves range from 430 to 1050mg/100g of DW (Joshi and Mehta, 2010; Teixeira *et al.*, 2014) similar to other plants rich in these compounds (Gupta *et al.*, 2005) while concentration of phytates range from 25 to 31g/kg of DW (Makkar and Becker, 1996) in dried leaves and from 21 and 23g/kg of DW in freeze-dried leaves (Makkar and Becker, 1997; Richter *et al.*, 2003).

4.10.5 Amphipathic compounds

4.10.5.1 Saponins

Saponins are amphipathic compounds possessing both polar and non-polar regions. These typically consist of a hydrophobic (non-polar) aglycone or sapogenin and a hydrophilic (polar) sugar moiety. This dual nature allows saponins to interact with both water and lipid molecules. The hydrophobic portion can interact with lipids while hydrophilic part can interact with water molecules. This property contributes to the surfactant or detergent-like characteristics of saponins which enable them to form foams or emulsify substances.

Saponins are a group of natural compounds that consist of an isoprenoidal-derived aglycone, designated genin or sapogenin, covalently linked to one or more sugar moieties (Augustin *et al.*, 2011). Even though some saponins have hemolytic side effects so they are studied for their anti-cancer properties (Man *et al.*, 2010; Tian *et al.*, 2013). *Moringa oleifera* leaves are a good source of saponins. Their concentration in dried leaves is approximately 50gDE/kg of DW (Makkar and Becker, 1996) while in freeze-dried leaves it ranges between 64 and 81 gDE/kg of DW (Makkar and Becker, 1997; Richter *et al.*, 2003).

4.10.6 Other compounds

Study conducted by Bhalla *et al.* (2021) delved into the identification of compounds in two different extracts of *M. oleifera*. In aqueous extract carbonic acid, butyl 2-pentyl ester (20.64%), 2-isopropoxyethyl propionate (16.87%), 4H-Pyran-4-one, 2,3-dihydro-3,5-dihydroxy-6-methyl- (8.99%) and 1,3-Dioxolan-2-one, 4,5-dimethyl- (6.16%) emerged as the predominant compounds based on their peak areas. The methanolic extraction revealed a more diverse compound profile due to differing polarities. In the methanolic extracts, 1,3-propanediol, 2-ethyl-2-(hydroxymethyl)- (21.19%), propionic acid, 2-methyl-, octyl ester (15.02%), ethanamine, N-ethyl-N-nitroso- (5.21%), and 9,12,15-octadecatrienoic acid were identified as the most abundant compounds. Antioxidant activity assays on free radicals demonstrated that methanolic extract exhibits a superior capacity for scavenging free radicals compared to the aqueous extracts. Bhalla *et al.* (2021) opine that this heightened antioxidant activity in the methanolic extract may be attributed to the presence of higher levels of polyphenolic molecules. Considering the increasing demand for compounds with potent antioxidant properties in both food and cosmetic industries, this finding holds significance. Such compounds are valued for their applications in skin treatment and as enriching agents in food products. These results underscore the potential utility of *M. oleifera* extracts, particularly the methanolic variant, in meeting the requirements of these industries.

4.11 Importance and Uses

4.11.1 Culinary uses

Moringa boasts an impressive nutritional profile, rich in essential vitamins such as A, C & E as well as being a noteworthy source of calcium that promotes bone health and helps prevent heart disease. Moringa serves as an excellent reservoir of potassium known for its ability to alleviate anxiety and stress. Moringa tree is a versatile resource with almost every part finding use in food, oil, fibre or medicine. All the components of this tree including leaves, pods, seeds, roots and flowers are suitable for both human and animal consumption. Particularly significant are its pods, leaves and seeds boasting rich nutritional profiles and serving diverse culinary and medicinal purposes. Widely cultivated across the tropics, the moringa tree is extensively promoted as a dietary supplement, health food or integral component in herbalism practices (Janick and Paull Robert, 2008). Extracted oils derived from the seeds offer a versatile application while leaves and roots can be transformed into powder. Typically, the desire for home consumption of pods and leaves can be readily satisfied by cultivating just one or two moringa trees in a backyard setting.

4.11.1.1 Pods

The fruit pods commonly referred to as 'drumsticks' have gained popularity as a food source worldwide with a notable increase in consumption observed, especially in South Asia (Somoza *et al.*, 2021) as these are delicious as well as nutritious. These are rich in vitamins, minerals and antioxidants. These are known for their potential to boost immunity, aid digestion and for promoting overall health. Enjoyed as a vegetable, young pods are commonly savored with very young *cum* fibreless ones offering the option to be prepared akin to string beans. Due to the minimal weight of very young pods, commercial production predominantly focuses on larger and more fibrous pods which are commonly employed in the preparation of soups, stews and curries (Radovich, 2011).

Preparation of drumsticks can vary widely across regions within India, Sri Lanka and Bangladesh. Each region might have its unique way of incorporating drumsticks into various dishes, utilizing different spices and cooking techniques. The preparation typically involves boiling young pods in a blend of coconut milk and spices like poppy or mustard seeds until they reach the desired tenderness. These versatile pods are commonly used in dishes like drumstick *dal* and *sambar* where these are pulped and simmered alongside various vegetables and spices such as turmeric and cumin. The pods are also cooked as vegetables along with potatoes. The mashed pulp of drumsticks is a staple in dishes like *bhurta*, a medley of lightly fried or curried vegetables (Lim, 2012). However, due to their tough and fibrous outer skin, consuming drumsticks involves a technique as some chew them to extract the flavourful juices and nutrients and discarding the fibrous remains while others prefer to suck out the flesh and tender seeds but discarding the fibrous tube-like skin. Apart from the mentioned curries and lentil-based dishes, drumsticks are also used in rice dishes, stir-fries and even pickles. Their unique flavour adds depth to soups, stews and vegetable-based gravies.

In addition to their culinary use, drumsticks hold cultural significance in some regions. These might be part of traditional festivities or rituals symbolizing fertility, prosperity or other cultural values.

4.11.1.2 Seeds

Moringa fruit seeds serve multiple purposes akin to green peas in various regions (Saha *et al.*, 2012). In Nigeria, their bitter taste is highly valued, often incorporated into sauces or consumed as a fried snack. The edible seed oil finds use in condiments or dressings (Lim, 2012). De-bittered and ground moringa seeds function effectively as a fortifying agent augmenting the protein, iron and calcium levels in wheat flours (Lim, 2012; Chinma *et al.*, 2014; Oyeyinka and Oyeyinka, 2018).

4.11.1.3 Leaves and flowers

Based on their hardiness whether consumed raw or cooked the moringa leaves offer a versatile culinary experience. Rich in nutrients, this plays a pivotal role in various dishes such as soups, stews and stir-fries ~ adding depth to flavours and taste. The young leaves and flowers when sautéed, transform into exquisite delicacies. These botanical elements are also favoured for brewing tea and infusing the beverage not just with flavour but with a wealth of nutritional benefits. Moringa leaves and flowers boast a wealth of antioxidants and hold a longstanding history in traditional herbal medicine.

Renowned for their high protein content, abundance of minerals, β-carotene and antioxidant compounds, moringa leaves serve as a nutritional powerhouse for both human and animal diets. Their significance extends to traditional medicine emphasizing their multifaceted role in enhancing overall well-being (Leone *et al.*, 2015a, b).

These prominently feature in dishes like *tinola* and *utan* ~ traditional Filipino clear broth-based soups. Finely chopped tender moringa leaves also serve as a garnish for various vegetable dishes and salads such as Kerala dish *thoran*. These are used as a substitute for or in combination with coriander leaves (cilantro) (Lim, 2012). Similar to spinach, these leaves are cooked and dried direct under sun, shade, freeze or through oven-drying at temperatures of 50–60°C and then the leaves are used though their efficacy in retaining specific micro- and macronutrients (Adenike, 2014; Ali *et al.*, 2017), often ground into a powder for soups, sauces and other culinary uses. The dried and powdered leaf powder is a popular addition to soups, sauces and smoothies (Lim, 2012) as leaf powder is highly valued as a dietary supplement and is often used to fortify diverse food products, spanning from dairy items like yogurt and cheese (Oyeyinka and Oyeyinka, 2018) to baked goods like bread and pastries thus maintaining a favourable taste profile (Lim, 2012; Oyeyinka and Oyeyinka, 2018).

4.11.1.4 Roots

The shredded roots serve as a condiment, boasting a sharp flavour profile attributed to their high polyphenol content (Atawodi *et al.*, 2010).

4.11.2 Oil and oil cake

M. oleifera seeds contain edible oil (Ramachandran *et al.*, 1980), flocculants for water purification (Jahn, 1984; Gassenschmidt *et al.*, 1995; Muyibi and Evanson, 1995; Ndabigengesere *et al.*, 1995), and antimicrobial substances (Jahn, 1984). Contrary to expectation, the seeds have garnered scientific attention due to the substantial oil content (38-40%) found in *M. oleifera* seed kernels, characterized by a high-quality fatty acid composition (with oleic acid surpassing 70%). The clear and odourless refined oil exhibits remarkable resistance to oxidative degradation and resist rancidity (Anwar *et al.*, 2005). The commercially recognized name for this oil is 'ben oil' or 'behen oil' named so for its high concentration of behenic acid. Its versatile properties make it suitable for both human consumption and for a variety of commercial applications. In fact, moringa oil stands as a promising alternative to olive oil in dietary use and it extends its utility to non-food sectors such as biodiesel production, cosmetics and serving as a lubricant for precision machinery.

Beyond its primary use in oil extraction, the residual seed cake serves dual purposes as both a water clarifying agent by serving as a natural coagulant and a fertilizer (Ndabigengesere and Subba Narasiah, 1998). Its application as a flocculent in water treatment is particularly noteworthy, given the presence of positively charged compounds within the seed cake. These compounds exhibit remarkable efficacy in precipitating suspended solids from water, a process known as flocculation. This phenomenon exploits the fact that most particles in aqueous solution carry a net negative surface charge. The international community has shown keen interest in the utilization of moringa-based flocculants as a sustainable and locally-produced alternative to aluminum sulphate ~ a common agent employed for water clarification. This interest is fueled by the biodegradable nature of moringa-based solutions, aligning with global efforts to reduce environmental impact (Radovich, 2011).

Moreover, the seed cake proves valuable as an organic fertilizer, playing a crucial role in enhancing agricultural productivity (Emmanuel *et al.*, 2011). Despite its versatile applications, the seed cake is not commonly used as livestock feed. This limitation stems from the presence of antinutritional compounds in mature seeds, prompting caution in its inclusion as a dietary component for animals. Consequently, the dual role of the press cake in agricultural and water treatment contexts underscores its potential as a valuable and eco-friendly resource (Radovich, 2011).

4.11.3 Medicinal uses

Moringa, hailed for its versatile medicinal properties finds extensive application across various health domains. The longstanding medicinal legacy of moringa, enriched by ongoing scientific validation, has led to a surge in the promotion and marketing of supplements and so-called 'superfoods' derived from moringa. Most components of the plant serve medicinal purposes, with its paramount contribution lying in its exceptional nutritional value.

Its leaves and bark are commonly employed in direct medical practices often fashioned into poultices and applied directly to wounds. This utilization not only harnesses the plant's antimicrobial attributes but also accelerates the healing process. A notable aspect of its medicinal prowess lies in its well-documented anti-fungal and anti-bacterial properties. These properties are believed to emanate, at least in part, from the presence of 4-(α-L-rhamnopyranosyloxy) benzyl isothiocyanate. This compound exhibits a remarkable efficacy against *Helicobacter pylori* ~ a bacterial pathogen afflicting human populations in medically underserved areas and impoverished communities globally. Moringa thus emerges as a promising natural resource, offering tangible health benefits, particularly in combating prevalent bacterial challenges in vulnerable populations (Fahey, 2005).

The mild horseradish smell in moringa roots and bark can be attributed to isothiocyanates giving the tree one of its common names the 'horseradish tree'. Moringa belongs to the same order (Capprales) as horseradish and other members of the cabbage family belong to. Isothiocyanates and related compounds found in the cabbage family have demonstrated anti-tumour and anti-carcinogenic effects. Research conducted at Johns Hopkins University and other institutions support the traditional use of moringa in cancer treatment. This underscores the potential therapeutic benefits of moringa aligning with its historical applications in addressing cancer-related concerns (Fahey, 2005).

4.11.3.1 Nutraceutical and pharmacological profile of various parts

Neutraceutical value and pharmacological properties of moringa on different tissues:

Plant parts	Neutraceutical significance	Pharmacological properties
Leaves	Rich in protein (arginine: 0.38g; histidine: 0.14g; lysine; 0.32g; tyrosine: 0.1g; phenylalanine: 0.29g; methionine: 0.11g; cysteine: 0.13g; threonine: 0.25g; leucine: 0.46g; isoleucine: 0.28g and valine: 0.35g). It contains magnesium: 24mg; potassium: 259mg as well as carotene, iron and ascorbic acid.	Anti-ulcer, hyperthyroidism, anti-diabetic, hypolipidemic, anti-helminthic, anti-oxidant, hypo-cholesterolemic, hepatoprotective, antifungal, antibacterial, nutritional supplement, *etc.* Foliage used as lactation enhancer.
Flowers	Copper: 0.62mg; sulphur: 137mg; chlorine: 423mg; oxalic acid: 101mg and phytin P:44mg. Lysine is also found in flowers.	Anti-arthritic
Pods	Rich in amino acid (alanine, arginine, glycine, serine, threonine, valine, glutamic acid, aspartic acid) and lysine. Sucrose also occurs in the fruit.	Pods and seeds: Hypotensive

Plant parts	Neutraceutical significance	Pharmacological properties
Seed	Benzylisothiocyanate derivative and benzoisothiocyanate	Analgesic, anti-spasmodic, diuretic, anti-allergic, antibacterial, larvicidal and anti-viral. Seed kernels: Anti-asthmatic, anti-inflammatory.
Stem	Hydroxymellein, vanillin, octacosanoic acid, beta-sitosterol and betasitostenone which have been reported for the first time from a plant species.	
Stem bark	White crystalline alkaloid, two resins, an inorganic acid and mucilage (gum).	Anti-urolithiatic
Root	Anthomine and pterygospermine (an antibiotic). The root yields a very pungent essential oil, which is offensive in odour.	Analgesic, anti-convulsive, anti-nociceptive, antiinflammatory, anti-cancer and antiurolithiatic.
Root bark	Moringinine and spirochine along with benzylamine and glucotropacoline	

(Saha *et al.*, 2012)

4.11.3.2 Ayurvedic properties of *Moringa*

Taste (rasa)	***Pungent**/katu, tikta/**bitter***
Virya	Heating/*ushna*
Post	Digestive metabolic state (*vipak*): pungent/*katu*
Guna	Light/*laghu*, dry/*ruksha*, sharp/*tikshana*, fluid/*sara*
Prabhava	Liver cleanser (*yakrit sodhana*); purifies blood (*rakta sodhaka*); enhances spleen/*pliha*; removes worms (*krimi*), acidic toxins from the blood (*amavishagni*); relieves from tumour (*gulma*); strengthens heart/*hridya*, fat metabolism and weight loss/*medovishahara* and regulates cholesterol.

In 'Bhava Prakash' part I (a 16th Century canonical textbook of Ayurveda, authored by Bhav Mishra and Rajnigantu), moringa is called sigru, or 'it moves like an arrow' in the body because it rapidly penetrates the tissues and has deep absorption and detoxification ability, making its effect on the deep bone marrow tissue swift and effectiveness (Agarwal, 2023).

4.11.4 Other traditional and contemporary uses

Apart from being used as vegetables, fodder, medicines, gum, food spices and water purification flocculants, the species is also used for production of rayon and paper pulp (Jahn *et al.*, 1986; Nautiyal and Venkataraman, 1987; Babu and Rajasekaran, 1991; Caceres *et al.*, 1991; Mayer and Stelz, 1993). Moringa serves a multitude of traditional and contemporary purposes including yielding a blue dye used in Jamaica and Senegal from its wood, serving as live fence posts, promoting crop plant growth through the mechanism of leaf extracts (though the exact mechanism is unknown), providing wood pulp suitable for crafting newsprint and writing paper, and offering bark that can be beaten to produce fibre for paper (Bwana-Simba, 2006).

4.11.5 Environmental services

Moringa tree serves various purposes including its use in windbreaks, live fences and as a trellis or shade tree. For optimal windbreak functionality, it is advised to propagate trees from seeds instead of cuttings. This method encourages the development of a robust taproot crucial for stability. To enhance wind resistance, it is recommended to pinch off the apical stem when the tree reaches 1m in height. This practice stimulates lateral branching so fortifying the tree against strong winds (Radovich, 2011).

4.12 Ecological Significance

The robust resilience of moringa extends beyond mere drought resistance as its dense root system plays a pivotal role in safeguarding against soil erosion. Functioning as a natural defense mechanism, these roots exert a cohesive force that binds soil particles together, effectively countering the threat of erosion. This is particularly crucial in regions characterized by loose or sandy soils where the risk of erosion is quite high.

Beyond erosion prevention, moringa tree actively contributes to enhancing soil fertility which further solidifies its ecological significance. The fallen leaves and root exudates release valuable organic matter into the soil which enriches its nutrient content. This dual benefit not only fortifies the landscape against erosion but also nurtures the soil promoting a sustainable and thriving ecosystem. In essence, the protective prowess of its root system coupled with its soil-enriching contributions, underscores the profound ecological importance of these trees in fostering and maintaining healthy and resilient ecosystems.

4.13 Toxicity

While human toxicity data regarding *M. oleifera* is limited, laboratory studies suggest that specific compounds found in its bark, roots or extracts could lead to adverse effects when consumed excessively (MSKCC, 2023). Supplementation using *M. oleifera* leaf extract might pose a risk of toxicity at doses above 3,000mg/kg of body weight, yet remains safe below 1,000mg/kg (Asare and Nyarko, 2012). *M. oleifera* could potentially interfere with prescription drugs that impact cytochrome P450, notably CYP3A4 and might hinder the antihyperglycemic effects of sitagliptin (MSKCC, 2023).

4.14 Cultivation

Moringa is a fast-growing drought-resistant tree that can reach 3 metres in just the first year. Moringa trees have a deep taproot system which means they need lots of space to stretch out their roots in the soil. The trees generally prefer loamy or sandy soils with a neutral pH. These will need to have full sun exposure year-round. Since the plant is native to subtropical and semi-arid regions, it can only tolerate light frost. Regular weather below 7.2^{0}C can be detrimental to these trees. Pruning moringa trees is an absolute necessity. Moringa can be propagated vegetatively through cutting and layering and sexually through seeds. Moringa is a deciduous tree, so it usually loses its leaves in the fall. Some areas with mild winters may experience leaves year-round.

4.14.1 Propagation

Moringa trees can be readily propagated using diverse methods such as seeds, stem cuttings (limb cutting) (Leone *et al.*, 2015a) and air layering. In India, direct-seeding emerges as the predominant method employed for commercial production. If direct-seeding is not used, 1–2 month old seedlings (about 30cm tall, 0.75cm in diameter) or well-rooted cuttings are transplanted into well cultivated soil. Size of transplants generated by cuttings is not important but the root system should be well developed. If grown in heavy soils, raised beds may be used to improve drainage. Moringa cross-pollinates readily (highly out-crossing) and variability within seedling populations can be high if multiple types are planted together. Cuttings should be used to propagate preferred individuals when other moringa trees are nearby (Radovich, 2011).

4.14.1.1 Propagation of annual types

Annual types are generally propagated by seeds (Eagri, 2023a, b). Usually the weight of single seed is 0.288g and the 10g seeds will contain some 35 seeds. Seeds exhibit no dormancy so these can be sown immediately upon extraction (Ramesh *et al.*, 2014). Some 625g seeds are required for planting an orchard of one hectare. However, before sowing the seeds should be treated with *Azospirillum* cultures @100g/625g of seeds as seeds require several years to fruit. This treatment helps early germination, *vis-a-vis* increased vigour in seedlings, their growth and yields (TNAU, 2023). These may directly be sown in the prepared pits or transplanted after raising seedlings in the polythene bags when these become about one month old, however some additional seedlings should also be raised together for the purpose of gap filling (Eagri, 2023a, b). Fully mature seeds with a hard outer shell are planted at a depth of about 2.5-5cm directly in well-draining soil either in pots or directly in prepared ground. The soil must be kept consistently moist but not waterlogged. Germination occurs in 1-2 weeks (Radovich, 2011).

4.14.1.2 Propagation of perennial types

Perennial types are usually propagated through limb cuttings (Eagri, 2023a, b) as seeds take several years to fruit. The cuttings some 1-1.5m long and 4-16cm in diameter must be taken from at least 1-year-old wood and after removing the lower leaves cuttings should be rooted in 30-60cm deep, loose and well-drained medium or sandy soil (Radovich, 2011). The soil should be kept consistently moist but not waterlogged. Roots develop within a few weeks.

4.14.1.3 Air layering

Air layering is a propagation technique employed to cultivate new plants from existing ones and it can be successfully applied even to *Moringa oleifera* trees. This method entails inducing the formation of a root system on a branch while it remains attached to the parent plant, allowing it to evolve into an independent plant once roots are established. The process involves selecting a healthy and mature branch, making a horizontal cut near a leaf node and removing a ring of bark to expose the inner wood. For good results, rooting hormone such as Rootex or indole butyric acid can be applied to expedite better root development. The exposed area is then packed with moist sphagnum moss or a similar rooting medium, covered with clear plastic wrap and sealed with string or twist ties to create an airtight environment. Aluminum foil is wrapped around the plastic-covered section to shield it from sunlight, maintaining a consistently dark and humid atmosphere conducive to root development. Regular monitoring and watering of the moss are essential, and after a few weeks to a couple of months, roots should begin to form.

Once a sufficient root system is established, the branch is cut just below the rooted section, and the rooted portion can be planted in a pot or directly in the ground. Providing proper care including regular watering, sunlight and nutrient-rich soil ensures the successful development of the newly separated plant. Air layering proves to be an effective method for propagating moringa trees, particularly useful for cloning specific varieties or when traditional propagation methods pose challenges.

4.14.2 Climate

Moringa thrives in sun-drenched environments and is well-suited for tropical and subtropical regions. Despite its aversion to freezing temperatures and frost, mature trees can endure brief exposures to 0°C, albeit at the expense of new growth. This versatile plant can flourish at elevations up to 2000m a.s.l. accommodating a broad spectrum of daily temperatures ranging from 15-35°C coupled with an annual rainfall spanning 250-4000mm. Moringa is particularly suitable for dry regions as it can be grown using rainwater without expensive irrigation techniques (Radovich, 2011; Eagri, 2023a, b; Wikipedia, 2023).

For optimal leaf and pod production, an elevation below 600m is ideal, demanding consistently high average daily temperatures of 25-30°C, well-distributed annual rainfall between 1000-2000mm, abundant solar radiation and well-drained soils. Growth experiences a notable deceleration when temperatures dip below 20°C. Heavy flower shedding occurs at high temperatures above 35°C, however, for optimum fruit production, a moderate day temperatures ranging between 30 to 35°C have been found ideal under Tamil Nadu conditions though temperatures above 40°C all the flowers tend to dry up and fall off so production is drastically reduced. Also when night temperature drops below 20°C the trees become vulnerable to fungal infection. In December in Tamil Nadu the plants become infected to powdery mildew. However, it is also observed that during rainy season yellow vein mosaic virus poses a serious threat to leaf production. Moringa is finely attuned to monsoon rainfall patterns, showcasing resilience against prolonged periods of drought in mature trees, though leaf production may be compromised during such arid spells (Radovich, 2011; Eagri, 2023a, b; Wikipedia, 2023). In Tamil Nadu, it was found that before harvesting of the leaves if there is heavy rainfall chlorophyll drainage occurs which reduces the market value.

These findings as reported by the National Horticultural Research and Development Foundation (NHRDF) in 2023, underscore the importance of understanding and managing these climatic factors to ensure a robust and thriving agricultural sector in the region.

4.14.3 Soil

While moringa exhibits resilience across various soil types including sandy, sandy-loam and loamy with a pH range of 4.5–9 (Wikipedia, 2023; Radovich, 2011), it thrives best in well-drained sandy loam soils rich in organic matter with a neutral pH (6-6.7) (Radovich, 2011; Eagri, 2023a, b). Despite its capacity to endure heavy clayey soils, it flourishes when these soils are prevented from excessive moisture. Light (sandy) soils are preferable when rooting branch cuttings directly in the ground (Radovich, 2011).

4.14.4 Season

Time of sowing of seeds for annual seed drumstick or planting of limb cuttings varies from region to region depending on receipt of monsoon (Eagri, 2023a, b). The stem cuttings are planted in pits in the months of June to August, so that the monsoon rains during the period facilitate easy rooting and further growth. The best suited season for sowing the seeds is September under Southern Indian conditions. The time of sowing has to be strictly adhered to because flowering phase should not coincide with monsoon rains which may result heavy flower shedding (TNAU, 2023).

4.14.5 Field preparation and planting

Soil erosion poses a significant challenge in tropical agriculture, underscoring the need for a delicate approach to soil treatment. Ploughing becomes essential primarily in scenarios marked by high planting density. However, for optimal root system penetration without exacerbating land erosion, a preferable alternative is the creation of pits filled with soil in moderate planting densities. Ideal dimensions for such pits are kept 20-40cm in width and 30-50cm in depth (Wikipedia, 2023). To optimize leaf production, the plants are spaced at 0.75 metre within rows and 1 metre between rows. For larger lots nearing 0.5 hectares, a wider spacing of 0.75×1m is more viable for leaf production as well as pod production. Only for pod production, the spacing maintained is 2.5×2.5m (Radovich, 2011) accommodating some 1600 plants per hectare. In case of single row planting alongside irrigation channels, the 2m spacing is deemed sufficient (TNAU, 2023).

Moringa leaves can be intensively cultivated in designated beds, spacing seeds at 10cm×10cm intervals akin to the cultivation method of an annual leafy vegetable. This focused leaf production technique within beds results in a remarkably high leaf yield within a confined area, presenting an optimal commercial application for home gardeners. However, due to substantial seed requirement, this intensive method might not be viable for larger-scale cultivation (Radovich, 2011).

4.14.6 Polyculture

Relatively open canopy of moringa lends itself well to intercropping. In small subsistence holdings, moringa trees are often interplanted with cassava, pumpkin and other vegetables. In commercial production, moringa is most often grown as a monocrop. For perennial polycultures, it is advisable to maintain spacings of at least 2×2m or 2×3m, taking into account factors such as pruning frequency of the trees, the shade tolerance and other specific requirements of the companion crops. Wider spacing may benefit pod production but will reduce production of moringa leaves per unit area. Wider spacing promotes pod production through precocious branching and flowering. In most situations, a spacing of 5×5m for largest trees is generally deemed appropriate (Radovich, 2011).

4.14.7 Manures and fertilizers

Moringa exhibits notable resilience to drought and nutrient-poor soils while displaying heightened productivity in response to adequate irrigation and fertilization. The crop requires 44:16:30g NPK/ tree at the time of pinching (75 days after sowing). Nitrogen @44g/tree must be applied as top dressing at first flowering stage, *i.e.* 150-160 days after sowing (Suthanthirapandian *et al.*, 1989).

To optimize the growth, it is advised to incorporate 300 grammes of complete fertilizer or 0.5–2 kilograms of manure per tree during planting. Significant improvements in yield have been observed with nitrogen fertilization rates reaching as high as 350kg per hectare. Trees demonstrate notable benefits when subjected to integrated fertilization methods combining organic and synthetic sources both. Subsequent fertilizer applications following coppicing are strongly recommended (Radovich, 2011).

4.14.8 Irrigation

Growing moringa plants may not require watering except during hot weather when they may be irrigated once a week. Annual moringa responds well to irrigation and the yield can be doubled by drip irrigation as compared to rain-fed crops. Drip irrigation at the rate of 4 litres/day can enhance yields by 57 per cent as compared to rainfed crop (Rajakrishnamoorthy *et al.*, 1994).

4.14.9 Growth regulator

Spray application of GA_3 @20ppm on 90th day of sowing increases all the pod characters such as length, specific gravity, number of seeds, flesh content and pulp more than untreated check (Ramesh *et al.*, 2014).

4.14.10 Intercultural operations

4.14.10.1 Pinching

Pinching of terminal bud on the central leader stem when it attains a height of 75cm will promote the growth of many lateral branches and reduce the height of the tree. It has been found that early pinching of growing tips carried out 60 days after sowing is better than pinching 90 days after sowing for obtaining a higher yield (Ramesh *et al.*, 2014). To enhance branching, boost production and to streamline harvesting it is advisable to employ methods such as pollarding, coppicing as well as lopping or pruning (Grubben, 2004). When harvesting leaves from mature trees, it is advisable to regularly trim branches, encouraging the development of fresh foliage (Radovich, 2011).

4.14.10.2 Weeding

Weeding is a crucial aspect of ensuring healthy plant growth and optimal yields. Moringa is known for its fast growth *cum* resilience and benefits significantly from a weed-free environment. Weeding requirements involve several key considerations to support development of plants. Since moringa plants have shallow root system so it is quite susceptible to weed competition as weeds grow quickly, rob essential nutrients, water and sunlight on the cost of main crop and work as host plants for certain insect-pests and diseases. Therefore regular and timely weeding is essential, especially during the initial growth stages. Manual or mechanised weeding is often employed to remove weeds without disturbing the fragile roots of main crop. Since moringa is grown under diverse climates and soil types so its weed types may vary. Understanding the local weed species and their growth patterns is crucial for effective management. Cultural practices such as mulching or cover cropping can also help suppress weed growth and maintain soil moisture which reduces the need for frequent weeding. Moreover, frequent and careful monitoring of the moringa plantation is essential to identify and address weed infestations promptly. Early intervention through regular inspection allows for timely weed removal so preventing them from establishing a stronghold in the plantation. Employing organic methods for weed control aligns well with principles of sustainable agriculture in moringa cultivation. Practices such as hand weeding, mulching with organic materials or using natural herbicides derived from plant extracts can effectively manage weeds without harming the moringa plants or the surrounding environment.

4.14.11 Harvesting and yield

Harvesting of leaves typically commences 75 days after initial sowing. Once ready, the leaves are delicately plucked from the branches, carefully packed into bags and subsequently transported to the market. The harvest process of pods involves manual techniques using knives, sickles and hooks with attached stabs. The timeline for harvesting green fruits varies with initiation occurring around 210 days (7 months) post-planting though dry fruit harvesting for seed collection occurs usually 252 days ($8^1/_2$ months) after planting as influenced by growing conditions and specific plant varieties.

Yields of moringa can significantly differ based on various factors such as climatic conditions of the area, variety planted, manuring *cum* fertilization and irrigation practices. For optimal results, moringa tends to thrive in warm dry conditions experiencing minimum and maximum temperature range of 22-35°C. It shows improved yields when supported by supplementary fertilizer and irrigation. Annual **leaf yield** on fresh weight basis ranges from 1-5kg/tree. This translates to an equivalent of 10,000 to 50,000kg per hectare per year with a planting spacing of 1×1m. It is worth noting that commercial leaf yield of moringa tends to decline during November to March, reportedly reaching 50 to 100% of the summer yields on the west coast of O'ahu, Hawai'i. This reduction in productivity is believed to be linked to lower solar radiation levels during winter months (Radovich, 2011). At 10×10cm spacing leaf yields are 7–8kg/m2 at the first cutting in well-drained irrigated and fertilized beds with up to seven cuttings a year (Radovich, 2011).

Pod yield in India reaches 19kg per tree per year which translates to an equivalent of 31,000kg per hectare annually with a spacing of 2.5m×2.5m, *i.e.* an average of 230 pods per tree, each pod weighing approximately 80–90g (Radovich, 2011). In Hawai'i, the Indian cultivar 'PKM-2' demonstrated 3-8 times more immature pod yield compared to local accessions within 6 months of transplanting the seedlings (Radovich and Paull 2008). When propagated from cuttings, the plant typically undergoes its initial pod harvest 6–8 months after. Frequently, fruit production may not occur in the first year and the yield may tend to be modest during the initial year but by second year the plant typically yields approximately 300 pods which increases 400–500 in the third year. A robust tree has the potential to produce 1,000 or more pods in a year (Booth and Wickens, 1988). In India, the agricultural yield of pods per hectare is estimated at 31 tonnes annually (Radovich, 2011). In North Indian regions, the fruit ripens primarily during summer as from October to March the temperatures may not be congenial though in South India there is occasional dual flowering and fruiting which may result into two harvest periods ~ one from July to September and another from March to April (Ramachandran *et al.*, 1980). Likewise, moringa **seed yield** is roughly 250kg/ha/year (Ramesh *et al.*, 2014). Moringa trees in Hawai'i yield approximately 3g of kernels per dry pod. The annual **oil production** per hectare can be estimated using this information. Assuming a relatively high oil yield of 20% by weight from the kernels with the specific gravity of 0.9 for oil, it can be anticipated that the trees might yield around 250 liters of oil per hectare (Radovich, 2011).

Polyculture crop yields generally tend to be lower for individual crops though overall production of the system often surpasses that of monocultures. For moringa, as the spacing between trees increases to accommodate companion crops, one can anticipate reduced leaf yields. However, opting for wider spacing such as 2-2.5×2-2.5m has the potential to enhance pod yields compared to densely planted crops (Radovich, 2011).

4.14.12 Postharvest handling and processing of moringa products

Fresh moringa products prized for their nutritional values are highly perishable and require careful postharvest handling *cum* processing to maximize their shelf life and value as elaborated below by Radovich (2011):

- Fresh moringa products especially the leaves should be consumed within 2 days of harvest, preferably within a day. Leaves being highly perishable should be stored in plastic bags under cool temperature of 10°C and 90-95% humidity so that wilting and leaflet abscission are avoided.

- Moringa fruits can be canned to extend their shelf life for later consumption.
- Leaves and flowers can be shade-dried or dehydrated and then pounded or ground for use as food additives to supplement protein and can also be used as tea.
- Converted into products like Zija™ which contains 30g of moringa leaf, pod and seed but due to high cost it is challenging in its retail sale though it can be prepared into local drinks and teas.
- Seed kernels of moringa yield precious edible oil containing 60-75% oleic acid which is valued for culinary and cosmetic applications as the oil is comparable to olive oil for cooking. Its rancidity action is quite slow due to its antioxidant contents. Oil is extracted either by grinding, boiling toasted seeds and then oil extracted through expellers. However, 'Komet Press' extracts 6.5 litres in 8 hour duration with 12% yield. Oil yield can be increased through 'Ram and Screw presses' along with dehulling which gives increased yield.

4.15 Physiological disorders and nutritional deficiencies

Physiological disorders affect plant growth and production drastically. These disorders often stem from environmental factors or imbalances in their nutition as defined here under:

i) **Leaf drop:** This occurs due to environmental stress such as excessive moisture, drought and/or sudden change in temperatures.

ii) **Stunted growth:** This occurs due to poor soil quality, inadequate nutrition, root compactness and overcrowding. Here root growth becomes utterly slow, leaves smaller and maturity delayed.

iii) **Fruit drop:** It is caused due to extremes of temperature and/or humidity, nutrient imbalance and water stress. Since fruits start falling prematurely so yield is negatively impacted.

iv) **Twisted or deformed growth:** This occurs at early stage of growth due to environmental stresses, *vis-a-vis* attack of diseases and/or pests. Its impact is seen as twisting of stems, irregular growth pattern and leaf deformity.

v) **Chlorosis/yellowing**: Leaf yellowing occurs due to nutrient deficiencies, soil alkalinity and/or improper soil pH levels. In such cases yellowing starts from the veins and spreads outward and in severe cases leaves may turn pale or white. Some common nutritional deficiencies are furnished hereunder:

- **N-deficiency** causes reduced and stunted growth and pale green to yellowing of leaves.
- **P-deficiency** causes development of purplish tint starting from tips and edges, slow plant growth and leaf deformity.
- **K-deficiency** causes yellowing or browning of leaf edges often progressing inward, along with weak stems and reduced fruiting.
- **Fe-deficiency** causes interveinal chlorosis (yellowing) though veins remain green, first appearing on younger leaves.
- **Mg-deficiency** causes interveinal yellowing while veins remaining green. In severe cases, leaves may develop reddish-purple colouration.
- **Ca-deficiency** is expressed in terms of new growth turning distorted and stunted with tips and margins of leaves showing necrosis. Young leaves may exhibit irregular shapes.

Soil testing is must to identify nutrient deficiencies accurately and to address remedies as following:

i) **Balanced fertilizers** should be applied or supplemented in the deficient soil or on the plants containing deficient nutrient(s) which will help restoring proper nutrient levels in the soil.

ii) **Foliar sprayings** should be followed directly on the leaves in case of any specific nutrient deficiency for rapid correction.

However, incorporation of sufficient **organic matters** into the soil at the time of field preparation will not allow deficiencies of any such nutrients. At the time of appearance of such symptoms the deficient organic matter should be applied to address any of such problems. Since symptoms can overlap or be indicative of multiple deficiencies, it is crucial to identify specific deficiency accurately through soil tests, by consulting the experts and through adjustment of soil pH accordingly through proper soil amendments, ensuring consistent and adequate watering, avoiding waterlogging as well as drought stress. The plants are also required to be protected from winds, extreme temperatures, creating sufficient spacing among the plants as well as inside the plants for proper aeration through pruning, and for conserving moisture, *vis-à-vis* regulating proper soil temperature and improving soil structure the beds are required to be mulched after planting

4.16 Insect-pests and diseases

4.16.1 Insect-pests

Cultivating moringa in kitchen gardens is often avoided due to persistent threat of caterpillar infestation. The presence of caterpillars is a common observation in the house-vicinity therefore people are discouraged growing moringa near their residence or in the kitchen garden.

Bark-eating caterpillar (*Indarbela quadrinotata* WLK.), hairy caterpillar (*Eupterote mollifera* WLK.), greenleaf caterpillar/defoliator caterpillar (*Noorda blitealis* WLK.) and budworm (*N. moringae* TAMS) infest on the crop. The impact of these pests often leads to significant defoliation so hampering overall health of moringa plants. Other lepidopterous larvae such as *Tetragonia siva*, *Metanastia hyrtaca*, *Helicoverpa armigera* and *Helopeltis antonii* also feed voraciously on moringa plants. The aphid (*Aphis caraccivora*), scale insects (*Ceroplastodes cajani* & *Diaspidotus* sp.), stem borers (*Indarbela tetraonis* Moore and *Diaxenopsis apomecynoides*) and a fruitfly (*Gitonia* sp.) contribute to the array of pests that threaten moringa cultivation (Kareem *et al.*, 1974; Verma and Khurana, 1974; Pillai *et al.*, 1979; Ramachandran *et al.*, 1980; Butani and Verma, 1981; Morton, 1991). Apart from these the crop is infested with certain other insects but these are of minor importance. In the Philippines, hairy caterpillar (*Streblote siva*) has earned notoriety as a significant pest. This voracious insect doesn't limit its impact to a single plant species. It poses a threat to various plants including *Acacia arabica*, cashew (*Anacardium occidentale*), jujube (*Zizyphus jujuba*), *Prosopis juliflora* and moringa. During early stages of development, the larvae exhibit a penchant for feasting on leaves, often congregating on tender shoots. As they mature, older larvae tend to congregate at the base of the main trunk as documented by Joshi *et al.* (2016). In Puerto Rico, moringa trees face notable vulnerability to termite attacks, further complicating efforts to maintain a healthy crop. Seed predation by an unidentified insect is a recurring issue, posing an additional challenge for those attempting to cultivate moringa.

Moderate use of nitrogen fertilization and avoiding waterlogged clay soils prevent most problems. Neem oil, horticultural soap and sulphur are low impact pesticides that can control mites, aphids and other pests. Though there are a few reports on variability in pest tolerance within moringa germplasm, local varieties are most likely to be best adapted for local conditions and should be included in new plantings.

4.16.1.1 Major pests

4.16.1.1.1 Fruit/pod fly (*Gitona distigma*): Moringa crop in South India is severely threatened by a formidable fruit fly with peak activity recorded from April to October. It is a diminutive yellowish fly with striking red eyes, the adult exhibiting wings that extend beyond its body and featuring a distinctive dark spot near the coastal margin. Its female lays cigar-shaped, sculptured and white-coloured eggs either individually or in clusters of 3-4 within the grooves of tender pods. The egg stage spans 3-4 days, followed by the maggot period lasting 18-25 days. Fully developed cream-coloured maggots pupate in the soil for 5-9 days (Eagri, 2023a, b). The maggots infiltrate tender fruits by making small holes at the terminal end and releasing the gummy exudation from the fruits which leads drying of the fruit from the tip. Thus the fruit undergoes internal rotting harbouring 20-28 maggots per fruit (Eagri, 2023a, b). These can be trapped through use of attractants such as citronella oil, eucalyptus oil, vinegar (acetic acid), dextrose or lactic acid. Regular collection and disposal of fallen and damaged fruits into a pit which is covered with a thick layer of soil is the best way to eliminate these. Soil maintenance practices including raking or ploughing to disrupt puparia prove beneficial. Chemical interventions include application of 0.04% Fenthion 80EC during the vegetative and flowering stage. The crop may be drenched with 5% NSKE at 2 litres/tree when the fruit set reaches 50%. Spraying 500ml of dichlorvos 76SC or 750ml of malathion 50EC in 500-750 litres of water per hectare when pods are 20-30 days old is quite effective. For further control, azadirachtin 0.03% or nimbecidine 0.03% can be applied at 50% fruit set and repeated 35 days later (Anjaneya Murthy and Reghupathy, 1992; Eagri, 2023a, b).

4.16.1.1.2 *Moringa* budworm (*Noorda moringae*): This is a major pest posing significant threat to moringa cultivation in South India. Its adult is characterized by its small size, featuring dark brown forewings and white hind wings with a dark brown border. Caterpillars exhibit a dirty brown hue with a distinct mid-dorsal stripe, accompanied by a black head and pro-thoracic shield. Once fully fed, the caterpillars emerge to pupate within minute brownish cocoons either in the soil or on the ground beneath dried leaves and debris. Entire life cycle including eggs, larvae and pupae last about 3-4, 8-16 and 6-10 days, respectively. Larvae inflict considerable damage by boring into flower buds so there is about 78% bud shedding. It remains very active during summer months in South India (Vikaspedia, 2023; Eagri, 2023a, b). To manage this pest, damaged buds along with caterpillars should be collected and destroyed, ploughing around the trees should be done to expose and eliminate the pupae and 1-2 light traps per hectare should be used to attract and kill the adults. Spraying of insecticides such as carbaryl (50 WP) at 1kg or malathion at 1 litre in 500-750 litres of water per hectare is advised for effective control (Eagri, 2023a, b).

4.16.1.1.3 Moringa/green leaf caterpillar (*Noorda blitealis*): Drumstick tree faces sporadic yet significant threat from this caterpillar, especially in South India. Medium-sized adult moths have uniformly dark forewings adorned with a small white streak near the base having hyaline hind wings accompanied with a broad black marginal band towards the anal side. Adult female lays eggs in batches on ventral surface of leaves. The duration of the egg, larvae and pupal stages last for 3, 7-15 and 6-9 days, respectively with pupation occurring in the soil (Eagri, 2023a, b). Its caterpillars feed on the leaf lamina in a manner that it transforms merely into a parchment-like structure. The peak infestation period is during March-April and December-January (Eagri, 2023a, b). It can be controlled by spraying with 0.2% Malathion.

4.16.1.1.4 Moringa hairy caterpillar (*Eupterote molifera*): Hairy caterpillar poses a significant threat to drumstick cultivation in South India. Its large adult moth has light yellowish-brown wings adorned with faint lines. Female moth lays clusters of eggs on tender plant parts with an incubation period of 5-7 days. Resulting caterpillar is dirty brown in colour, featuring whitish hairs in tufts on small warts which are not only unsightly but also poisonous and irritating. The larval and pupal stages span 12-14 weeks and 8-10 weeks, respectively, culminating in pupation in the soil within an earthen cocoon. The damage inflicted by these caterpillars is noticeable as they cluster on plant stems during hot daylight hours, becoming active at night. Their feeding behaviour involves scraping bark and gnawing foliage, leading to rapid defoliation of the tree (Eagri, 2023a, b; Vikaspedia, 2023). Its egg masses and caterpillars should promptly be collected and killed. Immediately after rains the light traps should be employed to attract and eliminate adults while a burning torch is effective in eliminating congregating larvae on the trunk. Chlorpyriphos 20EC or quinalphos 25EC at 0.2% should be sprayed on infested trees. Fish oil rosin soap solution at 2.5% can be sprayed on trunks and foliage immediately after rain and again 15 days later (Eagri, 2023a, b).

4.16.1.1.5 Wooly bear (*Olepa ricini*): This is known to attack a variety of crops including drumstick, banana, black gram, cotton, cucurbits, castor, cowpea, soybean, tea and yam. Robust larvae of this pest display a greyish black or blackish brown colouration with a distinctive red head and a thick tuft of hairs emerging from its body. Its adult is typically greyish brown or black featuring black spots on its wings while its hind wings exhibit a pink or red hue with additional black spots. Caterpillar feeds voraciously so defoliating entire tree. These caterpillars should be collected and eliminated. Dusting the infested crop with parathion 2D/malathion 5D or carbaryl 10D at 20-25kg/ha is quite effective to eradicate this pest.

4.16.1.1.6 Gristly citrus caterpillar (*Metanastria hyrtaca*): The caterpillar is prevalent throughout the Indian sub-continent which is a polyphagous pest with a particular affinity for various *Citrus* species and *Moringa*. Its stout caterpillars are cylindrical exhibiting a greyish-brown hue and are accompanied by a dense covering of hair. Its greyish-brown moths showcase sexual dimorphism ~ males exhibiting a chocolate-brown patch in the middle of their forewings and pectinate antennae. Its entire life cycle duration is 75-110 days. Their nocturnal caterpillars feed gregariously and voraciously though during daytime these cluster on the shaded side of tree trunks. Application of 0.2% chlorpyriphos 20EC or quinalphos 25EC should be sprayed on infested trees. Fish oil rosin soap solution at 2.5% can be sprayed on trunks and foliage immediately after rain and again 15 days later (Eagri, 2023a, b).

4.16.1.1.7 Jujube lappet moth (*Streblote siva*): This moth is widely distributed across Indian subcontinent and is known to infest drumstick and roses. Mature caterpillars display a pale ochreous-brown hue adorned with small black spots and extended lateral tufts of ochreous hair. The moth exhibits a greyish-white head and thorax, complemented by a whitish abdomen. Its forewings boast a captivating colouration, featuring a reddish-brown spot encircled by white. The hind wings are predominantly white with a subtle fuscous tint on the outer margin. Though its attack is not so serious as to that of roses but sometimes it becomes a problematic pest of leaves and pods. Chlorpyriphos 20EC 0.2% spraying will control this pest.

4.16.1.1.8 Bark-eating caterpillar (*Inderbela tetraonis* & *I. quadrrinotata*): This caterpillar poses a significant threat to various fruit-bearing trees including *Moringa*, *Zizyphus*, *Eugenia*, mango, litchi, guava, loquat, mulberry, pomegranate, orange, bauhinia, rose, *etc*. Its presence is widespread throughout India, Myanmar, Bangladesh and Sri Lanka. Its pale-brown adult moth is characterized by cream-white fore- and hind-wings alongwith brown spots and streaks on the forewings. Adult moths emerge during summer and lay 15-25 clustered eggs beneath loose bark on the tree. The eggs hatch in 8-10 days giving rise to larvae that feed on bark creating webs and forming zigzag galleries on the wood filled with frass and excreta. Subsequently, the larvae bore into the wood where these remain hidden during daytime but at night feed on the bark. The larval period spans 9-11 months, followed by pupation inside the stem for 3-4 months. In young trees its damage is severe as these bore into the trunk on branch junctions creating zigzag galleries which are filled with silk and frass. Its infestation stunts plant growth, hinders sap flow and reduces fruit formation. Caterpillars can be killed through inserting iron spikes into their tunnels and syringing a mixture containing 1 part ethylene glycol + 3 parts kerosene oil and sealing the opening with mud. Fumigants such as cotton-soaked chloroform, kerosene or petrol can be introduced into the tunnels and openings sealed with mud.

4.16.1.1.9 Long horn beetles (*Batocera rubus*): This beetle is widely distributed across Indian subcontinent. Its eggs are laid individually in cracks or crevices on the tree bark. Stout grubs some 10cm long display yellowish colouration with distinct segmentation. Pupation occurs within the tunnels. Yellowish-brown beetles are medium-sized with white spots on their elytra.

The egg, grub and pupal periods endure for 1-2 weeks, 24-28 weeks and 12-24 weeks, respectively. This species follows a single generation per year. While feeding the grubs creat zigzag burrows beneath the bark to feed on internal tissues & sapwood and causing death of affected branch or stem. Adults feed on the bark of young twigs and petioles. This pest is managed through cleaning all the webbed materials and excreta. In each hole cottonwool soaked with 0.5% monocrotophos 36WSC or any other effective fumigant such as carbon disulphide, carbon tetrachloride, chloroform or even petrol is inserted and the hole sealed with mud.

4.16.1.1.10 Aphids (*Aphis gossypii*): Nymphs and adults of *Aphis gossypii* are polyphagous pest posing a significant threat by rasping sap from twigs and other tender parts such as leaves, shoots, flowers and pods. Being parthenogenic, its population increases quite rapidly. To mitigate its impact, all the pods should be harvested and then dimethoate 30EC or malathion 0.2% is sprayed on full canopy of the trees for its control.

4.16.1.1.11 Scale Insects (*Ceroplastodes cajani*): While each individual scale insect extracts only a small amount of sap throughout its lifespan but cumulative impact of a vast population constantly feeding on the sap can weaken trees over time, ultimately influencing the size of pods. It also requires the same treatment as in case of aphid control.

4.16.1.1.12 Bud midge (*Stictodiplosis moringae*): Bud midge is identified as a minor pest affecting drumstick plants. Female midges deposit their clustered eggs inside the floral buds on anthers. Maggots start feeding on the internal tissues of the buds with particular focus on ovaries. The activity of this pest is prevalent from August to January. Infested buds exhibit premature falling. Fully-fed maggots emerge to pupate in the soil. Life cycle of the bud midge consists of three main phases ~ egg stage lasting 1-2 days, maggot stage 6-9 days and pupal stage extending for 5-8 days. This is controlled as to that of aphids.

4.16.1.1.13 Leaf eating weevils (*Myllocerus maculosus*): The weevil is known to consume a diverse range of crops. The reproductive process begins with the deposition of eggs in the soil. The grubs after emergence start feeding on the roots of cultivated crops and grasses. This undergoes pupation in the soil but upon reaching adulthood, these beetles emerge from the soil and engage in leaf nibbling resulting in minor damage to the vegetation. Using carbofuran in the soil @ 1g/m^2 immediately followed by watering will kill all the stages of weevils.

Indarbela quadrinotata Infested bark

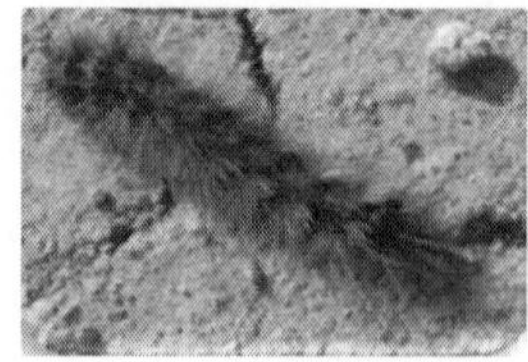

Larva - *Eupterote molifera*

Gitona distigma

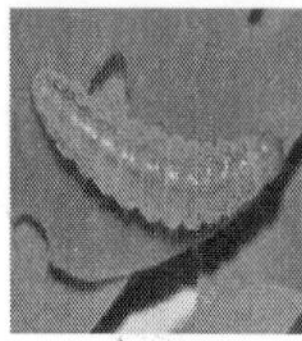

Noorda blitealis -Larva and adult

Bud worm-*Noorda moringae*

4.16.2 Diseases

While no severe afflictions have been noted, there have been reports of certain diseases causing minor harm to trees in southern India. Among these are root rot attributed to *Diplodia* sp. (Ramachandran *et al.*, 1980) and a fruit rot caused by *Cochliobolus hawaiiensis* (Kshirsagar and D'souza, 1989). Drumstick tree serves as a collateral host for *Leveillula taurica* ~ a powdery mildew known to inflict substantial damage to papaya (*Carica papaya*) nurseries in South India (Ullasa and Rawal, 1984). Preventive measures like proper sanitation, adequate spacing and avoiding waterlogged condition can significantly reduce the risk of *Moringa* diseases. Using disease-resistant varieties, implementing crop rotation or treatments as required can help manage these diseases. Regular monitoring and prompt action upon disease detection are keys to preserving the health. Consulting with local agricultural experts can provide specific guidance tailored to grower's conditions. Some of its other common diseases are damping off (*Pythium aphanidermatum*, *P. debaryanum* & *Rhizoctonia solani*), powdery mildew (*Leveillula taurica*), twig canker (*Fusarium pallidoroseum*), root rot (*Diplodia* sp.) and fruit rot (*Cochliobolus hawaiiensis*). Leaf spot may be caused due to fungal or bacterial infections that cause circular or irregularly shaped spots on leaves which may enlarge and merge affecting photosynthesis of plants. Twig canker is caused due to infection of *Fusarium pallidoroseum* which affects branches and stems leading to wilting. Root rot is caused due to waterlogging or through various fungi and due to this there is a decline in plant vigour, discolouration and root decay. Certain bacterial as well as fungal infections can cause sudden wilting or browning of leaves with ultimate plant death in severe cases. Rust appears as rusty spots on leaves which causes defoliation and reduced plant vigour. The diseases of special importance are:

4.16.2.1 Damping off (*Pythium aphanidermatum, P. debaryanum* and *Rhizoctonia solani*): Damping off disease in *Moringa oleifera* primarily stems from soil-borne pathogens, notably *Pythium, Phytophthora, Rhizoctonia* and *Fusarium*. These pathogens thrive in overly moist conditions, poor drainage or contaminated soil attacking the seeds or seedlings of *Moringa*. Excessive watering, high humidity and inadequate ventilation worsen the problem, creating an ideal environment for fungi to spread. Contaminated tools, containers or soil also introduce these pathogens and thus initiating the onset of the disease. The survival and spread mechanisms vary for different pathogens. *Pythium* relies on oospores in soil as its primary mode while *Rhizoctonia* uses sclerotia in the soil. *Pythium* spreads through zoospores *via* irrigation water whereas *Rhizoctonia* spreads through mycelial growth in soil and sclerotia *via* irrigation water. Certain conditions favour the growth of these pathogens. *Pythium* thrives in heavy rainfall, excessive and frequent irrigation, poorly drained soil and close-spaced cropping, particularly in high soil moisture at temperatures around 25-30°C. On the other hand, *Rhizoctonia* prefers high soil moisture at temperatures around 30–35°C. Identifying symptoms early is crucial. Damping off affects nursery beds and young seedlings causing reduced seed germination and poor seedling stand, often resulting in high seedling mortality ranging from 25-75%. In *Moringa oleifera*, affected seedlings may show poor germination rates or fail to emerge. Those that do appear may exhibit weakened and discoloured stems *cum* eventual wilting and collapsing at the soil level. Lesions or rotting may hinder growth leading to demise of the seedlings. *Rhizoctonia* attack results in thin and hard stems (wire stem symptoms) causing infected seedlings to topple. The disease tends to appear in patches both in nursery and field beds, significantly reducing the stand of young *Moringa* plants.

Implementing various control measures can help manage damping off disease. Sanitation practices like using sterilized soil and containers prevent pathogen introduction and spread. Adequate drainage, avoiding overwatering and ensuring proper ventilation reduce the moisture levels that is favourable for fungal growth. Using fungicide-treated seeds or applying biocontrol agents containing beneficial microorganisms can suppress damping off pathogens. Maintaining optimal environmental conditions including temperature and humidity supports healthy seedling growth by reducing susceptibility to the disease. Regular monitoring for early signs allows for prompt intervention and control measures to mitigate its impact on *Moringa* crops.

4.16.2.2 Twig canker (*Fusarium pallidoroseum*): Twig canker in *Moringa* is caused by fungal pathogens such as *Colletotrichum gloeosporioides* or *Phomopsis* spp. that affect the branches and stems which lead to severe damage and potential death of the plant. Initially its infection is visible through clearing of veinlets and chlorosis of the leaves. Subsequently, younger leaves may perish in succession leading to wilting and eventual death of entire plant within a few days. As the disease progresses, the petioles and leaves begin to droop and wilt. In young plants, symptoms primarily include veinlet clearing and petiole dropping. These symptoms persist and continue to appear in subsequent leaves. In the later stages, a browning of the vascular system becomes evident. Eventually, affected plants become stunted and succumb to disease. Twig canker disease in moringa can persist and spread through various channels and thrives in relatively high soil moisture and soil temperature. The disease may survive in the soil, leading to recurrent infections in subsequent plantings. Contaminated agricultural implements or tools can contribute to the spread of this ailment. Regular cleaning and maintaining the growing area by removing any infected plant debris as fungal spores can survive on fallen leaves and branches. Pruning and removing infected branches ensures preventing disease spread. Application of appropriate fungicides following manufacturer guidelines and local regulations can help manage its spread. Maintaining overall plant health by providing proper nutrition, adequate watering and ensuring optimal growing conditions will help maintaining plant health. If possible, growers should avoid planting *Moringa* in areas where disease has been prevalent previously. Crop rotation reduces the buildup of disease-causing agents in the soil. Tolerant to resistant moringa varieties should be found out for healthy crop. Consulting with local agricultural experts or extension services for region-specific advice is highly recommended for effective disease management.

4.16.2.2.1 Pest prevention and control (stagewise crop IPM)

Management	Activity
Pre-sowing	
	Common cultural practices: Deep summer ploughing; destroying the alternate host plants; avoid planting overlapping crops in adjacent area.
Nutrients	Apply FYM or compost @ 7-8kg per pit. In addition, basal dose of N:P:K in 350: 100: 200 g/pit should also be incorporated at the time of planting.
Weeds	Remove or incorporate previous crop residues before planting. Plan to grow suitable intercrops like legumes, ginger, tapioca, *etc.*
Soil borne fungus, resting stages of pests	Solarization is best option. Covering the beds with polythene sheet of 0.45mm thickness for three weeks before sowing will help in reducing the soil borne pests.
Damping off	Raise nursery in light soil with proper drainage; burning farm trash on the surface of the beds; sowing seeds on 15cm raised beds. Use low seed rate of 650g/40m^2 area.
Apply *Trichoderma viride/T. harzianum* and *Pseudomonas fluorescens* as seed/seedling/ planting material, nursery treatment and soil application (if commercial products are used, check for label claim. However, biopesticides produced by farmers for own consumption in their fields, registration is not required).	

Management	Activity
Vegetative stage	
	Common cultural practices: Collect and destroy crop debris; collect and destroy disease infected and insect damaged plant parts; provide irrigation at critical stages of the crop; avoid waterlogging; avoid water stress during flowering stage; judicious use of fertilizers; enhance parasitic activity by avoiding chemical spray, when 1-2 larval parasitoids are observed; field sanitation. Collection and destruction of eggs and early stage larvae; handpicking the older larvae during early stages (do not handpick hairy caterpillars with urticating hairs); the infested shoots and seed capsules may be collected and destroyed; collect and destroy the gregarious caterpillars and cocoons which are found on stem and destroy them in kerosene mixed water; use light trap @ 2/ha from 6 pm and 10 pm; erect the bird perches @ 50/ha for encouraging predatory birds such as king crow, common mynah *etc.* and set up bonfire around 7-8 pm. Conserve natural enemies through ecological engineering; augmentative release of natural enemies.
Nutrient	Under low fertile soils top dressing of N @ 100g/pit by soil incorporation should be done. Micronutrient deficiency should be corrected through foliar spraying of particular nutrient.
Weeds	Crop should be weed-free up to 3 months by interculture and hand weeding.
Hairy caterpillar	Collect caterpillars from plants and eliminate them. Utilise light traps to attract and eliminate adult insects. Apply 1kg of carbaryl or 1 litre of malathion 50 EC or endosulfan 35 EC in 500-750 litres of water per hectare through spraying.
Budworm bark caterpillar	Follow common cultural, mechanical and biological practices
Reproductive stage	
Nutrients	Micronutrient deficiency should be corrected by foliar spray of particular nutrient.
Weeds	All the weeds should be removed before seed formation to avoid further spread of weed seeds.
Pod fly	Collect and destroy fallen and damaged fruits; trap adults using attractants such as citronella oil, eucalyptus oil, vinegar (acetic acid), dextrose or lactic acid; rake up the soil under the trees or plough the infested field to destroy puparia. Spray NSKE 5% during 50% fruit set and later after 35 days.

(Vikaspedia, 2023)

4.17 Seed Production

Moringa produces seeds that are utilized for various purposes including planting, oil extraction and as a source of protein. Seed production in *Moringa* involves several key aspects:

- **Pollination:** Moringa flowers are typically small and white with a little of self-pollination and through insects such as bees. Adequate pollination is essential for seed development.
- **Seed formation:** After successful pollination, the flowers develop into seed pods. Moringa pods are slender and long from 20 to 90cm depending upon the variety. Pods contain seeds which mature as the pods dry and turn brown. Each pod can contain numerous seeds.
- **Harvesting:** Harvesting usually occurs when the pods are fully mature and have turned brown so at this stage the pods can be easily cracked open to retrieve the seeds. Harvesting at the right time ensures better seed quality and viability.
- **Seed processing:** After harvesting, the seeds are separated from the pods. Proper drying and storage are crucial to maintain seed viability. Once dried, the seeds can be stored for future planting or utilized for various purposes.
- **Storage:** Proper storage conditions, including cool and dry environments help maintaining the seed quality and viability for an extended period.
- **Germination and planting:** When planting *Moringa* seeds, it is beneficial to soak them in water for several hours or nick the seed coat slightly to enhance germination. Planting in well-draining soil with adequate moisture and sunlight promotes successful germination and growth.

Moringa seeds have a high oil content, making them valuable for oil extraction and as a source of nutrition. The ability of seeds to withstand various climates and soil conditions contributes to their popularity for cultivation and production.

4.17.1 Seed extraction and storage guidelines

- **Harvesting:** Ripe fruits are collected when they turn brown for optimal seed extraction. High-quality seeds known for strong germination are found mainly in the proximal and middle sections of the fruit rather than the distal part. Black and brown seeds show superior germination rates and vigour compared to white seeds.

- **Drying process:** Harvested pods require 1-2 days of shade-drying with adequate ventilation. Gentle manual opening of pods facilitates easy seed separation, allowing removal of any damaged or undesirable seeds.
- **Grading:** Specific gravity separators effectively grade seeds. Seeds from fractions 2 and 3 exhibit higher seedling emergence and vigour.
- **Storage conditions:** Optimum storage temperature is 20-25°C. Freshly harvested annual moringa seeds, dried to 8% moisture content, treated with captan (2g/kg) and packed in 700 gauge polythene bags can be stored for up to one year (Palanisamy *et al*, 1995). Black to brown seeds can also be treated with carbendazim (2g/kg) and then stored in 700 gauge polythene bags to maintain over 84% germination for 12 months (Sivasubramanian and Thiagarajan, 1997).
- **Seed testing:** For testing, sow seeds at 1cm depth in sandy medium with 80% moisture-retaining capacity for optimal conditions.

4.17.2 Seed certification standards

For certified seeds, only main crop is allowed and for labelled seeds main crop with two ratoons are allowed following 1000m isolation distance for breeder seeds and 500m for certified seeds. These seeds should have 70% minimum germination

References

1. Abbas, Z.A., N.M. Abu-Mejdad, Z.W. Atwan and N.A. Al-Masoudi, 2017. Synthesis and biological evaluation of new dipyridylpteridines, lumazines and related analogues. *J. Heterocyc. Chem.*, **54**(2): 895-903.
2. Adenike, A.B. 2014. The effects of drying on some nutrients of *Moringa oleifera* leaves (PDF). *Annals – Food Sci. Techn.*, **15**(2): 246–50.
3. Agarwal, V. 2023. The magical moringa. https://www.ayurvedacollege.com/blog/magicalmoringa/ [Accessed November 30, 2023].
4. Ali, A., Y.A. Yusof, N.L. Chin and M.N. Ibrahim, 2017. Processing of *Moringa* leaves as natural source of nutrients by optimization of drying and grinding mechanism. *J. Food Process Eng..* **40**(e12583): e12583. doi:10.1111/jfpe.12583.
5. Al-Ziaydi, A.G., M. Imran, A.M. Al-Shammari, H.S. Kadhim and M. Jabir, 2020. The anti-proliferative activity of D-Mannoheptulose against breast cancer cell line through glycolysis inhibition. *AIP Conf. Proc.*, 2307, 020023. HTTPS://DOI.ORG/ 10.1063/5.0032958.
6. Amaglo, N.K., R.N. Bennett, R.B. Lo Curto, E.A.S. Rosa, V. Lo Turco, A. Giuffrida, A. Lo Curto, F. Crea and G.M. Timpo, 2010. Profiling selected

phytochemicals and nutrients in different tissues of the multipurpose tree *Moringa oleifera* L. grown in Ghana. *Food Chem.*, **122**: 1047–54.

7. Anjaneya Murthy, J.N. and A. Regupathy, 1992. Seasonal incidence of moringa fruit fly, *Gitona* sp., *South Indian Horti.*, **40**:1: 43-8.

8. Anwar, F., M. Ashraf and M.I. Bhanger, 2005. Interprovenance variation in the composition of *Moringa oleifera* oilseeds from Pakistan. *J. Am. Oil Chem. Soc.*, **82**: 45–51.

9. Anzano, A., M. Ammar, M. Papaianni, L. Grauso, M. Sabbah, R. Capparelli and V. Lanzotti, 2021. *Moringa oleifera* Lam.: A Phytochemical and Pharmacological Overview. *Horticulturae*, **7(10)**:409.

10. Aparna, V., D. Vijayan, P. Mandal and P. Karthe, 2012. Anti-inflammatory property of n-hexadecanoic acid: structural evidence and kinetic assessment. *J. Chem. Bio.*, **80**(3): 434-9.

11. Asare, G.A. and A. Nyarko, 2012. Toxicity potentials of the nutraceutical *Moringa oleifera* at supra-supplementation levels. *J. Ethnopharmac.*, **139** (1): 265–72.

12. Atawodi, S.E., J.C. Atawodi, G.A. Idakwo, B. Pfundstein, R. Haubner, G. Wurtele, H. Bartsch and R.W. Owen, 2010. Evaluation of the polyphenol content and antioxidant properties of methanol extracts of the leaves, stem and root barks of *Moringa oleifera* Lam. *J. Medic. Food*, **13**(3): 710–6.

13. Augustin, J.M., V. Kuzina, S.B. Andersen and S. Bak, 2011. Molecular activities, biosynthesis and evolution of triterpenoid saponins. *Phytochem.*, **72**: 435–57.

14. Aziz, S., N. Jafarah, B. Yusof and Z. Zetty, 2019. Phytol-containing seaweed extracts as control for *Ganoderma boninense*. *J.Ooil Palm Res.*, **31**: 238-47.

15. Babu, S.C. and B. Rajasekaran, 1991. Agroforestry, altitude towards risk and nutrient availability: as case study of south Indian farming systems. *Agrofor. Syst.*, **15**: 1-15.

16. Bajpai, M., A. Pande, S.K. Tewari and D. Prakash, 2007. Phenolic contents and antioxidant activity of some food and medicinal plants. *Int. J. Food Sci. Nutr.*, **56**: 287–91.

17. Balamurugan, R., V. Duraipandiyan and S. Ignacimuthu, 2011. Antidiabetic activity of γ-sitosterol isolated from *Lippia nodiflora* L. in streptozotocin induced diabetic rats. *Eur. J. Pharmacol.*, **667**: 410–18.

18. Balamurugan, R., A. Stalin and S. Ignacimuthu, 2012. Molecular docking of γ-sitosterol with some targets related to diabetes. *Eur. J. Med. Chem.*, **47**: 38–43.

19. Bargah, R.K. and C. Das, 2014. Isolation and characterization of steroidal glycoside from chloroform extract of the stem bark of *Moringa pterygosperma* Gaertn. *Int. J. Innov. Res. Sci. Eng. Technol.*, **3**: 18319–22.

20. Bennett, R.N., F.A. Mellon, N. Foidl, J.H. Pratt, M.S. Dupont, L. Perkins and P.A. Kroon, 2003. Profiling glucosinolates and phenolics in vegetative and reproductive tissues of the multi-purpose trees *Moringa oleifera* L. (Horseradish tree) and *Moringa stenopetala* L. *J. Agric. Food Chem.*, **51**: 3546–53.

21. Bhalla, N., N. Ingle, S.V. Patri and D. Haranath, 2021. Phytochemical analysis of *Moringa oleifera* leaves extracts by GC-MS and free radical scavenging potency for industrial applications. *Saudi J. Biol. Sci.*, **28**(12): 6915-28.

22. Bhatnagar, A.S. and A.G.G. Krishna, 2013. Natural antioxidants of the Jaffna variety of *Moringa oleifera* seed oil of Indian origin as compared to other vegetable oils. *Grasas Y. Aceites.*, **64**(5): 537-45.

23. Bhatta, R., M. Saravanan, L. Baruah and K.T. Sampath, 2012. Nutrient content, *in vitro* ruminal fermentation characteristics and methane reduction potential of tropical tannin-containing leaves. *J. Sci. Food Agric.*, **92**: 2929–35.

24. Bhattacharya, A. and S. Mandal, 2004. Pollination, pollen germination and stigma receptivity in *Moringa oleifera* Lamk. *Grana*, **43**: 48–56.

25. Bizzarri, M., A.S. Laganà, D. Aragona and V. Unfer, 2020. Inositol and pulmonary function. Could myo-inositol treatment downregulate inflammation and cytokine release syndrome in SARS-CoV-2? *Eur. Rev. Med. Pharmacol. Sci.*, **24** (6): 3426-32.

26. Blondeau, N., R.H. Lipsky, M. Bourourou, M.W. Duncan, P.B. Gorelick and A.M. Marini, 2015. Alpha-linolenic acid: an omega-3 fatty acid with neuroprotective properties—ready for use in the stroke clinic?. *BioMed. Res.intern.*, **19**: 15-20

27. Booth, F.E.M. and G.E. Wickens, 1988. Non-timber Uses of Selected Arid Zone Trees and Shrubs in Africa, 98p. FAO, Rome. Retrieved 20-11-2013.

28. Brat, P., S.George, A. Bellamy, L. Du Chaffaut, A. Scalbert, L. Mennen, N. Arnault and M.J. Amiot, 2006. Daily polyphenol intake in France from fruit and vegetables. *J. Nutr.*, **136**: 2368–73.

29. Butani, D.K. and S. Verma, 1981. Insect-pests of vegetables and their control – drumsticks. *Pesticides*, **15** (10): 29–32.

30. Bwana-Simba, E. 2006. The Various Uses of *Moringa oleifera*. http://www.runetwork.de/html/en/index.html?article_id=3862. [accessed September 21, 2009].

31. Caceres, A., O. Cabrera, O. Morales, P. Mollinedo and P. Mendia, 1991. Pharmacological properties of *Moringa oleifera.* 1: preliminary screening for antimicrobial activity. *J. Ethnopharmacol.*, **33**: 213-6.

32. Cancelas, J., A. Acitores, M.L.V. Penacarrillo, W.J. Malaisse and I. Valverde, 2000. Activation of glycogen synthase a in hepatocytes exposed to alpha-D-glucose pentaacetate. *Intern. J. Mol. Med.*, **6**(2): 197-9.

33. Čechovská, L., K. Cejpek, M. Konečný and J. Velíšek, 2011. On the role of 2,3-dihydro-3,5-dihydroxy-6-methyl-(4H)-pyran-4-one in antioxidant capacity of prunes. *Eur. Food Res. Technol.*, **233**(3): 367-76.

34. Chelliah, R., S. Ramakrishnan and U. Antony, 2017. Nutritional quality of *Moringa oleifera* for its bioactivity and antibacterial properties. *Int. Food Res. J.*, **24**: 825–33.

35. Chinma, C., J. Abu and S. Akoma, 2014. Effect of germinated tigernut and moringa flour blends on the quality of wheat-based bread. *J. Food Proc. Preserv.*, **38** (2): 721–7.

36. Choi, J.-K., I. Carreras, A. Dedeoglu and B.G. Jenkins, 2010. Detection of increased scyllo-inositol in brain with magnetic resonance spectroscopy after dietary supplementation in Alzheimer's disease mouse models. *J. Neuropharm.*, **59**(4-5): 353-7.

37. Chuang, P.H., C.W. Lee, J.Y. Chou, M. Murugan, B.J. Shieh and H.M. Chen, 2007. Anti-fungal activity of crude extracts and essential oil of *Moringa oleifera* Lam. *Bioresour. Technol.*, **7**(98): 232–6.

38. Cieślik, E., A. Gręda and W. Adamus, 2006. Contents of polyphenols in fruit and vegetables. *Food Chem.*, **94**: 135–42.

39. Ciska, E., B. Martyniak-Przybyszewska and H. Kozlowska, 2000. Content of glucosinolates in cruciferous vegetables grown at the same site for two years under different climatic conditions. *J. Agric. Food Chem.*, **48**: 2862–7.

40. Coppin, J.P., Y. Xu, H. Chen, M.H. Pan, C.T. Ho, R. Juliani, J.E. Simon and Q. Wu, 2013. Determination of flavonoids by LC/MS and anti-inflammatory activity in *Moringa oleifera. J. Funct. Foods.* **5**: 1892–9.

41. Courtois, P., A. Sener and W.J. Malaisse, 2001. D-mannoheptulose phosphorylation by hexokinase isoenzymes. *Int. J. Mol. Med.*, **7**: 359-63.

42. Cushnie, T.P.T., B. Cushnie and A.J. Lamb, 2014. Alkaloids: An overview of their antibacterial, antibiotic-enhancing and antivirulence activities. *Int. J. Antimicrob. Agen.*, **44**: 377–86.

43. Dadamouny, M.A., M. Unterseher, P. König and M. Schnittler, 2016. Population performance of *Moringa peregrina* (Forssk.) Fiori (Moringaceae)

at Sinai Peninsula, Egypt in the last decades: Consequences for its conservation. *J. Nature Conserv.*, **34**: 65–74.

44. den Hartog, G.J., A.W. Boots, A. Adam-Perrot, F. Brouns, I.W. Verkooijen, A.R. Weseler, G.R. Haenen and A. Bast, 2010. Erythritol is a sweet antioxidant. *Nutrit.*, **26**(4): 449-58.

45. Dhifi, W., S. Bellili, S. Jazi, N. Bahloul and W. Mnif, 2016. Essential Oils' Chemical Characterization and Investigation of Some Biological Activities: A Critical Review. *Medicines*, **3**: 25.

46. Dinkova-Kostova, A.T. and R.V. Kostov, 2012. Glucosinolates and isothiocyanates in health and disease. *Trends Mol. Med.*, **18**: 337–47.

47. Duthie, G.G. and A.D. Wood, 2011. Natural salicylates: Foods, functions and disease prevention. *Food Funct.*, **2**: 515–20.

48. Eagri, 2023a. Development of e Courses for B.Sc. (Agriculture). Accessed on 9 December, 2023. http://www.eagri.org/eagri50/ENTO331/lecture26/Moringa/002.html.

49. Eagri, 2023b. Drumstick (*Moringa olerifra* Lam.). Development of e-Courses for B.Sc. (Agriculture). HORT 281: Lecture 32. Accessed on 29 November, 2023. http://eagri.org/eagri50/HORT281/lec32.html.

50. Ecoursesonline, 2023. Tropical and subtropical vegetables. http://ecoursesonline.iasri.res.in/mod/page/view.php?id=97994 [Accessed on December, 2023].

51. El-Seedi, H.R., A.M. El-Said, S.A. Khalifa, U. Göransson, L. Bohlin, A.K. Borg-Karlson and R. Verpoorte, 2012. Biosynthesis, natural sources, dietary intake, pharmacokinetic properties and biological activities of hydroxycinnamic acids. *J. Agric. Food Chem.*, **60**: 10877–95.

52. Emmanuel, S.A., B.S. Emmanuel, S.G. Zaku and S.A. Thomas, 2011. Biodiversity and agricultural productivity enhancement in Nigeria: Application of processed *Moringa oleifera* seeds for improved organic farming. *Biol. J. North Am.*, **2**: 867–71.

53. Fahey, J.W. 2005. *Moringa oleifera*: A Review of the Medical Evidence for Its Nutritional, Therapeutic, and Prophylactic Properties. Part 1. *Trees for Life J.*, **1**: 5.

54. Faujdar, E. and R.K. Singh, 2021. Methyl oleate derived multifunctional additive for polyol based lubricants. *Wear.*, **No.** 466, pp. 203550.

55. Foidl, N., H.P.S. Makkar and K. Becker, 2001. The potential of *Moringa oleifera* for agricultural and industrial uses. The Miracle Tree: The Multiple

Attributes of Moringa (ed. Fuglie, L.J.), pp. 45-76. Church World Service, Dakar, Senegal.

56. Forster, N., C. Ulrichs, M. Schreiner, C.T. Muller and I. Mewis, 2015. Development of a reliable extraction and quantification method for glucosinolates in *Moringa oleifera. Food Chem.*, **166**: 456–64.

57. Freeman, S. and T. Hudlicky, 2004. New oligomers of conduritol-F and muco-inositol. synthesis and biological evaluation as glycosidase inhibitors. *J. BMCL*, **14**(5).1209-12.

58. Fu, L., B.-T. Xu, X.-R. Xu, R.-Y. Gan, Y. Zhang, E.-Q. Xia and H.-B. Li, 2011. Antioxidant capacities and total phenolic contents of 62 fruits. *Food Chem.*, **129**: 345–50.

59. Gassenschmidt, U., D. Klans, D. Jarry, B. Taucher and H. Neebdergall, 1995. Isolation and characterization of a flocculating protein from *Moringa oleifera* Lam. *Biochem. Biophys. Acta*, **24**(093): 477-81.

60. Ge, L., M.M. Zhu, J.Y. Yang, F. Wang, R. Zhang, J.H. Zhang, J. Shen, H.F. Tian and C.F. Wu, 2015. Differential proteomic analysis of the anti-depressive effects of oleamide in a rat chronic mild stress model of depression. *Pharmacol. Biochem. & Behavior*, **131**: 77-86.

61. Ghosh A., K,X. Yu, H.L. Osswald, J. Agniswamy, Y.F. Wang, M. Amano, I.T. Weber and H. Mitsuya, 2015. Structure-based design of potent HIV-1 protease inhibitors with modified P1-biphenyl ligands: synthesis, biological evaluation and enzyme-inhibitor X-ray structural studies. *J. Med. Chem.*, **58**(13):5334-43.

62. Goplan, C., B.V.R. Sastri and S.C. Balasubramanian, 1999. Nutritive Value of Indian Foods. National Institute of Nutrition, Hyperabad, India.

63. GRIN, 2023. *Moringa oleifera. Germplasm Resources Information Network.* Agricultural Research Service, United States Department of Agriculture. Retrieved 19 December 2023.

64. Grubben, G.J.H. (ed.), 2004. Plant resources of tropical Africa: Vegetables, Vol. 2. PROTA, 394p. ISBN 978-9057821479. Retrieved 26 November 2023.

65. Gupta, S., A. Jyothi Lakshmi, M.N. Manjunath and J. Prakash, 2005. Analysis of nutrient and antinutrient content of underutilised green leafy vegetables. *LWT Food Sci. Technol.*, **38**: 339–45.

66. Hagiyama, K., K. Suzuki, M. Ohtake, M. Shimada, N. Nanbu, M. Takehara, M. Ue and Y. Sasaki, 2008. Physical properties of substituted 1,3-dioxolan-2-ones. *Chem. Let.*, **37**(2): 210-11.

67. Hamedi, A., A.P. Lashgari and A. Pasdaran, 2019. Antimicrobial Activity and Analysis of the Essential Oils of Selected Endemic Edible Apiaceae Plants Root from Caspian Hyrcanian Region (North of Iran). *Pharm Sci.*, **25** (2): 138-44.

68. Harnly, J.M., R.F. Doherty, G.R. Beecher, J.M. Holden, D.B. Haytowitz, S. Bhagwat and S. Gebhardt, 2006. Flavonoid content of U.S. fruits, vegetables and nuts. *J. agric. Food Chem.*, **54**: 9966–77.

69. He, Y., J.L. Johnson and S.H. Yalkowsky, 2005. Oral formulation of a novel antiviral agent, PG301029, in a mixture of Gelucire 44/14 and DMA (2⊠ 1, wt/wt). *AAPS PharmSci.Tech.*, **6**: E1-E5.

70. Heuze, V., G. Tran, P. Hassoun, D. Bastianelli and F. Lebas, 2016. Moringa (*Moringa oleifera*) [(accessed on 16 November 2016)]. Feedipedia, a Programme by INRA, CIRAD, AFZ and FAO. Available online: http://www.feedipedia.org/node/124.

71. Hiraman, 2020. Bhagya KDM & Sarpan 2: Next-Gen Moringa Varieties. Sunday Farmer. September 8, 2020. Accessed on 5 December, 2023. https://sundayfarmer.wordpress.com/2020/09/08/bhagya-kdm-sarpan-2-next-gen-moringa-varieties/.

72. Huang, A., Neil Brody and Tracey N. Liebman, 2017. Dihydroxyacetone and sunless tanning: knowledge, myths, and current understanding. *J. Am. Acad. of Dermat.*, **77**(I-5): 991-2.

73. Imad, H.H., J.H. Hussein, A.K. Muhanned and S.H. Nidaa, 2015. Identification of five newly described bioactive chemical compounds in methanolic extract of *Mentha viridis* by using gas chromatography – mass spectrometry (GC-MS). *J. Pharmacognosy Phytother.*, **7**(7): 107-25.

74. Iqbal, S. and M.I. Bhanger, 2006. Effect of season and production location on antioxidant activity of *Moringa oleifera* leaves grown in Pakistan. *J. Food Compos. & Anal.*, **19**(6–7): 544–551.

75. Islam, M.T., E.S. Ali, S.J. Uddin, S. Shaw, M.A. Islam, M.I. Ahmed, M. Chandra Shill, U.K. Karmakar, N.S. Yarla, I.N. Khan, M.M. Billah, M.D. Pieczynska, G. Zengin, C. Malainer, F. Nicoletti, D. Gulei, I. Berindan-Neagoe, A. Apostolov, M. Banach, A.W.K. Yeung, A. El-Demerdash, J. Xiao, P. Dey, S. Yele, A. Jóźwik, N. Strzałkowska, J. Marchewka, K.R.R. Rengasamy, J. Horbańczuk, M.A. Kamal, M.S. Mubarak, S.K. Mishra, J.A. Shilpi and A.G. Atanasov, 2018. Phytol: A Review of Biomedical Activities. *F. Chem. T.*, **121**: 82-94.

76. Jahn, S.A.A. 1984. Effectiveness of traditional flocculants as primary coagulants and coagulant aids for the treatment of tropical raw water with more than a thousand fold fluctuation in turbidity. *Water Supply*, **6**: 8-10.

77. Jahn, S.A.A., H.A. Mussad and H. Burdstaller, 1986. The tree that purifies water. Cultivating multipurpose Moringaceae in the Sudan. *Unasylva*, **152**: 23-8

78. Janick, J. and E. Paull Robert, 2008. The Encyclopedia of Fruit & Nuts. CABI, pp. 509–10. ISBN 978-0-85199-638-7.

79. Joshi, P. and D. Mehta, 2010. Effect of dehydration on the nutritive value of drumstick leaves. *J. Metabolomics Syst. Biol.*, **1**: 5–9.

80. Joshi, R.C., B.D. Vasantharaj and R. Kant, 2016. A review of the insect and mite pests of *Moringa oleifera* Lam. *Agric. for Developm.*, **29**: 1-5.

81. Kaiser, A., A. Gottwald and C. Wiersch, 2001. Effect of drugs inhibiting spermidine biosynthesis and metabolism on the *in vitro* development of *Plasmodium falciparum*. *Parasitol. Res.*, **87**: 963–72.

82. Kamenov, Z. and A. Gateva, 2020. Inositols in PCOS. *Molecul.*, **25**(23): 5566.

83. Kancheva, V.D. and O.T. Kasaikina, 2013. Bio-antioxidants-a chemical base of their antioxidant activity and beneficial effect on human health. *Curr. Med. Chem.*, **20**: 4784–805.

84. Kanhar, S. and A.K. Sahoo, 2018. Ameliorative effect of *Homalium zeylanicum* against carbon tetrachloride-induced oxidative stress and liver injury in rats. *J. Biopha.*, **111**: 305-314.

85. Kanthaswamy, 2006. Studies on phenology and floral biology in *Moringa oleifera*. *Inetrn. J. agric. Sci.*, **2**(2): 341-3.

86. Kareem, A.A., S. Sadakathulla and T.R. Subramanian, 1974. Note on the severe damage of moringa fruits by the fly *Gitona* sp. (Drospohilidae: Diptera). *South Ind. Hort.*, **22**(½): 71.

87. Kasolo, J.N., G.S. Bimenya, L. Ojok, J. Ochieng and J.W. Ogwal-Okeng, 2010. Phytochemicals and uses of *Moringa oleifera* leaves in Ugandan rural communities. *J. Med. Pl. Res.*, **4**: 753–7.

88. Khorassani, R., U. Hettwer, A. Ratzinger, B. Steingrobe, P. Karlovsky and N. Claassen, 2011. Citramalic acid and salicylic acid in sugar beetroot exudates solubilize soil phosphorus. *BMC Pl. Biol.*, **11**(1): 121.

89. Kshirsagar, C.R. and T.F. D'souza, 1989. A new disease of drumstick. *J. Mahar. Agric. Univer.*, **14**(2): 241–2.

90. Kumar, S. and A.K. Pandey, 2013. Chemistry and biological activities of flavonoids: An overview. *Sci. World J.*, **16**(2): 750.

91. Lahjie, A.M. and B. Siebert, 1987. Kelor or horseradish tree (*Moringa oleifera* Lam.). A report from East Kalimantan. German Forestry Group, Mulawarman Univ., *GFG Report*, **6**: 41–3.

92. Lalas, S. and J. Tasaknis, 2002. Characterization of *Moringa oleifera* seed oil variety "Periyakulum 1". *J. Food Compos. Anal.*, **15**: 65-7.

93. Lalas, S., V. Athanasiadis, I. Karageorgou, G. Batra, G.D. Nanos and D.P. Makris, 2017. Nutritional Characterization of Leaves and Herbal Tea of *Moringa oleifera* Cultivated in Greece. *J. Herbs Spices Med. Plants.*, **23**(4): 320–33.

94. Lee, J.H., Y.G. Kim, J.G. Park and J. Lee, 2017. Supercritical fluid extracts of *Moringa oleifera* and their unsaturated fatty acid components inhibit biofilm formation by *Staphylococcus aureus. Food Control.*, **80**: 74–82.

95. Leone, A., A. Spada, A. Battezzati, A. Schiraldi, J. Aristil and S. Bertoli, 2015a. Cultivation, Genetic, Ethnopharmacology, Phytochemistry and Pharmacology of *Moringa oleifera* Leaves: An Overview. *Intern. J. Molec. Sci.*, **16**(6): 12791–835.

96. Leone, A., G. Fiorillo, F. Criscuoli, S. Ravasenghi, L. Santagostini, G. Fico, A. Spadafranca, A. Battezzati, A. Schiraldi, F. Pozzi, S.D. Lello, S. Filippini and S. Bertoli, 2015b. Nutritional characterization and phenolic profiling of *Moringa oleifera* leaves grown in Chad, Sahrawi Refugee Camps and Haiti. *Int. J. Mol. Sci.*, **16**:18923–37.

97. Leone, A., A. Spada, A. Battezzati, A. Schiraldi, J. Aristil and S. Bertoli, 2016. *Moringa oleifera* Seeds and Oil: Characteristics and Uses for Human Health. *Int. J. Mol. Sci.*, **17**(12): 2141.

98. Li, Y., H. Wu, R. Zhang, G. Shu, S. Wang, P. Gao, X. Zhu, Q. Jiang and L. Wang, 2020. Diet containing stearic acid increases food reward-related behaviors in mice compared with oleic acid. *Brain Res. Bull.*, No.164, pp. 45-54.

99. Lim, T.K. 2012. *Moringa oleifera*. In: Edible Medicinal and Non-Medicinal Plants, Vol. 3: Fruits. Springer, Netherlands, pp. 453–85. ISBN 978-94-007-2534-8.

100. Liu, S., L. He and K. Yao, 2018. The antioxidative function of alpha-ketoglutarate and its applications. *Bio. Med. R.I.*, pp. 1-6.

101. Lopez-Rodriguez, N.A., M. Gaytán-Martínez, M. de la Luz Reyes-Vega and G. Loarca-Piña, 2020. Glucosinolates and Isothiocyanates from *Moringa*

oleifera: Chemical and Biological Approaches. *Plant Foods Hum. Nutr.*, **75**: 447–57.

102. Maharshi, D.P. 2020. Most Popular Moringa Seed Varieties. Building on Success. A Fresh Perspective. https://medium.com/@dpmaharshi/most-popular-moringa-seed-varieties-1e31e974aee4 [Accessed 26 November 2020].
103. Maiyo, F.C., R. Moodley and M. Singh, 2016. Cytotoxicity, antioxidant and apoptosis studies of quercetin-3-O-glucoside and 4-(beta-D-glucopyranosyl-1->4-alpha-L-rhamnopyranosyloxy)-benzyl isothiocyanate from *Moringa oleifera*. *Anticancer Agents Med. Chem.*, **16**: 648–56.
104. Makkar, H.P.S. and K. Becker, 1996. Nutrional value and antinutritional components of whole and ethanol extracted *Moringa oleifera* Leaves. *Anim. Feed Sci. Technol.*, **63**: 211–28.
105. Makkar, H.P.S. and K. Becker, 1997. Nutrients and antiquality factors in different morphological parts of the *Moringa oleifer*a tree. *J. agric. Sci.*, **128**: 311–22.
106. Man, S., W. Gao, Y. Zhang, L. Huang and C. Liu, 2010. Chemical study and medical application of saponins as anti-cancer agents. *Fitoterapia*, **81**: 703–14.
107. Manca, L., S.M. Palomo, C. Caddeo, A. Nacher, O.D. Sales, J.E. Peris, J.L. Pedraz, A.M. Fadda and M. Manconi, 2018. Sorbitol-penetration enhancer containing vesicles loaded with baicalin for the protection and regeneration of skin injured by oxidative stress and UV radiation. *J. Ijpharm.*, **555**: 175-183. HTTPS://DOI.ORG /10.1016/j.ijpharm.2018.11.053.
108. Mariano, B. and C. Gianfranco, 2014. Inositol: history of an effective therapy for polycystic ovary syndrome. *Eur. Rev. Med. Pharmacol. Sci.*, **18**: 1896-1903.
109. Markad, S.D., P. Kaur and B.K. Kishore Reddy, 2015. Novel lead generation of an anti-tuberculosis agent active against non-replicating mycobacteria: exploring hybridization of pyrazinamide with multiple fragments. *Med. Chem. Res.*, **24**: 2986–92.
110. Mark Olson, E. 2010. Moringaceae Martinov; Drumstick Tree Family. In: Flora of North America, North of Mexico, Vol. 7: Magnoliophyta: Dilleniidae, Part 2, pp. 167-9. Oxford University Press. ISBN 978-0195318227.
111. Mark, Olson, E. 2002. Combining Data from DNA Sequences and Morphology for a Phylogeny of Moringaceae (Brassicales) (PDF). *Syst. Bot.*, **27**(1): 55–73.

112. Mayer, F.A. and E. Stelz, 1993. *Moringa stenopetala* provides food and low cost water purification. *Agrofor. Today*, **5**: 16-9.

113. McBurney, R.P.H., C. Griffin, A.A. Paul and D.C. Greenberg, 2004. The nutritional composition of African wild food plants: From compilation to utilization. *J. Food Comp. & Anal.*, **17**: 277–89.

114. MIR, 2023. Market Intelligence Report: Moringa - Agri Exchange. Accessed on November 2023. https://agriexchange.apeda.gov.in/Weekly_eReport/Moringa_Report.pdf.

115. Morton, J.F. 1991. The horseradish tree, *Moringa pterygosperma* (Moringaceae) – a boon to arid lands? *Econ. Bot.*, **45**(3): 318–33.

116. MSKCC, 2023. *Moringa oleifera*: Purported Benefits, Side Effects & More. Memorial Sloan-Kettering Cancer Center. Last updated: Tuesday, June 20, 2023. Accessed on 26 December, 2023. https://www.mskcc.org/cancer-care/integrative-medicine/herbs/moringa-oleifera.

117. Muyibi, S.A and L.M. Evanson, 1995. Optimising physical parameters affecting coagulation of turbid waters with *Moringa oleifera* seeds. *Water Res.* **29**: 2689-95.

118. Nautiyal, B.P. and K.G. Venkataraman, 1987. Moringa (drumstick) - An ideal tree for social forestry: Growing conditions and uses - part 1. *MYFOREST*, **23**: 53-8.

119. ööNdabigengesere, A. and K.S. Narasiah, 1998. Quality of water treated by coagulation using *Moringa oleifera* seeds. *Water Res.*, **32**: 781–91.

120. Ndabigengesere, A,, K.S. Narasiah and B.G. Talbot, 1995. Active agents and mechanisms of coagulation of turbid waters using *Moringa oleifera. Water Res.*, **29**: 703-10.

121. NHRDF, 2023. Value Chain Study of *Moringa* Karnal. Tamilnadu – NHRDF. Accessed on December, 2023. Page 37. https://www.midh.gov.in/VCS%20Reports/16-Value%20Chain%20Study%20of%20Moringa%20Karnal%20Tamilnadu.pdf.

122. Nishio, T., Y. Yoshikawa, C.Y. Shew, N. Umezawa, T. Higuschi and K. Yoshikawa, 2019. Specific effects of antitumor active norspermidine on the structure and function of DNA. *Sci. Rep.*, **9**: 14971. HTTPS://DOI.ORG/10.1038/s41598-019-50943-1.

123. Nzikou, J.M., L. Matos, J.E. Moussounga, C.B. Ndangui, A. Kimbonguila, Th Silou, M. Linder and S. Desobry, 2009. Characterization of *Moringa oleifera* seed oil variety Congo-Brazzaville. *J. Food Techn.*, **7**(3): 56-65.

124. Ölgena, S., N. Altanlarb, E. Karataylıc and M. Bozdayıc, 2008. Antimicrobial and antiviral screening of novel indole carboxamide and propanamide derivatives. *Z. Naturforsch.*, **63c:** 189-95.

125. Oyeyinka, A.T. and S.A. Oyeyinka, 2018. *Moringa oleifera* as a food fortificant: Recent trends and prospects. *J. Saudi Soc. Agric. Sci.*, **17**(2): 127–36.

126. Palanisamy, V., K. Balakrishnan, T.V. Karivaratharaju and Arumugam, 1995. Influence of seed treatments and containers on the viability of annual moringa seeds, ***SIH***, 42-43. Via Ramesh *et al.*, 2014; www.arccjournals.com.

127. Panda, S., A. Kar, P. Sharma and A. Sharma, 2013. Cardioprotective potential of *N*,α-L-rhamnopyranosyl vincosamide, an indole alkaloid, isolated from the leaves of *Moringa oleifera* in isoproterenol induced cardiotoxic rats: *In vivo* and *in vitro* studies. *Bioorg. Med. Chem. Lett.*, **23**: 959–62.

128. Pandey, K.B. and S.I. Rizvi, 2009. Plant polyphenols as dietary antioxidants in human health and disease. *Oxid. Med. Cell Longev.*, **2**: 270–78.

129. Paquette, L.A., D. Crich, P.L. Fuchs and G.A. Molander, 2009. Encyclopedia of reagents for organic synthesis, 1017p. Wiley, New York.

130. Pillai, K.S., K. Saradamma and M.R.G.K. Nair, 1979. *Helopeltis antonii* Sign. as a pest of *Moringa oleifera*. Cur. *Sci.*, **49**(7): 288–9.

131. Pintaudi, B., G. Di Vieste and M. Bonomo, 2016. The Effectiveness of Myo-Inositol and D-Chiro Inositol Treatment in Type 2 Diabetes. I. *J. Endocrinol.*, **52**: 1-5. Article ID **9132052.** https://doi.org/10.1155/2016/9132052.

132. Prabu, M.J., 2009. A farmer's experimentation leads to a highly popular drumstick variety. *The Hindu* (Jan. 29, 2009). Updated September 16, 2010. http://www.thehindu.com/todays-paper/tp-features/tp-sci-tech-and-agri/A-farmerrsquos-experimentation-leads-to-a-highly-popular-drumstick-variety/article15951226.ece [Accessed November 30, 2013].

133. Prakash, D. and C. Gupta, 2012. Glucosinolates: The phytochemicals of nutraceutical importance. *J. Complement. Integr. Med.*, 9(1). https://doi.org/10.1515/1553-3840.1611.

134. Prakash, D., S. Suri, G. Upadhyay and B.N. Singh, 2007. Total phenol, antioxidant and free radical scavenging activities of some medicinal plants. *Int. J. Food Sci. Nutr.*, **58**: 18–28.

135. Radovich, T.J.K. and R. Paull, 2008. Early growth, leaf yield, protein content and pod yield of four *Moringa* accessions in Hawai'i. *HortSci.*, **43**: 1135.

136. Radovich, Ted (2011). C.R. Elevitch (ed.). Farm and Forestry Production and Marketing Profile for *Moringa* (PDF), In: Specialty Crops for Pacific Island Agroforestry (ed. Elevitch, C.R.). http://agroforestry.net/images/pdfs/Moringa_specialty_crop.pdf.

137. Rajakrishnamoorthy, V., S. Santhanabosu, V.K. Duraisamy and A. Rajagopal, 1994. Drip irrigation in annual moringa. *Madras agri. J.*, **81**(12): 678-9.

138. Rajamanickam, C. and S. Arokiamary, 2022. Assessment of *Moringa* Varieties for Growth and Yield Characters. *J. Krishi Vigyan*, **11**(1): 7-10.

139. Rajangam, J., R.S. Azahakia Manavalan, T. Thangaraj, A. Vijayakumar and N. Muthukrishan, 2001. Status of production and utilisation of moringa in Southern India. Development potential for Moringa products (Oct. 29th – Nov. 2nd), Dar es Salaam,Tanzania. www.arccjournals.com.

140. Ramachandran, C., K.V. Peter and P.K. Gopalakrishnan, 1980. Drumstick (*Moringa oleifera*) - a multipurpose Indian vegetable. *Econ. Bot.*, **34**(3): 276–83.

141. Ramesh, K.A., M. Prabhu, V. Ponnuswami, V. Lakshmanan and A. Nithyadevi, 2014. Scientific Seed Production Techniques in Moringa. Agricultural Research Communication Centre. *Agri. Rev.*, **35**(1): 69-73.

142. Rhew, Z.I., J.H. Lee and Y. Han, 2020. Aster yomena has anti-arthritic activity against septic arthritis induced by *Candida albicans*: its terpenoid constituent is the most effective and has synergy with indomethacin. *Adv. Tradit. Med. (ADTM)*, **20**: 213–21.

143. Richter, N., P. Siddhuraju and K. Becker, 2003. Evaluation of nutritional quality of moringa (*Moringa oleifera* Lam.) leaves as an alternative protein source for Nile tilapia (*Oreochromis niloticus* L.). *Aquacult.*, **217**: 599–611.

144. Rodriguez-Garay, B., G.C. Phillips and G.D. Kuehn, 1989. Detection of norspermidine and norspermine in *Medicago sativa* L. (Alfalfa). *Plant Physiol.*, **89**(2): 525-9.

145. Saeed, N.M., E.E. Demerdashb, H.M. Abdel-Rahmana, M.M. Algandabyc, F.A. Al-Abbasid and A.B. Abdel-Naim, 2012. Anti-inflammatory activity of methyl palmitate and ethyl palmitate in different experimental rat models. *J. TAAP.*, **264**(1): 84-93.

146. Sagiv, A.E., S. Dikstein and A. Ingber, 2001. The efficiency of humectants as skin moisturizers in the presence of oil. *PubMed. S.R&T*, **7**(1): **32**-5.

147. Saha, P., R.C. Jena, B. Sahoo, K. Sahoo and A. Lenka, 2012. Sufficing Neutraceutical Rich Multipurpose Leafy Vegetable on Earth : Moringa.

Odisha Rev., pp. 72-80. https://magazines.odisha.gov.in/Orissareview/2012/oct/engpdf/72-80.pdf.

148. Sahakitpichan, P., C. Mahidol, W. Disadee, S. Ruchirawat and T. Kanchanapoom, 2011. Unusual glycosides of pyrrole alkaloid and 4′-hydroxyphenylethanamide from leaves of *Moringa oleifera*. *Phytochem.*, **72**: 791–5.

149. Scruel, O., C. Vanhoutte and A. Sener, 1998. Interference of D-mannoheptulose with D-glucose phosphorylation, metabolism and functional effects: Comparison between liver, parotid cells and pancreatic islets. *Mol. Cell Biochem.*, **187**: 113–20.

150. Serafico, M.E., L.A. Perlas, C.R. Magsadia, J.A. Desnacido, R.V. Viajar, E.O. Rongavilla, G.P. Azana and T.P. Trinidad, 2017. Efficacy of Malunggay (*Moringa oleifera*) leaves in improving the iron and vitamins A and B status of Filipino schoolchildren. *Acta Hort.*, **1158**: 293–302.

151. Shendurse, A.M. and C.D. Khedkar, 2016. Encyclopedia of Food and Health, Elsevier, pp. 239-47.

152. Siddhuraju, P. and K. Becker, 2003. Antioxidant properties of various solvent extracts of total phenolic constituents from three different agroclimatic origins of drumstick tree (*Moringa oleifera* Lam.) leaves. *J. Agric. Food Chem.*, **51**: 2144–55.

153. Siddiqui, N., B. Azad, M.S. Alam and R. Ali, 2010. Indoles: role in diverse biological activities. *IJPCR*, **2**(4): 121-7.

154. Singh, B.N., B.R. Singh, R.L. Singh, D. Prakash, R. Dhakarey, G. Upadhyay and H.B. Singh, 2009. Oxidative DNA damage protective activity, antioxidant and anti-quorum sensing potentials of *Moringa oleifera*. *Food Chem. Toxicol.*, **47**: 1109–16.

155. Singh, P.H, 2010. Moringa: A Crop of Future. In: Brain Storming Session on Moringa. Organised at TNAU, Coimbatore on 23 September, 2010.

156. Sivasubramanian, K. and C.P. Thiagarajan, 1997. Storage potential of moringa seeds. *Madras Agric. J.*, **84**(10): 618-20.

157. Skoweranda, J., M. Bukowska-Strzyzewska, R. Bartnik and W. Strzyżewski, 1990. Molecular structure and some reactivity aspects of 2,6-diamino-4(3H)-pyrimidinone monohydrate. *J. Chem. Cryst.*, **20**(2): 117-21.

158. Somoza, V., P. Pirkwieser, S. Grosshagauer and K. Kraemer, 2021. The Future of Moringa Foods: A Food Chemistry Perspective. *Front. in Nutrit.*, **8**: 844.

159. Sreelatha, S. and P.R. Padma, 2009. Antioxidant activity and total phenolic content of *Moringa oleifera* leaves in two stages of maturity. *Plant Foods Hum. Nutr.*, **64**: 303–11.

160. Stadlmayr, B., U.R. Charondiere and V.N. Enujiugha, 2012. West African food composition table. The Food and Agriculture Organization of the United Nations, Rome.

161. Starostin, K.E., A.M. Lapitskaya, V.A. Ignatenko, K.K. Pivnitsky and L.G. Nikishin, 2000. Practical synthesis of hex-5-ynoic acid from cyelohexanone. *Russ. Chem. Bull.*, No. 49, pp.81-4.

162. Sultana, B. and F. Anwar, 2008. Flavonols (kaempeferol, quercetin, myricetin) contents of selected fruits, vegetables and medicinal plants. *Food Chem.*, **108**: 879–84.

163. Sultana, B., F. Anwar and M. Ashraf, 2009. Effect of extraction solvent/technique on the antioxidant activity of selected medicinal plant extracts. *Molecules*, **14**: 2167–80.

164. Sunkara, S.P., H.J. Zwolshen, J.N. Prakash and L.T. Bowlin, 1988. Mechanism of Antitumor activity of Norspermidine, a structural homologue of Spermidine. Progress in Polyamine Research, Plenum Press, New York, pp. 707-16.

165. sustainablebioresources.com, 2023. *Sustainable Bioresources*, LLC. Hawaii. **Varieties and Ecotypes of *Moringa oleifera.*** Last updated on May 16, 2023. Accessed on December 4, 2023. https://sustainablebioresources.com/plants/plant-families/moringaceae/moringa-oleifera-2/varieties-ecotypes-moringa-oleifera/.

166. Suthanthirapandian, I.R., S. Sambandamurthy and I. Iralappan, 1989. Variations in seedling population of annual moringa. *South Indian Hort.*, **35**(5): 301-2.

167. Teixeira, E.M.B., M.R.B. Carvalho, V.A. Neves, M.A. Silva and L. Arantes-Pereira, 2014. Chemical characteristics and fractionation of proteins from *Moringa oleifera* Lam. leaves. *Food Chem.*, **147**: 51–4.

168. Thampalayagowder, R. 2017. An ODC Moringa Variety – Nutritious and High Yielding Drumstick. Department of Agriculture, Government of Tamilnadu – VIRALIMALAI – PUDUKOTTAI. https://www.linkedin.com/pulse/odc-moringa-variety-nutritious-high-yielding-ramu-thampalayagowder-1 [Accessed 02 November 2020].

169. Tian, X., H. Tang, H. Lin, G. Cheng, S. Wang and X. Zhang, 2013. Saponins: The potential chemotherapeutic agents in pursuing new anti-glioblastoma drugs. *Mini. Rev. Med. Chem.*, **13**: 1709–24.

170. TNAU (Tamil Nadu Agricultural University). 2023. Advances in Production of moringa. *TNAU Agritech Portal* Pages:146. http://agritech.

tnau.ac.in/horticulture/pdf/Moringa%20English%20book.pdf [Retrieved on 25 November, 2023].

171. Tsakinis, J., V. Sapiliotis, S. Lalas, V. Gergis and V. Dourtoglou, 1999. Quality changes of *Moringa oleifera,* variety Mbololo of Kenya seed oil during frying. *Riv. Ital. Sost. Gras.,* **75**(1): 21-7.

172. Tsakinis, J., S. Lalas, V. Gergis and V. Spiliotis, 1998. A total characterisation of *Moringa oleifera* Malawi seed oil. *Riv. Ital. Sost. Gras.,* **75**(1): 21–7.

173. Ullasa, B.A. and R.D. Rawal, 1984. *Papaver rheoeas* and *Moringa oleifera,* two new hosts of papaya powdery mildew. *Curr. Sci.,* **53**(14): 754–5.

174. USDA, 2013. U.S. Department of Agriculture Agricultural Research Service. USDA national nutrient database for standard reference, release 25. https://ndb.nal.usda.gov/ndb/search/list Updated 2016.

175. van Rooijen, M.A., J. Plat, W.A. Blom, P.L. Zock and R.P. Mensink, 2021. Dietary stearic acid and palmitic acid do not differently affect ABCA1-mediated cholesterol efflux capacity in healthy men and postmenopausal women: A randomized controlled trial. *Clinic. Nutr.,* **40**(3): 804-11.

176. Vergara-Jimenez, M., M.M. Almatrafi and M.L. Fernandez, 2017. Bioactive components in *Moringa oleifera* leaves protect against chronic disease. *Antioxidants,* **6**: 91.

177. Verma, A.N. and A.D. Khurana, 1974. Further host records of *Indarbela* sp. (Lepidoptera: Matarbelidae). *Har. agric. Univ. J. Res.,* **4**(3): 253–4.

178. Verma, S., A. Singh and A. Mishra, 2013. Gallic acid: Molecular rival of cancer. *Environ. Toxicol. Pharmacol.,* **35**: 473–85.

179. Vikaspedia, 2023. Drumstick Insect Pests. Accessed on 9 December, 2023. https://vikaspedia.in/agriculture/crop-production/integrated-pest-managment/ipm-for-vegetables/ipm-strategies-for-drumstick/drumstick-insect-pests.

180. Waterman, C., D.M. Cheng, P. Rojas-Silva, A. Poulev, J. Dreifus, M.a. Lila and I. Raskin, 2014. Stable, water extractable isothiocyanates from *Moringa oleifera* leaves attenuate inflammation *in vitro. Phytochem.,* **103**: 114–22.

181. Waterman, C., P. Rojas-Silva, T.B. Tumer, P. Kuhn, A.J. Richard, S. Wicks, J.M. Stephens, Z. Wang, R. Mynatt, W. Cefalu and I. Raskin, 2015. Isothiocyanate-rich *Moringa oleifera* extract reduces weight gain, insulin resistance and hepatic gluconeogenesis in mice. *Mol. Nutr. Food Res.,* **59**(6): 1013-24.

182. Weill, P., C. Plissonneau, P. Legrand, V. Rioux and R. Thibault, 2020. May omega-3 fatty acid dietary supplementation help reduce severe complications in Covid-19 patients?. *Biochimie,* **179**: 275-80.

183. Wick, A.N., M.C. Almen and L. Joseph, 1951. The metabolism of sorbitol. *J. Amer. pharmac. Assoc.* (Scientific ed.), **40**(11): 542-4.
184. Wikipedia, 2023. Wikipedia – The Free Encyclopedia. https://en.wikipedia.org/wiki/Moringa_oleifera. Accessed on 25 November, 2023.
185. Yang, R.Y., R.Y. Yang, S. Lin and G. Kuo, 2008. Content and distribution of flavonoids among 91 edible plant species. *Asia Pac. J. Clin. Nutr.*, **17**: 275–9.
186. Yu, X., M. Zhao, F. Liu, S. Zeng and J. Hu, 2013. Identification of 2,3-dihydro-3,5-dihydroxy-6-methyl-4H-pyran-4-one as a strong antioxidant in glucose–histidine Maillard reaction products. *Food Res. int.*, **51**(1): 397-403.
187. Zaitseva, N.V., T.S. Ulanova and O.V. Dolgikh, 2018. Diagnostics of Early Changes in the Immune System Due to Low Concentration of N-Nitrosamines in the Blood. *Bull. Exp. Biol. Med.*, **164**: 334–8.
188. Zelikin, A.N. and D. Putnam, 2005. Poly(carbonate-acetal)s from the dimer form of dihydroxyacetone. *Macromolec.*, **38**(13): 5532-7.
189. Zhang, M., S.N. Hettiarachchy, R. Horax, A. Kannan, M.D.A. Praisoody, A. Muhundan and C.R. Mallangi, 2011. Phytochemicals, antioxidant and antimicrobial activity of *Hibiscus sabdariffa*, *Centella asiatica*, *Moringa oleifera* and *Murraya koenigii* leaves. *J. Med. Pl. Res.*, **5**: 6672–80.
190. Zhao, S. and D. Zhang, 2013. Supercritical fluid extraction and characterisation of *Moringa oleifera* leaves oil. *Sep. Purif. Technol.*, **118**: 497–502.
191. Zhao, Z. and M.H. Moghadasian, 2008. Chemistry, natural sources, dietary intake and pharmacokinetic properties of ferulic acid: A review. *Food Chem.*, **109**: 691–702.

Mahua [*Madhuca longifolia* (L.) J.F.Macbr.]

Family: Sapotaceae.

Synonyms (for Soth Indian one): *Bassia longifolia* L., *B. villosa* Wall. ex G. Don, *Illipe malabarorum* J.Koenig ex Gras subsp. *alphonsae* Dubard, (POWO, 2024c) nom. superfl.; *Vidoricum longifolium* (L.) Kuntze (POWO, 2024a) ~ with common names as 'South Indian mahua', Illupai', 'Ilupa', 'Irippa', 'Madhookam', 'Naatu billupai', 'Njannal', 'Poonam', *etc.* (BiodiversityIndia, 2024a); while **synonyms** of var. *latifolia* is *B. latifolia* Roxb., *Illipe latifolia* (Roxb.) F. Muell., *I. malabrorum* subsp. *latifolia* (Roxb.) Dubbard, *Madhuca indica* J.F. Gmel., *M. latifolia* (Roxb.) J.F. Macbr., *Vidoricum latifolium* (Roxb.) Kuntze (POWO, 2024b) ~ with common names 'Butternut tree', 'Bassia', 'Indian butter tree', 'Moatree', 'Moah tree', 'Honey tree', 'Mahuam', 'Mahuo', 'Mahuwa', 'Mauwa' though vernacular names in Hindi are 'Mahua'(महुआ), 'Mohwa', 'Gilaunda' (गिलौंदा), 'Gur' (गुड़), 'Gulu' (गुलू), 'Vanprasth' (वानप्रस्थ); **in Bengali**: 'Maul'; **in Kashmiri**: 'Mahwa-mova', 'Mahora'; **in Konkani**: 'Mohwa'; **in Kannada**: 'Irpe', 'Hunage', 'Kaadippe', 'Haalippe', 'Doddippe'; **in Malayalam**: 'Ilupa', 'Irippa', 'Njannal', 'Ilippa'; **in Tamil**: 'Illupei', 'Kattiruppai'; **in Telugu**: 'Ippa', 'Madhukamu', **in Marathi**: 'Kat-illipi', 'Moha', 'Mhomva'; and **in Sanskrit**: 'Madhuca' (मधूक), 'Madhucam' (मधुकम्), 'Moha' (मोंह) and 'Vānaprasthh' (वनप्रस्थः); *etc.* (Flowersofindia, 2024; Moolihai, 2024; BiodiversityIndia, 2024b; Suryawanshi and Mokat, 2021).

5.1 Introduction

The tree native to tropical regions of India, Central Nepal, Sri Lanka and Bangladesh (POWO, 2024b) grows throughout the greater part of India except the temperate and arid areas and southern parts of Indian peninsula. Flourishing in a variety of environments ~ it showcases remarkable adaptability thriving even in arid conditions, and stands out prominently in mixed tropical deciduous forests. It is a common tree of the deciduous forests of Madhya Pradesh, central India, Maharashtra, Gujarat, peninsular India, Chota Nagpur, Odisha and characteristically all the dry plains. It is extensively cultivated near villages because of its multi-utility. Two distinct varieties of mahua tree are recognized based on leaf morphology and geographical distribution. *Madhuca longifolia* var. *longifolia* characterized by its elongated leaves predominantly thrives in South India while *M.l.* var. *latifolia* (Roxb.) A. Chev. having broader leaves in rest of the Indian plains (TNAU, 2024a; efloraofindia, 2024; POWO, 2024a; BiodiversityIndia, 2024a, b; Suryawanshi and Mokat, 2021; Kendre and Wakte, 2021).

While it may not be coveted for its timber as it is sparingly cut for such purposes, the mahua tree holds a distinguished position among Indian forest trees due to its delectable and nutritive flowers. Its significance transcends the typical metrics of forestry due to tree's lush growth and notably its flowers which make it a vital source of sustenance for the inhabitants of Central India. The flowers, renowned for their exquisite taste, possess unique quality of being storable for extended periods, making them a staple and resilient component of local diet (Flowersofindia, 2024).

5.2 Botany

5.2.1 Tree size and shape

Trunk in Narsapur, Medak district, India

By J.M.Garg - Own work, CC BY 3.0, https://commons.wikimedia.org/w/index.php?curid=7766963

***Madhuca longifolia* tree in Narsapur, Medak district**

By J.M.Garg - Own work, CC BY 3.0, https://commons.wikimedia.org/w/index.php?curid=7766966

The mahua tree is characterized as a medium-sized to large deciduous tree, often reaching heights of up to 20 meters with a short bole and a large rounded crown. The tree exhibits a distinctive bark measuring approximately 1.3 centimeters in thickness. The grayish-brown and rough bark features vertical cracks with a tendency to exfoliate in thin scales. The blaze of the mahua tree is notable for its reddish-brown or dark red hue. Tree is known for its milky exudation. Its treewood is used as timber and for burning.

5.2.2 Root system

The mahua tree exhibits a superior taproot system characterized by a well-developed and extensive network of primary roots that delve deeply into the soil. This robust taproot system not only provides the plant with stability and anchorage but also facilitates efficient nutrient and water absorption from deeper soil layers. The depth and resilience of the mahua taproot contribute significantly to the overall health and vigour of the tree, enabling it to thrive in diverse environmental conditions and promoting long-term sustainability. Its dug out root wood is used as fuel wood.

5.2.3 Leaves

The leaves are characterized by their simple structure arranged alternately along the branchlets and clustered at the terminal ends. Lanceolate stipules located laterally shed soon. The petiole measuring 2-4cm is stout and smooth (glabrous). The lamina, or the flat part of the leaf is elliptic with dimensions ranging from 10-25cm in length and 6-12cm in width, var. *longifolia* having relatively narrower leaves (6-15×1.5-4cm) compared to var. *latifolia*. Lamina base is rounded, obtuse or cuneate while apex can be acute, acuminate or apiculate. Glabrous *cum* coriaceous (leathery) leaf has entire margin and exhibits a smooth texture. A leaf may have 9-14 pairs of prominent parallel lateral nerves and the intercostal areas between these veins display a faint reticulate patterning. Leaves are used as fodder, as dining plates for community dinners just by stitching 6-7 together and as community fire during winter.

5.2.4 Flowers

The bisexual flowers range in colour from cream to yellowish-white emerging from the axils of the leaves immediately after winter is over. The calyx consisting of four lobes arranged in pairs (2+2) is obovate and covered in fulvous tomentose giving it a densely fuzzy texture and this terminates in acute points. Corolla tube measuring about 8mm long has 6-12 oblong lobes. The stamens totalling 16 are arranged roughly in three series. The anthers ~ each measuring 3.5mm are sparsely hairy where connectives are apiculate. Ovary comprising of six cells is superior measuring 3mm long. Style extending to 2.5cm is exserted and pubescent below while stigma is obtuse in shape. This detailed description captures the intricate floral morphology of *Madhuca longifolia* illustrating the complexity and unique features of its reproductive structures.

Flowering pattern in mahua

Flowering pattern in mahua

(Singh *et al.*, 2008)

Flowers from Melghat Tiger Reserve, Maharashtra

By Dr. Raju Kasambe - Own work, CC BY-SA 3.0, https://commons.wikimedia.org/w/index.php?

Fresh mahua flowers for preparing sweet dishes

By Gypsypkd - Own work, CC BY-SA 3.0, https://commons.wikimedia.org/w/index.php?curid=7461237

Sun-dried mahua flowers for various sweet preparations

By Gurpreet Singh Ranchi - Own work, CC BY-SA 4.0, https://commons.wikimedia.org/w/index.php?curid=38007044

These corolla tubes are edible as fresh for various sweet preparations, used fresh by extracting its very sweet juice in various culineries and after perfect sun drying are stored for long duration for various preparations afterwards at special occasions. The dried ones are fried *cum* crushed and mixed either with fried *cum* crushed linseed seeds/sesamum seeds for using as food at home or when on a trip elsewhere. The dried corolla tubes may also be boiled with seeds either of pea/gram or pigeon pea and taken as food with or without curd ~ a very tasty food. Dried corolla tube is also served to lactating buffaloes and cows for increase of their milking capacity. Also the dried corolla tubes are used in distillation for preparing famous country liquor ~ 'Daaroo'.

5.2.5 Fruits

The fruit, manifested as fleshy berry is notable for its substantial size and ovoid shape. When mature, the fruit takes on a greenish hue and possesses a fleshy consistency. The outer surface of the fruit is often smooth and its colour can vary ranging from green when immature to a more subdued brown or reddish-brown when mature. The outer surface of the fruit is covered with a tawny-tomentose texture. The pedicel which serves as the stalk attaching the fruit to the plant is notably stout. Within the fleshy pulp the fruit contains one to four (generally 1-2) seeds. This distinctive fruit structure contributes to the reproductive biology of the mahua tree ~ playing a role in seed dispersal and the perpetuation of the species. On ripening, its pulp is very sweet and an attraction for wasps, bees, other insects and swines.

Fruit in Narsapur, Medak district, India

By J.M.Garg - Own work, CC BY 3.0, https://commons.wikimedia.org/w/index.php?curid=7766956

Raw fruits at matured stage are collected, peeled and then after removal of the seeds these are prepared into cooked vegetables with hot spices. Cooking may be carried out with or without potato. Its cooked vegetable is very tasty and a favourite for many.

5.2.6 Seeds

The seeds are encapsulated within hard and woody drupes which are the fruit structures of this species. The seeds are quite large, oval and brown in colour. Enclosed within the tough outer shell of the drupe the seeds are safeguarded until they reach maturity. Mahua seeds hold significant economic importance, primarily due to the extraction of edible oil from them. Seeds have brown peels (testa) which after removal are sun-dried and its oil is extracted for culinary purposes, though now as many alternatives are available so these are sold in the market to be used in soap industries. This oil is a valuable culinary ingredient in various regions and is also employed in traditional medicinal practices. Seeds serve as a potential food source contributing to the nutritional aspects associated with mahua utilization. The resilience of these seeds, nestled within protective drupes, showcases the tree's reproductive strategy and its adaptation to diverse environmental conditions, reflecting its ecological and economic significance. The harvesting of seeds typically takes place between May and July in northern part of India though little early in southern India, with villagers employing manual hand-picking techniques or bamboo sticks for collection. Subsequently, indigenous methods are commonly employed for extraction of oil from these seeds.

5.2.7 Phenology

Flowering typically commences in mahua trees when these are about 10 years old. The mahua tree exhibits a cyclic pattern in seed production with 1-2 prolific seed-years anticipated every three years. This periodicity in seed abundance could be influenced by various environmental factors.

From February to April, the mahua tree undergoes a distinct seasonal transition marked by the shedding of the majority of its leaves. This shedding coincides with emergence of musky-scented flowers arranged in clusters of a dozen or so dangling gracefully from the tips of the gnarled grey branchlets. Flowering of mahua tree occurs from March to April. During this period tubularly fleshy corolla which serves as the protective covering for the flower falls off either concurrently with or before the emergence of new leaves. The mahua tree blooms predominantly at night and as dawn breaks each ethereal flower with its brief existence delicately descends to the ground. Following this ephemeral floral display the tree progresses into fruiting (Oudhia and Paull, 2008; Khare *et al.,* 2018; Kendre and Wakte, 2021; TNAU, 2024a; BiodiversityIndia, 2024a, b; Flowersofindia, 2024).

5.3 Ecological and cultural significance

Mahua holds ecological significance by fostering biodiversity through its role as a nectar source for pollinators and as a food resource for various wildlife species. Furthermore, the tree's cultural importance is reflected in its traditional uses including extraction of edible oil from its seeds which holds both culinary and medicinal value in indigenous practices. Understanding the ecological and cultural dimensions of mahua enhances our appreciation of its multifaceted role in diverse ecosystems.

5.4 Nutritional value and utilization

Mahua is a versatile tree with various components finding applications in agriculture and animal husbandry.

5.4.1 Leaves

The leaves serve as nutritious fodder, while both flowers and fruits can be utilized as supplementary feed for cattle. The chemical composition of the leaves reveals a content of crude protein at 9.8%, crude fibre at 20.3%, N-free extract at 60.7%, ether extract at 4.1%, total ash content at 7.8%, calcium at 1.6% and phosphorus at 0.2% (TNAU, 2024a).

6.4.2 Seed oil

Seeds of mahua tree are particularly valuable containing approximately 39% oil. This oil is not only used for culinary purposes but also finds application in soap and detergent making, skin care and as a vegetable butter. It can also be used as a fuel oil. The residual seed cake, a byproduct of oil extraction serves as a nutrient-rich feed for cattle and as an excellent organic manure. On a dry matter basis the seed cake exhibits a composition of crude protein at 19.4%, ether extract at 12%, crude fibre at 62.2%, total ash content at 6.4%, calcium at 0.2%, phosphorus at 0.8%, digestible crude protein at 8.0% and total digestible nutrients at 60% (TNAU, 2024a).

The alkaloids found in the press-cake derived from mahua seeds are reportedly utilized for piscicidal purposes in certain regions of India, particularly in aquaculture ponds. This unique application involves using the cake not only for its piscicidal properties but also as a fertilizer for the pond. The process involves fertilizing the pond with the cake, followed by draining, sun drying and refilling it with water. Subsequently, the pond is restocked with fish fingerlings. This integrated approach not only leverages the piscicidal attributes of mahua seed alkaloids but also underscores their role in pond rejuvenation and sustainable aquaculture practices (Keenan, 1920; Pillay and Kutty, 2005).

Nutritional aspects of seed oil of mahua as reported by Marikkar and Yanty (2012) and Mishra and Pradhan (2013) is given here under:

S. No.	Parameters	Value
1.	Refractive index	1.452-1.462
2.	Iodine value	55-70
3.	Saponification value	187-197
4.	Unsaponifiable matter (%)	1-35
5.	Palmitic acid C 16:0 (%)	24.56
6.	Stearic acid C 18:0 (%)	22.77
7.	Oleic acid C C18:0 (%)	37.08
8.	Linolic acid C18:2 (%)	14.3

5.4.3 Flowers

Mahua flowers, a prolific product of a fully developed tree, can yield up to 90kg in a year. Mahua flowers boast a notable richness in total sugars, particularly characterized by a substantial presence of reducing sugars. The abundance of these reducing sugars underscores the significant carbohydrate content within the floral composition, contributing to the overall nutritional profile of mahua flowers. The chemical make up of the flowers includes crude protein 8%, ether extract 1.4%, crude fibre 30.4%, N-free extract 52.4%, total carbohydrate 83.8%, total ash content at 7.8%, calcium 0.22%, and phosphorus 0.16%. Protein content of the flowers is comparable to that of rice while the total carbohydrate aligns with maize and oats. Interestingly, mahua flowers also serve as a source of alcohol for tribal communities. This comprehensive utilization of mahua components highlights the tree's significance in sustainable agriculture and local livelihoods (Suryawanshi and Mokat, 2020; TNAU, 2024a). Mahua flowers and fruits are utilized in various culinary applications, showcasing the versatility and nutritional value of these plant components.

5.4.3.1 Culinary use of mahua flowers

i) **Mahua floral fritters:** The flowers can be coated in a light batter and deep-fried to make fritters. This preparation is a popular snack appreciated for its unique taste and crunchy texture.

ii) **Mahua floral pickles:** The flowers can be pickled to create a tangy and flavourful condiment. The pickling process involves marinating the flowers in a mixture of spices, vinegar and oil which results into a delightful accompaniment to meals.

iii) **Mahua floral saute:** Mahua flowers can be sautéed with spices and other ingredients to create a delicious side dish or topping for various dishes. The sautéing process enhances the flavour and aroma of the flowers.

iv) **Inclusion in sweets:** Mahua flowers are sometimes incorporated into the preparation of traditional sweets, adding a distinctive floral note to desserts like puddings, cakes or halwa.

5.4.3.2 Culinary use of mahua fruits

i) **Mahua fruit jam:** The fruits can be used to make jams or preserves, offering a delightful spread for bread, toast or as an accompaniment to desserts.

ii) **Mahua fruit chutney:** Mahua fruits are suitable for making chutneys, combining them with spices and seasonings to create a flavourful condiment that pairs well with a variety of dishes.

iii) **Mahua fruit beverages:** The fruits can be juiced or blended to create refreshing beverages. The juice can be consumed or used as a base for cocktails and mocktails.

iv) **Mahua fruit pulp in cooking:** The pulp of mahua fruits can be incorporated into various recipes, such as curries or sauces to impart a unique flavour and richness to the dishes.

5.5 Chemical Constituents and Uses

Plant parts	Chemical constituents	Uses	References
Leaves	B-carotene, xanthophylls, D-glucosides, erthrodiol, palmitic acid, protobassic acid, myricetin, quercetin, oleanolic acid, n-hexacosanol, 3-O-L-rhamnoside, 3β-caproxy and 3β-palmitoxy- olean-12-en- 28-ol, 3-O-β-D-glucoside, quercetin and its3-galactoside, myricetin and its 3-O-arabinoside, n- octacosanol, β-sitosterol and its 3-O-β-D-glucoside	Orchitis, Eczema, Chronic bronchitis, Cushing's diseases	Khare *et al.*, 2018; Singh *et al.*, 2014
Bark	α- and β- amyrin acetates, ethylcinnamate, betulinic acid, α-terpineol and sesquiterpene alcohol, 3β- monocaprylic ester of erythrodiol and 3β-capryloxy oleanolic acid.	Used to cure bleeding spongy gums, ulcer, tonsillitis, skin diseases, epilepsy, rheumatism, pneumonia and piles	Khare *et al.*, 2018; Sangeetha and Devi, 2016
Flowers	Vitamins like A & C, quercetin, β-amyrindecanate, betulinic acid, tannins, β- amyrin acetate, stigmasterol and β -amyrin cinnamate	Used as tonic, analgesic, (aphrodisiac, demulcent) and diuretic	Singh *et al.*, 2017; Mishra *et al.*, 2013
Seeds	Arachidic acid, oleic acid, linolenic acid, aspartic acid, protobassic acid, lysine, myristic, palmitic and stearic acids, isoleucine, cysteine, α-alanine, glycine, and leucine, methionine, proline, threonine, myricetin, serine, quercetin, mi-saponin A and saponin B, madlongiside A, B, C, D	Fat obtained from seeds is laxative and is used to treat skin diseases, rheumatism, headache, piles, galactagogue, *etc.*	Yadav *et al.*, 2012
Fruits	n-hexacosanol, β-sitosterol, and its 3β-D-glucoside, quercetin, dihydroquercetin, and α- and β- amyrin acetates	Used in tonsillitis, as an astringent and in pharyngitis	Kumar *et al.*, 2015

(Kendre and Wakte, 2021)

5.6 Pharmacological Properties

S.No.	Activities	Plant parts	Solvents	References
1	Anti-hyperglycemic	Bark	Methanol	Dahake *et al.*, 2010
		Leaves	Hydro-ethanol	Ghosh *et al.*, 2009
2	Anti-inflammatory	Bark	Ethanol	Jha and Mazumder, 2018
		Seed		
		Overall plant	Acetone	Chakma, 2011
3	Larvicidal and ovicidal	Seed cakes	Aqueous	Verma *et al.*, 2014; Lanjewar and Shukla, 1986
4	Spasmolytic activity	Leaves	Crude extract	Khare *et al.*, 2018
		Seeds		
5	Insecticidal and pesticidal activity	Seeds	Crude extract	Khare *et al.*, 2018
6	Antimicrobial activity	Flowers, leaves, stem and stem bark	Ethanol, acetone, water	Singh *et al.*, 2018; Jagram and Dhanraj, 2016
7	Antibacterial	Bark, stem, flowers, leaves	Alcoholic	Khare *et al.*, 2018
8	Antifungal	Leaves, flowers	Alcoholic	Kalaivani and Jegadeesan, 2013
9	Analgesic	Flowers, leaves	Methanol	Verma *et al.*, 2014
10	Hepatoprotective	Flowers	Methanol	Singh *et al.*, 2017
11	Anti-oxidant	Bark	Ethanol	Verma *et al.*, 2014
		Leaves		Khare *et al.*, 2018
12	Dermatological uses	Leaves	Ethanol	Khare *et al.*, 2018
		Bark	Ethanol	
13	Anti-convulsant/ anti-epileptic	Bark	Methanol	Palani *et al.*, 2010
		Leaves	Ethanol	
14	Anti-cancer	Leaves	Ethanol	Jagram and Dhanraj, 2016
15	Anthelmintic	Flowers	Ethanol Methanol	Singh *et al.*, 2017
16	Anti-ulcer	Flowers	Alcohol	Kalaivani and Jegadeesan, 2013
		seed	Ethanol	Seshagiri *et al.*, 2007
17	Anti-tumour	Leaves	Acetone	Sangameswaran *et al.*, 2012
18	Neuropharmacological response	Leaves	Methanol	Ingankal and Swamy, 2012

(Kendre and Wakte, 2021)

The Ayurvedic Pharmacopoeia of India has suggested many therapeutic formulations of *Madhuca longifolia.* These are Svasa, Daha, Ksaya, Trsna and Srama. Following are the formulations available in market:

Marketed formulations of *Madhuca longifolia* (Khare *et al.,* 2018; Mishra, 2019)

Name of the formulation	Dosage form	Cures
Madhukasava	liquid	Bleeding, Emaciation, Tiredness, Skin diseases
Abhayarishta	liquid	Constipation, Piles
Chandanasava	liquid	Burning sensation, Burning micturition, Spermatorrhoea
Nyagrodhadi churna	powder	Diabetes, Urinary disorders
Lakshmanarishta	liquid	Gynaecological disorders like heavy and irregular periods, metrorrhagia, menorrhagia
Pancha saarapanaka	liquid	Burning sensation, burning micturition, thirst
Stanyajanana rasayana	powder	Enhances milk production, provides strength to lactating mothers
Kutajarishta	liquid	Amoebiasis, bacterial dysentery, amoebic dysentery, blood diarrhea

(Kendre and Wakte, 2021)

5.7 Cultivation

5.7.1 Climate

Mahua thrives in regions characterized by a dry tropical and subtropical climate, showcasing its adaptability to specific environmental conditions. In its natural habitat, this resilient species experiences a broad temperature range, with the absolute maximum temperature fluctuating 41-48°C. Absolute minimum temperature in these regions almost varies from 8 to 10°C (TNAU, 2024a).

It thrives in locations with abundant sunlight, showcasing a preference for well-lit areas over shaded spaces. Typically, it is found at altitudes up to 1200 metres, indicating a versatile altitude range for successful growth (Singh *et al.,* 2017; Kendre and Wakte, 2021; TNAU, 2024a).

Mahua tree exhibits a remarkable ability to withstand drought conditions, making it well-suited for arid environments. It thrives in regions characterized by an average annual rainfall ranging from approximately 750 to 1875mm. Its natural habitat experiences a diverse range of relative humidity levels with values spanning from about 40 to 80% in January and extending to approximately 60 to

90% in July. These climatic conditions play a pivotal role in shaping the ecological niche of the mahua tree (TNAU, 2024a). This climatic versatility allows the mahua tree to establish itself successfully in diverse landscapes, contributing to its widespread distribution.

5.7.2 Soil

The mahua tree demonstrates a versatile adaptability to a diverse range of soils including shallow, sandy, rocky, clayey and calcareous soils with a particular inclination towards sandy soils. Flourishing in various soil compositions, it thrives notably in the alluvial soils of the Indo-Gangetic plains. Remarkably, even in Sal forests, the mahua tree showcases its resilience by flourishing in high-clay and calcareous soils. This adaptability across different soil types underscores the tree's ecological versatility, making it well-suited to a broad spectrum of environmental conditions (TNAU, 2024a).

5.7.3 Silvicultural characters

The species demonstrates robust regrowth through coppicing, particularly when selectively felled during the hot season. The efficiency of its regenerative capacity is underscored by a well-developed taproot system, contributing to the tree's resilience and adaptability.

5.7.4 Nursery techniques

Cultivating mahua involves specific nursery techniques to ensure optimal seed germination and successful transplantation. Harvesting ripe fruits involves shaking the branches to collect them, followed by a meticulous process of rubbing and washing to extract clean seeds. Approximately 450 seeds weigh one kilogramme. However, it is crucial to acknowledge that germination capacity of fresh seeds diminishes over time particularly during storage where susceptibility to fungi and insect infestation increases.

For propagation, fresh seeds are sown at a depth of 1.5-2.5cm. Early transplanting preferably when seedlings are one month old into nursery beds is crucial for achieving robust growth. Spacing in these transplant beds is maintained at 30×15cm and shading becomes imperative for seedling protection. Late transplanting may yield suboptimal results. Mahua seedlings can also be containerized. In this method, one-month-old seedlings are pricked into containers with shading remaining an essential practice (TNAU, 2024a).

5.7.5 Planting

Mahua planting involves the meticulous placement of saplings in 30cm^3 pits at a spacing of 4×4 metres. Method of planting stumps employs either crowbar holes or similarly sized 30cm^3 pits. This strategic approach aims to optimize the growth and development of mahua trees ensuring proper spacing for mature trees and facilitating efficient nutrient uptake from the soil. It is recommended to plant mahua saplings during proper planting season, taking into consideration local climatic conditions for optimal establishment. Planting is usually done during rainy season when atmospheric humidity is quite high. Adequate care, including regular watering and protection from pests, enhances the success rate of mahua cultivation (TNAU, 2024a).

5.8 Insect-Pests and Diseases

Stathmopoda basiplectra is a serious pest of seeds. Among the defoliators of the tree are *Achaea janata*, *Anuga multiplicans*, *Bombotelia nugatrix*, *Metanastria hyrtaca* and the larvae of *Acrocercops euthycolona* & *A. phaeomorphia* which mine the leaves. *Unaspis acuminata* is a sap sucker and *Indarbella quadrinotata* feeds on the bark. The fungi, *Polystictus steinheilianus* causes decay in felled timber, *Fomes caryophylli* causes heartrot, *Cercospora haticola* causes leaf spot and *Scopella echimulata* is a leaf rust. Leaf blight is caused by *Pestalotiopsis dichaeta*. *Loranthus* is a serious parasitic plant of trees in some localities.

5.8.1 Insect-pests

i) **Bark caterpillar/Bark feeder (*Indarbela quadrinotata*) (Singh *et al.*, 2008; TNAU, 2024b):** This inflicts damage, especially from February to March through the activities of its smooth-skinned larvae that exhibit dark chitinized patches on their segments. The adult form of this pest manifests as a moth with forewings characterized by rows of dark rusty-red spots. Manifestation of symptoms includes creation of short downward tunnels typically observed at the junction of a dead branch. Its feeding damage is noticed due to discolouration or peeling of the bark which weakens the outer protective layer (Orwa *et al.*, 2009). For effective control measures, it is recommended to insert in tunnels (boreholes) a 0.5% dichlorvos 76EC solution soaked in cotton and plugging the tunnel with mud. Tree's overall health should be maintained through proper watering, fertilization and disease control. Physical barriers or placing of deterrents on the trunk to ward off access of this pest, *via-a-vis* use of targeted insecticides will help controlling the pest. Use of petrol or 0.1% dichlorovos at 3-week intervals will control this pest effectively (Singh *et al.*, 2008).

ii) ***Miresa albipuncta* larvae (Meshram *et al.*, 1991):** Mahua tree nurseries and young plantations in Jabalpur (M.P.) were observed to be affected by larvae of *Miresa albipuncta*. The infestation primarily targeted the older leaves. Despite being of minor importance, the presence of these limacodid larvae signifies a potential threat to the overall health of the plants. Its infestation causes visible damage as feeding trails, skeletonization of leaves and formation of characteristic feeding patterns. Its infestation may be checked through introduction of natural predators/parasitoids, *vis-à-vis* through application of targeted insecticides.

iii) **Mahua fruit fly or peach fruit fly (*Bactrocera zonata*)** (Kumar *et al.*, 2021): This poses a significant threat to various fruit crops (banana, mango, guava, jamun, custard apple, pineapple, grapes, peach, apples, *etc.* and many of the vegetables) due to its reproductive behaviour and the subsequent damage inflicted on fruits. The adult female utilizes her ovipositor to lay eggs on the fruit's surface. Upon hatching, the larvae penetrate the fruit initiating feeding on the flesh or pulp so rendering the fruit unsuitable for consumption. Its infestation may be detected due to oviposition-puncturing symptoms on the fruits so fruits with high sugar content release a sugary liquid which solidifies near the site of oviposion. Proper orchard sanitation to destroy breeding sites and spraying malathion 0.2% on the crop will protect the plants.

iv) **Mahua seed borer (*Stathmopoda basiplectra*)** (Orwa *et al.*, 2009): Recognized as a significant threat as this bores into the seeds. Its presence can be noted through small entry holes on mahua seeds. Timely harvesting of mahua seeds can reduce some risk. Keeping seeds in dry and well-ventilated conditions can discourage the activity of seed borers. Introduction of natural predators/parasitoids of seed borers can be an effective biological control strategy.

v) **Defoliators** *(Achaea janata, Anuga multiplicans, Bombotelia nugatrix, Metanastria hyrtaca, Acrocercops euthycolona, A. phaeomorphia)* *(Orwa et al.,* 2009): These defoliators contribute to leaf loss, impacting the overall health of the tree. Larvae of leaf-mining moths such as *Acrocercops euthycolona* and *A. phaeomorphia* tunnel through the leaves. Severe defoliation reduces the photosynthetic activities. Presence of caterpillars or larvae on leaves, *via-a-vis* mining damage by leaf-mining moths can result in discoloured and weakened leaves. Natural predators such as birds and beneficial insects to control defoliator's population, judicious use of insecticides during early stages of infestation, removing severely infested branches and introduction of natural enemies specific to leaf-mining moths can be an effective control measure. Any systemic insecticide will also control their population.

vi) **Mahua leaf roller** (*Polychorosis cellifera*) (Singh *et al.*, 2008): The larvae exhibit a distinctive behaviour of webbing the leaves by folding the tip downwards along both margins parallel to the mid-rib. Feeding occurs inside the folded leaves. During severe infestations, up to $1/4^{th}$ of the leaf lamina may be consumed. Affected leaves should be regularly clipped and burnt to control the population of leaf rollers. In case of severe infestation, 0.2% Chlorpyriphos 20 EC is recommended.

vii) **Mahua scale insect** (*Unaspis acuminata*) (Orwa *et al.*, 2009) feeds on sap so over time weakens vigour of the trees. Its infestation causes yellowing and wilting of leaves *cum* appearance of sticky honeydew on the leaves. Application of horticultural oils can help smother and control scale insects. Introduction of predators/parasites that feed on scale insects will help reducing the population. Removing heavily infested branches will reduce scale population.

5.8.2 Diseases

A multitude of soil-borne pathogens wield the capability to instigate numerous diseases, with one of the prevalent afflictions being root rot, particularly rampant in nursery environments. Root rot caused by soil-borne pathogens poses a significant threat to health of nursery plants. Root system is affected thereby overall growth and development of plants are hampered (Verma *et al.*, 2015). A fungicide solution containing 0.2% Ridomil Gold by Syngenta is quite effective when administered at fortnightly intervals as soil drench. This formulation comprises metalaxyl (M) 4%, mancozeb 64%, sodium lignosulphonate 2% and dibutylnaphthalene sulphonic acid sodium salt 2% which is quite effective to mitigate the impact of the root rot pathogen on the seedlings. Simultaneously, careful monitoring of watering schedule should be followed to prevent excess moisture accumulation in and around root zone which will promote a healthier environment for seedling growth (Verma *et al.*, 2015).

i) **Leaf spot (*Pestalotia dichaeta*)** (Orwa *et al.*, 2009; Verma *et al.*, 2015): In this case small circular lesions develop on leaves often with darkened centre which causes leaf blight, leaf browning, necrosis, severe defoliation and wilting. Against *Pestalotia*-induced leaf spot, regular applications of 0.2% mancozeb, maintaining proper spacing for adequate ventilation and prompt removal *cum* destruction of infected leaves should be practiced.

ii) **Spotted anthracnose disease (*Sphaceloma madhucae*)** (Verma *et al.*, 2015): Its infections appear as irregularly-shaped sunken lesions on leaves which later on become covered in dark fungal spores. Effective control strategies include application of fungicides containing copper compounds, ensuring proper irrigation practices to avoid waterlogged conditions and pruning infected branches to minimize the spread of the pathogen.

iii) **Root rot (*Fusarium solani* (Mart.) Sacc.)** (Verma *et al.*, 2015): The threat of root rot is induced particularly in nursery settings in central India. Root rot symptoms encompass yellowing of leaves, general decline in plant health and wilting. To mitigate *Fusarium solani,* care should be taken that soil is well-drained, proper irrigation schedule is followed and seeds are treated with fungicidal solutions such as 0.2% thiride or captan prior to planting.

iv) **Timber decay (*Polystictus steinheilianus*)** (Orwa *et al.*, 2009)**:** Its infection causes decay in felled timber. Timber decay manifests as the softening and deterioration of felled timber where infected wood may exhibit discolouration, a musty odour and a spongy texture. After felling, the timber should properly be cured and dried and wood preservatives should be applied.

v) **Heartrot (*Fomes caryophylli*)** (Orwa *et al.*, 2009)**:** This pathogen affects central wood, potentially compromising the structural integrity of the tree. External symptoms may include fungal conks or brackets, wood discolouration and a hollowed-out central core. Infected branches should be pruned to prevent spread of the fungus and proper inspection of the tree or wood should be carried out. After cutting the fungicidal treatment of benlate or captan should be carried out time to time.

vi) **Leaf Spot (*Cercospora haticola*)** (Orwa *et al.*, 2009)**:** *Cercospora haticola* causes leaf spots characterized by dark lesions on the leaves. Severe infections may lead to defoliation, impacting the tree's photosynthetic capacity. Fungicidal sprays (0.2% mancozeb) at the first signs of infection, proper sanitation practices to remove and dispose of infected leaves and ensuring adequate spacing between trees for improved air circulation are some of the measures in mitigating this problem.

vii) **Leaf rust (*Scopella echimulata*)** (Orwa *et al.*, 2009)**:** Its infection induces leaf rust which can weaken the tree. To manage this disease, infected leaves and branches should be pruned and fungicides such as bupirimate with triforine, penconazole, mancozeb or myclobutanyl should be used. For proper air circulation in between and on the trees some pruning will have to be done. Fungicidal sprays during periods of high humidity; pruning and removing infected leaves to minimize disease spread; adequate spacing and ventilation to reduce conditions favouring fungal growth are recommended.

viii) **Parasitic plant (*Loranthus*)** (Orwa *et al.*, 2009)**:** These parasitic plants regularly draw their nutrients from the tree so affect seriously the health of mahua trees in certain locations. Physical removal of *Loranthus* plants by cutting its whole establishment is the best way to get rid of this oparasite. In that area, its population should be monitored and managed.

References

1. BiodiversityIndia, 2024a. *Madhuca longifolia longifolia*. India Biodiversity Portal. https://indiabiodiversity.org/species/show/230279. Accessed on 2 February 2024.
2. BiodiversityIndia, 2024b. *Madhuca longifolia latifolia* (Roxb.) A.Chev. India Biodiversity Portal. https://indiabiodiversity.org/species/show/230278. Accessed on 2 February 2024.
3. Chakma, C.S. 2011. Anti-microbial activity of fruits-seeds of *M. longifolia*. *Intern. Res. J. Pharm.*, **2**(9): 192-3.
4. Dahake, A.P., C.S. Chakma, R.C. Chakma and P. Bagherwal, 2010. The antihyperglycemic activity of methanolic extract of bark of *Madhuca longifolia*. *J. Diab., Endocrinol. & Metab. Dis.*, **39**(1): 1-44.
5. efloraofindia, 2024. eFloraofIndia. Accessed on 3 February 2024. https://efloraofindia.com/2011/03/13/madhuca-longifolia-var-longifolia/.
6. Flowersofindia, 2024. Mahua. Flowers of India. https://www.flowersofindia.net/catalog/slides/Mahua.html.
7. Ghosh, R., I. Dhande, V.M. Kakade, R.R. Vohra and V.J. Kadam, 2009. Antihyperglycemic activity of *Madhuca longifolia* in alloxan-induced diabetic rats. *Intern. J. Pharmacol.*, **6**(2): 1-2.
8. Ingankal, T.S. and P. Swamy, 2012. Neuropharmacological potential of methanolic extract and a triterpene isolated from *Madhuca longifolia* leaves in mice. *Ind. J. exper. Biol.*, **50**: 862-6.
9. Jagram, M. and M. Dhanraj, 2016. Medicinal and commercial potential of *Madhuca indica*: A review. *Inter. J. Medic. & Health Res.*, **2**(12): 23-6.
10. Jha, D. and P.M. Mazumder, 2018. Biological, chemical and pharmacological aspects of Asian *Madhuca longifolia*. *Pacif. J. Trop. Med.*, **11**(1): 9-14.

11. Kalaivani, M. and M. Jegadeesan, 2013. Evaluation of antiulcer activity of ethanolic extract of flowers of *Madhuca longifolia* in experimental rats. *Intern. J. Scient. & Res. Publicat.*, **3**(6): 6-9.

12. Keenan, G.I. 1920. The microscopical identification of mohraw meal in insecticides. *J. Amer. Pharmacy. Assoc.*, **9**(2): 144-7.

13. Kendre, N. and P. Wakte, 2021. A review on phytochemicals and biological attributes of *Madhuca longifolia. Asian J. Pharmacy & Pharmacol.*, **7**(2): 74-84.

14. Khare, P., K. Kishore and D.K. Sharma, 2018. Medicinal uses, phytochemistry and pharmacological profile of *Madhuca longifolia. Asian J. Pharmacy & Pharmacol.*, **4**(5): 570-81.

15. Kumar, A., B.K. Singh, N. Trivedi and A. Aggarwal, 2015. Mahua an important Indian species: A review. *Pharmacoarena: An Intern. J. Pharmaceut. Res.*, **1**(2): 13-21.

16. Kumar, P., S. Dwivedi and A. Jain, 2021. A new report of fruit fly infestation on fruits of *Madhuca indica* in central India. *J. Entom. & Zool. Stud.*, **9**(3): 318-22.

17. Lanjewar, R.D. and V.N. Shukla, 1986. The vulnerability of larvae and eggs of *Meloidogyne incognita* to some oilcake and fungicides. *Indian J. Nemat.*, **16**(1): 69-73.

18. Marikkar, N. and N.M. Yanty, 2012. Seed fat from *Madhuca longifolia* as raw material for halal alternative fats. *Borneo Sci.*, **31**: 97-106.

19. Meshram, P.B., Jamaluddin and S.C. Pathak, 1991. A new report of slug caterpillar, *Miresa albipuncta* Herr-Schaff (Lepidoptera: Limacodidae) as a pest of mahua, *Madhuca latifolia*. Indian *J. Appl. and Pure Biol.*, **6**(1): 79.

20. Mishra, A. 2019. Mahua (*Madhuca longifolia*) flowers: a review on processing and biological properties. *Nutr. & Food Sci.*, **49**(6): 1153-63.

21. Mishra, S. and S. Pradhan, 2013. *Madhuca longifolia* (Sapotaceae). A review of its traditional uses and nutritional properties. *Intern. J. Hum. & Soc. Sci. Inv.*, **2**(5): 30-36.

22. Moolihai, 2024. Mahua (Illupai): benefits, nutrition, uses, side effects. MOOLIHAI ORGANICS. https://www.moolihai.com/benefits-of-mahua/.

23. Orwa, C., A. Mutua, R. Kindt, R. Jamnadass and S. Anthony, 2009. *Madhuca latifolia*. Agroforestree Database:a tree reference and selection guide version 4.0 (http://www.worldagroforestry.org/sites/treedbs/treedatabases.asp) Retrieved on 3 February 2024 via https://apps.worldagroforestry.org/treedb/AFTPDFS/Madhuca_latifolia.PDF.

24. Oudhia, P. and R.E. Paull, 2008. Butter tree *Madhuca latifolia* Roxb.: Sapotaceae. *Encyclopedia of Fruit and Nuts* (eds: Janick, J. and R. E. Paull), pp. 827-8. CABI, Wallingford, United Kingdom.
25. Palani S., S. Raja, S. Karthi, Selvi Archana and B. Senthil Kumar, 2010. *In vivo* analysis of nephro and hepatoprotective effects and antioxidant activity of *Madhuca longifolia* against acetaminophen-induced toxicity & oxidative stress. *J. Pharm. Res.*, **3**(1) :9-16.
26. Pillay, T.V.R. and M.N. Kutty, 2005. Aquaculture: Principles and Practices (2nd ed.), 623p. Blackwell Publishing Ltd., Kerala.
27. POWO, 2024a. *Madhuca longifolia* (L.) J.F.Macbr. Plants of the World Online. Royal Botanic Gardens, Kew Science. Retrieved on 29 January 2024. (https://powo.science.kew.org/taxon/urn:lsid:ipni.org:names:787505-1). First published in Contr. Gray Herb., n.s., 53: 17 (1918).
28. POWO, 2024b. *Madhuca longifolia* var. *latifolia* (Roxb.) A.Chev. Plants of the World Online. Royal Botanic Gardens, Kew Science. Retrieved on 29 January 2024. (https://powo.science.kew.org/taxon/urn:lsid:ipni.org:names:77189991-1). First published in Rep. Bot. Appl. 23: 149 (1943).
29. POWO, 2024c. *Madhuca longifolia* var. *longifolia.* Plants of the World Online. Royal Botanic Gardens, Kew Science. Retrieved on 29 January 2024. (https://powo.science.kew.org/taxon/urn:lsid:ipni.org:names:77170637-1).
30. Sangameswaran, B., M.S. Saluja, A. Sharma and N. Manocha, 2012. Anti-tumor effect of acetone extract of *Madhuca longifolia* against Ehrlich Ascites Carcinoma (EAC) in mice. *Phytopharmacol.*, **3**(1): 130-36.
31. Sangeetha, R. and N. Devi, 2016. Sapotaceae: *Madhuca longifolia*. A review of its phytochemical and pharmacological profile. *Intern. J. Pharma and Bio Sci.*, 7(4): (B)106–14.
32. Seshagiri, M., R.D. Gaikwad, S. Paramjyothi, K.S. Jyothi and S. Ramchandra, 2007. Anti-inflammatory, anti-ulcer and hypoglycemic activities of ethanolic and crude alkaloid extract of *Madhuca indica* (Koenig) Gmelin seed cake. *Orient. Pharm. & exp. Med.*, **7**(2): 141-9.
33. Singh, S., A.K. Singh, B.G. Bagle and T.A. More, 2008. Mahua – A multipurpose tree for tribals. CIAH, Bikaner. Technical Bulletin No. 32. Accessed on 3 February 2024. https://ciah.icar.gov.in/publication/Technical_bulletin/Mahua_4_tribals.pdf.
34. Singh, P., A. Goel and M. Sehgal, 2018. Phytochemical analysis and antimicrobial activity of *Madhuca longifolia* and *Clerodendrum infortunatum*: Medicinal plants for application as textile finishes. *Intern. J. Chem. Stud.*, **6**(5): 2031-40.

35. Singh, S.K., R.V. Savadi and K.V. Akki, 2014. A review on *Madhuca longifolia. Asian J. Pharmac. Educat. & Res.*, **3**(2): 8-14.

36. Singh, V., J. Sinha, J. Singh and A.K. Rai, 2017. Phytochemistry, ethnomedical uses and future prospects of mahua (*Madhuca longifolia*) as a food: A review. *J. Nutr. & Food Sci.*, **7**:1.

37. Suryawanshi, Y. and D. Mokat, 2020. Variability studies in *Madhuca longifolia* var. *latifolia* flowers from northern Western Ghats of India. *Indian J. Hill Farm..* **33**(2): 261–6.

38. Suryawanshi, Y.C. and D.N. Mokat, 2021. Morphophysiological seed variability in mahua trees from Western Ghats and its impact on tribal life. *Proc. Nat. Acad. Sci. India: Sect B: Biol. Sci.*, **91**(1): 227–39. doi:10.1007/s40011-020-01223-w. S2CID 231876956.

39. TNAU, 2024a. Bio Fuels: Mahua. TNAU Agritech Portal. https://agritech.tnau.ac.in/bio-fuels/Biofuel_Mahua.html. Retrieved on 3 February 2024.

40. TNAU, 2024b. Mahua (Illupai). Pest management in tree crops. TNAU Agritech Portal. https://agritech.tnau.ac.in/forestry/forest_pest_Illupai%20mahua.html. Retrieved on 3 February 2024.

41. Verma, N., K.K. Jha, U. Kumar, D. Kanad, N.K. Singh, A.K. Singh and R. Sharma, 2014. Biological Properties, Phytochemistry, and Traditional Uses of Mahua. A *Madhuca longifolia* Review. *Intern. J. Adv. Res. & Innovat.*, **2**(3): 630-8.

42. Verma, R., F. Shirin and P. Verma, 2015. Root-rot disease of *Madhuca indica* (mahua): a new record. *Intern. J. Cur. Res.*, **7**: 16751-4.

43. Wani, D.D. and M.J. Thirumalachar, 1971. Studies on Elsinoe and Sphaceloma disease of plants from Maharashtra state (India) –VII. *SyIdowia, Annal. MIycol.* (Ser. 2): **25**: 47-50.

44. Yadav, P., D. Singh, A. Mallik and S. Nayak, 2012. *Madhuca longifolia* (Sapotaceae): A review of its traditional use, phytochemistry and pharmacology. *Intern. J. Biomed. Res.*, **3**(07): 36-40.

Bamboo

Bamboo forest in Arashiyama, Kyoto, Japan

By Naokijp - Own work, , https://commons.wikimedia.org/w/index.php?curid=124407938

6.1 Etymology

The etymological origins of the term 'bamboo' remain shrouded in uncertainty with plausible roots tracing back to either Dutch or Portuguese linguistic influences. The trajectory of its adoption suggests an initial incorporation from

Malay word 'mambu' or Kannada languages (Kramer, 2016; Kelchner, 2013; Ayer *et al.*, 2023). During the waning years of the 16th century precisely between 1590 and 1600, the Dutch bestowed upon this remarkable plant the moniker 'bamboo'. Subsequently, it garnered its neo-Latin nomenclature '*Bambusa*'. The etymological evolution of this designation is often traced back to an intriguing origin. Some scholars opine that the ancestral term may have been 'bambu' ~ a phonetic semblance to the resonating sound produced by bamboo when subjected to the fervent embrace of an open flame (Kramer, 2016).

According to this captivating hypothesis, the auditory resemblance arises during the transformative process of bamboo under heat. When exposed to elevated temperatures, the sealed hollow chambers within the internodes undergo a remarkable phenomenon. The incarcerated hot air within these compartments expands, culminating in an explosive crescendo reminiscent of a resounding boom-boom. This auditory spectacle akin to staccato beat of distant drums is said to have inspired the initial articulation of the term 'bambu', echoing the percussive symphony emanating from bamboo's reaction to the fervour of an open fire (Kramer, 2016).

6.2 Introduction

Bamboos, a diverse assemblage of towering tree-like grasses constitute the subfamily Bambusoideae within the expansive grass family Poaceae (McClure, 2013; Canavan *et al.*, 2016; Ayer *et al.*, 2023). This botanical cohort encompasses a staggering array, surpassing 115 genera and comprising over 1,400 distinct species. These resilient grasses manifest their ecological presence in tropical and subtropical to mild temperate zones. The epicentre of bamboo diversity lies in E&SE Asia, punctuated by a significant presence on the islands dotting the Indian and Pacific Oceans. Noteworthy exceptions include select species from the genus *Arundinaria* which find their native habitat in the southern United States. Here, they form dense canebrakes along riverbanks and in marshy terrain (Britannica, 2024).

India emerges as a pivotal player on the global stage securing the second-largest rank in bamboo production. The nation's annual bamboo output is estimated at a substantial 3.23 million tonnes, attesting to its significant contribution to the worldwide bamboo supply chain (Testbook, 2024).

6.3 National Bamboo Mission (NBM) of India

The Government of India initiated the National Bamboo Mission (NBM) in 2006-2007 under the Ministry of Agriculture and Farmers Welfare, allocating an initial budget of 500+ crores. This mission launched to recognize the significance of bamboo expanded in the 10th plan and concluded its first phase in 2015-2016. In 2018-19, a restructured NBM was introduced as part of the National Mission for Sustainable Agriculture (NMSA) encompassing market development, value addition, product development, skill enhancement and research *cum* development. The mission aims to foster holistic growth in the bamboo sector through a regionally differentiated area-based strategy focusing on increasing bamboo cultivation and marketing. To achieve these goals, the Mission promotes the establishment of new nurseries and strengthens existing ones to enhance the availability of quality planting material. It also addresses forward integration by boosting the marketing of bamboo products particularly handicraft items. Amendments to the Indian Forest Act in 1927 exempted bamboo grown outside the forest from being classified as a tree.

In consultation with industry and states, the Mission identified 10 priority bamboo species to be cultivated by farmers, aiming for self-sufficiency in raw material supply. Guidelines for bamboo nurseries accreditation and planting material certification have been released and states are in the process of accrediting all such nurseries whether private or in Govt. sector. As of November 30, 2021 Rs.324.08cr has been released to the states including Bamboo and Timber Sector Groups (BTSGs). NBM activities follow Annual Action Plans from State Governments and operational guidelines covering assistance for treatment, market establishment, incubation centres, value-added product development, processing, tools and equipment development and research *cum* development, including support for startups.

NBM is implemented in the following states: Andhra Pradesh, Arunachal Pradesh, Asom, Bihar, Chhattisgarh, Gujarat, Himachal Pradesh, Jharkhand, Karnataka, Kerala, Madhya Pradesh, Maharashtra, Manipur, Meghalaya, Mizoram, Nagaland, Odisha, Sikkim, Tamil Nadu, Telangana, Tripura, Uttarakhand and Uttar Pradesh (Bamboo Info, 2024).

6.4 Removal of Bamboo from 'Tree' Category (Testbook, 2024)

The significant amendment to the Indian Forest Act of 1927, in 2017 marked a pivotal shift in the classification of bamboo, separating it from the encompassing 'Tree' category. This legislative change holds profound implications as it emancipated bamboo from the stringent regulations associated with trees, fostering an environment conducive to individual cultivation and business activities centered around bamboo and its myriad products. Under revised legal framework, individuals are no longer burdened with the requirement of seeking permissions for felling and transitting of bamboos. This liberation from bureaucratic constraints serves a strategic purpose ~ to actively promote and facilitate its cultivation and utilization. The overarching objective is to harness bamboo as a sustainable resource, recognizing its multifaceted benefits and potential contributions to ecological balance and economic development.

This transformative legal amendment not only empowers individuals to engage in bamboo-related activities with greater ease but also aligns with broader environmental and economic goals. By endorsing bamboo as a viable and sustainable resource, the amended Indian Forest Act aims to stimulate its widespread cultivation, thereby promoting environmental conservation and bolstering economic opportunities for individuals involved in this sector.

6.5 Botany and Cytology

Bamboos characterized as remarkably fast-growing perennials exhibit an impressive growth rate where certain species capable of reaching up to 30cm per day (Britannica, 2024). Specific bamboo varieties can achieve remarkable growth of 91cm within a 24-hour timeframe, translating to an astonishing rate of almost 40mm per hour (equivalent to 1mm every 90 seconds) (GWR, 2024). Japanese giant timber bamboo (*Phyllostachys bambusoides*) has been documented to exhibit an extraordinary growth spurt reaching up to 1.56m in just one day (Austin and Ueda, 1970). This rapid growth coupled with bamboo's ability to thrive in marginal land conditions positions it as an excellent candidate for afforestation, carbon sequestration and climate change mitigation strategies (Singh *et al.,* 2020; Kumar *et al.,* 2021; Masisi *et al.,* 2022).

The distinctive features of bamboo stems referred to as 'culms' contribute to its unique structural characteristics. These culms being typically hollow between the ringed nodes form branching clusters emerging from a robust rhizome or underground stem. Herbaceous bamboos are usually small, resemble grasses and grow only a few centimetres tall. The height of bamboo culms varies widely ranging from 10 to15cm in the smallest species to an impressive 40m or more in the largest ones (Britannica, 2024) and reach a diameter of up to 20cm which is why they are often mistaken for 'trees'. Giant bamboos exemplified by *Dendrocalamus sinicus* can achieve extraordinary dimensions with individual culms reaching up to 46 metres in length, 36 centimeters in thickness and weighing up to 450 kilogrammes (forests.org, 2024).

The internodal regions of bamboos showcase considerable variability in length. Species like *Kinabaluchloa wrayi* boasts internodes up to 2.5 metres long (Ohrnberger, 2021) while *Arthrostylidium schombergkii* exhibits lower internodes extending up to 5 metres in length (Schombergk, 1841) ~ a length surpassed only by papyrus. In contrast, the diminutive *Raddiella vanessiae* ~ native to the savannas of French Guiana features culms measuring only 10–20 millimetres in length and about two millimetres in width (Judziewicz and Sepsenwol, 2007).

Akin to other grasses the bamboos display distinctive anatomical features in their stems. Unlike dicots and conifers, bamboo stems lack a cylindrical cambium layer between the phloem and xylem. The internodal regions of bamboo stems are typically hollow and vascular bundles are scattered throughout the culm walls. Bamboos lack the dicotyledonous woody xylem which contributes to its columnar rather than tapering stem structure (Wikipedia, 2024).

Flowering bamboo

By Mogens Engelund - Own work, CC BY-SA 3.0, https://commons.wikimedia.org/w/index.php?curid=3713086

Bunches of bamboo seeds

By Vijayanrajapuram - Own work, CC BY-SA 4.0, https://commons.wikimedia.org/w/index.php?curid=80591202

The slender leaves on juvenile culms typically emerge directly from the stem rings whereas mature culms frequently give rise to lateral branches adorned with horizontal leaf-bearing extensions. The flowering and seed production of most bamboo varieties occur only after a considerable period of growth which may span 12-120 years and this reproductive event transpires singularly in their entire lifespan so this is monocarpic in nature.

The floral biology of bamboo is a fascinating aspect of the reproductive cycle of these unique plants. Bamboo flowering is a complex and infrequent event, often occurring at long intervals ~ sometimes several decades. Although many bamboo species can bloom but most of them seldom bear fruits.

Here is an elaboration on the floral biology of bamboo:

- Bamboo species typically exhibit infrequent and synchronous flowering events often occurring at intervals of several years or even decades. This phenomenon known as gregarious or mast flowering is still not fully understood and the triggers for such synchronized flowering events remain a subject of research.

- Many bamboo species are monocarpic meaning these flower, set seeds and then die. This is a notable characteristic as bamboos usually have a long vegetative phase before transitioning to the reproductive stage. The simultaneous flowering of a population is thought to enhance cross-pollination and seed dispersal.
- Bamboo flowers are organized into spikelets which are small clusters of florets attached to a central axis or rachis. These spikelets are arranged in larger structures called inflorescences often in the form of panicles or racemes.
- Most bamboo species are wind-pollinated. The anemophilous nature of bamboo flowers means that they produce large quantities of lightweight pollen that is carried by wind. This adaptation is essential for effective pollination especially given the typically tall and dense growth habit of bamboo populations.
- Following successful pollination seeds develop within these florets. Once the seeds mature the bamboo culms that bore the flowers and seeds may start to decline or die.
- The synchronized, mast flowering strategy in bamboo populations is believed to confer evolutionary advantages. By flowering simultaneously, bamboos increase the chances of cross-pollination and genetic diversity, potentially enhancing ability of the species to adapt to changing environmental conditions.
- Bamboo flowering and subsequent seed production can have profound ecological impacts. The abundance of seeds provides a rich food source for various animals leading to population booms in certain species that feed on bamboo seeds. This can have cascading effects on ecosystems.

Understanding the floral biology of bamboo is crucial for conservation efforts and the management of bamboo resources, especially considering the economic and ecological importance of these plants in various cultures around the world.

Primarily, the propagation of these bamboos relies heavily on vegetative means. Certain species exhibit an aggressive spreading tendency forming dense undergrowth that can effectively exclude other plant varieties (Britannica, 2024). This unique combination of rapid growth, diverse culm characteristics and distinct stem anatomy underscores the remarkable nature of bamboo within the plant kingdom.

The fundamental chromosome composition varies among bamboo species ~ with most woody bamboo exhibiting a basic chromosome number of 12 (x=12) while herbaceous bamboo typically possesses 11 chromosomes (x=11). Within the realm of woody bamboo, there exist distinct polyploid subgroups. Tropical woody bamboo species are characterized by hexaploidy denoted as 2n=6x=72 while their temperate counterparts exhibit tetraploidy represented as 2n=4x=48 (Clark *et al.,* 1995).

6.6 Strength

Bamboo exhibits remarkable versatility and holds significant economic and cultural importance across S. Asia, SE Asia and E. Asia. Its multifaceted utility extends to various domains including construction materials, sustenance as a food source and as a raw material for various applications. Furthermore, bamboo has found a prominent place in the realm of arts, featuring prominently in bamboo paintings and intricate bambooworking.

In the context of structural applications, bamboo shares similarities with wood, being a natural composite material with an exceptional strength-to-weight ratio, rendering it highly suitable for constructing various structures (Lakkad and Patel, 1981). Its strength-to-weight ratio is comparable to timber with strength levels akin to robust softwood or hardwood timbers (Kaminski *et al.,* 2016a, b). The inherent strength of bamboo becomes evident in the testing of specific species such as *Bambusa tulda* from Bangladesh and adjoining regions which has demonstrated tensile strengths reaching an impressive 60,000 psi (400 MPa) (McClure, 1948).

Moreover, certain bamboo species such as *Bambusa tabacaria* from China exhibit exceptional hardness due to their high silica content. This characteristic is so pronounced that when struck by an axe, the bamboo can generate sparks (Farrelly, 1984). Such extraordinary properties contribute to the diverse and valuable applications of bamboo, making it a key resource in various cultural, economic, and artistic endeavours across the Asian continent.

6.7 Comparative Analysis of Bamboo and Traditional Trees

Though bamboo belongs to grass family, many of the larger woody bamboo species bear a striking resemblance to trees ~ often leading them to be colloquially termed 'bamboo trees'. Nevertheless, there exist notable distinctions between bamboo and conventional trees. Amidst all grasses, bamboo holds the distinction of being the largest standing alone in its ability to transform into a bonafide forest (Kramer, 2016).

In contrast to trees, bamboo lacks both meristem cells and a vascular cambium layer. The vascular cambium integral to trees is a perpetually growing stratum beneath the bark which contributes to the annual thickening of the tree trunk. Concurrently, meristem cells facilitate the upward growth of trees.

However, bamboo stems do not exhibit such prolonged growth patterns. Within a single growing season (typically spanning 6 to 8 months), a bamboo stem achieves its maximum height and thickness. Though culm fibres undergo maturation over the subsequent 3-4 years, the culm itself remains static ~ neither thickening nor elongating. A pivotal distinction lies in the absence of bark in bamboo ~ a feature ubiquitous in trees. Instead, during its initial developmental stage bamboo is enveloped by protective bracts surrounding the culm offering a unique protective layer during its formative years (Kramer, 2016).

Bamboo diverges significantly in its internal and external structures from conventional trees. Unlike the intricate network of branches and leaves seen in trees, bamboo exhibits a distinct cylindrical stem devoid of branches with leaves clustered at the top. This fundamental distinction sets the stage for further analysis into their respective growth dynamics. While both bamboo and trees share the common goal of reaching for the skies, their growth patterns diverge in noteworthy ways. Traditional trees characterized by a woody trunk and a branching canopy tend to grow incrementally over several years. In contrast bamboo known for its rapid growth can reach maturity in a matter of weeks or months. This accelerated growth is a result of its unique rhizome system which allows it to propagate swiftly and establish dense groves.

6.8 Uses (Britannica, 2024)

Bamboo a versatile plant finds an extensive array of applications particularly in E&SE Asia. The culinary domain benefits from certain bamboo species as their seeds are consumed as grains though flowering being quite rare, and the tender shoots when cooked become a popular vegetable in Chinese cuisines. Livestock thrive on the nutritional value provided by raw bamboo leaves which establishes these as valuable fodder.

Furthermore, bamboo serves as a pivotal resource in the paper industry with pulped fibres from select species such as *Dendrocalamus strictus* and *Bambusa bambos* which yield high-quality paper. The segmented structure of bamboo stems proves exceptionally valuable with the larger sections utilized for constructing durable planks for houses and rafts. Both sizeable and smaller stems are ingeniously bound together to create scaffolding indispensable in the realm of building construction.

Beyond construction, bamboo stems undergo various transformations ~ being split to craft buckets and pipes or fashioned into furniture, flooring, walking sticks, fishing poles, garden stakes and an assortment of utensils. Certain bamboo species double as ornamental elements which enhance the aesthetics of landscape gardens. The joints of bamboo stems produce a fine-grained silica known as 'tabasheer' which has served medicinal purposes in the Orient for centuries.

Not limited to practical applications, bamboo holds a revered place in East Asian culture where artists, poets and epicures have immortalized its beauty and utility in paintings and verses. The enduring legacy of bamboo extends beyond its physical uses, encapsulating a rich cultural tapestry that spans centuries.

6.9 Bamboo Importance in India

Bamboo stands as a remarkably versatile botanical marvel weaving its utility through diverse sectors within India's industrial landscape. Its significance extends across handicrafts, furniture, paper and construction industries. Notably, millions of individuals particularly those residing in rural regions rely on bamboo as a linchpin for their livelihoods. Beyond mere sustenance, bamboo emerges as a robust source of income and resources fostering employment opportunities and making substantial contributions to the nation's economic tapestry.

At its core, bamboo epitomizes a highly renewable resource owing to its rapid growth and adaptability to diverse climatic conditions. This intrinsic quality not only underscores its economic importance but also positions it as a potent force in environmental sustainability. With a minimal carbon footprint and the capacity to sequester significant amounts of carbon dioxide, bamboo assumes a pivotal role in climate change mitigation, rendering it an invaluable asset in the pursuit of environmental equilibrium.

Furthermore, Bamboo emerges as a beacon of sustainable practices presenting itself as an eco-friendly alternative to conventional materials like wood. Advocating for the increased use of bamboo aligns seamlessly with India's aspirations for a circular economy where resources are cyclically utilized and repurposed. Beyond conventional applications, bamboo transcends its utilitarian role and permeates into the realm of green infrastructure ~ contributing to the realization of eco-conscious initiatives such as green buildings and sustainable urban landscapes (Testbook, 2024).

Beyond its economic and environmental significance, bamboo also weaves itself into the intricate fabric of Indian culture and tradition. Wielding symbolic importance bamboo becomes an integral component of various festivals and rituals, enriching cultural practices across the nation. Moreover, the nutritional potency of bamboo shoots elevates these from mere sustenance to a gastronomic delicacy in specific regions of India ~ adding another layer to the plant's multifaceted role in the lives of its inhabitants (Testbook, 2024).

The narrative of bamboo in India transcends its botanical essence embodying a dynamic interplay between economic vitality, environmental stewardship, cultural resonance and nutritional sustenance. Recognizing and fostering the myriad dimensions of its utility stands not only as a testament to its versatility but also as a strategic step towards sustainable development and holistic well-being.

6.10 Bamboo Shoots (Sprouts)

Bamboo shoots, also referred to as bamboo sprouts, are the tender and edible shoots emerging from the ground of various bamboo species such as *Bambusa vulgaris* and *Phyllostachys edulis*. Harvesting is a delicate process requiring precise timing during the growing season to ensure optimal height and development. The prime time for collection falls in the middle of the growing season as shoots gathered at the beginning or end tend to be either underdeveloped or overgrown.

Post-harvest bamboo shoots undergo further processing for consumption available in both fresh and fermented forms. These culinary delights are integral to traditional dishes across numerous Asian countries including China, Korea, Japan, Thailand and Taiwan. The market offers a variety of processed shapes with options ranging from fresh and dried to canned bamboo shoots. Beyond their culinary appeal, freshly harvested bamboo shoots boast a rich nutritional profile. They are abundant in thiamine, niacin, vitamin A, vitamin B6 and vitamin E. Moreover, these shoots contain 17 different amino acids eight of which are essential for human health. However, it is worth noting that amino acid content in canned and fermented bamboo shoots is comparatively lower than in their freshly prepared counterparts (Nongdam and Tikendra, 2014). This underscores the importance of considering the method of preparation when assessing the nutritional benefits of bamboo shoots.

Bamboo shoot, *Phyllostachys parvifolia*, two year old division

By Jodarom - Own work, CC BY 3.0, https://commons.wikimedia.org/w/index.php?curid=7192063

Bamboo foliage probably *Phyllostachys aurea*

By MPF (talk) (Uploads) - photo User:MPF, CC BY-SA 3.0, https://en.wikipedia.org/w/index.php?curid=982246

Bamboo foliage probably *Phyllostachys nigra*

By by MPF - Own work, CC BY-SA 3.0, https://commons.wikimedia.org/w/index.php?curid=138533

Harvested whole bamboo shoots

By Joi Ito from Inbamura, Japan - Flickr, CC BY 2.0, https://commons.wikimedia.org/w/index.php?curid=2022556

Bamboo shoots for sale in a supermarket in Japan

By Chris 73 / Wikimedia Commons, CC BY-SA 3.0, https://commons.wikimedia.org/w/index.php?curid=26165

Bamboo shoot for sale in Kathmandu

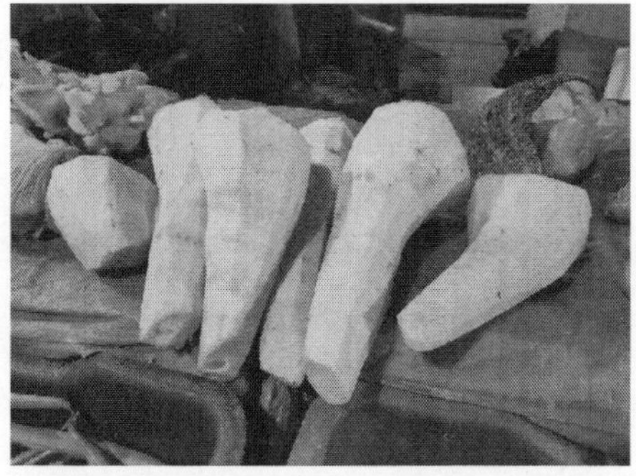

By Gaurav Dhwaj Khadka - Own work, CC BY-SA 4.0, https://commons.wikimedia.org/w/index.php?curid=126983051

Canned bamboo shoots

By Ferbr1 - Own work, CC BY-SA 3.0, https://commons.wikimedia.org/w/index.php?curid=19245416

6.11 Edible Bamboo Species

Majority of young bamboo shoots exhibit edibility potential upon undergoing a process of boiling to eliminate toxins (Dransfield and Widjaja, 1995). However, it is noteworthy that only a select group of approximately 100 species are routinely harvested for their edible shoots. Typically, these species are concurrently cultivated for various other purposes. The compilation of these edible species sourced from diverse references is presented below for reference (Akinlabi *et al.*, 2017; Hu *et al.,* 2015; Manan, 2023; Schroder, 2024).

Species	Common names and attributes	Origin	Edibility
Acidosasa is a native to South China, Taiwan and Vietnam. Its name is christened from Latin *Acidum* meaning 'sour' and *sasa* (another bamboo genus) referring to its edible shoots. Young shoots of the plants are preserved by the local population			
Acidosasa edulis	Yields up to 20,000kg/ha	China	Delicious
A. chinensis	Edible, *vis-a-vis* used for papermaking and weaving	Guangdong	Edible

Species	Common names and attributes	Origin	Edibility
A. chienouensis	Can grow up to 5m high with stem diameter of up to 2cm	Fujian, Hunan	Edible
A. Iingchuanensis		Guangxi	Edible
Bambusa ~ the most commonly harvested bamboo in tropical & subtropical Asia occurring from Philippines to India and from Sumatra to southern China			
Bambusa balcooa	Female bamboo. Edible and used for papermaking or wood chips.	NE India	Good
B. bambos	Giant thorny bamboo/Indian thorny bamboo	SE Asia	Edible
B. beecheyana	Beechey bamboo Large clump-forming bamboo	Southern China to Indo-China and Taiwan	Good
B. blumeana	Spiny bamboo/Thorny bamboo	Indonesia and Malaysia	Good
B. gibboides	Evergreen clumping bamboo	SE Asia, China	Good
B. polymorpha	Burmese bamboo (sweet-tasting edible shoots)	Mainland SE Asia, NE India & Bangladesh	Good
B. tulda	Bengal bamboo (mostly used as paper pulp in India)	Indian subcontinent, Indo-china, Tibet and Yunnan	Good
B. tuldoides	Punting pole bamboo Erect, evergreen, clump forming bamboo	Asia, China, Vietnam, Laos, Malaysia and Myanmar	Good
B. vulgaris	Common bamboo Open clump type bamboo species	Indochina and Yunnan province of southern China	Edible
Other edible species of *Bambusa: B. merrilliana* (endemic to the Philippines), *B. odashimae* (endemic to Taiwan), *B. oldhamii* (native to Taiwan and South China), *B. philippinensis* (endemic to Philippines), *etc.*			
Chimonobambusa – native to the Himalayas, mainland SE Asia, China and Japan			
Chimonobambusa communis	Running bamboo with swollen node	Himalayas,China, Vietnam, Japan and Myanmar	Good
C. macrophylla		–	Delicious
C. marmorea	Marbled bamboo	–	Delicious
C. pachystachys	Leaves all the year round	SC China	Delicious
C. puberula	Medium sized spreading bamboo	Assam, China and East Himalaya	Delicious
C. quadrangularis	Square bamboo Rounded square culms	–	Delicious
C. rigidula	Distinctive bulging nodes and shaped canes	China	Delicious

Species	Common names and attributes	Origin	Edibility
C. szechuanensis	Mostly used for medicinal purposes	E. Asia – W. China, Sichuan.	Delicious
C. tumidissinoda	Walking stick bamboo Used for walking sticks in Sichuan	–	Delicious
C. utilis		–	Good
Chimonocalamus –			
Chimonocalamus delicatus	Aromatic bamboo	China	Delicious
Dendrocalamus – native to tropical South Asia, SE Asia and South China			
Dendrocalamus asper	Rough bamboo	SE Asia	Good
D. brandisii	Velvet leaf bamboo, Teddy bear bamboo or Sweet dragon bamboo	–	Good
D. giganteus	Giant bamboo or Dragon bamboo	Myanmar, Bhutan, China and Thailand	Good
D. latiflorus	Taiwan giant bamboo	Southern China and Taiwan	Good
D. latiflorus 'Mei-Nung'	Taiwan giant bamboo	Southern China and Taiwan	Good
D. membranaceus	Waya/White bamboo	SE Asia	Edible
D. strictus	Male bamboo, Solid bamboo or Calcutta bamboo	SE Asia and India	Edible
Fargesia –			
Fargesia robusta	Umbrella bamboo Clumping bamboo with shiny deep olive-green leaves	–	Edible
Gigantochloa – native to tropical Asia			
Gigantochloa atter	Giant atter or Sweet bamboo	Malaysia	Good
G. levis	Smooth shoot *Gigantochloa* Evergreen and sympodial bamboo	Borneo, Malaysia	Delicious
G. ligulata	Perennial, evergreen	Malaya, Thailand	Good
G. nigrociliata	Black hair giant bamboo, Tropical bamboo		Good
G. pruriens	Perennial, evergreen, densely clump bamboo	North Sumatra, Indonesia	Good
G. robusta	Giant clumping bamboo	South and SE Asia	Good
G. thoii	Giant clumping bamboo	Southern China, SE Asia	Good

Species	Common names and attributes	Origin	Edibility
Guadua sarcocarpa	Fleshy fruit *Guadua*	–	Good
Himalayacalamus falconeri	Gold canes with multiple green stripes	–	Good
Nastus elatus	New Guinea green Delicately arching foliage	–	Edible
Oxytenanthera abyssinica	African lowland bamboo Drought-resistant species	Sub-Saharan Africa	Edible
Phyllostachys – native from Himalayas to East Asia			
Phyllostachys acuta			Delicious
P. angusta	Stone bamboo		Edible
P. arcana	Arcana bamboo Can grow up to 8m tall	China	Edible
P. atrovaginata	Incense bamboo, American bamboo Having wax on the surface	China, Myanmar and India	Edible
P. bambusoides	Madake, Giant or Japanese timber bamboo	China, Taiwan and Japan	Bitter
P. bambusoides f. *shouzhu*			Edible
P. bissetii	Dark gray-green canes	China	Edible
P. circumpilis			Edible
P. concava		China	Edible
P. incarnata		SE Asia	Delicious
P. decora	Beautiful bamboo	China	Edible
P. dulcis	Sweetshoot bamboo (displays cream-coloured stripes)	Massachusetts	Delicious
P. edulis	Moso bamboo or tortoise-shell bamboo	China and Taiwan	Delicious
P. edulis f. *edulis*			Delicious
P. elegans	Elegant bamboo (Culms are used as tools)	–	Delicious
P. erecta		–	Edible
P. fimbriata	Green bamboo	Mexico to northern Colombia	Edible
P. fimbriligula	Upright growth habit	Hunan and Jiangsu of China	Delicious

Species	Common names and attributes	Origin	Edibility
P. flexuosa	Zig-zag bamboo; Spring beauty	–	Delicious
P. glabrata	Woody culms	China (Fujian, Zhejiang)	Delicious
P. glauca	Remarkable evergreen bamboo	Hunan Province of China	Good
P. glauca f. *yunzhu*		–	Edible
P. glauca var. variabilis		–	Edible
P. heteroclada	Water bamboo	–	Edible
P. iridescens	Fastest growing strongest bamboo	Middle to eastern China and Japan	Delicious
P. makinoi	Makino bamboo Upright timber bamboo	SE China	Edible
P. meyeri	Meyer's bamboo Evergreen bamboo	Hunan (China)	Edible
P. mirabilis		–	Edible
P. nidularia	Tall and erect bamboo	Hunan province of China	Delicious
P. nidularia f. *farcta*		–	Edible
P. nidularia f. *mirabilis*		–	Edible
P. nidularia f. *speciosa*		–	Edible
P. nidularia f. *sulfurea*		–	Edible
P. nigella	Culm-sheath blade triangular	Mediterranean	Delicious
P. nigra f. *henonis*	Black bamboo	Hunan province of China	Delicious
P. nuda	Nude sheath bamboo Geat screening bamboo	Zhejiang in east China	Delicious
P. nuda 'Ink-finger'		–	Delicious
P. parvffolia	Bamboo with thick culms	Zhejiang province of China	Delicious
P. pingyangensis	Forest organic sliced bamboo shoots	–	Edible
P. platyglossa	Thin-walled bamboo	Jiangsu & Zhejiang (China)	Delicious
P. praecox	Running timber bamboo with yellow culms	–	Delicious
P. praecox f. *notata*		–	Edible

Species	Common names and attributes	Origin	Edibility
P. praecox f. *viridisulcata*	Striped green sulcus	–	Delicious
P. prominens	Grows around 10 metres tall	China	Good
P. propinqua	Straight dark green culms	–	Good
P. propinqua f. *lanuginosa*		–	Delicious
P. purpurata	Super cold hardy bamboo	Asia	Edible
P. rivalis		Subtropical SE China	Delicious
P. robustiramea	Distinctive grooves above pair of unequal branches		Edible
P. rubella		Hawaiian	Edible
P. rubromarginata	Red-margin bamboo; Red margins	Central China (Guangxi and Guizhou)	Edible
P. rutila		–	Edible
P. sapida	Herbarium specimens of bamboo	Anhui, Gansu, Jiangsu	Edible
P. sulphurea	Evergreen bamboo	–	Good
P. sulphurea f. *laqueata*		Anhui, Henan, Jiangxi provinces of China	Good
P. sulphurea f. *viridis*			Good
P. tianmuensis	Swollen culm-nodes	Anhui, Zhejiang (China)	Edible
P. viridiglaucescens	Greenwax golden bamboo Green glaucous bamboo	East Asia	Edible
P. vivax	Chinese timber bamboo	China	Delicious
P. vivax 'Huangwenzhu'	Yellow groove vivax and in Chinese called 'Huang Wen Zhu'.	China	Edible
P. vivax f. *aureocaulis*	Golden Chinese timber bamboo	–	Edible
P. yunhoensis		SE China	Delicious
Pleioblastus hindsii	Hardiest bamboo	Japan and East Asia California	Edible
Sasa kurilensis	Chishima Zasa Northern-growing bamboo	Korea, Japan and eastern Russia (Sakhalin)	Good
Sasaella masamuneana	Sam bamboo	Japan	Edible
Thamnocalamus aristatus	Clumping Himalayan bamboo	Himalayas	Edible
Thyrsostachys siamensis	Monastery bamboo, Thai bamboo or Umbrella bamboo	China, Myanmar, Laos, Vietnam and Thailand	Good
Yushania maling (*Arundinaria maling*)	Graceful open clumper	Nepal to Asom	Good

6.12 Nutritional value

Freshly harvested bamboo shoots boast an impressive nutritional profile, serving as a rich source of essential vitamins and minerals. These include thiamine, niacin, vitamin A, vitamin B6 and vitamin E. Bamboo shoots contain a noteworthy array of amino acids with 17 identified varieties including 8 essential ones crucial for human health. It is worth noting that amino acid content tends to be higher in freshly collected bamboo shoots compared to their canned or fermented counterparts (Qiu, 1992; Nongdam and Tikendra, 2014). In terms of protein content, bamboo shoots stand out as a valuable source with protein levels ranging from 1.49g/100g to 4.04g/100g fresh weight in their natural state (Sundriyal and Sundriyal, 2001; Chongtham *et al.*, 2011). However, it is essential to recognize that the protein content can vary depending on factors such as bamboo species, harvesting stage and processing methods like boiling, canning or fermenting. Bamboo shoots contribute significantly to the intake of essential minerals including potassium, phosphorus, sodium, calcium, magnesium and iron. These minerals play crucial roles in various metabolic activities within the body, supporting overall health and well-being. Carbohydrate levels in bamboo shoots are reported to be substantial with variations observed among different species. For instance, edible shoots of *Bambusa nutans*, *Bambusa vulgaris*, *Dendrocalamus strictus* and *Dendrocalamus asper* have carbohydrate contents of 3.3%, 3.4%, 0.6% and 2.9%, respectively (Kumbhare and Bhargava, 2007).

Bamboo shoots are remarkably low in fats. These emerge as an excellent dietary choice for individuals managing conditions such as diabetes and cardiovascular diseases. High fibre content in bamboo shoots contributes to reducing blood fat and cholesterol levels, positioning them as a sought-after health food among those grappling with modern lifestyle diseases. Pandey and Ojha (2013) found an increase in dietary fibre in the shoots of *Dendrocalamus asper*, *Dendrocalamus strictus* and *Bambusa tulda* with the maturation of bamboo shoots.

6.12.1 Nutritional value and per cent DV

Bamboo shoots, raw
Nutritional value per 100g

Energy	***115 kJ (27 kcal)***
Carbohydrates	5.2 g
Sugars	3 g
Dietary fibre	2.2 g
Fat	0.3 g
Protein	2.6 g
Vitamins	**Quantity (%DV[†])**

Bamboo shoots, raw
Nutritional value per 100g

Energy	***115 kJ (27 kcal)***
Thiamine (B1)	(0.15 mg) 13%
Riboflavin (B2)	(0.07 mg) 6%
Niacin (B3)	(0.6 mg) 4%
Pantothenic acid (B5)	(0.161 mg) 3%
Vitamin B6	(0.24 mg) 18%
Folate (B9)	(7 μg) 2%
Vitamin C	(4 mg) 5%
Vitamin E	(1 mg) 7%
Minerals	**Quantity (%DV†)**
Iron	(0.5 mg) 4%
Manganese	(0.262 mg) 12%
Phosphorus	(59 mg) 8%
Potassium	(533 mg) 11%
Zinc	(1.1 mg) 12%
Units: μg = micrograms; mg = milligrams; IU = International units Source: USDA, 2018	
†Percentages are roughly approximated using US recommendations for adults.	

6.12.2 Nutrient content of freshly harvested shoots of some bamboo species

(Bhatt** et al., **2005; Chongtham** et al., **2011; Bisht** et al., **2012; Nongdam and Tikendra, 2014)								
Nutrients	*B. bambos*	*B. nutans*	*B. polymorpha*	*B. tulda*	*B. vulgaris*	*D. asper*	*D. giganteus*	*D. strictus*
Amino acids (g/100g)	3.98	3.89	3.42	3.65	3.57	3.12	3.96	3.07
Protein (g/100g)	3.57	2.84	3.64	3.69	3.64	3.59	3.11	2.60
Carbohydrate (g/100g)	5.42	5.47	5.44	6.92	6.51	4.90	5.10	6.17
Fats (g/100g)	0.50	0.40	0.46	0.48	0.50	0.40	0.39	0.33
Fibres (g/100g)	4.49	2.28	3.81	3.97	4.24	3.54	2.60	2.26
Vitamin C (mg/100g)	1.90	1.19	2.60	1.42	4.80	3.20	3.28	2.43
Vitamin E (mg/100g)	0.61	0.47	0.49	0.61	0.52	0.91	0.69	0.58
Calcium (mg/100g)	0.36	1500	180.69	1300	320	5.51	26.93	139.5
Phosphorus (mg/100g)	30.12	900	15.06	700	220	40.95	12.57	58.13

(Bhatt et al., 2005; Chongtham et al., 2011; Bisht et al., 2012; Nongdam and Tikendra, 2014)								
Nutrients	B. bambos	B. nutans	B. polymorpha	B. tulda	B. vulgaris	D. asper	D. giganteus	D. strictus
Iron (mg/100g)	3.00	-	1.53	1.57	-	3.37	1.06	2.91
Sodium (mg/100g)	10.10	-	-	12.96	400	10.14	3.64	0.08
Potassium (mg/100g)	-	30.0	-	20.0	920	464	275	-
Magnesium (mg/100g)	5.38	40.0	-	40.0	100	10.14	9.57	0.17
Note: Hyphen (-) in table indicates authentic data not available for the particular bamboo species.								

6.12.3 Nutrient composition of raw, boiled and fermented shoots of different bamboo species

(Bhargava et al., 1996; Bhatt et al., 2005; Kumbhare and Bhargava, 2007; Nirmala et al., 2008; Nongdam and Tikendra, 2014)							
Nature of shoots	*Bamboo species*	*Carbohydrate (g/100g)*	*Protein (g/100g)*	*Fats (g/100g)*	*Vitamin C (mg/100g)*	*Vitamin E (mg/100g)*	*Fibres (g/100g)*
Raw shoots	*Dendrocalamus strictus*	85.98	21.51	0.10	5.80	-	0.98
	Dendrocalamus asper	2.90	25.80	3.59	3.30	0.91	0.70
	D. giganteus	5.10	-	0.39	3.28	0.69	-
	Bambusa vulgaris	3.40	25.70	0.20	13.70	-	0.70
	Bambusa nutans	3.30	21.10	1.00	5.30	-	0.76
Boiled shoots	*Dendrocalamus strictus*	5.00	-	-	-	-	0.96
	Dendrocalamus asper	3.10	17.10	-	-	-	0.70
	D. giganteus	-	11.66	-	-	-	-
	Bambusa vulgaris	5.00	13.50	-	-	-	0.96
	Bambusa nutans	5.10	17.30	-	-	-	0.75
Fermented shoots	*Dendrocalamus strictus*	-	-	-	-	-	-
	Dendrocalamus asper	-	-	-	-	-	-
	D. giganteus	-	-	-	0.32	1.09	0.21
	Bambusa vulgaris	-	-	-	-	-	-
	Bambusa nutans	-	-	-	-	-	-
Note: Hyphen (-) in table indicates authentic data not available for the particular bamboo species.							

6.12.4 Comparison of nutrient composition of young bamboo shoots of *Bambusa tulda* and *Dendrocalamus hamiltonii* with some common vegetables

(**Chongtham** *et al.*, **2011**)									
Nutrients	*Bambusa tulda*	*Dendrocalamus hamiltonii*	*Daucus carota*	*Raphanus sativus*	*Spinacea oleracea*	*Solanum tuberosum*	*Abelmoschus esculentus*	*Curcuma sativus*	*Cucurbita maxima*
Amino acids (g/100 g)	3.65	3.18	0.20	0.40	0.30	0.20	0.30	0.10	0.20
Protein (g/100 g)	3.69	3.72	0.90	0.70	2.00	1.60	1.90	0.60	1.40
Carbohydrate (g/100 g)	6.92	5.50	10.60	3.40	2.90	22.6	6.40	2.50	6.50
Fats (g/100 g)	0.48	0.41	0.10	0.40	0.70	0.10	0.20	0.10	0.40
Fibres (g/100 g)	3.97	3.90	1.20	0.60	2.00	0.40	1.20	0.40	1.10
Vitamin C (mg/100 g)	1.42	2.45	15.00	1.60	0.60	0.40	1.20	0.70	0.07
Vitamin E (mg/100 g)	0.61	0.71	3.00	15.00	28.10	19.70	13.00	3.20	0.90
Potassium (mg/100 g)	408	416	108.00	393.00	558.00	424.00	103.00	135.00	340.00
Iron (mg/100 g)	3.19	2.69	1.03	1.00	2.70	0.80	0.35	0.90	0.80
Calcium (mg/100 g)	4.06	3.00	80.00	35.00	9.90	12.00	56.00	14.00	21.00
Cupper (mg/100 g)	0.44	0.29	0.10	0.02	0.10	0.16	0.11	0.09	0.10
Sodium (mg/100 g)	19.96	9.32	35.60	39.00	79.00	11.00	6.90	2.00	5.60
Zinc (mg/100 g)	0.72	0.70	0.36	0.30	0.50	0.30	0.42	0.23	0.30
Source: (Nongdam and Tikendra, 2014)									

6.12.5 Phytosterol

Phytosterols possessing a molecular structure akin to cholesterol are abundantly present in various plant sources with a pronounced presence in both fresh and fermented bamboo shoots (Srivastava, 1990; Miettinen and Gylling, 2003). The significance of phytosterols in enhancing quality of life has been extensively documented by researchers such as Kritchevsky and Chenm (2005), Phillips *et al.* (2005) and Ostlund (2007). In a comprehensive study conducted by Lu *et al.* (2009), the phytosterol content in bamboo shoots of *Pleioblastus amarus, P. pubescens, P. praecox* and *Dendrocalamus latiflorus* was meticulously evaluated using ultraperformance liquid chromatography. Their findings revealed a notable prevalence of β-sitosterol compared to other sterols like campesterol and stigmasterol.

The consumption of diets rich in phytosterols has been linked to a reduction in the incidence of colon, breast and prostate cancers (Bradford and Awad, 2007; Choi *et al.*, 2007; Woyengo *et al.*, 2009). The inhibitory effects on tumour growth progression are attributed to phytosterols' interference with the cell cycle, induction of apoptosis and inhibition of tumour metastasis. Moreover, the presence of phytosterols hinders the absorption of dietary cholesterol and cholesterol esterification in the intestinal mucosa, thereby reducing blood cholesterol levels. Sitosterol has been identified as the most abundant phytosterol in bamboo shoots as reported by Sarangthem and Singh (2003). Noteworthy is the potential of extracting a substantial amount of phytosterols from freshly harvested bamboo shoots for production of steroidal drugs. This prospect holds promise in alleviating the reliance on *Dioscorea* and *Solanum* species as primary natural sources of steroidal drugs.

6.12.6 Phenols

Bamboo leaves are recognized for their substantial antioxidant capacity attributed to the elevated presence of phenolic compounds (Rice-Evans *et al.*, 1997; Lu *et al.*, 2005; Oboh and Ademosun, 2012). Park and John (2010) identified eight phenolic acids in bamboo shoots of *Phyllostachys pubescence*, among which protocatechuic acid, p-hydroxybenzoic acid and syringic acid were identified as the most abundant. Velioglu *et al.* (1998) established a statistically significant correlation between the antioxidant properties of the plant and content of phenolic compounds. Extracts from stems and leaves of *Phyllostachys* spp. exhibited robust antibacterial activities (Kim *et al.*, 2001).

Impact of various cooking methods on phenolic compounds in bamboo shoots was investigated (Zhang *et al.*, 2011). Boiling and stir-frying resulted in a slight reduction in the total phenolic content compared to fresh shoots. However, steamed bamboo shoots exhibited an increase in the level of phenolic compounds.

The reduction observed in boiled or stir-fried shoot samples was suggested to be linked to the decomposition of phenolic compounds during heat treatment. Yamaguchi *et al.* (2003) opined that heat treatment might lead to the inactivation of existing phenolic oxidases responsible for polyphenol decomposition.

The positive correlation between higher level of total phenolic compounds and increased antioxidant activity was underscored by Baardseth *et al.* (2010). Meanwhile, Pandey and Ojha (2013) reported variations in phenolic compounds in bamboo shoots of *Dendrocalamus asper*, *D. strictus* and *Bambusa tulda* at different harvesting times. Gallic acid concentration increased in *Dendrocalamus asper*, while caffeic acid increased in all bamboo species with shoot age. The level of vanillic acid in *Bambusa tulda* was found decreased but increased level was recorded in *Dendrocalamus asper* and *D. strictus*. Pandey and Ojha (2011) observed a decrease in the total phenolic content in shoots of *Bambusa bambos*, *B. tulda* and *Dendrocalamus asper* when subjected to boiling in higher salt concentrations for longer durations.

6.12.7 Toxins

Raw bamboo shoots contain varying amounts of cyanogen glycosides, specifically taxiphyllin (Schwarzmaier, 1977; Nartey, 1980; Vetter, 2000). Cyanogen glycosides ~ the natural toxins also found in cassava (cfs.gov.hk, 2024) pose a potential health risk. When bamboo shoots are disrupted β-glycosidase is released which acts on taxiphyllin to generate harmful hydrogen cyanide. It is crucial to ensure that the level of hydrogen cyanide does not surpass the toxic threshold for human consumption (Seigler, 1991).

To mitigate this risk, thorough cooking is essential as the toxins can be effectively destroyed through boiling process. Consequently, fresh bamboo shoots are routinely boiled before being incorporated into various culinary preparations. Canning process also serves as a means of toxin destruction, rendering the bamboo shoots safe for consumption. Therefore, it is imperative to follow appropriate cooking methods to eliminate harmful substances and guarantee the safety of bamboo shoots for human consumption.

6.12.8 Culinary uses

Bamboo shoots stand out as a noteworthy inclusion in the realm of beneficial health foods, owing to their substantial concentrations of proteins, carbohydrates, vitamins, fibres and minerals coupled with minimal fat content even then acceptance of bamboo shoots remains predominantly confined to SE Asian and East Asian countries. This limited popularity can be attributed to the pronounced pungent smell and bitter acidic taste associated with bamboo shoots. In India, utilization of bamboo as a culinary ingredient is predominantly observed in NE regions of country where it holds an indispensable role in various traditional specialty dishes. Different ethnic communities in this region highly favour fresh or fermented bamboo shoots as a key component of their traditional culinary repertoire such as *ushoi, soibum, rep, mesu, eup, ekhung, hirring* and many more. Occasional reports of bamboo shoots being used as food in other parts of India such as Odisha, Gujarat and the Western Ghats (Karnataka) but these instances are relatively less widespread (Satya *et al.*, 2009). The Kandha tribe in Western India practices a traditional culinary method known as *kardi* where shredded bamboo shoots are soaked in water for a day to eliminate bitterness before cooking (Panda and Padhy, 2007). Bal *et al.* (2009) claimed that bamboo shoots are consumed only in five Odishi villages after undergoing a meticulous boiling process to eliminate the acrid taste and any potential toxic element. In remote and hilly areas of the Western Ghats, these processed bamboo shoots find application in preparation of pickles, snacks, papads and various other fried food items.

It is crucial to emphasize the importance of proper processing before consuming bamboo shoots as freshly harvested shoots are known to contain high levels of toxic cyanogenic glycosides which could pose serious health risks (Nongdam and Tikendra, 2014). Hence, ensuring that bamboo shoots undergo appropriate boiling at varying temperature conditions and durations is imperative to mitigate potential health concerns associated with their consumption.

6.12.9 Meditional values

Bamboo shoots boast a rich composition of amino acids including tyrosine, arginine, histidine and leucine. Tyrosine ~ a pivotal component of adrenal glands serves as a precursor for adrenaline, an essential catalyst for robust metabolic activities in the body. Moreover, tyrosine plays a crucial role in the optimal functioning of thyroid and pituitary glands, integral in the production and regulation of hormones within human body. The nutritional profile of bamboo shoots extends beyond amino acids encompassing high levels of dietary fibre and phytosterols. This combination contributes to a reduction in blood fat and cholesterol levels, positioning bamboo shoots as highly coveted health foods especially for individuals contending with lifestyle-related disorders. Dietary fibre, with its myriad health benefits aids in controlling blood pressure, mitigating hypertension and addressing obesity concerns. It serves as a safeguard against coronary diseases and potential carcinogens (George *et al.*, 1982; Anderson and Strong, 1983).

Kalita and Dutta (2012) revealed that certain ethnic tribes in NE India leverage bamboo shoots to manage high blood pressure and cardiovascular ailments. The substantial dietary fibres coupled with low fat content contribute to reduction of arterial thickening thereby maintaining healthy blood pressure levels. Park and John (2009) underlined the positive impact of bamboo shoot-enriched diets on lowering serum cholesterol and low-density lipoprotein, along with promoting regular bowel movements, indicative of its potential role in cholesterol reduction and diabetes prevention. The medicinal properties of bamboo shoots extend to their anticancer properties attributed to the presence of lignans and phytosterols. Phytosterols play a pivotal role in inhibiting the production of carcinogens, impeding the growth of cancer cells and preventing cell invasion *cum* metastasis (Meric *et al.,* 2006). Karbi Anglong tribes in India have traditionally used bamboo shoots to combat early stages of cancer (Kalita and Dutta, 2012).

Moreover, bamboo shoots play a role in enhancing reproductive health, particularly in females. Local tribes, including Bodo, Thadau, Mosang and Tiwa utilize bamboo shoots for addressing issues such as irregular menstrual cycles, excessive post-delivery bleeding, infertility, alleviating labour pain, and inducing puberty in young females. While not scientifically proven, some tribes caution against bamboo shoot consumption during the first trimester of pregnancy due to suspicion that it causes abortion (Nongdam and Tikendra, 2014).

6.13 Cultivation (Bamboo Info, 2024)

6.13.1 Climatic conditions

Bamboo plantations thrive in tropical to warm temperate climates with an optimal temperature of 16°C for growth. Avoiding temperatures below 15°C in summer is advisable. Planting is feasible throughout the year in mild climatic zones. Bamboo adapts to diverse soil and climatic conditions flourishing in valleys, lower hill slopes and near water bodies. Most bamboo species prefer temperatures between 7 to 40°C with clump-forming species found in humid tropics having above 20°C temperatures. Indian bamboos are vulnerable to cold and frost. Rainfall generally supports bamboo growth, with the best results with a combination of high temperature and rainfall. The ideal rainfall range is 1200-4000mm per year. Drought-resistant species like *Dendrocalamus strictus* thrive in semi-dry forests tolerating temperatures -5-45°C and rainfall as low as 750mm to 4000mm annually.

Monopodial bamboos such as *Phyllostachys* prefer cool temperate zones (18-26°C) at elevations of 1000-2000m or even higher and exhibit cold resistance. Large-scale bamboo plantation success depends on factors such as selection of species, rhizome nature, agroclimatic features and spacing. While most species thrive in plains and at altitudes of 400–500 meters, certain such as *Sinarundinaria falcata*, *Phyllostachys aurea*, *Phyllostachys manni*, *Kuruna densifolia* and *Kuruna floribunda* prefer higher altitudes of 1500 meters or above.

6.13.2 Soil

Bamboos exhibit remarkable adaptability to a wide range of soil conditions, from organically poor to mineral-rich and can thrive in diverse moisture levels spanning drought to flooding. These are well-suited for reclaiming degraded lands. While generally favouring soil rich in organic matter, moist, porous, well-drained and with a neutral to marginally acidic pH but most bamboo species can grow on various soils excluding rock-strewn ones. Preferably cultivated in well-drained sandy loam to clay loam soils with a pH range of 4.5-6, bamboo species may tolerate pH levels as low as 3.5. Optimal conditions for high yield include loose, well-aerated and thick soil, moderate moisture, slightly acidic pH and ample organic matter. The presence of N, P_2O_5, K_2O, CaO, and SiO_2 in the soil enhances bamboo culm growth with nitrogen content being particularly crucial.

Generally bamboos have better stand in non-saline soil but there are species tolerating well even to saline soils to a certain level. However, prolonged waterlogged conditions and swamps can lead to root and rhizome rot and gradual drying. Bamboos such as *Ochlandra travancorica, O. scriptoria* and *O. wightii* found in the flood plains in Kerala, Tamil Nadu, Goa and Karnataka can endure short seasonal floods. For riverbank stabilization programmes, reeds such as *Ochlandra* with fibrous root systems serve as effective options forming a network that prevents topsoil erosion.

6.13.3 Propagation

Bamboos propagate through seeds and vegetative means (including tissue culture). Vegetative methods encompass conventional techniques such as offset planting, rhizome planting and cuttings (suitable for clump-forming species with rooting induced in culm, branch, and nodal cuttings). Layering may also be employed in certain cases whether thin-walled or thick-walled. Each method carries distinct advantages and disadvantages, ensuring varying survival rates.

6.13.3.1 Propagation through seeds

The size, shape and seed quantity per kilogram vary among bamboo species. To identify seed maturity, frequent visits are essential. Maturity is determined by a cutting test on representative seed samples. Mature seeds are collected under bamboo clumps on a polythene sheet which are processed through winnowing, sieving and handpicking impurities and kept in bags after proper drying and labelling. These seeds have little dormancy so are sown soon after harvesting. Before sowing these are water-soaked overnight, the floating ones are removed and sown in seedbeds or in trays.

On seed production behaviour, the bamboos have following classification:

i) Abundant seed forming, *e.g.*, *Bambusa bambos* and *Dendrocalamus hamiltonii.*

ii) Sparse seed forming, *e.g.*, *Dendrocalamus giganteus* and *Bambusa nutans.*

iii) Sterile group: *e.g.*, *B. vulgaris, B. balcooa, Pseudoxytenanthera monadelpha, Dendrocalamus stocksii.*

Since seeds have no dormancy so after harvesting these require immediate sowing. These have epigeal germination. Their viability can be tested through tetrazolium and hydrogen peroxide agents under laboratory condition. Seeds should be sown at well-drained location directly in nursery beds or polybags under partial sunlight. Soil for sowing should be a mixture of soil, sand and compost. After sowing the seeds are covered with sand and mulched with organic materials and irrigated daily over the mulch by each evening until germination. The seedlings when attain 4-6-leaf stage are gently uprooted and transplanted in polybags and afterwards when properly grown then to the place earmarked for this. After transplanting these are immediately irrigated. Keep in partial shade before transferring to direct sunlight.

6.13.3.2 Propagation using culm cuttings

In case of thick-walled culms, propagation through culm cuttings is particularly well-suited. Such cuttings boast prominent nodal buds or branches with aerial roots. The optimal time for such propagation is pre-monsoon to early monsoon season. It is advisable to extract mature culms (1–2 years old) from healthy and mature mother clumps without splitting them. The ideal portion is the middle segment with trimmed branches and the top portion discarded. While removing leaves and side branches, caution should be exercised to avoid injuring axillary buds on nodes. Certain species such as *Bambusa balcooa, B. vulgaris, Dendrocalamus stocksii, etc.* don't produce viable seeds so these are propagated through culm cuttings or by splitting the rhizomes. Seed-producing species have long flowering intervals varying from 30 to 70 years (*Bambusa bambos, B. polymorpha, Dendrocalamus strictus, D. brandisii, D.giganteus, etc.,*), with some flowering gregariously, leading to clump death, while others sporadically flower with or without parent plant deaths (such as *Ochlandra setigera, O. wightii, etc.*). Culm cuttings may have one, two or three internodes with buds or branches. Using a hacksaw or sharp knife 1- or 2-noded cuttings can be crafted. A 2×1cm opening or two holes of approximately 7mm diameter can be made in the centre of the internodes. For thin-walled species such as *Ochlandra travancorica*, it is crucial to avoid splitting the culm at the cut surface. Cutting at the half internodal point on each side is recommended. After extraction, culms should be swiftly transported to the nursery site, taking care to prevent drying by wrapping cut ends with moist gunny bags or covering these with wet rice sacks.

To enhance rooting response, treatment with growth-regulating substances such as IBA and NAA are recommended. For bamboos with hollow internodes, a 100ppm IBA/NAA hormone solution can be poured into the cavity followed with plugging while for solid bamboos the dip method is more proper (immersing overnight in the solution). The dip method involves immersing the basal part in solution for 12-15 hours before planting. For preparing hormonal solution, 1g IBA/NAA powder is dissolved in 50-100ml of 90% ethyl alcohol with addition of water slowly to avoid coagulation. The solution is then made up to 1000ml while stirring to prevent precipitation. The stock solution can be stored for up to one week in dark containers. Solutions can be prepared at desired concentrations (100 ppm, 200 ppm, 250 ppm, 500 ppm) by diluting the stock solution with water.

Treated cuttings are planted horizontally in nursery beds, spaced 15-30cm apart. One week before planting, the beds are drenched with insecticides and fungicides to prevent termite and fungi attacks. Sprouting occurs within a week, slender roots develop within a month and rhizome development starts in 3-6 months. Rooted cuttings can then be transferred to polybags or planted directly in the field. The success rate is high for thick-walled species (60-95%) but less encouraging for thin-walled species (0-40%).

6.13.3.2 Propagation using branch cuttings

Propagation through branch cuttings is a straightforward and accessible method, particularly suited for species characterized by swollen branch bases, aerial roots and prominent branches. While both primary and secondary branches can be employed, the success rate of rooting is not as consistently high as with culm cuttings. Despite this, a 50% rooting success is considered satisfactory. The optimal period for employing this propagation technique is during early monsoon, especially following pre-monsoon showers or within three months of the monsoon's onset. This technique has proven effective for species such as *Bambusa balcooa, B. vulgaris, Dendrocalamus stocksii, etc.*, predominantly thick-walled species with stout branches and prominent primary branches from the culm base. However, it exhibits a notably low survival percentage for species like *Dendrocalamus giganteus* and *Thyrsostachys oliveri.*

To prepare branch cuttings, sharp knives or saws are used to excise branches, followed by trimming leaves and small branches. Cuttings should have 4-8 nodes with their basal parts immersed in a rooting hormone solution for 24 hours. Dipping selected branches in rooting hormone solution (200 ppm) for a few hours, preferably overnight, can also aid in root formation and increase in survival rates. Rooting process in this case typically spans 4-8 months while rhizome formation up to a year. A sand bed in partial sunlight is recommended for its planting after dipping of the cut ends in fungicide for five minutes. To protect the exposed cut ends, wax or cow dung should be applied. Regular watering 2-3 times a day is essential but depending on climatic conditions. Sprouting of branch buds usually occurs within 10 to 20 days. These post-rooting cuttings can be transplanted into polybags (15×20cm or larger size) filled with a potting medium (soil, sand and manure @ 1:1:1 or soil+compost 3:1). Preparing pre-rooted and pre-rhizomed branch cuttings can expedite the rooting and rhizome development process. In some species, natural aerial rooting and rhizome formation can be observed. Inducing pre-rooting is possible by chopping off the top part of the culm and covering nodal buds with moist media such as moss or coir. Sprouting and rooting time ranges from 30 to 70 days, contingent upon the season. Once rooting is successful, the cuttings should be carefully removed from nursery beds and transplanted into poly pots.

6.13.3.3 Propagation using rhizome cuttings

Propagation through rhizome cuttings is a specialized technique particularly well-suited for monopodial bamboos characterized by runner-type rhizomes. This method involves utilizing rhizomes in three distinct forms: i) rhizomes with roots, ii) rhizomes with both culms *cum* roots, and iii) rhizomes with culm-stocks and roots. The success of this propagation technique hinges on the careful selection of rhizomes that are no more than one year old, devoid of any physical damage and equipped with attached roots essential for both successful rooting and subsequent sprouting. In order to ensure optimal outcome, it is imperative that rhizomes are shielded from damage during transportation akin to the precautions taken with offsets. Specifically, if transportation is part of the propagation process, adequate measures must be taken to protect the rhizomes from drying out. This nuanced approach not only safeguards the integrity of the rhizomes but also enhances the overall success rate of the propagation method.

6.13.3.4 Propagation by layering

Propagation by layering is a sophisticated horticultural technique employed to reproduce plants by utilizing a segment of a culm or branch while it remains attached to the parent plant and is in direct contact with a moist rooting medium, typically soil. This method is particularly effective for species such as *Bambusa polymorpha, B. textilis, B. vulgaris, Dendrocalamus giganteus, D. longispathus* and *Guadua angustifolia*.

Various layering techniques exist, including ground layering (simple layering), stump layering, air layering (marcotting) and seedling layering. In case of ground layering, entire culm or a branch section is gently bent to the ground, placed in a prepared shallow trench, secured in the soil with pegs and covered with soil. For more rigid culms, a partial cut can be made at the base, and the trench size is adjusted based on the culm's dimensions. To promote successful layering mulching with a coir mat or moist straw coupled with regular watering is desirable. However, precautions must be taken to prevent waterlogging. Rooting and sprouting primarily occur at the nodes and if rhizome formation is observed, nodes can be separated using a sharp knife or saw.

Stump layering involves cutting the culm leaving 2-3 basal nodes and covering the stump with a rooting medium. Addition of 200ppm IBA has been identified as beneficial in enhancing both sprouting and rooting, with visible sprouting occurring approximately six months after cutting the culms. Marcotting represents a modified layering technique wherein the swollen base of a branch located in the mid-culm zone is covered with a propagating medium and secured in place using a suitable receptacle. This nuanced approach further expands the possibilities for successful plant propagation through layering.

6.14 Soil and Land Preparation

Establishing a bamboo nursery holds significant merit when embarking on large-scale bamboo plantations. Optimal location for nursery should be in close proximity to the intended plantation area but terrain should have comprehensive irrigation system and well-drained. Bamboos thrive in degraded lands, marginal land, expansive areas, slopes, river banks and at farm boundaries and from coastal regions at sea level to high mountains reaching up to 4000 metres above mean sea level. Rocky soils are not suitable, however, the soil should be some 40cm deep which can be added with fly ash as a soil enricher and conditioner. Moreover, the area should have access to transportation, *vis-à-vis* near to the markets for specific bamboo species. Large-scale plantations cater to various purposes such as timber production, edible shoots, pulp, bio-energy, for various edible preparations and for manufacturing furniture and handicrafts. Initially, the area requires adequate fences and afterwards there should be even fire protection measures.

6.14.1 Spacing

Effective bamboo plant spacing is crucial for successful cultivation, considering factors such as purpose for which planting is to be done, rhizome nature, culm diameter and growth parameters. Adequate spacing minimizes competition for resources such as light, water and soil nutrients. However, quite wide spacing can lead to canopy exposure, soil moisture loss and weed growth which may impact productivity. Recommended spacings for large bamboo plants are 8×8m and 8×6m with 150 and 200 clumps, respectively. Sympodial species such as *Dendrocalamus latiflorus* can be planted at 4×5m or 5×5m apart. For pulp and paper purposes the plants can be spaced 3×4m, 4×4m or 4×5m. Timber stands of *Dendrocalamus strictus* suggest a density of 700 clumps per hectare.

Taller bamboos such as *Dendrocalamus giganteus* require low planting density with wider spacing though for small and medium-sized bamboos it is otherwise as furnished here under.

i) 2×2 to 3×3m for erosion prevention or riverbank stabilization.

ii) 3×3m or 2.5×2.5m for erosion control with interspersed timber species.

iii) 4×4m for small-diameter bamboos.

iv) 5×5 to 7×7m for medium-diameter bamboos.

v) 7×7 to 10×10m for large-diameter bamboos.

Rectangular or triangular spacing can be employed, with triangular spacing recommended for large-scale commercial planting. Aligning rows North-South ensures optimal sun exposure and air circulation.

Before planting, pit-spots are marked with stakes in a North-South orientation. Pit sizes range from 60×60×60 to 100×100×100cm for rhizomes and 30×30×30 to 45×45×45cm for seedlings. Coastal areas may require enriched pits.

Ideal planting time is early rainy season. Manure and fertilizer (5kg FYM/compost) are mixed with topsoil. While planting polybags are removed, pits filled with topsoil and after planting soil mounds around the plants are created. Post-planting operations include casualty replacement, clump and culm maintenance, soil care, mulching, weeding, pest control and putting of guards against each plant.

6.14.2 Manures and fertilizers

Effective bamboo cultivation requires careful attention to manures and fertilizers. Bamboo plants are known as heavy feeders and even nutrient-rich soils can deplete over time without proper fertilizer application. To attain optimal quality and yield, it is essential to understand the vital components of fertilizers, *viz.* potassium, phosphorous and nitrogen. For a comprehensive approach, it is recommended to incorporate green manures as pre-plant, compost or FYM and wood ash in conjunction with traditional fertilizers. However, it is essential to test the soil before incorporating any fertilizer in the standing field. Under standard soil conditions, the ideal composition of nitrogen, phosphorous and potassium should follow a ratio of 5:2:1 or 4:2:1.

Tailoring the fertilizer application to the specific needs of bamboo cultivation is crucial. In plantations focused on culm production, chemical fertilizer application may reach 1,500kg per hectare annually while shoot production could demand up to 4,000kg per hectare per year. A general guideline recommends the dose of 150 grammes each of N, P and K per plant. Bamboo exhibits positive responses to a combination of NPK and farmyard manure or vermicompost. To foster growth, 15–20kg of manure or compost can be added to each clump. At on-farm or high-intensity managed matured bamboo plantations, 0.5–1kg of NPK fertilizer per well-grown clump may be suitable.

The first fertilizer application 1-2 months after planting along with 5kg of well-rotted compost and 50g of NPK should be applied to each hill. Subsequent applications should align with the beginning of the rainy season with 10kg of manure or compost and 100–150g of NPK in the second year and 30kg of manure in addition to some 500g of NPK in the third year. This meticulous approach ensures sustained and healthy bamboo growth over time.

6.14.3 Irrigation

Ensuring optimal water conditions is crucial for its successful cultivation. The chosen planting site should experience well-distributed rainfall while areas prone to waterlogging should be avoided to prevent water-related issues. Both bamboo nurseries and plantations necessitate irrigation for optimal growth. In instances where proper drainage infrastructure is lacking artificial means must be implemented to prevent water accumulation. During dry periods, it is recommended to provide irrigation at regular intervals with each plant receiving 10-12 litres of water per week for initial two years. To maximize water efficiency a drip irrigation system stands out as the most effective solution.

For situations allowing consistent refilling, an alternative method involves burying clay pots up to their necks in the bamboo's root zone. An inverted plastic bottle with a wick inserted in the neck can be employed to facilitate a gradual outflow of water. To retain moisture effectively, creating a soil mound around the plant base and digging a trench (with a radius of at least 50cm) around the bamboo plant can be instrumental.

It is crucial to acknowledge that the water requirements and preferred irrigation methods are species-specific. For example, *Dendrocalamus strictus* thrives in rainfed areas whereas species such as *D. hamiltonii, D. stocksii* and *Pseudoxytenanthera ritcheyi* are better suited for minimal irrigation conditions. Understanding the specific needs of each bamboo species ensures a tailored and efficient approach to irrigation practices.

6.14.4 Intercultural practices

6.14.4.1 Mulching, mounding, pruning and coppicing

Mulching is an integral practice for ensuring production of high-quality bamboo shoots. Not only does mulching retain soil moisture but it also suppresses weed growth effectively. It plays a very crucial role in plantations maintained for their edible shoots as this practice shields young shoots from direct sunlight, helps retaining the soil moisture levels, does not permit any pathogen or insect to feed upon, provides warmth during winter and promotes optimal growth without compromising their edible quality. Uniform mulching involves spreading a substantial layer (10cm deep) of leaf litter or organic matter around the bamboo clump. This process contributes, apart from facilitating the main crop in many ways, enriches the soil with organic nutrients, and enhances overall soil fertility and organic carbon content. Retaining bamboo leaves under the plant serves as on-site mulching material.

In older plantations of sympodial bamboos where rhizomes become exposed due to soil erosion and minimal mulching, **soil mounding** becomes imperative. When rhizomes grow horizontally beneath the soil, exposing themselves as the clump ages, their growth can be hindered by sunlight exposure. This can lead to reduced availability of nutrients and moisture so soil mounding addresses this issue by protecting the rhizome system and promoting healthy growth.

Vigilant **weeding** during the initial years of plantation is essential to prevent competition for sustenance with young bamboo shoots. Biannual weeding after rains and towards the end of the wet season aids in quick establishment and accelerated growth of bamboo rhizomes. Weed clearing and soil loosening can be achieved through ploughing the entire plantation or employing spot weed control and soil loosening around planted bamboo.

During early maturity stage of 3-5 years, **pruning** to maintain clump congestion, *vis-a-vis* coppicing activities are crucial for efficient harvesting. Pruning lower branches prevents congestion, promotes air circulation and facilitates fertilization and harvesting, especially in short-necked clumping bamboos. Pruning can commence from second and third years with intensive pruning starting from the fourth year during December and January.

To ensure a healthy bunch-structure, weak or deformed culms are removed through thinning and pruning. Dead culms pose a risk of pathogens and act as potential fire hazards. Thinning involves removal of dead, old or damaged culms/poles from the 3rd or 4th year to maintain an optimal bunch structure. New shoots typically emerge on the outer side while older ones to the inner portion.

6.14.5 Intercropping

Bamboo cultivation offers a sustainable and versatile approach that extends beyond mere timber production. The gestation period of a bamboo plantation spans five years, presenting an opportunity to maximize land utilization through intercropping during the initial three years to generate an additional income.

Various crops such as ginger, chillies, various vegetables and shade-loving aromatic or medicinal plants can be strategically planted in the interspace to generate an additional income, *vis-a-vis* will reduce the weed and pest population to a minimum apart from contributing to overall health and growth of main crop automatically due to management of intercrops.

In humid zone of NE India, bamboo cultivation harmoniously coexists with crops like siris, aonla, bakain, banana, neem, semul, betel nut and coconut. The cultivation spectrum extends to higher altitudes where temperate bamboo species are paired with crops growing at that altitude. Shade-tolerant crops such as pineapple, ginger, turmeric, colocasia and dioscorea find potential synergy with bamboo in various regions.

Madhya Pradesh showcases successful combinations with maize and soyabean thriving in *Bambusa nutans* and *Dendrocalamus strictus* plantations. Cultivation of isolated clumps of potentially significant species such as *D. stocksii* enriches farm bunds, homesteads, boundaries and block plantations particularly in the Konkan belt of Maharashtra. Medicinal plants such as *Aloe vera*, *Catharanthus roseus*, *Curcuma domestica*, *Plantago ovata* and *Withania somnifera* find a compatible growth environment as intercrops due to their adaptability.

The innovative integration of bamboo with other trees represents a departure from conventional cultivation practices. For instance, farmers in south Konkan region embrace a holistic approach by cultivating bamboo alongside native trees such as *Tectona grandis, Phyllanthus emblica, Aegle marmelos, Spondiyas pinnata* and *Gmelina arborea*. The balance between bamboo and broadleaf trees is crucial, typically maintained at ratios like 7:3 or 8:2 in semi-naturally mixed stands. Trees with deep umbrageous crowns such as *Mangifera indica* and *Artocarpus heterophyllus* are judiciously avoided.

In essence, the diversified cultivation practices outlined above exemplify a sustainable and resilient approach to maximize the productivity and economic benefits of bamboo plantations. This harmonious integration not only ensures financial gains for farmers but also contributes to environmental conservation and enhanced agricultural sustainability.

6.14.6 Harvesting

Harvesting bamboo is a nuanced process that depends on the species being cultivated. Typically, commercial plantations become ready for harvest 4-6 years after establishment. Scientific management practices, including fertilization, pruning, thinning and weeding are crucial for optimal yield. For non-structural applications such as those not requiring peak physical and mechanical properties, 2-3-year old culms from mature clumps are harvested.

Harvesting should be avoided during growing season, *i.e.* during monsoon to prevent damage to the emerging shoots as this will hinder future growth of clumps. For timber production harvesting during dry season is ideal unless congestion control is necessary. All the mature culms which have attained 3-years of age and with a bit of yellowing of culms should be harvested as low as possible but a few of such culms should be left from harvesting to protect the new ones. However, the debris and the dead bamboos if any should be removed at this time.

Harvesting of the healthy **edible shoots** should be done when the culms are 15-50cm high depending on the species but mother rhizomes should be spared from damage. Different harvesting ages cater to various applications due to changing mechanical strength, *vis-a-vis* moisture and starch content. Young shoots up to 30 days old having high water content have soft tissues suitable for food, juvenile shoots up to 1-year old with immature culms having high starch and moisture contents are suitable for basketry and rope production, adult ones 1-2 years old with high starch content but prone to decay and insect attacks are suitable for basketry and mat formation, late adult ones with 2-3 years old are suitable for pulping *cum* paper industry, for furniture making and for non-structural applications and old aged ones being 3-4 years old are ideal for structural applications and for industrial products with high structural stability. Culm age is determined based on features of culm sheath, branch and leaf development, external culm colour and new culm positions.

6.14.7 Stacking and storage

Harvested culms should be vertically stacked for seasoning. Proper covering and shading on concrete or cement flooring or the use of tarps or polythene sheets protect against termites and borer pests, however, the floor may be treated with chlorpyriphos. These should not be dried directly under sun as this may cause cracking on the culms.

6.14.8 Yield

The productivity of bamboo plantations is intricately tied to various factors including management practices, site conditions and quality of planting materials. Yield also fluctuates significantly across different species. A meticulously managed mature plantation has the potential to produce an impressive 25-30 tonnes (2500-3000) of culms per hectare, *i.e.* 5-6 culms per clump in case of bamboos of medium diameter.

However, the yield of *Bambusa tulda* in unmanaged natural forest patches in Asom is documented at 12.5 tonnes (dry) per hectare on a four-year cycle. In village settings, this figure increases to a range of 20-23 tonnes. Scientifically well-managed plantations, particularly those of species like *B. bambos, B. balcooa,*

B. nutans and *D. strictus* boast annual yields varying from 25 to 40 tonnes per hectare.

For *Dendrocalamus brandisii*, a spacing of 6×6 metres accommodates around 278 bamboo plants per hectare with harvestable yields starting from an average of 2 culms/clump from the third year onwards but from 8th to 40th year an average of 10 or more culms/clump may be anticipated. In case of *D. stocksii* with spacing of 4×4 metres (625 plants/ha), 10 culms/plant in 4th year and about 15 culms/clump from the 5th year onwards can be obtained. *Guadua angustifolia* from Coorg (Karnataka) usually provides an average of 12 extractable culms per clump starting from the 4th year which account to 4320 culms per hectare. These insights underscore the nuanced dynamics influencing bamboo yield across diverse species and cultivation methodologies.

6.11 Insect-Pests and Diseases

6.11.1 Insect-pests (Mohan and Berry, 2021)

Bamboo cultivation faces a significant challenge in the form of pest-related impediments, posing a major obstacle to achieving optimal production levels. The susceptibility of bamboo to various insects from the orders Coleoptera, Lepidoptera, Orthoptera and Hemiptera is a prominent concern. Among the defoliators, the greater leaf roller (*Pyrausta coclesalis*) emerges as the most detrimental particularly during post-monsoon season when it reaches epidemic proportions. Defoliators such as *P. bambucivora, Heiroglyphus banian* and *Poecilo cerus pictus* contribute to the pest burden.

The bamboo weevils (*Cyrtotrachelus dux* & *Cyrtotrachelus longimanus*) play a pivotal role as shoot borers impacting bamboo growth. Hispine beetle (*Estigmena chinensis*) inflicts substantial damage to *Dendrocalamus strictus* culms. Silvicultural thinning is employed as a control measure against borers affecting standing bamboos. Presence of sap-suckers including the bamboo aphid (*Oregma bambusae*) and seed bug (*Ochrophara* sp.) can reach epidemic proportions.

Threat posed by various leaf-biting and sucking insects ~ aphids should not be underestimated as these can cause minor damage, primarily affecting young bamboo plants. Collins and Keilar (2005) have documented these observations emphasizing the need for comprehensive management strategies to safeguard bamboo crops from their infestation.

To mitigate these challenges, a comprehensive approach to pest management for bamboo involves a range of preventive and remedial measures. These encompass physical, chemical and biological techniques aimed at curbing insect infestations. IPM concept is pivotal in this context as it leverages all available options to minimize the environmental impact of insecticides and ensures sustainable bamboo cultivation practices.

6.11.1.1 Seed pests

One prominent pest is *Ochrophara montana* which poses a significant threat during flowering of bamboo species. Its nymphs and adults extract sap from developing seeds which lead to infertile seed production. To counteract this effective management strategies involve bringing adult bugs and nymphs to the ground through smoke application. Once on the ground, these pests can be eliminated through mechanical destruction or by employing 0.002% cypermethrin spray.

Another formidable pest in this context is larvae of *Sitotroga cerealella* moth, a notorious threat to stored grains that also causes 100% seed loss in storage. To address this issue, a recommended management approach includes closed space fumigation of infested seeds. This process entails using 6ml of carbon disulphide per 100kg of seeds and allowing these to remain in this treated environment at least for 48 hours. This meticulous strategy ensures the effective control of this pest and safeguards stored bamboo seeds from potential devastation.

6.11.1.2 Insect-pests of bamboo nursery

In bamboo nurseries, the threat posed by insect-pests is of significant concern, particularly impacting the vitality of young seedlings. *Agrotis ipsilon* larvae have been identified as formidable adversaries capable of causing fatal damage to these fledgling bamboo plants. Simultaneously, mature seedlings, which have developed rhizomes face a different peril in the form of chafer grubs specifically those belonging to the *Holotrichia serrata* species. The insidious attacks of termites (*Odontotermes* spp. & *Microtermes* spp.) further compound the challenges faced by bamboo nurseries. To combat cutworms and chafer grubs in nurseries, a recommended approach involves the application of thimet at a rate of 200g per bed. This intervention serves as a proactive measure to safeguard the vulnerable seedlings from potential devastation. In case of termite infestation, a targeted response is necessary with application of chlorpyriphos at a concentration of 0.2%. To ensure comprehensive coverage, a judicious quantity of fifty litres of emulsion per bed should be meticulously administered.

By adopting these management practices, bamboo nurseries can fortify their defenses against the diverse array of insect-pests, thereby promoting the robust growth and development of bamboo seedlings.

6.11.1.3 Green culm borers

Green culm borers pose a significant threat to bamboo plants affecting both young and mature culms. Several species such as *Cyrtotrachelus dux*,

C. longimanus, *Myocalandra exarata*, *Argyroploce paragramma*, *Chelyophora ceratatina* and *Estigmina chinensis* being particularly noteworthy. Of these, *Cyrtotrachelus* spp. and *Estigmina chinensis* stand out as the primary culprits responsible for causing maximum damage.

One recommended approach involves preventing clump congestion by systematically removing culms at regular intervals employing a horse-shoe-shaped thinning method. This practice significantly diminishes the likelihood of borer infestation.

To address existing infestations, an internodal injection of 0.02% Dimethoate is advocated. The injection should be administered at the middle of the basal internode with an optimal dose of 20-25ml emulsion per culm. This targeted application of insecticide is most effective during monsoon season when larval activity of these borers is at its peak. By adopting these proactive measures and implementing timely interventions, bamboo cultivators can safeguard their crops against the detrimental impact of green culm borers.

6.11.1.4 Defoliators

Defoliators belonging to Lepidoptera and Orthoptera group pose a significant threat to various bamboo species. Noteworthy lepidopteran defoliators include *Algedonia* (*Pyrausta*) *coclesalis*, A. *bambusivora* and *Massepha absolutalis*. Meanwhile, orthopteran defoliators such as *Hieroglyphus banian*, *Poecilo cerus pictus* and *Schistocerca gregaria* are also known for damaging bamboo foliage.

Greater bamboo leaf roller (*Algedonia coclesalis*) feeds on a variety of bamboo species including *Arundinaria* sp., *Bambusa vulgaris*, *B. nutuns*, *Cephalostachyum pergracile*, *Dendrocalamus strictus*, *D. giganteus*, *D. latiflorus*, *Phyllostachys pubescens*, *P. viridis* and *Schistostachyumper gracile*. Outbreaks of this defoliator are a recurrent issue in bamboo forests and plantations across India. The larvae of *Algedonia coclesalis* roll on bamboo leaves into the protective structures feeding on the upper leaf tissue. Despite its prevalence, natural predators such as tachinid and ichneumonid parasitoids, as well as carabid and mantid predators help regulate the population. Management strategies involve spraying with 0.01% dimethoate, monocrotophos or thimet @ 200g per bed.

Another lepidopteran defoliator is 'lesser bamboo leaf roller' [*Algedonia* (*Pyrausta*) *bambucivora*]. This species is particularly problematic in bamboo forests of NW Himalaya, especially in Punjab from July to October. The eggs are laid in clusters and the larvae feed between rolled-up leaves eventually pupating inside a cocoon within the roll. Bamboo species such as *Bambusa nutans*, *B. vulgaris*, *Cephalostachyum pergracile*, *Dendrocalamus strictus*, *D. giganteus* and *Schistostachyumper gracile* are susceptible to this defoliator.

Other defoliators include *Hieroglyphus banian, Massepha absolutalis, Poecilo cerus pictus* and *Schistocerca gregaria*. Effective management strategies involve spraying 0.01% dimethoate, monocrotophos or thimet at a rate of 200g per bed. Termite control can be achieved through application of 0.2% chlorpyriphos solution with 50 litres of emulsion applied per bed. These comprehensive measures aim to mitigate the impact of defoliators on bamboo plantations, forests and nurseries.

6.11.1.5 Shoot *cum* culm borer

Bamboos face the threat of infestation by various shoot and culm borers. Among these, the 'hispine bamboo beetle (*Estigmena chinensis*) is particularly noteworthy which is distributed across India, Bangladesh, Myanmar and Sri Lanka. This beetle targets bamboo species such as *Bambusa bambos, B. burmanica, Dendrocalamus strictus, Gigantochloa scortechinii* and *Schizostachyumper gracile* during initial stages of culm growth. Its voracious attack results in weakened and deformed bamboos not worth for any thing. **Hispine bamboo beetle** has annual life cycle laying eggs in concealed batches, often covered with chewed bamboo leaf fragments. Upon hatching, the larvae scratch tissue from the culm and leaf sheath creating small tunnels that eventually develop into irregular chambers. Pupation occurs within these tunnels with emergence of adult beetles coinciding with the next rainfall. The extensive larval tunnelling degrades the culm, sometimes causing bending at point of damage and in severe cases entire clumps of culms can be affected.

Bamboo weevil (*Cyrtotrachelus dux*) is found in Northern India, Bangladesh, Sri Lanka and Myanmar. This weevil poses a threat to young sprouting culms of *Dendrocalamus hamiltoni, D. stictus* and certain other bamboo species. With an annual life cycle, adult weevils become active at the onset of the monsoon. Mating leads to the biting of small pits on one-metre-tall culms, followed by feeding and egg-laying. The hatched grubs bore through internodal walls, creating irregular galleries that weaken the culm, making it prone to breakage. Management strategies involve exposing pupae in the soil or using 0.04% dimethoate on sprouting culms. Attacked culms can be treated by internodal injection of 0.2% monocrotophos.

Phloeobius crassicollis causes damage to felled and green standing bamboos. With a life cycle spanning a year, this beetle feeds on the outer surface of bamboo culms, primarily targeting nodes or culm sheaths. Hosts include *Bambusa bambos, B. nutans, B. polymorpha, B. tulda, B. vulgaris, Dendrocalamus calostachyus, D. giganteus, D. strictus* and *Gigantochloa atroviolacea*. Management entails internodal injection of 0.05% chlorpyriphos.

Various **other culm borers** such as *Cyrtotrachelus longimanus*, *Chelyophora ceratatina*, *Argyroploce paragramma* and *Myocalandra exarata* pose threats to bamboo. Managing these culm borers involves using of dimethoate or monocrotophos at a concentration of 0.04% through internodal injection. Vigilant monitoring and timely intervention are crucial to safeguarding bamboo ecosystems from the detrimental impacts of these borers.

6.11.1.6 Sap-sucking pests

Sap-sucking pests primarily belonging to families of Aleyrodidae, Aphididae, Coccidae and Membracidae exhibit a diverse array of species that have been documented on various bamboo species. Nymphs and adults of bamboo aphid (*Oregma bambusae*) are sometimes highly damaging as these are polyphagous. To effectively manage the infestation, 0.01% dimethoate or monocrotophos should be sprayed to the affected areas. Another category of pests that poses a threat to bamboo is 'ghoon borer' group (*Chlorophopus annularis*, *Dinoderus brives*, *D. minutus*, *D. ocellaris* and *Lyctus africanus*) so to manage bamboo products from these borers these are required to be subjected to a kiln with a temperature of 50-60°C for 48 hours which will eliminate them effectively. As a preventive measure, a thorough application of cypermethrin at 0.5% in diesel is recommended, and a sticker such as triton (2ml/l of emulsion) may be added to enhance treatment adherence. Preservation of bamboo can be achieved through treatment with a 5% aqueous solution of CCA. Protection against dry bamboo borers can be achieved through treatment with a mixture of 3% boric acid and borax (1:1) or 3% boric acid combined with zinc chloride. These comprehensive measures aim to mitigate the impact of sap-sucking pests and borers on bamboo, ensuring the health and longevity of these invaluable plant resources.

6.11.1.7 Subterranean pests

Bamboo cultivation faces challenges from subterranean pests that can adversely affect the health and growth of the plants. One such pest is the **white grub** (*Holotrichia serrata*). The infestation period of these insects occurs between June and September leading to wilting of transplanted culms. To manage this pest, it is advisable to avoid sandy soil for raising seedlings, refraining from using semi-decomposed FYM and avoid soil working during the monsoon season (June-July). An effective preventive measure involves soil mixing with Phorate 10G @ 200g/10m^{2}bed.

Another subterranean pest affecting bamboo is the **termite** (*Microtermes* spp. & *Odontotermes* spp.) These termites have a broad host range including teak, bamboo, neem, shishum, babul, siris, semul, anola, eucalyptus, and many others. While *Odontotermes obesus* and *Microtermes obesi* have been observed to feed on bamboo roots, the resulting damage is usually negligible. Termites pose a threat throughout the year, causing harm to plants in their early growth stages by feeding on roots, rhizomes and stems. Their consumption of decaying matter in FYM leads to a loss of valuable nitrogen for plant growth. Soil drenching with a water emulsion of chlorpyriphos (0.05%) or soil application of Imidacloprid (17.8% SL) at a concentration of 0.05% is quite effective.

6.11.2 Diseases

Bamboo confronts various diseases primarily induced by fungal pathogens that target critical components such as the rhizomes, roots, culms, foliage, flowers and seeds. The bamboo culms, in particular fall prey to an extensive array of both saprophytic and parasitic fungi. Among the myriad of afflictions, one of the most significant and pervasive diseases is the bamboo blight. *Bambusa vulgaris* in particular exhibits heightened susceptibility to bamboo blight caused by *Sarocladium oryzae*, as well as physiological disorders when grown in areas with impeded drainage (Boa and Rahman, 1987). This particular ailment holds the potential to spread like an epidemic, posing a substantial threat to bamboo plantations.

In addition to these threats, bamboo plants are susceptible to various diseases including leaf spot (*Diplozythiella bambusina*), rust (*Puccinia* spp.), *Ustilago shiraiana* and culm rot induced by *Pterulicium xylogenum* (Boa and Rahman, 1987). To address the challenge posed by *P. xylogenum*, Harsh *et al.* (2005) suggest effective control measures such as spraying and drenching with copper oxychloride and carbendazim.

The bamboo rust diseases affecting bamboo species worldwide, are attributed to at least 29 rust fungi that fall under the purview of six distinct genera: *Kweilingia, Puccinia, Uredo, Phakospora, Stereostratum* and *Tunicopsora*. Prior to 2007, the extent of reported rust pathogens on bamboo in Hawai'i was limited to two: *Puccinia phyllostachydis* responsible for leaf rust and *Stereostratum corticiodes* which induces culm rust. In 2007, a new bamboo rust disease emerged due to infection of *Kweilingia divina* (syn. *Dasturella divina*) (Nelson and Goo, 2011).

6.11.2.1 Bamboo blight (*Sarocladium oryzae*)

Bamboo blight between 1985 and 1988 in *Bambusa vulgaris* was caused by *Sarocladium oryzae* in the coastal regions of Odisha. The disease resulted in extensive mortality of mature clumps within village groves and traced its origins back to Bangladesh spreading through West Bengal before affecting Odisha. The onset of the blight primarily targets young culms, occurring either during or shortly after the elongation process. Distinctive symptoms manifest as premature death of the culm sheath, distinguishable by its easy removal compared to healthy ones, along with a partial collapse of the apical region. Wet-rot patches develop on the internodes creating a conducive environment for various insect larvae, beetles and ants exacerbating the damage. As the internodes rot the affected areas rapidly spread within the tender culm resulting into death of apical region. This progressive deterioration culminates in the demise of the entire clump within a span of 3 to 4 years. Treating soil, specifically drenching bamboo clumps with copper oxychloride and Dithane M45 has demonstrated promising results in controlling the spread of bamboo blight (odishabamboo.org, 2024).

6.11.2.2 Culm rot (*Pterulicium xylogenum*)

A novel culm rot disease in bamboo caused by *Pterulicium xylogenum* has been recorded. This disease manifests in three bamboo species, *viz. Bambusa vulgaris* var. *waminii*, *Dendrocalamus giganteus* and *Gigantochloa* sp. Alarming observations revealed that approximately 45% of *B. vulgaris* var. *waminii* culms and 36% of *Gigantochloa* culms exhibited signs of morbidity or had succumbed to the disease. A combination of 0.05% copper oxychloride and carbendazim (expressed in active ingredient weight) emerged as a potent mixture for spraying and drenching to control this problem even at 0.025% concentration (Harsh *et al.*, 2005).

6.11.2.3 Bamboo rust (*Dasturella divina*)

Bamboo rust (*Dasturella divina*, syn. *D. oxytenantherae* Sathe; and *B. bambustina*; Mundkur and Kheswala, 1943; Sathe, 1965; Bakshi and Singh, 1967; Rangaswami *et al.*, 1970) exerts its detrimental effects primarily on the leaves of affected plant so disrupting the photosynthetic activities. Notable distinctions between *D. divina* and *D. bambusina* include the number of teliospores per chain and the depth of the telial column.

Typically manifesting in August, the infection becomes visible on mature leaves as greyish-brown minute flecks due to high density display of uredinia though juvenile leaves exhibit resistance. As the small flecks merge, spindle-shaped dark brown pustules with pale halos emerge. These yellowish-brown uredinia form on the adaxial surface of leaves though with rare occurrences on the abaxial surface. The rust infection persists until late May, leading to the development of dark brown teliosori in mature uredinial sori or separately on the adaxial surface in linear rows during January ultimately resulting in necrosis and withering of mature leaves (Mohanan, 1897; Mishra *et al.*, 2015). It is challenging to manage this problem due to production of vast rust spores and spread through winds. Mishra *et al.* (2015) advocated proper plant nutrition, increased plant spacing for better aeration, maintenance of proper hygiene and spraying effective fungicides. Effective fungicides are triforine with bupirimate, mancozeb, mycobutanil or penconazole.

6.11.2.4 Bamboo mosaic

Bamboo mosaic virus (BaMV) is a plant pathogenic virus belonging to the genus Potexvirus within the family Flexiviridae. The BaMV virus particles exhibit a distinctive filamentous morphology, characterized by flexuous rods measuring 490nm in length and 15nm in width. The genomic composition of BaMV is monopartite, featuring a positive-sense single-stranded RNA.

The primary natural host of BaMV is bamboo, with susceptibility observed in at least 10 different bamboo species (Lin *et al.,* 1993). Other plant families including Amaranthaceae, Chenopodiaceae, and Poaceae are known to harbour susceptible hosts. Geographically, BaMV has been identified in regions such as Hawaiʻi, various Pacific Islands, Australia (Dodman and Thomas, 1999), the Philippines, Taiwan, Brazil (Brunt *et al.,* 1996) and the United States (Elliot and Zettler 1996).

No insect vector has yet been recorded as a transmitter of BaMV. The virus primarily spreads through mechanical inoculation. Manifestations of BaMV infection include chlorotic mottling, mosaic patterns parallel to leaf veins, necrotic streaks on shoots and culms, vascular discolouration, aborted stems and in severe cases death of plant. Some susceptible hosts may exhibit no visible symptoms (Nelson and Borth, 2011).

Diagnostic methods for BaMV involve PCR (polymerase chain reaction), ELISA (enzyme-linked immunosorbent assay), inoculation of assay hosts and observation of symptom expression. BaMV represents the sole mosaic disease affecting bamboo and its sequence is available in GenBank. PCR primers for virus detection are also accessible (Nelson and Borth, 2011).

Once established, BaMV-infected plants cannot be eradicated without destroying them. To prevent infection and disease spread, the following disease-management practices are recommended (Nelson and Borth, 2011):

i) Avoid planting of BaMV-infected bamboo plants displaying mosaic symptoms on leaves, however, suspected species should be got tested before planting.

ii) Heat-sterilization of blades is pre-requisite between cuts though some infected plants may exhibit only mild or subtle symptoms.

iii) To ensure virus-free planting material, use only tissue cultured plants.

6.12 Physiological Disorders

Bamboos like any other plant also show various physiological disorders that affect their growth and health as shown below:

- **Yellowing/Chlorosis:** It occurs due to deficiencies of nitrogen, iron and magnesium which is due to poor soil conditions or improper fertilization.
- **Leaf rolling:** Leaf's curling or inward rolling mainly occurs due to overwatering or drought conditions, and also due to certain insect or disease attack.
- **Tip burn:** This appears as browning or necrosis at leaf tips which results due to waterlogged condition, salt accumulation in the soil or due to excessive fertilizer application.
- **Culm rot**: Fungal or bacterial infection which is often facilitated due to waterlogged and poorly drained conditions, *vis-à-vis* wounds on the culms.
- **Shoot abnormalities**: Deformities of new shoots occur due to insect-pests and disease affecting the growing tips. However, poor soil conditions may also cause such abnormalities.
- **Bamboo mosaic virus (BaMV):** The plants may show chlorotic mottling, necrotic streaks, mosaic pattern on leaves, vascular discolouration and in certain cases even plant death. BaMV is transmitted mechanically and can spread through contaminated tools or pruning practices.
- **Flowering disorders:** Irregular or abnormal pattern of flowering occurs including early, delayed or mast flowering due to environmental factors, stress or genetic predisposition.
- **Dieback:** Occurrence of gradual or sudden death of entire culms or branches due to certain environmental stresses, physical damage, incorrect pruning, inadequate spacing, *vis-à-vis* disease problems.

Management of physiological disorders includes adjusting watering practices, improving soil conditions, implementing proper fertilization and adopting disease and pest control measures. Regular monitoring and timely intervention can help maintain the health and vitality of bamboo stands.

References

1. Akinlabi, E.T., K. Anane-Fenin and D.R. Akwada, 2017. Bamboo: The Multipurpose Plant, pp. 205-13. Springer.
2. Anderson, J. and M.F. Strong, 1983. The effect of fiber on nutrition of man. *Ind. J. Nutrit. & Dietet.*, **81**: 279–85.
3. Austin, R. and K. Ueda, 1970. Bamboo, 193p. Walker/Weatherhill, New York.

4. Ayer, S., S. Timilsina, A. Aryal, A.K. Acharya, A. Neupane and K.P. Bhatta, 2023. Bamboo forests in Nepal: Status, distribution, research trends and contribution to local livelihoods. *Adv. Bamboo Sci.*, **4**: 100027. doi:10.1016/j.bamboo.2023.100027. ISSN 2773-1391. S2CID 259134632.

5. Baardseth, P., F. Bjerke, B.K. Martinsen and G. Skrede, 2010. Vitamin C, total phenolics and antioxidative activity in tip-cut green beans (*Phaseolus vulgaris*) and swede rods (*Brassica napus* var. *napobrassica*) processed by methods used in catering. *J. Sci. Food & Agric.*, **90**(7): 1245–55.

6. Bakshi, B.K. and S. Singh, 1967. Rusts on Indian Forest Trees. Indian Forest Records (NS). *Forest Pathol.*, **2**: 139–98.

7. Bal, L.M., S.N. Naik and S. Satya, 2009. Scientific validation of indigenous knowledge system for bamboo shoot processing and utilization. *Abstract Proc Nat. Sem. Bamboo Plantat. Manag. & Its Utiliz.*, p. 70. Held on March 2009 at Jodhpur (India).

8. Bamboo Info, 2024. Bamboo Technical Support Group (BTSG - KFRI). Retrieved on 2 February 2024. https://www.bambooinfo.in/cultivation/.

9. Bhargava, A., V. Kumbhare, A. Srivastava and A. Sahai, 1996. Bamboo parts and seeds for additional source of nutrition.*J. Food Sci. & Techn.*, **33:**(2): 145–6.

10. Bhatt, B.P., K. Singh and A. Singh, 2005. Nutritional values of some commercial edible bamboo species of the North Eastern Himalayan region, India. *J. Bamboo & Rattan*, **4**(2): 111–24.

11. Bisht, M.S., C. Nirmala and P. Nongdam, 2012. Bamboo shoot as resource for health food and socio-economic development in North-East India. In *Proc. 9th World Bamboo Cong.* (WBC '12), held in April 2012 in Antewerp, Belgium, pp. 393–402.

12. Boa, E.R. and M.A. Rahman, 1987. Bamboo blight and the bamboos of Bangladesh. The results and conclusions of an investigation into a serious new disease of bamboos. *Bulletin - Forest Pathology Series*, Forest Research Institute (Chittagong), No. 1, pp. 43.

13. Bradford, P.G. and A.B. Awad, 2007. Phytosterols as anticancer compounds. *Molec. Nutr. & Food Res.*, **51**(2): 161–70.

14. Britannica, 2024. The Editors of Encyclopaedia. Bamboo. In: *Encyclopedia Britannica*, 5 Jan. 2024, https://www.britannica.com/plant/bamboo. Accessed 1 February 2024.

15. Brunt, A.A., K. Crabtree, M.J. Dallwitz, A.J. Gibbs, L. Watson and E.J. Zurcher (eds.), 1996. Plant Viruses Online: Descriptions and Lists from

the VIDE Database. Version: 20 August 1996. http://www.agls. uidaho. edu/ebi/vdie/refs.htm#authors.

16. Canavan, S., D.M. Richardson, V. Visser, Le J.J. Roux, M.S. Vorontsova and J.R.U. Wilson, 2016. The global distribution of bamboos: assessing correlates of introduction and invasion. *AoB Plants*, **9**(1): plw078. doi:10.1093/aobpla/plw078. ISSN 2041-2851. PMC 5499700. PMID 28013249.

17. cfs.gov.hk, 2024. *Naturally Occurring Toxins in Vegetables and Fruits Food Contaminants,* Hong Kong Government Centre for Food Safety. Retrieved on 2 February 2024. https://www.cfs.gov.hk/english/programme/programme_rafs/programme_rafs_fc_02_05.html.

18. Choi, J.M., E.O. Lee, H.J. Lee. K.H. Kim, K.S. Ahn, B.S. Shim, N.I. Kim, M.C. Song, N.I. Baek and S.H. Kim, 2007. Identification of campesterol from *Chrysanthemum coronarium* L. and its antiangiogenic activities. *Phytother. Res.*, **21**(10): 954-9.

19. Chongtham, N., M.S. Bisht and S. Haorongbam, 2011. Nutritional properties of bamboo shoots: potential and prospects for utilization as health food. *Compreh. Rev. Food Sci. & Food Saf.*, **10**(3): 153–68.

20. Chongtham, N., M.S. Bisht and S. Haorongbam, 2011. Nutritional properties of bamboo shoots: potential and prospects for utilization as health food. *Compreh. Rev. Food Sci. & Food Saf.*, **10**(3): 153–68.

21. Clark, L.G., W. Zhang and J.F. Wendel, 1995. A phylogeny of the grass family (Poaceae) based on ndhF sequence data. *Syst. Bot.*, **20**:436–60.

22. Collins, R. and S. Keilar, 2005. The Australian bamboo shoot industry: a supply chain approach. In: *RIRDC* (*Rural Industries Research and Development Corporation*) Publ. No. 05/022, 92p. Barton, Australia.

23. Dodman R.L. and J.E. Thomas, 1999. The first record of bamboo mosaic potexvirus from Australia. *Aust. Pl. Path.*, **28**: 337–9.

24. Dransfield, S. and E.A. Widjaja (eds), 1995. *Bamboos.* Plant Resources of South-East Asia No. 7. Prosea Foundation, pp. 20–21. ISBN 90-73348-35-8.

25. Elliot, M.S. and F.W. Zettler, 1996. Bamboo mosaic virus detected in ornamental bamboo species in Florida. *Proc. Fda St. Hort. Soc.*, **109**:24–5.

26. Farrelly, D. 1984. The Book of Bamboo, 143p.. Sierra Club Books, San Francisco.

27. forests.org, 2024. World's Biggest Bamboo Plant Found in Southwest China. Forest Conservation Portal. 28 August 2003. https://web.archive.org/web/20031110222335/https://forests.org/articles/reader.asp?linkid=25225. Retrieved 30 January 2024.

28. George K., M.G. Noordhoff and J. Slagan, 1982. Dietary fiber used in the management of hypertension and obesity. *J. Sci. Food & Cantwell Agric..*, **32:** 494–7.

29. GWR, 2024. Fastest growing plant. *Guinness World Records*. https://www.guinnessworldrecords.com/world-records/fastest-growing-plant/. Retrieved 30 January 2024.

30. Harsh, N.S.K., Y.P. Singh, H.K. Gupta, B.M. Mushra, D.J. Mclaughlin and B. Dentinger, 2005. A new culm rot disease of bamboo in India and its management. *J. Bamboo & Rattan*, **4**(4): 387–98.

31. Hu, H., J. Zhang, K. Lu and Y. Tian, 2015. Characterization of *Acidosasa edulis* shoot shell and its biosorption of copper ions from aqueous solution. *J. environ. Chem. Eng.*, **3**(1): 357–64.

32. Judziewicz, E.J. and S. Sepsenwol, 2007. The World's Smallest Bamboo: *Raddiella vanessiae* (Poaceae: Bambusoideae: Olyreae), a new species from French Guiana. *J. bot. Res. Inst. Texas*, **1**(1): 1–7.

33. Kalita, T. and U. Dutta, 2012. A comparative study on indigenous usage of bamboo shoot in the health care practices in NE India, *The Clarion*, **1**(2): 130–41.

34. Kaminski, S., A. Lawrence and D. Trujillo, 2016a. Structural use of bamboo. Part 1: Introduction to bamboo. *Struct. Eng.*, **94**(8): 40–43.

35. Kaminski, S., A. Lawrence, D. Trujillo, I. Feltham and L. Felipe López, 2016b. Structural use of bamboo. Part 3: Design values. *Struct. Eng.*, **94**(12): 42–5.

36. Kelchner, S.A. 2013. Higher level phylogenetic relationships within the bamboos (Poaceae: Bambusoideae) based on five plastid markers (PDF). *Mol. Phylogen. & Evol.*, **67**(2): 404–13. doi:10.1016/j.ympev.2013.02.005. ISSN 1055-7903. PMID 23454093. Archived from the original (PDF) on 5 June 2015.

37. Kim, N.K., S.H. Cho, S.D. Lee, J.S. Ryu and K.H. Shim, 2001. Functional properties and antimicrobial activity of bamboo (*Phyllostachys spp.*) extracts. *Korean J. Food Proc.*, **8**: 475–80.

38. Kramer, L. 2016. Interesting facts about bamboo. Bamboo, Import Europe. https://bambooimport.com/en-at/blogs/everything-about-bamboo/interesting-facts-about-bamboo. Retrieved on 3 February 2024.

39. Kritchevsky, D. and S.C. Chen, 2005. Phytosterols-health benefits and potential concerns: a review. *Nutr. Res.*, **25**(5): 413–28.

40. Kumar, R., M.M. Thangaraju, M. Kumar, S.T. Thul, V.C. Pandey, S. Yadav, L. Singh and S. Kumar, 2021. Ecological restoration of coal fly ash–dumped

area through bamboo plantation. *Environ. Sci. & Pollut. Res.*, **28**(25): 33416–32.

41. Kumbhare, V. and A. Bhargava, 2007. Effect of processing on nutritional value of central Indian bamboo shoots. Part I. *J. Food Sci.*, **44**(3): 935–6.
42. Lakkad, S.C. and J.M. Patel, 1981. Mechanical properties of bamboo, a natural composite. *Fibre Sci. & Techn.*, **14**(4): 319–22.
43. Lin, N.-S., J.-Y. Chai, T.-Y. Huang, T.-Y. Chang and Y.-H. Hsu, 1993. Incidence of bamboo mosaic potexvirus in Taiwan. *Pl. Dis.*, **77**: 448–50.
44. Lu, B.Y., J.F. Bao, L. Shan and Y. Zhang, 2009. Technology for supercritical CO_2 extraction of bamboo shoot oil and components of product. *Trans. Chinese Soc. Agric. Eng.*, **25**(8): 312–6.
45. Lu, B., X. Wu, X. Tie, Y. Zhang and Y. Zhang, 2005. Toxicology and safety of anti-oxidant of bamboo leaves. Part 1: acute and subchronic toxicity studies on anti-oxidant of bamboo leaves. *Food & Chem. Toxicol.*. **43**(5): 783–92.
46. Manan, A. 2023. 107 Edible Bamboo Shoot Species: Attributes and Edibility. *CropForLif*, 2021 (Updated on May 6, 2023). https://cropforlife.com/107-edible-bamboo-shoot-species-attributes-and-edibility/.
47. Masisi, B., A. Zabel, J. Blaser and S. Augustino, 2022. Fighting climate change with bamboo in Africa: The case of Kyela, Rungwe and Mufindi districts – Tanzania. *Adv. Bamboo Sci.*, **1**: 100009. doi:10.1016/j.bamboo.2022.100009. ISSN 2773-1391. S2CID 253535691.
48. McClure, F.A. 2013. The Bamboos: A Fresh Perspective. Harvard University Press, Cambridge, MA and London, England. https://doi.org/10.4159/harvard.9780674428713.
49. McClure, F.A. 1948. 1948 Yearbook of Agriculture – Grasses, 738p. USDA, Washington, D.C.
50. Meric, J.B., S. Rottey, K. Olaussen, J.C. Soria, D. Khayat, O. Rixe and J.P. Spano, 2006. Cyclooxygenase-2 as a target for anticancer drug development. *Critic. Rev. Oncol./Hematol.*, **59**(1): 51-64.
51. Miettinen, T.A. and H. Gylling, 2003. Non-nutritive bioactive constituents of plants: phytosterols. *Intern. J. Vit. & Nutr. Res.*, **73**(2): 127–34.
52. Mishra, V., M.K. Thakur and R.P. Mishra, 2015. *Dasturella* rust of bamboo in India. *Int.J.Curr.Microbiol.App.Sci.*, **4**(6): 467-70.
53. Mohan, C. and N. Berry, 2021. Insect-pests of bamboos and their management. *Biotica Res. Today* (An International eMagazine), **3**(10): 874-7. https://biospub.com/index.php/biorestoday/article/view/1149.

54. Mohanan, C. 1997. Diseases of bamboos in Asia. International Network for Bamboo and Rattan, New Delhi, 228p.

55. Mundku, B.B. and K.F. Kheswala, 1943. *Dasturella* - a new genus of the Uredinales. *Mycologia*, **35**: 201–6.

56. Nartey, F. 1980. Toxicological aspects of cyanogens in tropical foods. Toxicology in the Tropics (eds Smith, R.L. and E. R. Bababummi), pp. 53–73. Taylor & Francis, London, UK.

57. Nelson, S. and W. Borth, 2011. Bamboo Mosaic. College of Tropical Agriculture and Human Resources (CTAHR). University of Hawai'i at Mänoa. https://www.ctahr.hawaii.edu/oc/freepubs/pdf/pd-76.pdf.

58. Nelson, S. and M. Goo, 2011. Kweilingia Rust of Bamboo in Hawai'i. College of Tropical Agriculture and Human Resources (CTAHR). University of Hawai'i at Mänoa. https://www.ctahr.hawaii.edu/oc/freepubs/pdf/PD-74.pdf.

59. Nirmala, C., M.L. Sharma and E. David, 2008. A comparative study of nutrient component of freshly harvested, fermented and canned bamboo shoots of *Dendrocalamus giganteus* Munro. *J.Amer. Bamboo Soc.*, **21**(1): 33–9.

60. Nongdam, P. and L. Tikendra, 2014. The Nutritional Facts of Bamboo Shoots and Their Usage as Important Traditional Foods of Northeast India. *Intern. Schol. Res. Notic.* https://doi.org/10.1155/2014/679073.

61. Oboh, G. and A.O. Ademosun, 2012. Characterization of the antioxidant properties of phenolic extracts from some citrus peels. *J. Food Sci. Techn.*, **49**(6): 729–36.

62. odishabamboo.org, 2024. Diseases and Pests. Odisha Bamboo Development Agency. https://www.odishabamboo.org/Bamboo/diseases-of-pests. Retrieved on 2 February 2024.

63. Ohrnberger, D. 2021. *Kinabaluchloa wrayi*. Bamboos of Thailand. Published on the internet from 1 January 2021 onwards. https://sites.google.com/view/bamboos-of-thailand/home. Retrieved 30 January 2024.

64. Ostlund, R.E., Jr. 2007. Phytosterols, cholesterol absorption and healthy diets. *Lipids*, **42**(1): 41–5.

65. Panda, T. and R.N. Padhy, 2007. Sustainable food habits of the hill-dwelling Kandha tribe in Kalahandi district of Orissa. *Indian J. Trad. Knowl.*, **6**(1): 103–5.

66. Pandey, A.K. and V. Ojha, 2011. Precooking processing of bamboo shoots for removal of anti-nutrients. *J. Food Sci. Techn.*, **51**(1): 43–50.

67. Pandey, A.K. and V. Ojha, 2013. Standardization of harvesting age of bamboo shoots with respect to nutritional and anti-nutritional components. *J. For. Res..*, 24(1): 83–90.

68. Park, E.J. and D.J. John, 2009. Effects of bamboo shoot consumption on lipid profiles and bowel function in healthy young women. *Nutrit.*, **25**(7-8): 723–8.

69. Park, E.J. and D.Y John, 2010. The antioxidant, angiotensin converting enzyme inhibition activity, and phenolic compounds of bamboo shoot extracts. *Food Sci. & Techn.*, **43**(4): 655–9.

70. Phillips, K.M., D.M. Ruggio and M. Ashraf-Khorassani, 2005. Phytosterol composition of nuts and seeds commonly consumed in the United States. *J. agric. & Food Chem.*, **53**(24): 9436–45.

71. Qiu, F.G. 1992. The recent development of bamboo foods. *Proc. intern. Symp. on Indust. Use of Bamboo*, pp. 333–7. Intern. Trop. Timber Organ. & Chinese Acad., Beijing, China.

72. Rangaswami, G., V.S. Sheshadri and K.K. Lucy Channamma, 1970. Fungi of South India, 193p. *Univers. Agric. Sci.*, Bangalore.

73. Rice-Evans, C.A., N.J. Miller and G. Paganga, 1997. Antioxidant properties of phenolic compounds. *Trends in Pl. Sci.*, **2**(4): 152–9.

74. Sarangthem, K. and T.N. Singh, 2003. Transformation of fermented bamboo (*Dendrocalamus asper*) shoots into phytosterols by microorganisms. *J. Food Sci. & Techn.*, **40**: 622–5.

75. Sathe, A.V. 1965. A new species of *Dasturella* (uredinales) from India. *Sydowia*, **19**: 148–9.

76. Satya, S., P. Singhal, V.G. Prabhu, L.M. Bal and P. Suhakar, 2009. Exploring the nutraceutical potential and food safety aspect of bamboo shoot of some Indian species. *Proc. 8th World Bamboo Conf.*, **6**: 78–88, Bangkok, Thailand, September 2009.

77. Schombergk, S.R.H. 1841. Some Accounts of the Curata....etc. *Trans. Linnean Soc. London.*, **18**: 559–60.

78. Schroder, S. 2024. Guadua Bamboo. 101 Edible Bamboo Species You Can Eat. Retrieved 25 January 2024. https://www.guaduabamboo.com/blog/edible-bamboo-species?rq=bamboo%20shoots.

79. Schwarzmaier, U. 1977. Cyanogenesis of *Dendrocalamus*: taxiphyllin. *Phytochem.*, **16**(10): 1599–1600.

80. Seigler, D.S. 1991. Cyanide and cyanogenic glycoside. In: *Herbivores: Their Interactions with Secondary Plant Metabolites*, vol. I (eds Rosentahel, G.A. and M.R. Berenbaum). *The Chemical Participants*, pp. 35–77, Academic Press, San Diego, California, USA.

81. Singh, L., S. Sridharan, S.T. Thul, P. Kokate, P. Kumar, S. Kumar and R. Kumar, 2020. Eco-rejuvenation of degraded land by microbe assisted bamboo plantation. *Indust. Crops & Prod.*, **155**: 112795. doi:10.1016/j.indcrop.2020.112795. ISSN 0926-6690. S2CID 225025086.

82. Srivastava, R.C. 1990. Bamboo: new raw materials for phytosterols. *Curr. Sci.*, **59**(24): 1333–4.

83. Sundriyal, M. and R.C. Sundriyal, 2001. Wild edible plants of the Sikkim Himalaya: nutritive values of selected species. *Econ. Bot.*, 55(3): 377–90.

84. Testbook, 2024. Bamboo Cultivation in India: Cultivation Requirements, Cropping Pattern & National Bamboo Mission! https://testbook.com/ias-preparation/bamboo-cultivation-in-india. Retrieved on 30 January 2024.

85. USDA, 2018. Bamboo shoots, raw. FoodData Central. Released in April 2018. Accessed on 11 January 2024. https://fdc.nal.usda.gov/fdc-app.html#/food-details/169210/nutrients.

86. Velioglu, Y.S., G. Mazza, L. Gao and B.D. Oomah, 1998. Antioxidant activity and total phenolic in selected fruits, vegetables and grain products. *J. agric. & Food Chem.*, **46**(10): 4113–7.

87. Vetter, J. 2000. Plant cyanogenic glycosides. *Toxicon.*, **38**(1): 11–36.

88. Wikipedia, 2024. Bamboo. https://en.wikipedia.org/wiki/Bamboo. Retrieved on 30 January 2024.

89. Woyengo, T.A., V.R. Ramprasath and P.J.H. Jones, 2009. Anticancer effects of phytosterols. *Europ. J. Clin. Nutr.*, **63**(7): 813–20.

90. Yamaguchi, T., M. Katsuda, Y. Oda, J. Terao, K. Kanazawa, S. Oshima, T. Inakuma, Y. Ishiguro, H. Takamura and T. Matoba, 2003. Influence of polyphenol and ascorbate oxidases during cooking process on the radical-scavenging activity of vegetables. *Food Sci. & Techn. Res.*, **9**(1): 79-83.

91. Zhang, J., R. Ji, Y. Hu, J. Chen and X. Ye, 2011. Effect of three cooking methods on nutrient components and antioxidant capacities of bamboo shoot (*Phyllostachys praecox* C.D. Chu et C.S. Chao). *J. Zhejiang Univ.: Sci. B. Biomed. & Biotechn.*, **12**(9): 752–9.

Curry tree or Curry leaf tree [*Bergera koenigii* (L.)]

Family: Rutaceae. **Chromosome number**: 2n = 18.

Synonyms: *Camunium koenigii* (L.) Kuntze, *Chalcas koenigii* (L.) Kurz, *Chalcas siamensis* (Craib) Tanaka, *Murraya foetidissima* Teijsm. & Binn., *M. koinigii* (L.) Spring (POWO, 2023a), *M. koenigii* (L.) Spreng (POWO, 2023a) *M. siamensis* Craib, *Nimbo melioides* Dennst. (POWO, 2023a)

Other common Indian names: As described by various authors, especially in Hindi (Parmar and Kaushal, 1982; Henry, 2015; etymonline.com, 2023) are 'barsanga', 'curry bush', 'curry leaf tree', 'gandhela', 'kathnim', 'kurry patta', 'meethi neem' and 'sweet neem'; in **Sanskrit** are 'Krshnaneembapatram', 'mahanimb' & 'surabhinimba'; in **Asomese** are 'Bishahari' & 'Narasingha'; in **Andhra Pradesh** is 'karepaku'; in **Bengal** are 'barasanga' & 'kartaphulli'; in **Gujarat** are 'gorenimb' & 'kadhilimbdo'; in **Himachal Pradesh** 'gandhela', 'gandhelu', 'gandhla' & 'mitha neem'; in **Karnataka** 'karibeva'; in **Kerala** 'kariveppilei'; in **Kumaon** 'gandhela', 'gandla' & 'gani'; in **Odisha** 'basango', 'bassan' & 'bhursanga' and in **Tamil Nadu** are 'karivempu', 'karuveppilai' & 'karuveppilei'.

7.1 Introduction

Curry tree stands as an aromatic perennial, thriving in tropical and subtropical climates. Not only it is a robust source of bio-available calcium and various essential nutrients, it also holds a significant place in Indian diet ~ offering numerous direct and indirect health advantages to those who consume it. Recognized as curry leaves, its leaves boast an amicably mildly bitter pungency and delicately acidic taste, retaining their distinct flavour and properties even after the drying process.

Being a leafy perennial, the curry leaf not only assures small and marginal farmers of a steady income when cultivated on a commercial scale but also contributes substantially to foreign returns through exports, notably to regions such as the Gulf and European Unions. This dual role as a dietary staple and a valuable export commodity underscores its importance within both local and international markets, creating economic stability for farmers while meeting global demand for its unique qualities (Joseph and Peter, 1985).

7.2 Etymology

Botanical species *Bergera koenigii*, initially documented by Carl Linnaeus in Mantissa Plantarum (1767), vol. 2, page 563 (POWO, 2023a), was formerly recognized as *Murraya koenigii* (L.) Spreng (POWO, 2023b). While some sources still refer it under previous designation (George *et al.*, 1993), the genus name '*Murraya*' originated from Johan Andreas Murray, a student of Carl Linnaeus who later became a medicine professor focusing on medicinal plants at the University of Göttingen, Germany (MBG, 2023). The specific epithet '*koenigii*' honours botanist Johann Gerhard König. The popular term 'curry leaf' stems from its frequent use in flavouring curries (Vasanthkumar *et al.*, 2023). 'Curry' itself is derived from the Tamil word 'kari,' linked to the tree's leaves' perceived blackness. This association is evident when comparing it to the Sanskrit name for the tree, 'krshnaneembapatram' or 'black neem leaf' (etymonline.com, 2023).

Already a plant of the family Asteraceae named as 'curry plant' (*Helichrysum italicum*) exists but neither that is edible nor in any way related to the present 'curry leaf plant' (*Bergera koenigii*) though aroma of both the plants are almost similar (MBG, 2023).

7.3 Origin and distribution

The origins of the curry leaf can be traced back to the Tarai tract in Uttar Pradesh (India) from where it has spread extensively across various regions including Myanmar, Ceylon, China, Pakistan, Australia and the Pacific Islands (Joseph and Peter, 1985; Vasanthkumar *et al.*, 2023). This versatile plant is widely distributed across India, flourishing in most regions except the higher altitudes of the Himalayas. It thrives in the outer Himalayas, spanning from the Ravi eastwards and ascending to approximately 1500 meters. Its presence extends to

Asom, Chittagong, Upper and Lower Myanmar, and it is a common sight in both evergreen and deciduous forests of peninsular India, often growing as underwood plant. In Himachal Pradesh, it can be found in areas situated between 800 and 1,450m a.s.l. Southern India witnesses its ubiquitous presence in the homestead gardens of almost every household (Parmar and Kaushal, 1982).

7.3.1 Germplasm rich regions of curry leaf in India

Drawing from numerous ethno-botanical reports and floral distribution studies, the regions rich in curry leaf germplasm have been delineated into six distinct zones within India for prospective exploration and genetic enhancement (Raghu, 2020) as given under:

i) **Region-I:** Encompassing sub-tropical forests tracing the sub-Himalayan foothills from Jammu & Kashmir, Himachal Pradesh, Uttarakhand to the Terai regions of Uttar Pradesh and Bihar.

ii) **Region-II:** Embracing tropical evergreen, semi-evergreen, moist and dry deciduous forests in the NE states. This zone spans from the foothills of South Sikkim, Darjeeling, Jalpaiguri to Cooch Behar districts in West Bengal. Notably, the Curry leaf features prominently in tribal medicinal practices and is commonly cultivated in homestead gardens in this region.

iii) **Region-III:** Encompassing central India, extending from the Sundarban delta to the Satpura range. This area covers the Chotanagpur plateau, Hazaribagh plateau, Ramgarh hills, Malayagiri, Dandakaranya and Vindhyan ranges. Differing from regions I & II, this zone loosely spans the SW parts of West Bengal, Jharkhand, Northern Odisha, Chhattisgarh and Northern Maharashtra to Madhya Pradesh including Southern Uttar Pradesh.

iv) **Region-IV:** This is limited to Eastern parts of India, primarily encompassing the Eastern Ghats. This area commences from the southern parts of Odisha, traverses Andhra Pradesh and extends up to Northern Tamil Nadu.

v) **Region-V:** It is restricted to the western parts of India, primarily covering the Western Ghats. This zone commences from the southern parts of Gujarat, extends through Maharashtra, Goa, Karnataka and culminates in Kerala.

vi) **Region-VI:** It is centered on the Andaman & Nicobar Islands, predominantly covering the Andaman semi-evergreen forests, Andaman moist deciduous forests and Andaman secondary moist deciduous forests.

7.4 Production Area and Export

India stands as the leading producer and consumer of curry leaves, predominantly cultivated in the S&SW states including Sikkim, Asom, Bengal and Madhya Pradesh. Recently it has gained importance as a commercial crop. Substantial commercial cultivation occurs in Tamil Nadu, Andhra Pradesh and Karnataka (Dharwad) as highlighted by Vasanthkumar *et al.* (2023). Districts like Coimbatore, Periyar, Madurai, Salem and Trichy in Tamil Nadu emerge as the significant hub for curry leaf production within the country. Moreover, other countries such as Myanmar, Malaysia and South Africa also contribute to the global production of curry leaves.

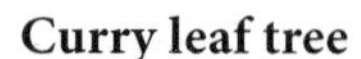

Curry leaf tree

By Rantingsteve at English Wikipedia - Transferred from en.wikipedia to Commons., Public Domain, https://commons.wikimedia.org/w/index.php?curid=3518477

According to Volza's Global Export data, curry leaves witnessed a total export volume of 66.4 thousand units from various countries across the world. This substantial volume was contributed by 2,421 exporters reaching out to 4,995 buyers. The primary destinations for these exports are Qatar, Oman and the United Arab Emirates. India has emerged as the foremost contributor to this export with a significant shipment count of 62,579 units, followed by Sri Lanka with 1,102 units and Cameroon securing the third spot with 368 shipments. In terms of product categorization, the top three categories for curry leaves in the world are identified by their HSN codes such as i) HSN Code 07099990, ii) HSN Code 09109100, and iii) HSN Code 09109990. Apart from the major exporters mentioned earlier, other countries also participate in export of curry leaves and these are Kenya (349), Thailand (224), Bangladesh (216), Vietnam (198), Fiji (191), Ghana (155), South Africa (153), United Arab Emirates (137) and several more. Global distribution and involvement in curry leaf export are widespread, encompassing numerous countries such as United States, China, Nigeria and many others ~ each contributing to varying extents which indicates a diverse market for this product. The data demonstrates the widespread interest and participation of countries across the globe for export of curry leaves, showcasing a broad market reach and diverse sourcing origins. The data mentioned above underscores the global significance and diverse distribution of curry leaves in international trade (Volza, 2023).

7.5 Botany

Curry leaf plant, emitting a delightful aroma, embodies a small, semi-deciduous nature. This tropical to sub-tropical shrub or small tree with its pubescent and unarmed characteristics, possesses the remarkable ability to reach heights of up to 6 metres. Its slender yet sturdy woody stem matures into a densely clustered shaded crown which adds to its visual appeal. The branches are covered with dark-grey bark.

Renowned for its pleasantly pungent and aromatic leaves, curry leaf is an indispensable ingredient in Indian and Asian cuisines. The leaves, arranged alternately, exhibit bipinnately compound structures, typically 15-20 centimetres in length. Initially glabrous, these leaves may occasionally display a pubescent nature in their early stages. Leaves are adorned with glandular dots. Bursting with an intense aroma due to numerous volatile compounds, each leaflet is thin, shiny, dark green and varied in shape from ovate to ovate-lanceolate. A single leaf can contain anywhere from 9 to 21 or even more leaflets, each slightly pungent, bitter and faintly acidulous in taste.

Curry leaf plant predominantly undergoes self-pollination. Its delicately fragrant white bisexual flowers cluster in terminal pedunculate inflorescences forming compact, corymbiform and cymose panicles, each flower measuring around 8mm across. These flowers feature five deeply cleft and pubescent calyces, five free and spreading petals and 10 short and long stamens arranged freely and alternately each with small anthers. The elongated thick style culminates in a capitate or grooved stigma. Flowering primarily occurs from mid-March to mid-May reaching its zenith during final week of April.

Curry leaf flowers appearing during spring

By Muhammad Mahdi Karim - Own work, CC BY-SA 2.5, https://commons.wikimedia.org/w/index.php?curid=3078616

The plant produces petite ovoid to oblong berries in a captivating bluish-black hue. These berries typically range from 1 to 2cm in diameter and boast a delicate pericarp enveloping a mucilaginous pulp that safeguards 1-2 seeds within. The enchanting fruiting spectacle commences in the heart of July, extending its vibrant display until the close of August. As the ripening process unfolds, the berries undergo a breathtaking transformation, transitioning from a vibrant red hue to a striking bluish-black. The seeds

Fruits – ripe and unripe

By Muhammad Mahdi Karim - Own work, CC BY-SA 2.5, https://commons.wikimedia.org/w/index.php?curid=3078616

nestled within are non-endospermic, adorned with a membranous and smooth testa encapsulating small embryonic life. While the sweet flavoured fruits themselves are edible and offer a delectable experience, it is crucial to note that the seeds are not fit for consumption.

Propagation of this splendid plant occurs through two primary means ~ either through dispersal or by cultivation of rooted suckers ~ both methods contributing to the proliferation of this magnificent botanical gem (Raghu, 2020; MBG, 2023; Vasanthkumar *et al.*, 2023).

7.6 Growth, Development and Life-Cycle

The growth and development of curry leaf trees follow a typical pattern influenced by various factors such as climate, soil, watering and care as mentioned here under:

- **Germination and seedling stage:** Curry leaf trees can be grown from seeds or rooted suckers. Seeds usually germinate in a fortnight under warm and moist conditions. Seedlings develop a few sets of true leaves within the first few months. Once sprouted, the seedling stage lasts approximately 6 to 12 months, depending on care and environmental factors.
- **Early growth:** During first or second year the tree focuses on establishing its strong root system and developing a sturdy trunk. Adequate watering and proper care during this stage are crucial for healthy growth.
- **Tree growth:** From seedling stage this tree progresses through a young tree phase during which it continues growing and developing branches and foliage. This phase can last anywhere from 2 to 5 years, again contingent upon factors such as soil quality, climatic factors and further maintenance.
- **Leaf and canopy development:** As the tree matures, it starts producing compound leaves. These leaves consist of numerous small leaflets arranged along a central stem. The canopy expands and the tree forms a bushy growth habit.
- **Flowering and fruiting:** Curry leaf trees typically start flowering in a few years, about three years. Fruits start developing soon after appearance of clustered pleasantly fragrant small white flowers. These fruits become green to red to shining black on maturity.
- **Maturity and maintenance:** Once it reaches maturity it starts producing leaves and may begin flowering. It can take 3-4 years for

maturing when these bear an abundant amount of leaves suitable for regular harvest. On maturity, these require regular pruning to maintain shape and promote healthy growth. Proper watering, occasional fertilization and protection from extreme weather conditions contribute to their continued development and productivity.

- **Environmental factors:** Curry leaf trees thrive in warm tropical to subtropical climates. These prefer well-draining soil and are relatively drought-tolerant once established. However, consistent moisture is essential for optimal growth.
- **Life span:** These trees when properly cared can have a relatively long life span, say 20-30 years.

7.7 Diversity and Classification of Curry Leaves in India

Curry leaves originating from India exhibit an astonishing spectrum of diversity evident across multiple levels ~ morphological, biochemical and molecular (Raina *et al.* 2002; Rao *et al.*, 2011; Raghu, 2020; Vasanthkumar *et al.*, 2023). These variations manifest in the form of distinct chemo-types and morpho-types, each unveiling a unique facet of this botanical wonder. In Indian landscape, the curry leaves boast four genetically diverse chemo-types: β-pinene, α-pinene, β-caryophyllene and β-phellandrene chemotypes as per essential oils extracted in plants obtained from different localities of Western Ghats (India).

Not limited only to their chemical makeup, the leaves also display three discernible morphotypes ['brown'/'gamthi' (GM)), 'regular' (RE) and 'dwarf' (DF)] characterized by variations in leaf colour, size, habitat and flavour. Highly aromatic and pungent brown type entices with its intense fragrance, slow growth and small *cum* thick serrated-edged dark brown leaves. On the other hand, the ubiquitous regular variant stands out for its rapid growth resembling a tree adorned with exstipulate and bipinnately compound leaves featuring reticulate venation and a dark hue thriving across the country. Meanwhile, the container-friendly yet less aromatic dwarf variety assumes a shrub-like appearance, moderately growing with spreading branches. Its leaves, resembling those of the regular type albeit in a lighter shade of green, possess their unique aromatic essence. Each morphotype distinguishes itself not only through appearance but also in the intensity of flavour they offer.

Further delving into their properties, the methanolic extracts of these three morpho-types unveil an order in radical scavenging capacity: Gamthi>Dwarf>Regular, evidenced by their respective IC50 values of 171, 365 and 471 (g/ml). Moreover, Gamthi exhibits a higher phenolic content of 532.8mg/ml compared to Dwarf (168.2mg/ml) and Regular (111.6mg/ml) types. The GM,

DF and RE morpho-types display varying levels of flavonoids with 6.01, 4.82 and 3.58 mg/ml, respectively.

Variability doesn't halt here as curry leaves showcase substantial diversity concerning the chemical composition of their essential oils across different geographical locations and seasons, contributing to their multifaceted nature and potential applications.

7.8 Genetic Improvement

The advancement of curry leaf genetics, focusing on germplasm collection, characterization and conservation is in its early stages within India. Most commercial-scale curry leaf farmers currently rely on locally available genotypes whose full genetic potential for yield, resistance to pests and diseases and quality attributes remains largely unknown.

However, there have been notable developments with introduction of two distinct and improved varieties, *viz.* DWD-1 and DWD-2 (Suwasini) which were released for commercial cultivation by UAS, Dharwad. Alongside these, a local landrace known as 'Senkaampu' has gained popularity in various regions of Tamil Nadu due to its exceptional aroma and high oil content.

Recent research initiatives at the ICAR- IIHR, Bengaluru have embarked on genetic enhancement efforts for curry leaves. These efforts aim to boost yield and quality while bolstering resistance against prevalent pests and diseases. Institute amassed successfully over 100 curry leaf germplasms from diverse regions across the country, establishing a comprehensive field genebank at ICAR-IIHR at Bengaluru.

Furthermore, rigorous endeavours are underway to characterize these collected germplasms, with plans to register them with the national genebank. This registration is envisioned to facilitate their widespread utilization in various research programmes across India (Raghu, 2020).

7.9 Cultivars

(TANAU, 2023; Vasanthkumar *et al.*, 2023; Eagri, 2024)

i) **Sen Kaampa/Senkaampu:** It is a local cultivar found in various parts of T.N., particularly in the Karamadai tract of Coimbatore district. It is known for its purplish-red petioles and leaves with excellent aroma and flavour.

ii) **DWD-1 (Suwasilli)**: Originating from Dharwad's University of Agricultural Sciences (UAS), DWD-1 is a clone of root suckers from a single plant. Cultivated mainly in Karnataka, it features glossy, dark green leaves rich in chlorophyll content (0.1629 mg/g of fresh leaf) and high fragrance. However, it is susceptible to low winter temperatures, resulting in poor bud burst. Its leaves, with a 5.22 per cent oil content, retain quality when desiccated at 50°C and ground into a powder.

iii) **DWD-2:** It was developed at UAS (Karnataka) mainly for cultivation in Karnataka from an unidentified seedling progeny. While its leaves are less fragrant and exhibit a modest pale green hue, it surpasses DWD-1 in bud burst count, internodal length and shoot growth. It grows eight times faster than DWD-1 so giving higher yield even. It displays less sensitivity to low temperatures. It has an oil content of 4.09 per cent.

7.10 Nutritional Value

7.10.1 Leaves

The leaves offer a significant nutritional profile, serving as a rich source of various essential elements, notably carotenoids and beta-carotene (Drisya *et al.*, 2015), alongside calcium and iron (Drisya *et al.*, 2015; Vasanthkumar *et al.*, 2023). The leaves provide a fair amount of vitamin A and contain notably a high concentration of oxalates ~ comprising 1.35% of total oxalates and 1.15% soluble oxalates, which significantly influence their nutritional availability. Furthermore, upon hydro-distillation, these leaves yield 0.5% of essential oil based on their fresh weight. This essential oil is characterized by its dark yellow hue, spicy fragrance and pungent taste (Vasanthkumar *et al.*, 2023; Raghu, 2020).

Comparative nutrient content of fresh and dehydrated curry leaves			
Sr.No.	**Nutrients**	**Value of fresh curry leaves (100g)**	**Value of dehydrated curry leaves (100g)**
1.	Proteins	6.1 g	12 g
2.	Fat	1.0 g	5.4 g
3.	Carbohydrate	18.7 g	64.31 g
4.	Calcium	830 mg	2040 mg
5.	Iron	0.93 mg	12 mg
6.	β carotene	7560 μg	5292 μg
7	Vitamin C	4 mg	4 mg

Source: Singh and Mohan (2011); Singh (2023)

7.10.2 Fruits

The fruit pulp has a moisture content of 64.9%. The juice extracted from the fruit has a TSS content of 16.8%. The pulp contains 9.76% total sugars, 9.58%

reducing sugars, 0.17% non-reducing sugars and almost a negligible amount of tannins and acids, besides containing 13.35% of vitamin C that is better than that of many fruits (Parmar and Kaushal, 1982).

The ash representing mineral content in the edible part of the fruit is 2.162%. Moreover, in every 100g of this portion there is a noteworthy presence of protein alongside trace amounts of minerals: 1.97% phosphorus, 0.082% potassium, 0.811% calcium, 0.166% magnesium and 0.007% iron (Parmar and Kaushal, 1982; Bonde *et al.*, 2011).

7.11 Phytochemicals

The curry leaf tree boasts a wealth of bioactive phytochemicals encompassing alkaloids, essential oils, phenolics, minerals, proteins (Singh *et al.*, 2011), terpenoids, tocopherol, β-carotene and lutein (Patterson and Verghese, 2015). The compounds, namely sabinene, pinene, cadinol, caryophyllene and cadinene contribute significantly to its distinctive aroma and flavour (Raghu, 2020). Rajvanshi and Mittal (2019) conducted a comprehensive phytochemical analysis of curry leaves, exploring their aqueous, methanol and ethanol extracts. Among the 12 tested phytochemical constituents, alkaloids, glycosides, steroids, tannins and terpenoids were consistently present in all the extracted samples. Cardiac glycosides were exclusively detected in the methanol and ethanol extracts. Phenolic compounds exhibited positive results in the aqueous and ethanol extracts, while quinone was solely found in the aqueous medium. Flavonoids, saponins and amino acids showed negative results across all the three solvents, *i.e.*, aqueous, methanol and ethanol (Rajvanshi and Mittal, 2019).

Phytochemical analysis on the leaf extracts			
Phytochemical constituents	**Aqueous**	**Methanol**	**Ethanol**
Alkaloids	moderately present	present	present
Flavonoids	-	-	-
Glycosides	moderately present	present	moderately present
Steroids	present	present	present
Cardiac glycosides	-	appreciable amount	appreciable amount
Saponins	-	-	-
Phenols	moderately present	-	moderately present
Tannins	present	present	present
Terpenoid	present	moderately present	present
Quinone	moderately present	-	-
Amino acids & protein	-	-	-

Source: Rajvanshi and Mittal (2019)

Rajendran *et al.* (2014) through a study revealed that curry leaf oils primarily consist of oxygenated monoterpenes. Through gas chromatography (GC) and gas chromatography-mass spectrometry (GC-MS) analysis, they identified 33 constituents. Predominant compounds in the oil were found to be linalool (32.83%), elemol (7.44%), geranyl acetate (6.18%), myrcene (6.12%), allo-ocimene (5.02%), α-terpinene (4.9%), (e)-β-ocimene (3.68%) and neryl acetate (3.45%).

Despite linalool (32.83%) and the monoterpenic ester linalyl acetate (16%) dominating the oil, a distinct composition was observed in the same plant species from Bangladesh. There, 3-carene (54.22%) was the predominant compound while linalool (0.19%) appeared as a minor component. The identified compounds were categorized into four groups: oxygenated monoterpenes (72.15%), monoterpene hydrocarbons (11.81%), oxygenated sesquiterpenes (10.48%) and sesquiterpene hydrocarbons (03.12%) (Rajendran *et al.*, 2014).

Chemical composition (%) of the essential oil of curry leaves from south region of T.N. (India)

Compound	Oil (%)	KI
A- Eudesmol	0.2	1657
β-Pinene	0.3	974
Myrcene	6.12	991
p-Cymene	0.23	1023
Limonene	0.61	1027
1,8-Cineole	1.56	1030
(Z)-β-Ocimene	0.36	1036
(E)-β-Ocimene	3.68	1047
γ-Terpinene	0.21	1057
α-Terpinolene	0.3	1087
Linalool	32.83	1104
1-Octen-3 yl acetate	0.24	1108
3-Octanyl acetate	0.26	1112
Allo-Ocimene	5.02	1129
α-Terpineol	4.9	1191
Nerol	1.32	1228
Carvone	0.16	1244
Linalyl acetate	16	1258
Lavandulyl acetate	0.23	1293
Myrtenyl acetate	0.1	1325
Neryl acetate	3.45	1359

Compound	Oil (%)	KI
Geranyl acetate	6.18	1383
Β-Elemene	0.26	1393
Z-Jasmone	0.3	1402
α-Gurjunene	0.1	1411
β-Caryophyllene	1.3	1422
Germacrene D	0.2	1479
α-Amorphene	0.7	1483
δ-Cadinene	0.26	1522
Elemol	7.44	1552
Viridiflorol	0.92	1595
γ-Eudesmol	0.77	1635
Β- Eudesmol	1.15	1654
Monoterpene Hydrocarbons 11.81 %		
Oxygenated Monoterpenes 72.15 %		
Sesquiterpene hydrocarbons 03.12 %		
Oxygenated Sesquiterpenes 10.48 %		
*Identification of volatile components is based on mass spectra value in reference to NIST 08 and standard libraries. Composition of grouped volatile compounds (%)		

Source: (Rajendran *et al.*, 2014)

Essential oils notably rich in α-pinene (39.93%), sabinene (13.3%) and trans-caryophyllene demonstrate antibacterial effects against *Bacillus subtilis*, *Proteus vulgaris* and *Corynebacterium pyogenes* (Kumar *et al.*, 1999). These oils primarily consist of monoterpenoids and their oxygenated derivatives, constituting the key chemical components of the curry leaf essential oil (Raghu, 2020). Fresh leaves collected from various regions of Western Ghats yielded essential oils with significant levels of monoterpene hydrocarbons such as sabinene (6.90-40.59%), β-phellandrene (1.39-45.89%) and α-pinene (1.93-63.66%). These have been found containing the sesquiterpene hydrocarbon β-caryophyllene 6.68-18.46% (Raghu, 2020).

GS-MS analysis of its essential oil revealed significant antibiofilm agents, including spathulenol (5.85%), cinnamaldehyde (0.37%) and linalool (0.04%) (Ganesh and Vittal, 2015). Moreover, diverse carbazole alkaloids such as mahanimbine, girinimbine, koenimbine, koenine, koenioline, mahanine, bismurrayafoline, murrayanol, isomahanine, bismahanine, euchrestine, bispyrafoline, murrafoline, murrayacine, murrayazolidine, xynthyletin and kurryam among others have been identified in various parts of the curry leaf tree (Tachibana *et al.*, 2003; Bhandari, 2012; Bhusal and Thakur, 2021; Mani *et al.*, 2021; Vasanthkumar *et al.*, 2023; PubChem, 2024a, b).

The leaves of this plant yield various compounds upon isolation including crystalline glycosides 'ksenigin' in the leaves and murayin in flowers. Carbazole alkaloids such as koenigin, girinimbin, iso-mahanimbin, koenine, koenidine, koenimbine, murrayastine, murrayaline, pyrayafoline, mahanine, mahanimbine and murrayanol have been identified. Triterpenoid alkaloids like cyclomahanimbine and tetrahydromahanimbine have also been recorded. The bark houses carbazole alkaloids including murrayacine, murrayazolidine, murrayazoline, mahanimbine, girinimbine, koenioline and xynthyletin as well (Bhandari, 2012; Bhusal and Thakur, 2021; Vasanthkumar *et al.*, 2023).

The N-hexane seed extract is rich in three bioactive carbazole alkaloids ~ kurryam, koenimbine and koenine (Mandal *et al.*, 2010). Meanwhile, acetone extraction from fresh leaves contained mahanine, mahanimbine and murrayanol ~ all bioactive carbazole alkaloids (Bhusal and Thakur, 2021). On the other hand the aqueous extracts revealed presence of tannins and carbazole alkaloids (Shivashankara *et al.*, 2012).

7.12 Importance and Uses

Curry leaves esteemed for their distinct aroma, owe their popularity to the presence of volatile oils, making them a prized leaf spice used in minute quantities. Renowned for enhancing digestion, these vibrant leaves hold an indispensable role within Indian cuisine and traditional medicine.

The nomenclature of these leaves is rooted in their frequent inclusion in curry flavourings, yet their significance transcends this as these contribute profoundly to augmenting the aroma and taste of diverse dishes. With a nuanced profile encompassing slight pungency and a subtly acidic undertone, these leaves maintain their distinct flavour even after drying, retaining their essence.

Beyond culinary applications, curry leaves find extensive use in numerous Indian Ayurvedic and Unani prescriptions, showcasing multifaceted industrial values and medicinal properties (Vasanthkumar *et al.*, 2023).

7.12.1 Culinary uses

Curry leaves, renowned for their intense aroma when crushed, are best used fresh in cooking to maximize their flavour; however, dried leaves while less potent can still be utilized. Frying these leaves in oil or butter enhances aroma and taste of the meal. Predominantly used in Indian and Sri Lankan cuisine, these subtly elevate vegetable, meat, seafood, chutneys, coconut sauces, relishes, marinades and even omelets. These leaves aid digestion being rich in vitamins A & C. Commonly found in southern and west coast Indian dishes, they are fried with mustard seeds and onions in the initial cooking stages. These are integral to dishes like *thoran, vada, rasam* and *kadhi*. Often dry-roasted and ground, they are a key component in spice blends such as South Indian *sambar masala*, a crucial element in the famous vegetable stew ~ *sambar*. Moreover, they lend their flavour to *masala dosa* ~ a South Indian favourite made from fermented lentil and rice batter. Highly regarded in South and SE Asian cuisines, these leaves significantly enhance the overall taste of a dish (Parmar and Kaushal, 1982; Verghese, 1989; MBG, 2023; Vasanthkumar *et al.*, 2023).

7.12.2 Use as traditional medicine

Curry leaves are historically utilized in both culinary and medicinal practices, offer a rich array of benefits. Integral to Indian Ayurvedic and Unani remedies, these leaves have long been recognized for their role in enhancing appetite, aiding digestion and serving as a potent stimulant.

Their medicinal properties extend widely: from managing diabetes mellitus and exhibiting antiemetic, anthelmintic, and analgesic actions to alleviating kidney-related discomfort and addressing various skin conditions such as bruises and eruptions.

Ayurvedic medicine attributes curry leaves with an impressive spectrum of benefits, including but not limited to reducing blood pressure, displaying antibacterial, antiviral, antifungal and antiprotozoal activities, aiding in digestion, promoting wound healing, combating cancer, controlling diabetes, reducing inflammation, acting as an antioxidant, lowering cholesterol and exhibiting anti-ulcer and anti-tumour properties.

Moreover, beyond the leaves' medicinal use, curry fruits offer nutritional value while the branches are employed to strengthen gums and are traditionally utilized for dental hygiene.

These diverse attributes have been substantiated by various studies (Gupta *et al.*, 2011; Reddy *et al.*, 2018; Bhusal and Thakur, 2021; Supe *et al.*, 2021), positioning curry leaves as a multifaceted resource in both traditional medicine and modern research.

7.12.3 Pharmacological properties

Curry tree boasts a rich medicinal profile that has been extensively researched, revealing a spectrum of therapeutic properties. Its bark, roots and leaves are recognized for their carminative, stomachic and tonic attributes. The essential oil derived from it showcases remarkable antibacterial and antifungal qualities. Furthermore, the seeds exhibit insecticidal and repellent properties.

Upon distillation, fresh curry leaves yield a vibrant yellow volatile oil with a robust, spicy aroma and a pungent clove-like flavour. The decoction of its leaves and stalks serves as a remedy for ailments such as rheumatism, leprosy, epilepsy, internal poison, fever and worms. Moreover, it acts as a laxative, enhances brain function and addresses eye diseases (Vasanthkumar *et al.*, 2023).

The fruit of the Curry tree contains the alkaloid koenigin which displays antifungal properties while the root juice is known to alleviate kidney pain (Vasanthkumar *et al.*, 2023). A group of carbazole alkaloids isolated from the curry plant exhibits a range of beneficial activities including antioxidant, anti-inflammatory, anti-tumour, antidiabetic, anticancer, antibacterial and anti-viral effects (Shyamala, 2005; Knolker and Reddy, 2008; Xie *et al.*, 2020).

Studies have highlighted the extensive therapeutic potential of the curry tree across various medical realms:

- **Antioxidant:** Studies comparing curry tree to other green leafy vegetables found its total antioxidant activity to be exceptionally high (2691 micromol of ascorbic acid/g sample), surpassing methanol extracts of *Amaranthus* sp., *Centella asiatica* and *Trigonella foenum graecum* (Gupta and Prakash, 2009). The alcohol-water extract of curry tree displayed superior antioxidant and free radical scavenging activity, potentially benefiting individuals exposed to cadmium (Ningappa *et al.*, 2008; Mitra *et al.*, 2012).
- **Anti-nociceptive and anti-inflammatory effects:** Petroleum-ether extracts and alkaloids from curry tree showed promising anti-nociceptive effects, hinting at its analgesic potential (Patil *et al.*, 2012). Alcohol extracts from the stem bark and crude root exhibited anti-inflammatory activities (Darvekar *et al.*, 2011). Moreover, methanolic extracts from dried leaves demonstrated analgesic and anti-inflammatory effects comparable to the drug diclofenac (Gupta *et al.*, 2010).
- **Lipid-lowering and metabolic benefits:** Dichloromethane and ethylacetate extracts from curry tree leaves significantly reduced body weight gain, total cholesterol and triglyceride levels in obese rats. Moreover, the aqueous extract showed promise in reducing diabetes severity and associated nephropathic complications (Kesari *et al.*, 2007; Birari *et al.*, 2010).
- **Diabetes management and islet protection:** Specific carbazole alkaloids found in curry tree, like mahanimbine, have shown potential in decreasing blood sugar levels and reducing total cholesterol. Curry tree extracts demonstrated islet protective roles and insulin

production in studies involving Swiss mice. They also exhibited inhibitory effects on post-prandial hyperglycemia (Dineshkumar *et al.*, 2010; Bhandari, 2012; Dusane and Joshi, 2012; Saied *et al.*, 2015).

- **Immunomodulatory and nephroprotective effects:** Leaf extracts not only possessed antidiabetic properties but also showcased regulatory effects on oxidative stress metabolism. Aqueous leaf extracts demonstrated significant reductions in serum urea and creatinine levels, indicating nephroprotective potential (Paul *et al.*, 2011; Yankuzo *et al.*, 2011).
- **Neuroprotection and anti-aging properties:** Mahanimbine could be a potential substitute for treating aging-related conditions and could improve spatial learning and memory. Curry tree extracts displayed neuroprotective benefits and potential implications in managing conditions like Alzheimer's disease. These improved cognitive functions and reversed memory impairment in mice. Moreover, these exhibited anti-amnesic potential and decreased oxidative stress in the aged mouse brain (Vasudevan and Parle, 2009; Tembhurne and Sakarkar, 2010; Bhandari, 2012; Mani *et al.*, 2012).
- **Gastrointestinal health:** Research indicates that kurryam and koenimbine active carbazole alkaloids present in curry leaves show promise in reducing diarrhoea and gastrointestinal motility in rat models (Mandal *et al.*, 2010).
- **Anticancer properties:** Girinimbine, a carbazole found in curry tree bark demonstrated inhibition of tumour growth and induced apoptosis in human hepatocellular carcinoma cells *in vitro* (Syam *et al.*, 2011). Curry leaf extracts exhibited potential in reducing cancer cells in mice (Ghosh *et al.*, 2012).
- **Hepatoprotective effects**: Aqueous extracts from curry plant containing tannins and carbazole alkaloids show hepatoprotective activity against ethanol-induced liver damage comparable to standard drugs (Sathaye *et al.*, 2011; Syam *et al.*, 2011; Shivashankara, 2012).
- **Antibacterial action:** Essential oils from curry tree exhibit robust inhibition against various bacteria including *Staphylococcus aureus* and *Klebsiella pneumoniae* (Rajendran *et al.*, 2014). Specific components like α-pinene, sabinene and trans-caryophyllene show antibacterial effects against other strains (Kumar *et al.*, 1999). Bioactive carbazole alkaloids like mahanine, mahanimbine and marrayanol display significant antimicrobial activities (Bhusal and Thakur, 2021).

7.12.4 Value-added products

i) **Dried curry leaves:** The process of drying curry leaves helps retain their aromatic essence, making them a convenient and long-lasting alternative to fresh leaves in culinary endeavours. They add a distinct flavour to dishes, retaining much of their original taste.

ii) **Curry leaf powder:** Once dried, the leaves are finely ground into a powder form. This powdered form of curry leaves serves as an excellent flavour enhancer for a wide array of dishes including curries, soups, rice and various other recipes, offering a concentrated burst of flavour.

iii) **Curry leaf extracts:** Extracts derived from curry leaves are valued for their potential health benefits, boasting antioxidant and anti-inflammatory properties. These extracts are commonly used in production of dietary supplements, promoting overall well-being and health maintenance.

iv) **Curry leaf oil:** Through steam distillation, essential oil is extracted from curry leaves, yielding a highly aromatic and potent oil. This oil finds diverse applications in aromatherapy for its soothing and relaxing properties. It is a sought-after ingredient in cosmetics and skincare products, adding fragrance and potential skin benefits.

v) **Infused vinegar and oils:** Utilizing curry leaves, vinegars and oils are infused with their aromatic essence. These infused products serve as flavourful bases for dressings, marinades and cooking, providing an added depth of aroma and taste to various culinary preparations.

vi) **Spice blends and seasonings:** Combining curry leaves with an assortment of other spices results in unique seasoning blends. These blends elevate the overall flavour profile of a multitude of dishes, offering a distinctive and nuanced taste experience.

Each value-added product derived from Curry leaves encapsulates their aromatic essence and distinct flavour, catering to various industries and culinary preferences, ranging from traditional cooking applications to health and wellness products.

7.12.5 Curry leaf oils

The Curry leaf plant's essential oils serve various industrial purposes including soap, perfume, cosmetics and food processing (Joseph and Peter, 1985). Through steam distillation under pressure, fresh leaves yield 1.6-3.7ml of volatile oils ~ referred to as curry leaf oil per kilogram of biomass (Syamasunder *et al.*, 2012).

This leaf oil boasts a specific gravity of 0.9748 at 25°C, a saponification value of 5.2 and an acid value of 3.8. Similarly, mature fruits produce a yellow-coloured volatile oil with a specific gravity of 0.872 at 130°C and a boiling point of 173.74°C. It exhibits a neroli-like odour and a peppery taste, imparting a pleasant cool sensation on the tongue. Extracting oil from the seeds results in a yellow-coloured oil known as limbolee oil.

The leaf oil acts as a fixative in heavy soap perfumes. Essential oils from other parts find use in the cosmetics industry and aromatherapy. Berries' extracts prevent oxidative damage in meat and meat products. Incorporating curry leaves in functional poultry meat finger sticks enhances lipid stability and antimicrobial quality, proving beneficial as an alternative to synthetic food preservatives in functional food industries.

Furthermore, India exports curry leaf oil. The oil composition changes with leaf maturity, showing a gradual decrease in volatile oils and oleoresin (acetone-extracted). Curry leaf oil has a specific gravity of 0.97, a refractive index of 1.50, and an optical rotation of +48 at 25°C. Its saponification value stands at 5.2, increasing to 54.6 after acetylation (Raghu, 2020).

7.13 Toxicity

Curry leaf tree is generally not known to cause any severe toxicity issues in humans or animals. However, there are a few considerations regarding its toxicity:

- **Sap irritation:** Handling its leaves or any other organ might cause skin irritation or give an allergic reaction in some individuals due to compounds present in the sap. It is advisable to handle the plant with care and wash hands thoroughly after contact.
- **Ingestion concerns:** While curry leaves are commonly used as a culinary herb and are generally safe when consumed in moderate amounts in cooking, however, excess consumption might lead to digestive discomfort or allergic reactions in some sensitive individuals.
- **Pesticides or chemicals:** In case of some pesticides, herbicides or other chemicals are applied to the curry leaf tree, consumption of their leaves soon after application could potentially pose health risks. Always follow recommended waiting periods after chemical applications before harvesting or consuming any part of the plant.
- **Pets and animals:** Ingestion of large quantities of curry leaves by pets or livestock might lead to gastrointestinal upsetting or allergic reactions so excessive feeding should be avoided.

7.14 Cultivation

(TANAU, 2023; Vasanthkumar *et al.*, 2023)

7.14.1 Propagation

i) **Seed propagation challenges:** Traditionally, propagation *via* seeds has been common but often proves unreliable due to low seed viability and the vulnerability of young seedlings to natural conditions. Commercially, seeds are harvested from well-ripened fruits gathered from high yielding mother trees, typically available between July and August. Each fruit contains approximately 2-3 seeds. These seeds immediately after harvesting are extracted and placed in nursery beds or polybags filled with a suitable mixture of soil and farmyard manure to encourage successful germination. When these become some 10cm high these can be lifted and planted in pots or in the field after preparing the mixture or the field as the case may be.

ii) **Alternative propagation methods:**

- **Root suckers:** One-year-old root suckers serve as another viable planting material, offering an alternative to seed propagation so these are severed gently from the plant during monsoon and planted in the prepared field or in pots filled with soil and FYM.
- **Micropropagation:** Utilizing axillary shoot multiplication and shoot bud regeneration *via* standardized techniques, micropropagation presents a reliable means of vegetative propagation. This involves using nodal segments from mature trees on a specific medium supplemented with growth regulators like 6-benzylaminopurine, Kinetin, NAA and IBA.

7.14.2 Climate

Curry leaf trees exhibit robust growth in warm climates, displaying adaptability to a diverse range of soil conditions. Optimal growth and development occur in regions with consistent warmth and ample sunlight. Its ability to thrive within various temperature ranges and its adaptability to different elevations make it versatile for wide cultivation.

- **Temperature tolerance:** Curry leaf trees flourish within a temperature range typically spanning from 26-37°C. However, when the temperature drops below 13°C, the growth of shoots can be adversely affected, potentially stunting the development of the plant.
- **Altitude suitability:** Curry leaf trees demonstrate resilience and can successfully be cultivated at elevations of up to 1500 meters above sea level (MSL). This adaptability to varying elevations widens the geographical regions suitable for their growth.

7.14.2 Soil

The ideal soil composition significantly influences the growth, productivity, and overall health of curry leaf trees. Red sandy loam soil characterized by good drainage is considered highly conducive to its optimal growth. This specific soil type facilitates proper aeration and moisture distribution, fostering robust root development and subsequent healthy foliage. However, other soils are also good provided these are sandy having better composition of organic matter, well-drained and the location is quite sunny but with ample water for irrigation.

7.14.3 Field preparation and planting

Achieving an optimal tilth is crucial for successful curry leaf cultivation. Field is ploughed meticulously 3-4 times to ensure thorough soil aeration, breaking down of larger soil clods and for making a fine tilth so that it becomes crumbly in texture, *vis-a-vis* ideal for root penetration and nutrient absorption. Soil should be enriched through incorporating 20 tonnes per hectare of well-rotten farmyard manure per hectare before last ploughing. Organic matter contributes essential nutrients, enhances soil structure and fosters a conducive environment for healthy root development. Before one month of planting, the field is dug out for making trenches of the dimension of 30×30×30cm spaced at 1.2-1.5 metres apart. However, the planting distance mostly depends on the height and canopy growth of the variety, *vis-a-vis* climate, fertility status of the soil and intended management practices. Ample space between trees ensures that the plants receive adequate sunlight, nutrients and airflow. In case the curry leaf trees are part of an agroforestry system or intercropped with other plants, the spacing might be adjusted accordingly to accommodate overall layout and management practices of the combined crops. Pits are prepared by mixing excavated soil with farmyard manure before planting. Seedlings are now planted singly in each pit by ensuring their appropriate depth. After planting, these plants are watered generously to settle the roots. Regular monitoring of soil moisture levels, especially during initial growth phase is crucial. Adequate irrigation practices should be implemented to support the newly planted seedlings. Afterwards, post-planting care such as timely weeding, watering and protection of plants from grazing animals, adverse weather conditions and disease and insect-pest management are essential.

7.14.4 Manures and fertilizers

Cultivation of curry leaf plants requires a substantial supply of organic nutrients for optimal growth and yield as furnished herewith. As already advocated, FYM should have been mixed in the soil before planting, however, at the time of pit-filling some 5kg FYM should again be mixed with planting mixture. However, after each pruning again the FYM is added in the pit. Requirement of various nutrients per tree of curry leaf tree is 150g of nitrogen, 25g of P_2O_5 and 50g of K_2O. TNAU recommended 90:10:10kg of N:P:K per hectare of land annually. The crop requires to be fed every 3-month after leaf harvesting.

7.14.5 Aftercare

7.14.5.1 Irrigation and weed management

During summer when there is no rain, this requires to be **watered** twice every week though during winter fortnightly. However, always the soil should remain consistently moist but not waterlogged. Weed-free environment is vital for healthy and optimal growth therefore any **weed** coming up if in the field should immediately be removed as these rob the main crop of their due light, nutrients and moisture supplied to the plants, *vis-a-vis* harbour insect-pests and pathogens injurious to the main crop.

7.14.5.2 Training and pruning

To encourage robust growth and ease of management curry plants benefit from proper training and pruning. First pruning is usually done after 15 months of planting leaving some 50cm from the ground to force developing lateral branches and to foster a bushier and most productive growth pattern. To keep the plants at manageable height roughly at one metre these are maintained through pruning in most of the cases though there may be clones growing to tree-size by nature to produce more leafy harvest so in such cases these are accordingly pruned akin to their requirement but in any case not more than six branches are allowed to grow to monitor proper access of sunlight and aeration between the trees as well as inside a tree. Every year, these require pinching of the terminal buds to encourage growth of the side shoots (lateral branching) so that leaf can be harvested to the maximum level. After first 2-3 leaf harvsts the plants are again pruned to initial height. Regular pruning and maintenance prolong the life of the bushes apart from more leaf yield. Normally four harvests are taken annually with an average leaf yield of 8-10t/ha annually.

7.14.6 Post-harvest management

After harvesting of leaves, immediately a meticulous sorting process starts where discoloured, damaged and dust or debris-accumulated leaves are relegated and only high quality leaves are selected and cleaned. These leaves are now either sun-dried for home consumption or dried through regulated mechanical drying to maintain consistent quality. Dried leaves maintaining their distinctive flavour are marketed as freshly dried ones, and fried or ground into a powder form. However, the fresh ones can be dried in an oven at a termperature not exceeding 50°C to produce a fine dry powder. Curry leaf oil is extracted through steam distillation at specific pressures which results into 2.6% yield of volatile oil or the oil can also be extracted through hydro-distillation of fresh leaves around 70°C.

7.14.7 Yield

Age of plantation	*Yield of leaves (kg/ha)*	*Interval of harvest*
First year	250-400	Once at the end of first year
Second and third year	1800-2200	Once in four months
Fourth year	2500	Once in 3 months
Fifth year onwards	3500 to 5000	Once in 2.5 to 3 months

7.15 Physiological Disorders

Siumilar to other plants, curry leaf trees also exhibit various nutrient deficiency symptoms that affect plant growth and overall health as elaboratred here under:

i) **Nitrogen deficiency** is expressed in terms of leaf yellowing (chlorosis) of older leaves starting from the tips and spreading inward, stunted growth and reduced leaf size.

ii) **Phosphorus deficiency** is expressed in terms of darker leaf greening with reddish or purplish tint. In such cases leaves may become brittle and show signs of necrosis, especially around the edges.

iii) **Potassium deficiency** is expressed as yellowing *cum* browning of leaf edges (necrosis). Leaves may appear scorched or burnt starting from the leaf tips and progressing inward.

iv) **Magnesium deficiency** symptoms appear as interveinal chlorosis where leaf tissues between veins turn yellow while the veins remain green. Leaves may also show curling or cupping.

v) **Iron deficiency** symptoms appear as yellowing of young leaves while the veins remain green (interveinal chlorosis). In severe cases, the leaves might turn almost white.

vi) **Zinc deficiency** symptoms appear in the form of reduced leaf size, distortion and mottling or chlorosis between leaf veins. Leaves may become narrow and cup-shaped.

vii) **Calcium deficiency** symptoms appear on new leaves which exhibit distorted growth or show necrotic spots. Tip burn or dieback might also occur.

Addressing nutrient deficiencies involves proper diagnosis and corrective measures such as soil testing, proper fertilization in the soil or through leaves, adjusatment of soil and water pH and organic mulching which helps retaining soil moisture and provides well-rotten FYM in the soil for amendment and for better nutrition. Leaf yellowing, **browning or tip *cum* edge burn** may result due to water stagnation in the field, *via-a-vis* excessive fertilizer application. **Leaf drop** may occur due to occurrence of environmental stressors such as drastic temperature fluctuations, drought and sudden change in light exposure. **Blossom fall** or premature **fruit drop** is generally caused due to inconsistent watering, extreme temperatures during flowering or at the time when fruits have set or due to inadequate pollination. **Scalding** on leaves and stems occurs due to intense sunlight exposure, especially during summer.

7.16 Insect-Pests and Diseases

Curry leaf tree is susceptible to various insect-pests and diseases that affect its growth and health. Mentioned below are some of the prominent insect-pests and diseases (Vikaspedia, 2023).

7.16.1 Insect-pests

- **Citrus butterfly (*Papilio demoleus*) and common mormon (*P. polytes polytes*)**: Apart from infesting on citrus plants, these also feed on curry leaf plants. Yellow *Papilio demoleus* females having black markings lay yellowish-white round eggs individually on tender leaves and shoots. Eggs hatch within 3-8 days releasing dark brown larvae with irregular white markings on their bodies. These feed voraciously on leaves and tender shoots preferring midribs and defoliate the plants completely. *Papilio polytes* displays sexual dimorphism ~ males being black while females exhibit variations in their wing patterns. These are controlled by releasing parasitoids in the garden. *Trichogramma* spp. lays the eggs inside the butterfly eggs. *Cotesia* and *Chelonus* wasps lay its eggs inside the larvae. *Brachymeria* wasp lays the eggs on or inside the larvae and *Telenomus* tiny wasp lays eggs inside the eggs so such eggs do not hatch. Sparing the beneficial insects, 0.1% Malathion spraying will also kill all these larvae of feeding butterflies.
- **Psyllid bugs/citrus psylla/citrus psyllid (*Diaphorina citri*)**: These are quite tiny insects whose nymphs and adults continue rasping the juice of leaves and producing honeydew ~ a sticky substance which attracts sooty mould. This lays its eggs on tender plant parts which after hatching release nymphs. Both nymphs and adults feed on the tender parts of the plants so leaves become curled *cum* twisted and turn brown. Their feeding may also cause invasion by bacteria. Affected areas should be pruned to maintain plant health.

Some prominent parasitoids used in biological control programmes against the citrus psyllid include *Tamarixia radiata, Diaphorencyrtus aligarhens, Coccidoctonus clavis* and *Psyllaephagus bliteus* which are tiny natural parasitoid wasps as natural enemies of citrus psyllids. These wasps lay eggs inside the nymphs so killing them soon. Certain species of wasp *Apanteles* sp. have shown potential in parasitizing citrus psyllids. So all these can be used to control citrus psyllids. In severe infestation, 0.1% dimethoate or *neem* oil can be sprayed but not when parasitic wasps are present.

- **Scales (*Unaspis citri*)**: These tiny insects are characterized by their small, flat and immobile bumps that are visible on the surface of leaves. Scales undergo a life cycle that typically involves stages such as egg, nymph and adult. Eggs are often laid beneath the protective covering created by adult female scale. The crawling nymphs emerging from eggs start moving to other parts of the plant to settle before maturing into adults. This cycle can vary in duration depending on environmental conditions but generally progresses relatively quickly contributing to rapid population growth. These pierce the plant tissue and extract sap by which plants are weakened. This sap-sucking behaviour results into yellowing, stunting and wilting of plants. Affected branches should be pruned and natural enemies should be introduced. *Aphytis melinus, Encarsia citrina, Coccophagus lycimnia* and certain species of *Metaphycus* are parasitic wasps that lay eggs inside the scale insects which after hatching start feeding on scales. *Neem* oil spraying is also effective. Horticultural oils and soaps are also effective in suffocating scale insects. Dimethoate at 0.1% can also be sprayed in severe cases but chemical control should be last resort.
- **Mealy bugs (*Planococcus citri*)**: Mealy bugs (*Planococcus citri, P. Pacificus* and *Icerya purchasi*) are characterized by their soft and pinkish-white bodies covered in a waxy substance. The 1.2-5mm long wingless insects where females lay eggs in several hundreds on leaf surface that hatch into yellow nymphs which feed on plant sap. In clusters the nymphs and adults may be observed on the branch-tips rasping plant sap. Their infestation may cause weakening of the plant, stunting, yellowing, shrivelling and falling of leaves and also secreting honeydew so attracting sooty mould. To protect the crop in the field, there should be proper sanitation below the trees. Initially the affected shoots should be pruned, any colony below or nearby such trees should be destroyed and raking the tree basins during April-May exposes the egg masses to quite hot climate and enemies.

Spraying of 2% *neem* or *Pongamia* oil or basin application of 0.125% Buprofezin will keep this insect at bay. Introduction of parasitoid wasps (*Leptomastix dactylopii, Anagyrus pseudococci, Aenasius bambawalei, Pseudaphycus angelicus* and *Coccidoxenoides perminutus*) on the infested plants can serve as an effective biological control as these lay their eggs on or inside the body of various stages of mealy bugs and kill them. In severe cases, insecticidal soaps or neem oils, as well as specific pesticides labeled for mealy bug control can be used. However, caution and adherence to label instructions are crucial to minimize adverse effects on beneficial insects and the environment.

- **Aphids/orange spirea aphid** (*Toxoptera aurantii*)**:** These pear-shaped diminutive insects exhibit a spectrum of hues ranging from yellow, green, brown to white. Typically small in size, these cluster on plants particularly on the underside of leaves as these reproduce asexually numerous generations during favourable weather conditions. These cluster around tender plant parts including flowers and fruits and keep on rasping plant sap, *vis-a-vis* enticing sooty mould due to secretion of honeydew so plants become very weak. Use of insecticidal soap, *neem* or other horticultural oils and through predatory insects their population may drastically be reduced. Spraying of chemical insecticides should be the last resort. There are several parasitoids infesting on eggs or nymphs of this aphid. *Aphelinus* sp., *Lysiphlebus testaceipes, Aphidus colemani* and *Diacretiella rapae* wasps lay their eggs inside the aphid bodies which after hatching start feeding on them. Spraying with 1% *neem* oil, 0.2% Dimethoate or 0.03% Imidacloprid will also control aphids but chemical insecticides should be the last option.
- **Tortoise beetle/curry-leaf tortoise beetle/curry leaf mite (*Silana farinosa*)**: Adults of this beetle are reddish-brown and the grubs are hued black with distinctive forked posterior and flattened body. Adult beetle poses a significant threat to curry leaf tree due to its nature of defoliating the tree while feeding. Presence of larvae is felt due to extensive leaf defoliation. To minimise population and menace of beetle, intercropping with other crops is a must in such cases so that its life cycle is disrupted. Certain mites such as *Amblyseius* spp., *Neoseiulus* spp., *Euseius* spp. and *Cheyletus eruditus* predate over this beetle so these should be introduced in the planting. *Eucarsia* and *Eretmocerus* wasps also predate over these insects. Chemicals used in controlling mealy bugs will control even beetles.

- **Two-spotted carmine spider mites (*Tetranychus* spp.)**: These are minute arthropods barely visible to the naked eye and typically measuring about 0.4mm in length. These mites often exhibit two distinctive dark spots on their bodies. Its female lays eggs in considerable number on various plant parts such as buds, leaves, twigs, stems and trunks. Nymphs appear after hatching which before maturity undergo multiple moltings. Initially the first-instar nymphs are 6-legged but at subsequent stages these have a full set of eight legs. Overwintering occurs in the soil and the generation time varies with temperature, potentially as brief as four days in warmer conditions. Their damage occurs as stippling, scarring and bronzing of leaves and calyces within the first 2-5 months of transplanting but during summer or fall. To minimise its attack, there should be proper aeration inside the planting, plants should be properly fed and there should be proper sanitation. Natural predators such as *Phytoseiulus persimilis, Neoseiulus californicus, Amblyseius andersonii, A. swirskii* and *Euseius ovalis* mites should be introduced in such plantings. If the population of two-spotted mite exceeds the threshold levels acaricides as per manufacturer instructions should be used. Azadirachtin (Nemactin 0.15EC) @ 2.5ml/l spraying soon after mite infestation should be done. Spraying of *Lecanicillium* (*Verticillium*) *lecanii* containing 10^9 cfu/ml at a concentration of 1ml/litre of water is quite useful in controlling the mites.
- **Citrus leaf miner (*Phyllocnistis citrella*)**: It is a lepidopterous insect whose larvae invade the crop through making serpentine-like winding mines through the leaf tissue. These mines initially appear whitish to silvery lines but turning soon to brownish. This pest has four stages ~ egg, larva, pupa and adult. Adult female moths lay eggs individually on young leaf undersides. The larvae on hatching tunnel winding mines into the leaf tissues as these feed and grow and there itself these on maturity pupate. Adults appear from mature pupa. In single growing season these complete multiple generations. Its infestation is symptomised mining of leaves leading to distortion, curling and malformation, *vis-a-vis* leaf drop and ultimate plant death in severe infestations. In getting rid of this pest, one should maintain good farm hygiene, initial pruning of infested parts, proper irrigation and fertilizing. Moreover, infested parts should be pruned away and destroyed. Natural parasitoid wasps such as *Goniozus nephantidis* and *Citrostichous phyllocnistoides* ~ both parasitize the larvae; and *Ageniaspis phyllocnistoides* the pupae. Use of pheromone traps can aid

in monitoring and disrupting the mating behaviour of adult moths. In cases where infestations reach damaging levels, targeted and judicious use of insecticides following IPM principles can be considered as a last resort, ensuring minimal impact on non-target organisms and the environment.

7.16.2 Diseases

- **Leaf spot disease (*Phyllosticta* spp.)**: The primary mode of infection occurs through soil and rain splash while secondary spread occurs through air and rain during moist conditions. This cycle facilitates the rapid infection of the plants. *Phyllosticta* leaf spot is characterized by its distinct symptoms appearing as round lesions on leaves. Initially the lesions are small and circular but these often expand into irregular yellowish-brown spots. Favourable conditions lead to development of tiny black fruiting bodies in circular formations at the centre of these lesions resulting in dead tissue that easily breaks away creating holes. The symptoms of leaf spot vary in severity ranging from a few round spots to extensive lesions that can cause premature leaf loss. Severe infestations can weaken and debilitate the affected trees. Carbendazim spraying at 0.1% satisfactorily controls its spread, however, no use of sulphur compounds should be allowed. Moreover, to permit good airflow among the planting, proper pruning is also required.
- **Powdery mildew (*Acisporium* spp.)**: It is a fungal disease commonly identified by the presence of a white powdery growth on the surfaces of leaves. This growth often appears as a fine flour-like substance and can spread rapidly across foiage. Though primarily it affects leaves but manifest on stems, flowers and even fruits. The invasion under favourable conditions, *i.e.* when temperature is moderate, humidity is high and air circulation is poor, the spores of this pathogen germinate and penetrate the plant tissues and start producing new spores. With progress of the disease the leaves start distorted, curling and turning yellow. In severe cases the leaves may wither and drop prematurely. Care at planting should be taken to ensure proper spacing and after full growth proper pruning to ensure good air circulation, *vis-a-vis* reduced humidity levels. Overhead watering should not be done in any case. Fungicidal sprays containing sulphur or potassium bicarbonate can help suppressing its growth and preventing its spread. Spraying the crop with 0.1% Indocar/Bayleton/Systhane or 0.2% Sulfex/Wettasul/Thiovit can control this problem.

7.17 Stagewise IPM

Management	Activity
Pre-sowing	
Nutrients	The field is ploughed 3-4 times to get a fine tilth. Before last ploughing well decomposed farmyard manure is applied @ 20t/ha. Pit size of 30×30×30cm is dug one to two months before planting at a spacing of 1.2×1.5m (in case of sole crop) or 1.5×3m or 1.5×2m. Application of 65:22:20kg N:P:K/ha at the time of planting.
Weeds	Plan to grow suitable intercrops like legumes, ginger, tapioca, *etc.* so that intercultural operations done in one case may benefit other crops too.
Resting stages of pests, soil-borne fungus	Solarization in summer with deep ploughing and covering the field with 0.45mm polythene sheets.
* Applying *Trichoderma* as seed and nursery treatment and *Pseudomonas fluorescens* as seed, nursery treatment and soil application (in case of commercial products are used check for label claim (biopesticides produced by farmers for own consumption does not require registration).	
Vegetative stage	
Nutrients	Apply 4 kg of FYM + 5:10:10g N:P:K/plant and mix with soil after every harvest.
Weeding and other intercultural operations	Periodical hoeing or hand weeding in pits should be done following every irrigation. In the first year the crop can be intercropped with pulses. After attaining 1m height, the terminal bud is cut off to encourage basal branching. In total 5-6 branches are maintained per bush.
Leaf spot	**Cultural control** Disease-free seed and seedlings should always be used. **Biological control** Preventive spray of tobacco decoction could be done. Citrus oil at 1360ppm inhibits maximum growth of the fungus followed by lemongrass oil at 1720ppm and peppermint at 2260ppm, respectively Spray NSKE 5%
Tortoise beetle	Larvae can be handpicked. To kill the beetle stages in the field, parasitoids such as eggs of *Trichogramma evanescens* & *Telenomus* sp., larvae of *Distatrix papilionis* & *Brachymeria* sp., and pupae of *Pteromalus* sp. are introduced in the field. Spray NSKE 5% or neem oil @ 1-2 %.

Management	Activity
Leaf eating caterpillar (Citrus butterfly)	Yellow sticky traps @10-12/ha for monitoring Conserve predators such as predatory wasps, lady bird beetles, lacewings and syrphid fly larvae. Horticultural mineral oil spraying @ 0.5 -1%. Dusting of cow dung ash. Spray NSKE 5% or Azadirachtin 0.03 %. Application of fish oil rosin soap at one part in 25 parts of water is highly effective.
Citrus psylla or psyllid	Close monitoring and pruning of the infested plant parts Conserve natural predators such as predatory wasps, lacewings, ladybugs (*Chilocoris nigritus*), predatory ants, predatory mites *etc.* Spraying of horticultural mineral oils @ 0.5-1%.
Scale insects	**Biological control** Predatory wasps, syrphid/hover flies, ladybugs or mealybug destroyers (a conccinellid, *Cryptolaemus montrouzieri*), *etc.*
Mealybug	**Biological control** Predatory midges, green lacewings, lady bird beetles, hover flies, wasps, *etc.*
Citrus leafminer	**Biological control** Conserve parasitoids such as *Closterocerus* spp., *Cirrospilius* spp., *Pnigalio* spp., *Chrysocharis* spp., and *Sympieses* spp. Spray NSKE 5%
Aphid	**Biological control** Conserve parasitoids such as *Aphidius colemani*, *Diaeretiella* spp. *Aphelinus* spp. *etc.* Conserve predators such as anthocorid bugs/pirate bugs (*Orius* spp.), mirid bugs, syrphid/hover flies, green lacewings (*Mallada basalis* and *Chrysoperla carnea*), predatory coccinellids (*Stethorus punctillum*), staphylinid beetle (*Oligota* spp.), predatory cecidomyiid fly (*Aphidoletis aphidimyza*) and predatory gall midge (*Feltiella minuta*), earwigs, ground beetles, rove beetles, spiders, wasps, *etc.*
Reproductive stage (in subsequent seasons)	
Nutrient management	Apply 4kg of FYM, 5:10:10g N:P:K/plant and mix with soil after every harvest.
Weed management	Need-based hoeing and weeding should be done around the plant.
Citrus psylla/psyllid, aphid, mealy bug, scale & citrus leaf miner	Same as in vegetative stage

Source: Vikaspedia (2023)

7.18 Seed Production

Curry leaf trees produce seeds through natural flowering and pollination processes. They are typically self-pollinating but cross-pollination by insects or wind can occur leading to seed variation. After fertilization the fruits in the form of berries (drupes) develop which contain seeds. Ripe fruits are collected which on extraction give 1-2 seeds per fruit. These as such or just after drying can be sown for new plants.

To maintain genetic purity, isolation distance of 300-500m should be maintained where seeds will be set through pollinating insects or *via* wind so genetic purity will remain almost the same. Through regular monitoring and removal of any off type plant will help maintaining the purity.

References

1. Bhandari, P. 2012. Curry leaf (*Murraya koenigii*) or Cure leaf: Review of its curative properties. *J. Med. Nutr. & Nutraceut.*, **1**(2), 92.
2. Bhusal, D. and D.P. Thakur, 2021. Curry leaf: A Review. *Rev. Food & Agr.*, **2**(1): 36-8.
3. Birari, R,, V. Javia and K.K. Bhutani, 2010. Antiobesity and lipid lowering effects of *Murraya koenigii* (L.) Spreng leaves extracts and mahanimbine on high fat diet induced obese rats. *Fitoterapia*, **81**:1129-33.
4. Bonde, S.D., L.S. Nemade, M.R. Patel and A.A. Patel, 2011. *Murraya koenigii* (Curry leaf): Ethnobotany, phytochemistry and pharmacology - A review. *Int. J. Pharm. Phytopharmacol. Res.*, **1**:23-7.
5. Darvekar, V.M., V.R. Patil, A.B. Choudhari and D. Road, 2011. Anti-inflammatory activity of *Murraya koenigii* Spreng on experimental animals. *J. Nat. Prod. Pl. Res.*, **1**(1): 65–9.
6. Dineshkumar, B., A. Mitra and M. Manjunatha, 2010. Antidiabetic and hypolipidemic effects of mahanimbine (carbazole alkaloid) from *Murraya koenigii* (Rutaceae) leaves. *Int. J. Phytomed.*, **2**: 22–30.
7. Drisya, C.R., B.G. Swetha, V. Velu, D. Indrani and R.P. Singh, 2015. Effect of dried *Murraya koenigii* leaves on nutritional, textural and organoleptic characeteristics of cookies. *J. Food Sci. & Techn.*. **52**(1): 500–506.
8. Dusane, M.B. and B.N. Joshi, 2012. Islet protective and insulin secretion property of *Murraya koenigii* and *Ocimum tenuflorum* in streptozotocin-induced diabetic mice. *Can. J. Physiol. Pharmacol.*, **90**: 371-8.
9. Eagri, 2024. Origin, area, production, varieties, package of practices for drumstick, coccinea and curry leaf. Hort. 281: Lecture 32. Development

of e-Courses for B.Sc. (Agriculture). Accessed on December, 2024. http://eagri.org/eagri50/HORT281/lec32.html.

10. etymonline.com, 2023. Curry and Curry tree. Online Etymology Dictionary. 2023. Retrieved 30 December, 2023. https://www.etymonline.com/search?q=Curry+tree.

11. Ganesh, P.S. and R.R. Vittal, 2015. *In vitro* antibiofilm activity of *Murraya koenigii* essential oil extracted using supercritical fluid CO_2 method against *Pseudomonas aeruginosa* PAO1. *Nat. Prod. Res.*, **29**(24): 2295–8.

12. George, A.S., A.E. Orchard and H.J. Hewson, 1993. Oceanic islands 2. *Flora of Australia* (50 ed.), 606p. Australian Gov. Pub. Serv., Canberra.

13. Ghosh, D., S.B. Firdaus, E. Mitra, M. Dey and D. Bandyopadhyay, 2012. Protective effect of aqueous leaf extract of *Murraya Koenigii* against lead induced oxidative stress in rat liver, heart and kidney: A dose response study. *Asian J. Pharmac. & Clin. Res.*, **5**(Suppl. 4): 54–8.

14. Gupta, P., A. Nahata and V.K. Dixit, 2011. An update on *Murraya koenigii* Spreng: A multifunctional Ayurvedic herb (eds Zhong, Xi, Yi Jie and He Xue Bao, **9**:824-33.

15. Gupta, S. And J. Prakash, 2009. Studies on Indian green leafy vegetables for their antioxidant activity. *Pl. Foods Hum. Nutr.*, **64**: 39-45.

16. Gupta, S,, M. George, M. Singhal, G.N. Sharma and V. Garg, 2010. Leaf extract of *Murraya koenigii* Linn for anti-inflammatory and analgesic activity in animal models. *J. Adv. Pharm. Technol. Res.*, **1**:68-77.

17. Henry, A.B.T. 2015. A Hand-book to the Flora of Ceylon, 1893p. *Dulau and Co.*, Sri Lanka.

18. Joseph, S. and K.V. Peter, 1985. Curry leaf (*Murraya koenigii*), perennial, nutritious, leafy vegetable. *Econ. Bot.*, **39**(1): 68–73. *JSTOR*, http://www.jstor.org/stable/4254709. Accessed 2 Jan. 2024.

19. Kesari, A.N., S. Kesari, S.K. Singh, R.K. Gupta and G. Watal, 2007. Studies on the glycemic and lipidemic effect of *Murraya koenigii* in experimental animals. *J. Ethnopharm.*, **112**: 305-11.

20. Knolker, H.J. and K.R. Reddy, 2008. Biological and pharmacological activities of carbazole alkaloids. *Alkaloids Chem. Biol.*, **65**: 181–93.

21. Kumar, V.S., A. Sharma, R. Tiwari and S. Kumar, 1999. *Murraya koenigii*: A review. *J. Medic. & Aromat. Pl. Sc.*, **2**(1): 1139-44.

22. Mandal, S., A. Nayak, M. Kar, S.K. Banerjee, A. Das, S.N. Upadhyay, R.K. Singh, A. Banerji and J. Banerji, 2010. Fitoterapia Antidiarrhoeal activity

of carbazole alkaloids from *Murraya koenigii* Spreng (Rutaceae) seeds. *Fitoterapia*, **81**(1): 72–4.

23. Mani, V., N.S. Mohd. Azahan, K. Ramasamy, S.M. Lim and A.B. Abdul Majeed, 2021. Mahanimbine improved aging-related memory deficits in mice through enhanced cholinergic transmission and suppressed oxidative stress, Amyloid levels and neuroinflammation. *Brain Sci.*, **12**(1):12. doi: 10.3390/brainsci12010012. PMID: 35053756; PMCID: PMC8773876.

24. Mani, V., K. Ramasamy, A. Ahmad, M. Parle, S.A. Shah and A.B. Majeed, 2012. Protective effects of total alkaloidal extract from *Murraya koenigii* leaves on experimentally induced dementia. *Food Chem. Toxicol.*, **50**: 1036-44.

25. MBG, 2023. *Murraya koenigii. Missouri bot.* Gdn., St. Louis, MO, USA. http://www.missouribotanicalgarden.org/PlantFinder/PlantFinderDetails.aspx?kempercode=d441. Retrieved 30 December, 2023.

26. Mitra, E., A.K. Ghosh, D. Ghosh, D. Mukherjee, A. Chattopadhyay and S. Dutta, 2012. Protective effect of aqueous curry leaf (*Murraya koenigii*) extract against cadmium-induced oxidative stress in rat heart. *Food Chem. Toxicol.*, **50**: 1340-53.

27. Ningappa, M.B., R. Dinesha and I. Srinivas, 2008. Antioxidant and free radical scavenging activities of polyphenol-enriched curry leaf (*Murraya koenigii* L.) extracts. *Food Chem.*, **106**(2): 720–28.

28. Parmar, C. and M.K. Kaushal, 1982. *Murraya koenigii.* In Wild Fruits, p.45–8. Kalyani Publishers, New Delhi

29. Patil, R.A., P.M. Langade, P.B. Dighade and Y.A. Hiray, 2012. Antinociceptive activity of acute and chronic administration of *Murraya koenigii* L. leaves in experimental animal models. *Indian J. Pharmacol.*, **44**:15-9.

30. Patterson, J. and M. Verghese, 2015. Anticancer and toxic effects of curry leaf (*Murraya koenigii*) extracts. *J. Pharmacol. & Toxicol.*, **10**(2): 49–59.

31. Paul, S., T.K. Bandyopadhyay and A. Bhattacharyya, 2011. Immunomodulatory effect of leaf extract of *Murraya koenigii* in diabetic mice. *Immunopharmacol. Immunotoxicol.*, **33**: 691-9.

32. POWO, 2023a. *Bergera koenigii* L. Plants of the World Online. Royal Botanic Gardens, Kew Science. Retrieved on 29 December, 2023. (https://powo.science.kew.org/taxon/urn:lsid:ipni.org:names:771522-1). First published in Mant. Pl. 2: 563 (1767).

33. POWO, 2023b. *Murraya koenigii* (L.) Spreng. Plants of the World Online. Royal Botanic Gardens, Kew Science. Retrieved on 29 December, 2023.

(https://powo.science.kew.org/taxon/urn:lsid:ipni.org:names:774433-1). First published in Syst. Veg., ed. 16. 2: 315 (1825).

34. PubChem, 2024a. National Center for Biotechnology Information (2024). PubChem Compound Summary for CID 167963, Mahanimbine. Retrieved January 1, 2024 from https://pubchem.ncbi.nlm.nih.gov/compound/Mahanimbine.

35. PubChem, 2024b. National Center for Biotechnology Information (2024). PubChem Compound Summary for CID 96943, Girinimbine. Retrieved January 1, 2024 from https://pubchem.ncbi.nlm.nih.gov/compound/Girinimbine.

36. Raghu, B.R. 2020. Diversity and distribution of curry leaf in India. *J. hort. Sci.*, **15**(1): 1-8.

37. Raina, V.K., R.K. Lal, S. Tripathi, M. Khan, K.V. Syamasundar and S.K. Srivastava, 2002. Essential oil composition of genetically diverse stocks of *Murraya koenigii. Flav. & Fragr. J.*, **17**: 144-6.

38. Rajendran, M.P., B.B. Pallaiyan and N. Selvaraj, 2014 Chemical composition, antibacterial and antioxidant profile of essential oil from *Murraya koenigii* (L.) leaves. *Avicenna J. Phytomed.*, **4**(3): 200-14.

39. Rajvanshi, R. and K. Mittal, 2019. Phytochemical Analysis of curry Leaves. *Intern. J. Sci. & Res.*, (IJSR), **8**(9): 366-7.

40. Rao, B.R.R., D.K. Rajput and G.R. Mallavarapu, 2011. Chemical diversity in curry leaf (*Murraya koenigii*) essential oils. *Food Chem.*, **126:** 989–94.

41. Reddy, B.M., C.K. Dhanpal and B.V.S. Lakshmi, 2018. A review on curry leaves (*Murraya koenigii*): versatile multi-potential medicinal plant. *Intern. J. Adv. in Pharmacy Med. & Bioallied Sci.*, **1**: 31-41.

42. Saied, S., F. Hassan, S. Naz and H. Siddiqi, 2015. Carbazole alkaloids from stem bark of *Murraya koenigii* (L.) Spreng. *J. Basic Appl. Sci.*, **11**: 146–8.

43. Sathaye, S., Y. Bagul, S. Gupta, H. Kaur and R. Redkar, 2011. Hepatoprotective effects of aqueous leaf extract and crude isolates of *Murraya koenigii* against *in vitro* ethanol-induced hepatotoxicity model. *Exp. Toxicol. Pathol.*, **63**: 587-91.

44. Shivashankara, A.R., A. Azmidah, R. Haniadka, M.P. Rai, R. Arora and M.S. Baliga, 2012. Dietary agents in the prevention of alcohol-induced hepatotoxicity: preclinical observations. *Food & Funct.*, **3**(2): 101–9.

45. Shyamala, M. 2005. Hepatoprotective activity of *Murraya koenigii* leaf extract on paracetamol induced hepatic damaged rats. *Indian J. Pharmacol.*, **37**: 184-5.

46. Singh, A.P., T. Wilson, D. Luthria, M.R. Freeman, R.M. Scott, D. Bilenker, S. Shah, S. Somasundaram and N. Vorsa, 2011. LC-MS-MS characterisation of curry leaf flavonols and antioxidant activity. *Food Chem.*, **127**(1): 80-85.

47. Singh, R. 2023. Curry leaves (kadi patta): Uses, benefits, side effects and more! Last updated: Dec 20, 2023. In *PharEasy*. https://pharmeasy.in/blog/ayurveda-uses-benefits-side-effects-of-Curry-leaves.

48. Singh, S. and S.M. Mohan, 2011. Curry leaves (*Murraya koenigii* Linn. Sprengal)- A miracle plant. *Indian J. Nat. Prod. & Res.*, **2**(4): 508-11.

49. Supe, D., P.B. Deshpande, P. Kandhare and S. Swami, 2021. Review on *Murraya koenigii*: Multipurpose role in management of human health. Available from: https://www.ijppr.humanjournals.com/wp-content/uploads/2021/09/6.Dipak-Supe-Padmanabh-B.-Deshpande-Priyanka-Kandhare-Sandeep-Swami.pdf.

50. Syam, S., A.B. Abdul, M.A. Sukari, S. Mohan, S.I. Abdelwahab and T.S. Wah, 2011. The growth suppressing effects of girinimbine on HepG2 involve induction of apoptosis and cell cycle arrest. *Molecules*, **16**(8): 7155–70.

51. Syamasundar, K.V., B. Srinivasulu, A.P.L. Gowda, R. Srinivasaiyer and R.R. Rao, 2012. Chemo variations of wild curry leaf (*Murraya koenigii* spreng.) from Western Ghats of India. *J. Pharmacognosy,* **2**: 126-30.

52. Tachibana, Y., H. Kikuzaki, N.H. Lajis and N. Nakatani, 2003. Comparison of antioxidative properties of carbazole alkaloids from *Murraya koenigii* leaves. *J. agric. Food Chem.*, **51**: 6461–7.

53. TNAU, 2023. Curry leaf (*Murraya koenigii* Linn. Sprengal). Horticulture, TNAU Agritech Portal. Retrieved on 29 December, 2023. https://agritech.tnau.ac.in/horticulture/horti_vegetables_Curry%20leaf.html.

54. Tembhurne, S.V. and D.M. Sakarkar, 2010. Influence of *Murraya koenigii* on experimental model of diabetes and progression of neuropathic pain. *Res. Pharm. Sci.*, **5**:41-7.

55. Vasanthkumar, S.S., U.K. Pooja, and M.H. Kumaresan, 2023. Production technology of curry leaf (Chapter 4). Trends in Vegetable Science, pp. 52-8. ISBN 978-81-19821-41-9.

56. Vasudevan, M. And M. Parle, 2009. Antiamnesic potential of *Murraya koenigii* leaves. *Phytother. Res.*, **23**: 308–16.

57. Verghese, J. 1989. Indian curry leaf. *Perf. & Flavourist*, **14**(3): 69-70.

58. Vikaspedia, 2023. Curryleaf pests and diseases. Accessed on 9 December, 2023. https://vikaspedia.in/agriculture/crop-production/integrated-pest-

managment/ipm-for-vegetables/ipm-strategies-for-curryleaf/Curry-leaf-pests.

59. Volza, 3023. Curry leaves exports from world. Volza Crow Global. Retrieved on 31 December, 2023. https://www.volza.com/p/Curry-leaves/export/.

60. Xie, H., T. Zhang, N. Yang, Z. Li and Y. Liu, 2020. Anticancer effects of mahanimbine alkaloid on the human bladder cancer cells are due to the induction of G0/G1 cell cycle arrest, apoptosis and autophagy. *JBUON*, **25**: 1166–71.

61. Yankuzo, H., Q.U. Ahmed, R.I. Santosa, S.F. Akter and N.A. Talib, 2011. Non-beneficial effect of the leaves of *Murraya koenigii* (Linn.) Spreng (Rutaceae) on diabetes-induced renal damage *in vivo*. *J. Ethnopharmacol.*, **135**: 88-94.

Indian Bauhinia or Kachnar (*Bauhinia variegata* L.)

Family: Fabaceae. Chromosome number: $2n = 28$.

Synonyms: *Bauhinia variegata* var. *purpurascens* Voigt, *Perlebia variegata* (L.) A.Schmitz, *Phanera variegata* (L.) Benth. (POWO, 2024).

Other common names: Camel's Foot Tree, Mountain Ebony or Orchid Tree.

8.1 Introduction

Kachnar, distinguished by its striking orchid-like flowers and unique bilobed leaves, stands as a medium-sized deciduous tree esteemed for its ornamental allure and multifaceted significance. Revered culturally, ecologically and medicinally, it holds a notable place. In certain regions, its blossoms find application in traditional medicine while serving as a vital nectar source for pollinators such as bees and butterflies. Moreover, it acts as a host plant for specific butterfly species while its seeds serve as sustenance for birds and small mammals.

Blooming flowers of *Bauhinia variegata* tree in the month of March, Delhi, India.

Kachnar demonstrates ecological adaptations, thriving in various soil types and environmental conditions from dry to moist habitats. As a leguminous plant, it can fix nitrogen in the soil so contributing to soil fertility. Culturally, it holds significance as an ornamental tree often planted in gardens, parks and along roadsides. Various parts of tree are used in traditional medicine systems for its purported medicinal properties.

8.2 Etymology

The etymology of '*Bauhinia variegata*' traces back to the Bauhin brothers, Gaspard and Johann who were Swiss botanists in 16th century. The genus *Bauhinia* is named after them due to their significant contributions to botany. The specific epithet '*variegata*' in *Bauhinia variegata* refers to the variegated or multicoloured appearance of flowers or leaves. '*Variegata*' is derived from the Latin word '*variegatus*' which means 'having various colours' or 'varied in appearance'.

The etymology of the word 'kachnar' is rooted in several languages, particularly Hindi and Sanskrit.

1. Hindi: 'Kachnar' (कचनार) is derived from the word 'Kachnar' or 'Kachanar' (कचनार / कचनार) itself. This word has its origins in Sanskrit.
2. Sanskrit: The Sanskrit term for this plant is 'Kanchnara' (कञ्चनारा) which is a combination of two words: 'Kanchana' (कञ्चना) meaning 'gold' or 'golden' and 'Nara' (नारा) meaning 'man' or 'person'. This name possibly refers to the golden- coloured flowers of *Bauhinia variegat*a.

The term 'Kachnar' essentially means a plant with golden or golden- coloured flowers which is an apt description for the beautiful blooms of the *Bauhinia variegata* tree.

8.3 Origin and distribution

Bauhinia variegata, native to South Asia, particularly India, Nepal and Myanmar boasts a rich history of cultivation and admiration in these regions. It graces gardens, parks and roadsides for its captivating beauty. Distributed in the sub-Himalayan and outer Himalayan regions such as Sikkim and Punjab (India) it has also been introduced to tropical and subtropical areas globally including parts of Africa, SE Asia, Australia and the Americas. Appreciated for its ornamental value, it often embellishes botanical gardens, urban landscapes, and private gardens in these introduced regions.

In its native habitat it thrives naturally in various environments from forests to scrublands and riverbanks. Known for its adaptability, it flourishes in diverse soil types and climatic conditions from dry to moist environments.

8.4 Botany

Bauhinia variegata is a medium-sized deciduous tree that typically grows to heights of 10-12 meters. It has a spreading canopy with a rounded or irregular shape. The bark is smooth and grayish-brown when young, becoming rougher and darker with age. It may develop shallow fissures and ridges as the tree matures. The tree typically has a deep and extensive root system, allowing it to anchor firmly in the soil and access water and nutrients from deep within the ground.

The leaves are simple and alternate displaying palmate venation. These are obcordate shaped, long *cum* broad, rounded and deeply bilobed at the base & apex resembling the shape of a camel's foot. Each leaf is typically 7-10 centimeters long and wide with a smooth texture and a dark green colour. The lobes of the leaves are rounded or slightly pointed at the tips and the leaf margin is entire.

Kachnar produces large showy flowers that are reminiscent of orchids. The bisexual and zygomorphic flowers have 8–12 centimetres diameter and are typically pink to purplish in colour though white and red varieties also exist. Each flower has five sepals, five petals, 10 stamens and a single pistil. Pollens are elongated, approximately 75 microns in length. The petals are arranged in a characteristic butterfly-like shape. The flowers are borne in clusters at the ends of branches and bloom profusely during spring season.

One of the distinguishing characteristics of *Bauhinia variegata* is that it tends to be nearly leafless or has significantly reduced foliage when it flowers. In contrast, species like *Bauhinia purpurea* and *Bauhinia blakeana* typically retain their leaves while flowering. This difference in leaf retention during flowering can help differentiate between these *Bauhinia* species and avoid confusion, especially for those familiar with their appearance.

Reproduction occurs sexually through seed production with pollination primarily facilitated by insects like bees and butterflies. The resulting fruits are long *cum* flat seed pods that are brown or reddish-brown in colour and 15–30cm long containing multiple seeds, dispersed by wind or animals. The seedpod dries completely on the tree, and when mature begins to twist into a helix or corkscrew shape, ultimately exploding open with a very audible clack to deliver its seeds into the environs.

8.4.1 Anatomy of stem

The stem's anatomy through transverse section analysis shows distinct structures such as the periderm and cortex. Secondary phloem appears wide and forms a continuous cylindrical layer, characterized by thin, narrow straight rays as well as 3-4 cylinders of discontinuous fibre masses and randomly distributed sieve elements. Secondary xylem exhibits diffuse porosity and comprises vessels, fibres, xylem rays and xylem parenchyma. Xylem fibres display thick lignified walls while some exhibit gelatinous walls. Abundant xylem parenchyma cells are found within the xylem. Xylem rays are narrow, straight and consist of radially elongated thick-walled lignified walls. Predominantly prismatic crystals and druses type of calcium oxalate crystals are observed. Powder microscopical examination reveals the presence of fibres, parenchymatous cells, periderm and vessel elements. Histochemical analysis indicates the presence of protein, tannin, lignin and cellulose within the stem tissues (Patel and Golwala, 2012).

8.4.2 Root anatomy

The root anatomy examined through transverse section analysis revealed distinct secondary phloem and secondary xylem structures. The secondary phloem exhibit broad rays, densely packed phloem fibres and radial arrangements of phloem elements. Meanwhile, the secondary xylem displays notably wider vessels, often with thin walls and occurring either singly or in radial clusters. Xylem fibres are observed in both gelatinous and normal forms. Prismatic calcium oxalate crystals are prevalent. Microscopic powder examination further identifies xylem parenchyma cells, xylem fibres and vessel elements.

8.5 Nutritional value

Kachnar has the following nutritional values.

Constituents (%)	*Dried leaves*	*Flowers*	*Seeds*	*Buds*
Protein	15.19	3.24-5	41.9 ± 1.6	3.7
Carbohydrates	66.82	16.01	28.4 ± 1.6	6.4
Fats	4.15	0.15-2.5	0.6 ± 0.1 (FFA)	2.44
Fibres	4.26	8.66	6.9 ± 0.8	6.8
Moisture	8.83	77.80	6.7 ± 0.46	84.51
Ash	4.9	2.81	4.8 ± 0.1	4.33
Total oils	–	–	18 ± 0.1	–

Nutritional value of Kachnar (Sharma *et al.*, 2021)

8.6 Phytochemicals

The plants of the genus *Bauhinia* exhibit a rich chemical composition characterized by a significant presence of flavonoids, terpenes, steroids and alkaloids. Studies have reported the isolation of compounds such as lupeol, 5,7-dimethoxyflavonone-4–O–α– L–rhamnopyranosyl-β-D-glucopyranoside, β-sitosterol and kaempferol-3-glucoside from the stem of these plants (Mali *et al.*, 2007). Furthermore, phytochemical screening of *B. variegata* has revealed the presence of flavonoids, terpenoids, reducing sugars, steroids, saponins, tannins, and cardiac glycosides (Al-Snafi, 2013).

Sharma *et al.* (2019) investigated the reported potential and cytotoxic activity of pure fractions isolated from methanol extract (MEB) using silica gel column chromatography. These fractions, identified as kaempferol, stigmasterol, protocatechuic acid-methyl ester (PCA-ME) and protocatechuic acid (PCA) were subjected to analysis against various cell lines.

Kaempferol exhibits potent antioxidant activity as reported by Wang *et al.* (2018). This flavonoid is commonly found in plant-based foods and various plant parts utilized in traditional medicine. Stigmasterol, also known as the Wulzen anti-stiffness factor, is a plant sterol similar to animal cholesterol. Its role extends to the synthesis of vitamin D3 and semi-synthetic progesterone, a valuable human hormone (Kanimozhi and Bai, 2012). Protocatechuic acid (PCA) is widely distributed in nature and shares structural similarities with well-known antioxidants such as caffeic acid, vanillic acid, gallic acid and syringic acid (Kakkar and Bais, 2014). Both PCA and PCA-ME exhibit significant antioxidant and cytotoxic activities, with PCA-ME showing superior potential in scavenging free radicals compared to PCA (Sharma *et al.*, 2019).

8.7 Uses

For many years, various parts of kachnar tree such as its leaves, flowers and floral buds have served as vegetables in multiple regions of India. The stem of the kachnar tree yields fibre, and it is also known to produce a type of gum (Kansal *et al.*, 2020).

Kachnar, a local name in the Indian subcontinent refers to the edible buds harvested from the tree, widely incorporated into numerous regional recipes. A traditional dish, Kachnar curry combines these buds with yogurt, onions and indigenous spices. The buds are stir-fried as a vegetable and used to prepare achaar, a pickle enjoyed across the subcontinent. Kachnar exhibits notable antioxidant and anticancer properties (Golwala *et al.*, 2020). In Nepal, the plant is known as koiralo, with its flower termed as koiralo ko phool in Nepali. Nepalese cuisine utilizes both the flowers and buds in pickling, a dish prepared by boiling them and mixing with mashed potato, green chilies, onions, peas and various spices like timmur, turmeric and cumin. Mustard oil, fenugreek seeds and lemon juice are added for flavour with the achar serving as a significant component of meals during the Ghode Jatra Festival.

8.8 Other species

The genus *Bauhinia* is large and diverse, comprising numerous species distributed across various regions of the world. While exact numbers may vary due to taxonomic revisions and ongoing botanical research, these are estimated to be around 300 to 500 species. New species are occasionally discovered and the classification of existing species may change based on further study and analysis.

Bauhinia acuminata flower

By Joydeep, CC BY-SA 3.0, https://commons.wikimedia.org/w/index.php?curid=21015780

Bauhinia purpurea ~ flowers

By J.M.Garg - Own work, CC BY-SA 4.0, https://commons.wikimedia.org/w/index.php?curid=5924058

While *Bauhinia variegata* is one of the more commonly known species whose buds are consumed as vegetables, there are several other *Bauhinia* species whose buds or other parts are also edible in certain cultures, e.g.

i) ***Bauhinia acuminata:*** *Bauhinia acuminata,* also referred to as the 'Dwarf White Bauhinia' or 'White Orchid Tree', is indigenous to South Asia, particularly India and Sri Lanka. Unlike its larger relative *Bauhinia variegata, Bauhinia acuminata* typically grows as a shrub or small tree. It bears fragrant white flowers with yellow centres, adding a touch of elegance to its compact stature. The leaves of *Bauhinia acuminata* are deeply divided into two lobes, reminiscent of butterfly wings. Cultivated for its ornamental value, this species is prized for its attractive flowers and manageable size in landscaping and garden settings.

ii) ***Bauhinia purpurea:*** *Bauhinia purpurea*, commonly known as the 'Purple Orchid Tree' or 'Butterfly Tree' hails from SE Asia but has found widespread cultivation in tropical and subtropical regions. It is renowned for its striking purple or pink flowers with vibrant yellow centres which adorn the tree in abundance. Like other *Bauhinia* species, *Bauhinia purpurea* features bilobed or cloven leaves that resemble butterfly wings. This species is highly valued as an ornamental tree often planted in parks, gardens and along roadsides for its showy flowers and attractive foliage, adding a splash of colour to the landscape.

iii) ***Bauhinia galpinii***: *Bauhinia galpinii*, also known as the 'Pride of De Kaap', 'Red Orchid Bush' or 'Red Bauhinia' is native to southern Africa, particularly South Africa and Zimbabwe. It stands out for its bright red-orange flowers with contrasting yellow centres which provide a striking display against its green foliage. Similar to other *Bauhinia* species, *Bauhinia galpinii* features bilobed or cloven leaves, adding to its aesthetic appeal. This species is prized for its colourful flowers and drought tolerance, making it a popular choice for ornamental planting in gardens and landscapes across its native range.

iv) ***Bauhinia variegata candida:*** This cultivar of *Bauhinia variegata* produces white flowers instead of the typical pink to purplish flowers. While primarily grown for its ornamental value, the buds of this cultivar may also be edible in certain cuisines. It retains the same general habitat preferences and growth requirements as the species it was derived from.

Flower of Bauhinia galpinii

work,http://www.flowersview.com/Bauhinia-galpinii-/02-IMG_7663.jpg.html, CC BY-SA 4.0, https://commons.wikimedia.org/w/index.php?curid=43139234

var. candida in Hyderabad, India

By J.M.Garg - Own work, CC BY-SA 4.0, https://commons.wikimedia.org/w/index.php?curid=6624249

Among the above mentioned species, only *Bauhinia variegata* is recognized for its edible parts, particularly the buds. These buds are widely used in various culinary traditions in the Indian subcontinent and Nepal. These are commonly incorporated into dishes such as kachnar curry and stir-fried vegetables. These are utilized in preparation of pickles and other culinary delights. The other species including *Bauhinia acuminata*, *Bauhinia purpurea* and *Bauhinia galpinii* are primarily cultivated for ornamental purposes but are not known for their edible qualities.

It is important to note that edibility of *Bauhinia* species can vary depending on factors such as local culinary traditions, preparation method and individual tolerances. Before consuming any part of a *Bauhinia* plant, it's advisable to verify its edibility and ensure that it has been properly prepared.

8.9 Cultivation

Cultivating kachnar requires attention to several factors to ensure its healthy growth and development.

8.9.1 Climate and Location

Ideally suited to warm, tropical and subtropical climates, this species thrives in temperatures ranging from 20°C to 35°C and is sensitive to frost. Adequate sunlight is crucial with the plant preferring full sun exposure while tolerating partial shade. Moderate to high levels of rainfall during growing season are beneficial though kachnar can withstand short dry spells with proper watering. Well-draining soil such as sandy loam or loamy soil with a neutral to slightly acidic pH supports healthy growth. While the plant can tolerate a range of humidity levels, moderately humid conditions are preferred. Ensuring adequate air circulation around the plant helps preventing fungal diseases and promoting vigorous growth. By providing these optimal climatic conditions, cultivators can foster the healthy development and ornamental beauty of *Bauhinia variegata* plants.

8.9.2 Propagation

Propagation of kachnar can be achieved through several methods including seeds, cuttings and air layering. Seeds are collected from mature healthy plants, scarified to break the seed coat, soaked in water to soften the coat and then planted in pots or directly in the ground during warm season. Cuttings, typically 15-20cm long are taken from healthy mature plants with lower leaves removed, the cut end dipped in rooting hormone and then planted in a well-draining rooting medium. These cuttings are kept in a warm humid environment with indirect sunlight, ensuring the medium remaining consistently moist until roots develop within a few weeks to months. Air layering involves selecting a healthy branch, making a shallow incision, dusting it with rooting hormone, wrapping it with moist sphagnum moss and securing it with plastic wrap. Roots should form within a few weeks to months. Once roots are established, the cuttings or air-layered sections can be transplanted into pots or directly into the ground, providing proper care and attention for successful growth and establishment.

8.9.3 Sowing season

The sowing season for *Bauhinia variegata* depends on the climate and growing conditions of the region. In general, the best time to sow kachnar seeds is during warm season when temperatures are consistently above 20°C and the risk of frost has passed.

For tropical and subtropical regions where kachnar thrives, the sowing season typically falls in early spring or late winter after the coldest months have passed and temperatures begin to rise. In these regions, seeds can be sown directly in the ground or in pots from February to April.

In temperate regions with colder winters it is advisable to wait until the danger of frost has passed before sowing kachnar seeds. This may mean waiting until late spring or early summer to sow seeds directly in the ground or in pots.

Regardless of the region, it's essential to consider the local climate and weather conditions when determining the best time to sow kachnar seeds. Providing optimal growing conditions during the sowing season will help ensure successful germination and establishment of *Bauhinia variegata* plants.

8.9.4 Planting

When planting kachnar, the process begins by digging a hole twice as wide and deep as the root ball of the plant. This ensures ample space for root expansion and establishment. The hole is then backfilled with a mixture of compost and soil, enriching the planting area with nutrients and improving soil structure. Carefully placing the plant in the centre of the hole, ensuring the top of the root ball is levelled with the surrounding soil surface, is essential. Backfilling follows, with the soil-compost mixture gently pressed down to eliminate air pockets and for providing firm contact between the roots and the soil. Thorough watering immediately after planting helps settle the soil around the roots and provides essential moisture. Proper planting techniques are crucial for successful establishment and future growth of *Bauhinia variegata*, ensuring it thrives in its new environment.

It is recommended to space trees 3-4.5m apart to accommodate their mature size, allowing ample room for growth. Ensure they are positioned at least 5-6m away from structures or other trees to prevent overcrowding and to promote airflow. For hedge planting, space individual plants 1.5-2.5m apart to achieve a dense and uniform appearance. In containers, select pots with a diameter of at least 45-60cm and space them 1.5-2.5m apart for optimal growth. Adjusting spacing based on specific landscape conditions and preferences is essential to achieve desired results.

8.9.5 Watering

Careful attention to the watering needs of kachnar is essential for its optimal growth and health. This species prefers moderate watering which entails watering the plant deeply but infrequently allowing the soil to dry out slightly between waterings. This watering regimen encourages the development of a robust root system while helping to prevent issues such as root rot due to waterlogging.

Monitoring the soil moisture level is crucial and watering should be given when the top inch or two of soil feels dry to the touch. During the plant's dormant season, typically in winter, watering should be reduced to account for its decreased growth activity. By following these watering guidelines, cultivators can ensure that their *Bauhinia variegata* plants receive the appropriate moisture levels to thrive and flourish.

8.9.6 Fertilization

There are various options for both organic manures and commercial fertilizers to consider. For organic options, well-rotted compost and aged livestock manure such as cow or poultry manure are excellent choices. These organic amendments not only enrich the soil with essential nutrients but also improve its structure and fertility over time. Bone meal and fish emulsion are organic amendments that provide specific nutrients like phosphorus and nitrogen, respectively to support root development and flowering.

Alternatively, commercial fertilizers such as balanced formulations with equal proportions of nitrogen, phosphorus and potassium or slow-release fertilizers can provide a convenient and effective way to supply nutrients to the plants. Regardless of the chosen fertilizer, it is essential to follow recommended application rates and guidelines to avoid over-fertilization which can lead to nutrient imbalances or adverse effects on plant health. By selecting the appropriate manures or fertilizers and applying them judiciously, gardeners can ensure the healthy growth and vibrant flowering of their kachnar plants.

8.9.7 Pruning

Regular pruning is crucial to maintain desired shape of the plant and eliminate dead or damaged branches that could compromise its overall vigour. It is advisable to prune after flowering as this encourages branching and promotes compact growth allowing the plant to allocate its energy towards new growth and flower production during the growing season. When pruning, it is essential to use sharp and clean pruning shears and make cuts just above a node or lateral bud to encourage new growth in the desired direction. While pruning, gardeners should also consider the plant's flowering habits as *Bauhinia variegata* blooms on new wood. Therefore, pruning should be performed judiciously to avoid removing potential flowering branches. Rejuvenation pruning can be undertaken if the plant becomes overgrown or leggy, involving the removal of one-third of the oldest branches to ground level in late winter or early spring to promote new growth from the base of the plant.

8.9.8 Support

Young plants can be supported with stakes to establish strong roots and prevent leaning or toppling over. Taller specimens should be secured to stakes using soft ties or straps to minimize damage from strong winds. This support promotes stability and encourages healthy growth, ensuring the plant's long-term success in the landscape.

8.10 Pest and disease management for kachnar

8.10.1 Pests

i) **Aphids**: These small insects feed on the sap of kachnar causing leaf yellowing and stunted growth. Aphids are controlled by spraying plants with insecticidal soap or neem oil.

ii) **Scale insects:** Scale insects attach themselves to the stems and leaves of kachnar, sucking sap and causing wilting. Infested areas should heavily be pruned and applied with horticultural oil to suffocate scales.

iii) **Caterpillars:** Caterpillars may feed on kachnar foliage, leading to defoliation. Caterpillars when possible should be hand picked or organic insecticides containing *Bacillus thuringiensis* (Bt) should be used to control infestations.

8.10.2 Diseases

i) **Powdery mildew:** Powdery mildew appears as a white powdery substance on full area of leaves affecting plant vigour and flowering. Improved air circulation and avoidance of overhead watering will keep powdery mildew at bay. Fungicidal sprays containing sulphur or potassium bicarbonate as a preventative measure are quite effective.

ii) **Leaf spot:** Leaf spot diseases cause dark lesions on kachnar foliage leading to leaf drop and reduced growth. Removal and destroying the infected leaves and application of fungicidal sprays containing copper-based compounds will control this problem.

iii) **Root sot:** Root rot caused by fungal pathogens leads to yellowing of foliage, wilting and eventual plant death. Improving soil drainage and avoiding overwatering will prevent root rot. Infected plants should be treated with fungicidal drenches containing thiophanate-methyl or potassium phosphonate.

8.10.3 Cultural practices

i) Maintaining proper plant hygiene by removing fallen leaves and debris will reduce pest and disease pressure.

ii) Avoiding overhead watering will reduce the disease to the minimum. Watering at the base of the plant early in the day will allow foliage to dry before evening.

iii) Plants are monitored regularly for signs of pests and diseases, in case of any incident prompt action will check the incident reaching to alarming lavel.

By implementing integrated pest and disease management strategies including cultural practices and targeted treatments, gardeners can effectively manage pest and disease issues in kachnar plants.

References

1. Al-Snafi A.E. 2013. The pharmacological importance of *Bauhinia variegata*—A review. *Int. J. Pharm. Sci. Res.,* **4**:160–64.
2. Golwala, D.K., S.K. Vaidya, K. Dholwani, D.S. Patel and S. Sahoo, 2020. Antioxidant and antimutagenic (anticlastogenic) activity of alcoholic extract of *Bauhinia Variegata* Linn root. *Europ. J. Medic. Pl.,* **31**(2) :32-9.
3. Kakkar, S. and S. Bais, 2014. A review on protocatechuic acid and its pharmacological potential. *ISRN Pharmacol. (*doi: 10.1155/2014/952943).
4. Kanimozhi D. and V. Bai, 2012. Evaluation of phytochemical antioxidant antimicrobial activity determination of bioactive components of ethanolic extract of aerial and underground parts of *Cynodon dactylon* L. *Int. J. Sci. Res. Rev.,* **1**:33–48.
5. Kansal, M., P. Shukla and P. Shukla, 2020. *Bauhinia variegata* a boom to human health *Int. J. Pharmacogn.,* **7**(7): 155-61.http://doi.or/10.13040/IJPSR.0975-8232.
6. Mali, R.G., S.G. Mahajan and A.A. Mehta, 2007. Rakta Kanchan (*Bauhinia variegata*): Chemistry, traditional and medicinal uses-a review. *Pharmacogn. Rev.,* **1**:314.
7. Patel, L.D. and D.K. Golwala, 2012. Pharmacognostical studies of *Bauhinia variegata* Linn. stem. *Intern. J. Pharmac. Res.,* **4**(3): 127-30.
8. POWO, 2024. *Bauhinia variegata* L. Plants of the World Online. Royal Botanical Garden, Kew. https://powo.science.kew.org/taxon/urn:lsid:ipni.org:names:481756-1. Accessed on 16 February 2024.
9. Sharma, K., V. Kumar, S. Kumar, R. Sharma abd C.M. Mehta, 2021. *Bauhinia variegata*: a comprehensive review on bioactive compounds, health benefits and utilization. *Adv. Tradit. Med.,* **21**:645–53.

10. Sharma, N., A. Sharma, G. Bhatia, M. Landi, M. Brestic, B. Singh, J. Singh, S. Kaur and R. Bhardwaj, 2019. Isolation of phytochemicals from *Bauhinia variegata* L. bark and their *in vitro* antioxidant and cytotoxic potential. *Antioxidants* (Basel), **8**(10):492.

11. Wang, J., X. Fang, L. Ge, F. Cao, L. Zhao, Z. Wang and W. Xiao, 2018. Antitumor, antioxidant and anti-inflammatory activities of kaempferol and its corresponding glycosides and the enzymatic preparation of kaempferol. *PLoS ONE.*, **13**:e0197563. doi: 10.1371/journal.pone.0197563.

Our other publications by same authors

1. **Misra, R.L. and Sanyat Misra, 2023. Commonly Used Ornamental Plants (Rs. 7995/-).**
2. Misra, R.L. and Sanyat Misra, 2023. *Commercial Ornamental Crops* **Cut Flowers** (Rs. 5995/-).
3. Misra, R.L. and Sanyat Misra, 2023. *Commercial Ornamental Crops* **Traditional and Loose Flowers** (Rs. 5995/-).
4. Sanjay Mishra, Sanyat Misra and R.L. Misra, 2017. Soilless Crop Cultivation (Rs. 2395/-).
5. Sanyat Misra, Sanjay Mishra, R.L. Misra, Pragya Ranjan and J.K. Ranjan, 2023. **Herbs, Spices and Medicinal plants** (2 volumes) (Rs. 7995/- for both the volumes).
6. Sanyat Misra, Sanjay Mishra and R.L. Misra, 2004. **Tuber Crops** (Rs. 4995/-).
7. Sanyat Misra, Sanjay Mishra and R.L. Misra, 2004. **Root Vegetables** (Rs. 3995/-).
8. Sanyat Misra, Sanjay Mishra and R.L. Misra, 2004. **Edible Alliums** (Rs. 3995/-).
9. Sanyat Misra, Sanjay Mishra and R.L. Misra, 2004. **Cole Crops** (Rs. 2995/-).

Index

Family

Scientific Names

Common Names

Colour Plates

Young banana plant
(*pg.* 11)

Banana rhizome
(*pg.* 11)

Plant showing fruit and inflorescence
(*pg.* 12)

Female flowers with petals at the tip of ovary (*pg.* 12)

Partially opened inflorescence

(*pg.* 12)

Fruits of four cultivars

(*pg.* 13)

Fruits of wild-type bananas with numerous large, hard seeds (*pg.* 14)

Cv. Nendran
(*pg.* 23)

Jackfruits on main stem of the tree
(*pg.* 102)

Jackfruit tree laden with clustered fruits
(*pg.* 102)

Jackfruit tree trunk showing texture and colouration (*pg.* 106)

Flower buds and leaves (*pg.* 107)

(a)

(b)

(c)

(d)

(e)

Footstalk initiation (the emergence of floral bud covered with stipule) and elongation (a-c); footstalk in maximum elongation (d); the opening of stipules and emergence of inflorescences (e). (*pg.* 108)

(a)

(b) (c)

(d)

(e)

Developmental stages of male inflorescences: The emergence of male inflorescences from the opening stipules of footstalk (a); pollen maturity (pollen are sticky and yellow); anthers dehiscence; pollination and fertilization (b-c); end of pollen viability; the dry up and abscission of anthers (d); and the dry up and abscission of male inflorescences (e). (*pg.* 109)

Developmental stages of female inflorescences: the emergence of female inflorescences from the opening stipules of footstalk (a); stigma receptivity (stigma are sticky, straight and white in colour); pollination and fertilization (b-c); end of stigma receptivity; the drying up and abscission of stigma; the formation of young fruits (syncarps) (d); and the development of syncarps until maturity (e-f). (*pg.* 110)

Jackfruit flesh
(*pg.* 111)

Jackfruit seeds
(*pg.* 113)

Artocarpus altilis
(*pg.* 200)

Tree showing form of the buttress root
(*pg.* 203)

Male inflorescence
(*pg.* 204)

Female inflorescence
of ovary (*pg.* 2074)

By US Pacific Basin Agricultural Research Center - US Pacific Basin Agricultural Research Center (http://pbarc.ars.usda.gov/pages/research/tpgrmu/Hart-49.jpg; http://pbarc.ars.usda.gov/pages/research/tpgrmu/HART49.shtml; http://pbarc.ars.usda.gov/pages/research/tpgrmu/artocarpus.shtml), Public Domain, https://commons.wikimedia.org/w/index.php?curid=131476 (*pg.* 205)

Moringa foliage
(*pg.* 250)

Moringa Tree Flowers
(*pg.* 250)

Moringa flowers}
(*pg.* 250)

Moringa pods in Panchkhal (Nepal)
(*pg.* 252)

Moringa tree and dry seed pods
(*pg.* 252)

Indarbela quadrinotata Infested bark
(*pg.* 318)

Larva - Eupterote molifera
(*pg.* 318)

Gitona distigma
(*pg.* 318)

Noorda blitealis -Larva and adult (*pg.* 318)

Bud worm-*Noorda moringae*
(*pg.* 318)

Trunk in Narsapur, Medak district, India (*pg.* 342)

Madhuca longifolia tree in Narsapur, Medak district (*pg.* 342)

Flowering pattern in mahua
(*pg.* 344)

Flowers from Melghat Tiger Reserve, Maharashtra (*pg.* 344)

Fresh mahua flowers for preparing sweet dishes (*pg.* 344)

Sun-dried mahua flowers for various sweet preparations (*pg.* 344)

Fruit in Narsapur, Medak district, India (*pg.* 345)

Bamboo forest in Arashiyama, Kyoto, Japan (*pg.* 361)

Flowering bamboo (*pg.* 365)

Bamboo shoot, p*hyllostachys parvifolia*, two year old division (*pg.* 361)

Bamboo foliage probably *Phyllostachys aurea* (*pg.* 365)

Bamboo foliage probably *Phyllostachys nigra* (*pg.* 365)

Bunches of bamboo seeds
(*pg.* 365)

Bamboo shoots for sale in a supermarket in Japan
(*pg.* 371)

Harvested whole bamboo shoots (*pg.* 371)

Bamboo shoot for sale in Kathmandu
(*pg.* 371)

Canned bamboo shoots
(*pg.* 371)

Curry leaf tree
(*pg.* 414)

Curry leaf flowers appearing during spring
(*pg.* 415)

Fruits – ripe and unripe
(*pg.* 415)

Blooming flowers of Bauhinia variegata tree in the month of March, Delhi, India.
(*pg.* 447)

Bauhinia acuminata flower
(*pg.* 452)

Bauhinia purpurea ~ flowers
(*pg.* 452)

Flower of Bauhinia galpinii
(*pg.* 453)

var. *candida in Hyderabad, India*
(*pg.* 453)